5th International Symposium on Marine and Fisheries Research

IOP Conference Series: Earth and Environmental Science
Volume 1289

Yogyakarta, Indonesia
24 July 2023

Editors:

Muhammad Prima Putra
Eko Hardianto
Desy Putri Handayani
Wee Hin Boo
Andreas Bauer
Siriyupa Netramai
Al'Ain Shoufi Rahmah Arifin

ISBN: 978-1-7138-8754-6
ISSN: 1755-1307

Printed from e-media with permission by:

Curran Associates, Inc.
57 Morehouse Lane
Red Hook, NY 12571

Some format issues inherent in the e-media version may also appear in this print version.

Institute of Physics
Dirac House, Temple Back
Bristol BS1 6BE UK

Phone: 44 1 17 929 7481
Fax: 44 1 17 920 0979

techtracking@iop.org

Additional copies of this publication are available from:

Curran Associates, Inc.
57 Morehouse Lane
Red Hook, NY 12571 USA
Phone: 845-758-0400
Fax: 845-758-2633
Email: curran@proceedings.com
Web: www.proceedings.com

TABLE OF CONTENTS

Preface

Peer Review Statement

Consumption Rate of Stingrays in the Southern Coastal Area of Flores Island 1
Dati Nawastuti, Dwidjono Hadi Darwanto, Jangkung Handoyo Mulyo, Suadi

Optimising Empowerment of Fostered Villages for Increased Income of Tambak Farmers in Bulu Cindea Village 12
R Rahmi, A Syarif, J Ibrahim, F Indah Yani, A Chadijah, A R Nanda, A S Syamsuri, F Fitriyani, S Syamsia

A Case Study of Anchovy Diurnal Migration Patterns in Teluk Bakau, Bintan Regency, Riau Island, Indonesia 21
Try Febrianto, Asep Ma'mun, Arif Pratomo, Esty Kurniawati, Mario Putra Suhana, Donny Apdillah, Ahmad Zahid

Business Analysis of Processed Baby Squid Chili Products (Micro, Small and Medium Enterprises MSMEs "Dapoer Ikan Diana"): Case Study in Samarinda City, East Kalimantan, Indonesia 31
Eko Sugiharto, Elly Purnamasari, Komsanah Sukarti

Nutritional Value of Sea Cucumber Acaudina Molpadiodes from Delta Wulan Estuary, Demak Regency, Central of Java, Indonesia 36
W Widianingsih, R Hartati

Echinoderm Diversity in the Soft Bottom of Wulan Estuary, Demak, Indonesia 41
Retno Hartati, Widianingsih Widianingsih, Muhammad Zainuri, Edy Supriyo

Variation in Eastern Little Tuna (Euthynnus Affinis) Catches Related to El Niño Southern Oscillation (ENSO) Events in the Makassar Strait 49
M L Syamsuddin, A R Puspita, F Syamsudin, Y N Ihsan, Sunarto, M Zainuddin

Estimation of Carbon Storage of Tropical Mangroves in the North Maluku Islands Region 63
Najamuddin, I Tahir, M I Pujiati, Z A Harahap

Tropical Cyclone Saudel Impact on Ocean Dynamic Over the South China Sea 75
S R Arum, A Wicaksono, A D Siswanto

First Record and Phylogenetic Relationship of the Endangered Species Betta Foerschi (Vierke, 1979) (Anabantiformes : Osphronemidae) from Belitung Island, Indonesia 85
AF Syarif, FS Valen, M Herjayanto, S Aisyah

A New Distributional Record of Spanner Barbs, Striuntius Lateristriga (Valenciennes, 1842) (Cypriniformes: Cyprinidae) in Belitung Island, Indonesia 93
AF Syarif, AM Hafidz, FS Valen, M Herjayanto, R Robin

Sustainable Marine Ecotourism Governance Based on Tourism Preferences 99
Atikah Nurhayati, Titin Herawati, Asep Agus Handaka, Wahyuniar Pamungkas, Nora Akbarsyah, Agnes Sudarmono

COVID-19 Pandemic and the Momentum of Seaweed Industry Development in Indonesia Towards Sustainable Food System: An Overview 109
B Wardono, A S Prabakusuma, A Zulham, R Yusuf, E S Luhur, I Muliawan, R Deswati, F Y Arthatiani

Profile of Fisheries Capture in Cikidang Fish Landing Base (PPI), Pangandaran, West Java Study 2017-2021 and Its Potential Utilization for Fisheries Product Processing Industry 119
T Yuniarti, M Mulyono, M Mardiana, I Nurmala, I J Puspadewi, H Triyono, M H Alauddin, Y P Handoko, M Maulita, A Maulani

Prelimenary Study Technology Readiness Level and Techno-Economic Analysis of Seaweed-Based Soft Capsule Prototypes 134
T. Hidayat, K. Syamsu, T.C. Sunarti, M. Nurilmala, L.P. Manalu

The Resilience of Fishery Production Supply Chain in West Java Province, Indonesia 143
A. Nurhayati, A.A. Handaka, T. Herawati, A. Yustiati, Iskandar

The Growth Rate and Antibacterial Activity of Lactic Acid Bacteria GMH2 and GMH3 in Various Salt Concentration 155
A N Rosanto, R I Lestari, M M P Putra

Growth Pattern of Yellow Finned Barb (Mystacoleucus Marginatus, Valenciennes 1842) in the Lau Belawan River, Pancur Batu District, North Sumatera Province 161
Desrita, J S Hasibuan, A Fadhilah, D S Mulia, Nurfadillah

Molecular Analysis of the Blood Cockle, Tegillarca Granosa (Linnaeus 1758) from Indonesia 171
J I Castaneda, E Hardianto, E Setyobudi, MD R Islam

The Effect of Continuous Lighting on the Rearing of Yellowfin Tuna (Thunnus Albacares) Larvae 180
Bejo Slamet, Jhon H. Hutapea, Gunawan, Ananto Setiadi

Survival Rate and Growth of Red Tilapia (Oreochromis Sp.) Seed Hybridized from Four Different Strains 192
Farkhana Laily, Afida Sabilil Hikma, Rustadi

Leaf Extract Effectiveness of Pecut Kuda (Stachytarpheta Jamaicensis L. Vahl) on Barramundi (Lates Calcarifer) Eggs Hatchability 201
P Hartami, Mahdaliana, Mainisa, S Adhar, M Hatta, E Ayuzar, DF Hanum, Rasidi

Performance of Maintaining Brood Stock of Clarias Gariepinus with Different Densities 214
G Prabowo, M G E Kristiany, E Marlina, A Suhermanto, C A Adi

Towards a Customizable and Cost-Effective Monitoring System for Vannamei Shrimp Cultivation 221
Aynuddin, B Triyatmo, A Rakhmanda

Gelling Characteristics of Fish Surimi Gel Added with Calcium Hydroxyapatite (HAp) 228
M R Hasan, C A C Abdullah, N A Mustapha, M S M Ghazali, M Noranizan

Pathogenic Bacteria Contaminant and Heavy Metal Content in Peeled Green Mussels Sold in the Traditional Market in Yogyakarta Indonesia: A Case Study in 2022 238
W. Suwito, Andriani, G. R. Barokah

Quality of Skipjack Surimi (Katsuwonus Pelamis) with Addition of Calcium Carbonate (CaCO3) 245
M F Hidayat, Ustadi

Consumer Preference and Nutritional Value of Catfish Nuggets Substituted with Jellyfish 258
M T P Haji, Ustadi

Screening of Protease-Producing Lactic Acid Bacteria from Indonesian Fermented Fish Products and Its Application in Chitin Extraction from Vannamei Shrimp (Litopenaeus Vannamei) Shell 263
A N Aisyah, M Gavi, M M P Putra

Ultraviolet Irradiation Effect on Characteristics of Protease Enzyme from African Catfish (Clarias Sp.) Intestines 271
A Tamaya, A Suwanto, Hernawan, M Anwar, C D Poeloengasih, B Juligani, W Apriyana, A Fauzi, P Yuliyanto, R Suryani, Y Khasanah, S M Syarifah, D J Prasetyo

Seasonal Effect on the Amino Acid and Fatty Acid Profiles of Ulva Spp. Collected from Sepanjang Beach, Yogyakarta, Indonesia 279
CD Poeloengasih, R Pangestuti, EA Siahaan, Y Putra, TB Bardant, DJ Prasetyo, Hernawan, TH Jatmiko, Suratno, Y Khasanah, L Ali, DA Christyandari, W Widiastuti, R Suryani, A Windarsih, B Kumayanjati, N Wahyudin

Consumer Preference, Quality, and Financial Analysis of Tekwan Made from Snakehead, Yellow Tail, Wolf Herring, and Spanish Mackerels 288
F Y Pratama, Ustadi

Effect of Adding Egg White Powder on Surimi Quality Based on Pangasius and Catfish Flesh 298
A N F Rahma, Ustadi

Valorization of Marine By-Products Using Green Technology 310
S Netramai, F Saleah, C Kawinkamraksa, A Wipatanawin, T Kijchavengkul, F Courtes, C K Rosales

Characteristics of Ready to Cook Product from Dried Pempek with Freezing Treatment Before Drying 315
A S Fahmi, Romadhon

The Characteristics and Consumer Acceptance of Jelly Drink Fortified with Ulva Lactuca Fatty Acid Microemulsion 322
D R S Kusumajati, S A Budhiyanti

High Temperature Short Time Blanching Enhances the Antioxidative Properties of Sea Grapes (Caulerpa Racemosa) Powder 336
E Windrayani, S A Budhiyanti, N Ekantari

Effect of Carboxymethyl Cellulose Addition on the Characteristic of Chitosan-Based Bioplastic 345
L N Putranti, P S Nugraheni

Effect of Carrageenan Addition on the Characteristic of Chitosan-Based Bioplastic 360
E Favian, P S Nugraheni

Cinnamon and Amylum Mucilage Alter the Physicochemical, Sensory, and Antioxidant Activity of Arthrospira Platensis Granules 375
A W Elmaghriba, M D Candraningsih, N Ekantari

Physical, Chemical and Sensory Properties of Cookies Added with Arthrospira Platensis Granules and Stevia Sugar 389
S Raharjo, N Ekantari

Author Index

IOP Conf. Series: Earth and Environmental Science 1289 (2023) 011001 doi:10.1088/1755-1315/1289/1/011001

Preface

Pandemic COVID-19 has hardly impacted global fisheries and aquaculture, as FAO reported in the 34th session of the Committee on Fisheries (COFI) in February 2021. FOA further suggest that this pandemic impact catalyzes the fisheries sector to be more innovative, socially, and environmentally responsible. Fish and fisheries product are well-known healthy food. It contains high quality and quantity of protein and provide essential vitamins and minerals to maintain human health status. The innovation of science and technology in the fisheries sector is a crucial point to improve and assure the efficiency and sustainability of the production and utilization of the resource. Research activities and development should be synergically conducted to implement technologies that benefit the communities.

Department of Fisheries, Faculty of Agriculture, Universitas Gadjah Mada, holds a biennially international symposium to provide access and exchange of research data and fisheries experience to support the transfer of knowledge and technology to strengthen the world's fisheries sector. The 5th International Symposium on Marine and Fisheries Research (The 5th ISMFR) aims to bring together academic scientists, researcher, government institutions, private parties, and stakeholders to share and exchange progress information, experiences, and research results in all aspects of marine and fisheries sciences. The 5th ISMFR theme was "Transformative Adaptation: An Approach in Building Resilience into Sustainable Fisheries". It covered a broad spectrum of fisheries-related topics, including aquaculture, fish disease, fish genetics, biotechnology, marine natural product, seafood processing technology, seafood safety, fisheries biology, fisheries resources management, fisheries socio-economics, oceanography, climate sciences, and marine ecotoxicology. The 5th ISMFR was held on July 24nd 2023 at Grand Rohan Hotel, Yogyakarta. The seminar's organizing committee was located in the Department of Fisheries, Faculty of Agriculture, Universitas Gadjah Mada, Yogyakarta, Indonesia. Management of symposium was carried out using the website (https://ismfrugm.faperta.ugm.ac.id/) and for the proceeding submission using the Morresier platform.

The symposium consists of plenary session and parallel presentation sessions. The plenary session presented four keynote speakers, namely Professor Dr. Rockmin Dahuri the Head of Indonesian Aquaculture Society (Indonesia), Professor Dr. Sc. Amir Husni form Department of Fisheries, Universitas Gadjah Mada, Indonesia (Indonesia), Noranizan Mohd Adzahan, Ph.D from Universiti Putra Malaysia (Malaysia), and Dr. Krishna R. Salin the President World Aquaculture Society Asia Pacific Chapter (India). A total of six parallel class was conducted with 7 invited speakers and presenters from five countries, namely Egypt, USA, Thailand, Japan, and Malaysia. Presentations in each parallel class were divided into presentation panels consisting of approximately five presenters. Each presenter was given 12 minutes for presentation and discussion. A total of 144 scientific papers has been presented at the 5th ISMFR. The 5th ISMFR was attended by 136 participants. Presenters and attendances join the symposium in person at Grand Rohan Hotel, Yogyakarta. Presenters and attendances who were unable to attend in person took part in symposium virtually via Zoom Meeting platform.

This proceeding provides an opportunity for readers to gain more information from the reviewed papers that have been presented in the 5th ISMFR. The articles published in this proceeding were selected gorm the papers presented in the symposium. The reviewed from four countries (Austria, Japan, Malaysia, and Indonesia) and editors from four countries (Malaysia, Thailand, Austria, and Indonesia) have participated in the abstract screening, improving, and finalizing the manuscript. The proceedings divided into three sections, namely aquaculture, aquatic resource management, dan fish product technology. From this proceeding, readers will find recent research finding on broad aspect of fisheries and marine sciences to come up with new knowledge and idea to promote sustainable fisheries through technology and research innovation for a healthy community.

We want to thank all parties for the success of the 5th ISMFR. Our gratitude is presented to organizing committee, keynote and invited speakers, reviewers, editors, and editing staff for the dedication, hard work, and tireless efforts in implementing the symposium and publication process. We express our acknowledgment to the Rector of Universitas Gadjah Mada, the Dean of the Faculty of Agriculture, the Head of Fisheries Department, and the Publication Agency of Universitas Gadjah Mada, who provided continuous support on the symposium. Our thank you also conveyed to the speakers and participants, who have given their best effort to disseminate, discuss, and publish papers. We also thank to all parties who have contributed to the success of the 5th ISMFR and the publication. We sincerely hope readers will be find notable pieces of knowledge on fisheries and marine science from different points of view.

Chief Editor

IOP Conf. Series: Earth and Environmental Science 1289 (2023) 011001 doi:10.1088/1755-1315/1289/1/011001

List of committees of the 5th ISMFR

Steering Committee	: Prof. Dr. Ir. Alim Isnansetyo, M.Sc. (Universitas Gadjah Mada, Indonesia)
Chairman of the committee	: Mgs. Muhammad Prima Putra, S.Pi., M.Sc., Ph.D (Universitas Gadjah Mada, Indonesia)
Secretary	: Dr. Senny Hemiati, S.Pi., M.Sc. (Universitas Gadjah Mada, Indonesia)
Treasurer	: Candra Aryudiawan, S.Pi., M.Sc. (Universitas Gadjah Mada, Indonesia) Indrianto (Universitas Gadjah Mada, Indonesia)
Secretarial section	: Theresia Adven Dea K. (Universitas Gadjah Mada, Indonesia) Afif Whelly Artissandi (Universitas Gadjah Mada, Indoesia) Nahla Alfiatunnisa, S.Pi. (Universitas Gadjah Mada, Indonesia) Anisah Luthfi Purboasih (Universitas Gadjah Mada, Indonesia) Ma'rufa Nurul Latifa (Universitas Gadjah Mada, Indonesia) Dinda Laksmi Paramita (Universitas Gadjah Mada, Indonesia) Tsania Fara Zafira, S.Pi. (Uniersitas Gadjah Mada, Indoesia) Al 'Ain Shoufi Rahmah Arifin, A.Md. (Universitas Gadjah Mada, Indonesia)
Event and Conference section	: Dr. Prihati Sih Nugraheni, S.Pi., M.P (Universitas Gadjah Mada, Indoensia) Indun Dewi Puspita, S.P., M.Sc., Ph.D. (Universitas Gadjah Mada, Indonesia) Prof. Dr. Sc. Amir Husni, S.Pi., M.P (Universitas Gadjah Mada, Indonesia) Dr, Susilo Budi Priyono, S.Pi, M.Si., (Universitas Gadjah Mada, Indonesia) Indah Istiqomah, S.Pi., M.Si., Ph.D (Universitas Gadjah Mada, Indonesia) Dr. Ir. Ign. Hardaningsih, M.Si. (Universitas Gadjah Mada, Indonesia) Dr. Ir. Triyanto, M.Si. (Universitas Gadjah Mada, Indonesia) Ir. Sukardi, M.P. (Universitas Gadjah Mada, Indonesia) Drs. Namastra Probosunu, M.Si. (Universitas Gadjah Mada, Indonesia) Dr. Ratih Ida Adharini, S.Pi., M.Si. (Universitas Gadjah Mada, Indonesia) Tony Budi Satriyo, S.Pi., M.Sc., Ph.D. (Univesitas Gadjah Mada, Indonesia) Dr. Suwarman Partosuwiryo, A.Pi., M.M. (Universitas Gadjah Mada, Indonesia)
Article section	: Dr. rer. nat. Riza Yuliratno Setiawan, S.Kel., M.Sc. (Universitas Gadjah Mada, Indonesia) Prof. Dr. Ir. Djumanto, M.Sc. (Universitas Gadjah Mada, Indonesia) Prof. Dr. Ir. Alim Isnansetyo, M.Sc. (Universitas Gadjah Mada, Indonesia) Dr. Eko Setyobudi, S.Pi., M.Si. (Universitas Gadjah Mada, Indonesia) Prof. Dr. Ir. Ustadi, M.P. (Universitas Gadjah Mada, Indonesia) Dr. R.A. Siti Ari Budhiyanti, S.TP., M.P. (Universitas Gadjah Mada, Indonesia) Prof. Dr. Ir. Murwantoko, M.Si. (Universitas Gadjah Mada, Indonesia) Prof. Dr. Ir. Rustadi, M.Sc. (Universitas Gadjah Mada, Indonesia) Dr. Dini Wahyu Kartika Sari, S.Pi., M.Si. (Universitas Gadjah Mada, Indonesia) Dr. Ir. Bambang Triyatmo., M.P. (Universitas Gadjah Mada, Indonesia)
Fittings section	: Nur Ari Purnomo, S.Pi. (Universitas Gadjah Mada Indonesia) R. Budisantoso (Universitas Gadjah Mada, Indonesia)
Publication and documentation section	: Dr. Ega Adhi Wicaksono, S.Pi. (Universitas Gadjah Mada, Indonesia) Sebastian Dani Wiraredya (Universitas Gadjah Mada, Indonesia) Nokhirra Rimulso (Universitas Gadjah Mada, Indonesia)
Proceeding section	: Dr. Desy Putri Handayani, S.Pi. (Universitas Gadjah Mada, Indonesia)

IOP Conf. Series: Earth and Environmental Science 1289 (2023) 011001 doi:10.1088/1755-1315/1289/1/011001

	Eko Hardianto, S.Pi., M.Si., M.Sc., Ph.D (Universitas Gadjah Mada, Indonesia)
Consumption section	: Dr. Nurfitri Ekantari, S.Pi., M.P (Universitas Gadjah Mada, Indonesia) Murwani (Universitas Gadjah Mada, Indonesia) Rima Sabtiana (Universitas Gadjah Mada, Indonesia) Wagiran (Universitas Gadjah Mada, Indonesia)
Sponsorship section	: Ir. Hery Saksono, M.A. (Universitas Gadjah Mada, Indonesia) Suadi, S.Pi., M.Agr.Sc., Ph.D (Universitas Gadjah Mada, Indonesia) Dr. Ir. Latif Sahubawa, M.Si. (Universitas Gadjah Mada, Indonesia)

IOP Conf. Series: Earth and Environmental Science 1289 (2023) 011001 doi:10.1088/1755-1315/1289/1/011001

List of Editors of the 5th ISMFR Proceeding

Name	Institution	Email	Position
Mgs. Muhammad Prima Putra, S.Pi., M.Sc., Ph.D	Universitas Gadjah Mada, Indonesia	primaputra@ugm.ac.id	Editor In Chief
Eko Hardianto, S.Pi., M.Si., M.Sc., Ph.D	Universitas Gadjah Mada, Indonesia	ekohardianto@ugm.ac.id	Co Editor
Dr. Desy Putri Handayani, S.Pi.	Universitas Gadjah Mada, Indonesia	desy.putri.h@mail.ugm.ac.id	Co Editor
Wee Hin Boo	Universiti Kebangsaan Malaysia	weehinboo@ukm.edu.my	International Invited Editor
Andreas Bauer	Universitat fur Bodenkultur Wien	andreas.bauer@boku.ac.at	International Invited Editor
Siriyupa Netramai	Mahidol University	siriyupa.net@mahidol.edu	International Invited Editor
Al ‘Ain Shoufi Rahmah Arifin, A.Md	Universitas Gadjah Mada, Indonesia	al.ain.shoufi@mail.ugm.ac.id	Managing Editor

IOP Conf. Series: Earth and Environmental Science 1289 (2023) 011001 doi:10.1088/1755-1315/1289/1/011001

Documentation of the 5th ISMFR

Report of the
Chairman of the
ISMFR Committee
Mgs. Muhammad Prima
Putra, S.Pi., M.Sc., Ph.D.
Organized by:
Sponsored by:
ISMFR
The 5th International Sym
on Marine and Fisheries
Sustainable Fisheries and Aquac
for Global Food Resilience

Acknowledgement
STEAMBOAT
Ministry of Education, Malaysia
FRGS, MOHE
TRGS, MOHE
ScienceFund, MOSTI
Geran Putra Berimpak UPM
Geran Putra Berkumpulan UPM
Research, Publish, Travel and Talk
THANK YOU

Acknowledgement

Establishment of detection and identific
method for gram-negative histamine-
producing bacteria and research on the
behavior in fish
Hajime Takahashi Ph.D.(Prof
Lab. of Food Microbiology,
Dept. of Food Science and Technology
Tokyo University of Marine Science and Tech

VIP

SCREENING OF PROTEASE-PRODUCING LACTIC ACID BACTERIA AND
ITS APPLICATION IN DEPROTEINIZATION AND DEMINERALIZATION
PROCESSES OF CHITIN FROM VANNAMEI SHRIMP
SHELL
CONCLUSION
REFERENCES

Peer Review Statement

All papers published in this volume have been reviewed through processes administered by the Editors. Reviews were conducted by expert referees to the professional and scientific standards expected of a proceedings journal published by IOP Publishing.

- **Type of peer review:** Single Anonymous
- **Conference submission management system:** Morressier
- **Number of submissions received:** 61
- **Number of submissions sent for review:** 47
- **Number of submissions accepted:** 41
- **Acceptance Rate (Submissions Accepted / Submissions Received × 100):** 67.2
- **Average number of reviews per paper:** 1.14
- **Total number of reviewers involved:** 18
- **Contact person for queries:**
  Name: Eko Hardianto, Ph.D
  Email: ekohardianto@ugm.ac.id
  Affiliation: Department of Fisheries, Faculty of Agriculture, Universitas Gadjah Mada, Indonesia

Consumption rate of stingrays in the southern coastal area of Flores Island

Dati Nawastuti[1,3]*, Dwidjono Hadi Darwanto[1], Jangkung Handoyo Mulyo[1] and Suadi[2]

[1]Social Economy of Agriculture Department, Faculty of Agriculture, Gadjah Mada University, Indonesia
[2]Fisheries Department, Faculty of Agriculture, Gadjah Mada University, Indonesia
[3]Fishery Technology Department, Faculty of Technology, Institut Keguruan dan Teknologi Larantuka, Indonesia

* Corresponding author mail: datinawastuti@mail.ugm.ac.id

Abstract. Stingrays are an elective wellspring of creature protein for individuals living in the southern coastal area of Flores Island. Considered to have low economic value, the presence of stingrays in the Chondrichthyan meat trade is still often ignored. This study seeks to determine the influencing factors on the consumption rate of stingrays among the people of the southern coastal area of Flores Island. The analytical method used is descriptive analysis and multiple linear regressions with the SPSS 23 program. Essential information was gotten through perception, circulating polls by means of a google form, and inside and out interviews. The answers of respondents who entered amounted to 132 people and were selected based on the research area amounted to 46 people. Results stated that the variables of income level, education level, place of residence, number of dependents, product knowledge, fish products, and fish species had a significant simultaneous effect on consumption levels with an R^2 value of 0.764 percent, while 23.6% of the dependent variable is influenced by other variables that are not part of the research model. Intensive socialization among consumers, especially those who have a high level of education regarding the types of stingrays in protected regional languages and should not be consumed is very necessary. Knowledge gaps which, if not addressed immediately, will disrupt conservation area management.

1. Introduction

Nowadays there are approximately 1,160 *Elasmobranch* divided, namely sharks totaling 510 species, rays as many as 650 species [1], and several families of stingrays found in coastal areas namely wedge fishes, sleeper rays, stingrays, and guitarfishes [2]. The increasing catch of various species of rays is evidenced by the decreasing number of populations [3] which is followed by slow growth with low productivity, as well as a fairly high risk of catching [4]. In the *Chondrichthyan* meat trade, the presence of rays and skates is still often unobserved [5].

The production of stingrays in Nusa Tenggara Timur (NTT) Province, in 2020 is 1,660.06 tons [6]. In several locations in NTT Province there are a small number of fishermen who target sharks and rays, especially Manta and Mobula rays as the main catch objects, including the Island of Seraya Besar (Northern Labuan Bajo, NTT), Village of Nangaroro (NTT), Village Bola (NTT), Ende (NTT) and Lamakera (NTT). Each body component of stingrays has a different economic value and stingrays that have low economic value are only marketed and consumed locally [7].

Flores Island is administratively located in the province of East Nusa Tenggara and has an area of 14,300 km^2 divided into 8 districts. East Flores Regency is located between 08^{0}04'-08^{0}40'SL and 122^{0}38'-123^{0}57'E. The northern part of the district is bordered by the Flores Sea, to the south by the Savu Sea, to the west by Sikka Regency, and to the east by Lembata Regency. Its area is 5,983.38 km^2 with a land area of 1,812.85 km^2 and a water area of about 4,170.53 km^2 which is divided by Flores Island, Adonara Island, and Solor Island as well as 24 other small islands [8]. In the opinion of Tiffany [9] the patterns in each region depend on natural resources, culture, and tastes.

Consumers in the southern coastal area of Flores Island, especially those in East Flores Regency, have long known and consumed dried stingray meat called Belélang. Belélang products originating from Lamakera can be found in traditional markets and online sales, while fresh stingrays of various species can be found throughout the waters of Flores Island. Global data on consumption trends, consumer attitudes, and public knowledge about culinary origins from *Elasmobranch* are still very limited [10]. Research on consumption levels of stingrays is still rare and remains a major challenge due to data limitations. The objective of this research is to ascertain the factors that influence the rate of stingray consumption in the southern coastal region of Flores Island, Indonesia.

2. Materials and Methods

2.1. Data source and collection

The exploration information was taken from a survey utilizing a google form which was conveyed for a month from August 4, 2021, to September 4, 2021. The information entered were 132 respondents from Java Island, Timor Island, Sumba Island, Adonara Island, Lembata Island, Belitung Island, and Flores Island. The research focuses on the East Flores Regency, Flores Island's southern coast; the information that can be utilized are 46 pieces. Considering the information that has been acquired, semi-organized interviews were led, both face to face and phone interviews with 26 individuals. The informants are consumers who have consumed both fresh and dried stingrays from 2014 to 2021. Interviews were conducted on 17 men, and 9 women. The location of this research is the Pasar Inpres and Pasar Senja in the Larantuka District, as well as the Pasar Oka which operates once a week to accommodate the needs of consumers from two sub-districts, namely Ile Mandiri District and Lewolema District.

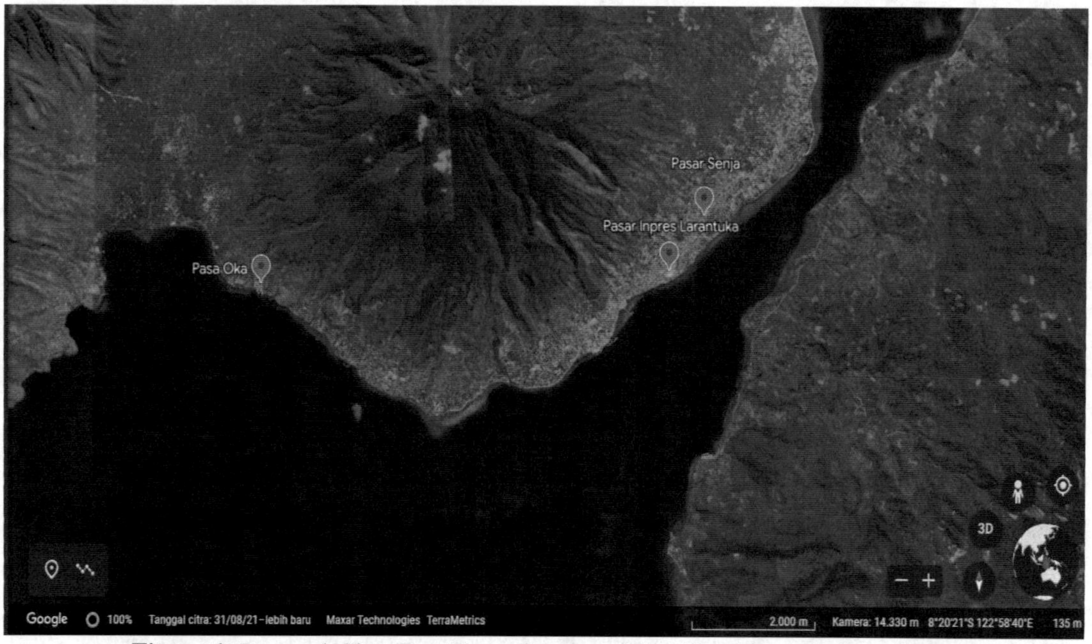

Figure 1. Research Site, East Flores Regency, East Nusa Tenggara Province

The method used to determine the factors that influence the consumption of stingrays is a multiple linear regression analysis. The independent variables needed include income, education level, place of residence, number of dependents in the family, product knowledge, consumption goals, and consumer tastes. Multiple linear regression model as follows [9]:

$$Y = \alpha + \beta_1 X_1 + \beta_2 X_2 + \beta_3 X_3 + \beta_4 X_4 + \beta_5 X_5 + d_1 + d_2 + \mu \text{ .... } (1)$$

Description:

Y = Stingray consumption (IDR/Kg)
α = Constant
β = Regression Coefficient
d_1 = Dummy variable stingray product (0 = fresh stingray; 1 = dried stingray)
d_2 = Dummy variable protection status of stingrays (0 = not protected; 1 = protected)
μ = Error Term (interference error)
X_1 = Income (IDR)
X_2 = Education level (head)
X_3 = Residence (head)
X_4 = Number of family dependents (head)
X_5 = Product knowledge (head)

Variables in the study:

1. Family income:
a. < IDR 1,500,000
b. IDR 1,500,000 to 2,500,000
c. IDR 2,500,000 to 3,500,000
d. > IDR 3,500,000
2. Last education level:
a. Elementary School
b. Junior high school
c. Senior High School
d. College
3. Place of residence:
a. Larantuka Sub Regency
b. Ile Mandiri Sub Regency
c. Lewolema Sub Regency
d. In addition to Larantuka, Ile Mandiri, and Lewolema sub-Regency
4. Number of family dependents:
a. 1-3 people
b. 4-6 people
c. More than 6 people
5. Product knowledge (protected stingray species):
a. Know
b. Do not know

2.2. Data analysis

Normality, multicollinearity, heteroscedasticity, and autocorrelation using the Cochrane Orchut method were the traditional assumption tests applied to the data. Comparison between the variables of income level, education level, place of residence, number of family dependents, product knowledge, fish products, and types of fish to consumption level was carried out by t-test. Variations between the dependent variable, independent variable, and dummy variable were compared using one-way ANOVA. Decision-making is done at the value of 10%. Statistical analysis using SPSS 23 tool.

3. Result and Discussion

3.1. Result

3.1.1. Partial testing. Partial test results to make decisions using the value of = 10%. The partial test shows a significance value (sig.) on the income level variable (X_1) of $0.070 < 0.1$, indicating that the level of income variable (X_1) has a partial effect on the consumption level variable (Y) (Table 1). Consumer income which ranges from less than IDR 1,500,000 up to more than IDR 3,500,000 choose to consume stingrays either in fresh or dried form which is sold in traditional markets or sold by walking around because the price of stingrays is quite affordable.

Table 1. Coefficients[a]

Model	Unstandardized Coefficients B	Unstandardized Coefficients Std. Error	S. Coeff. Beta	t	Sig.
1 (Constant)	59316.60	6440.80		9.20	<.001
Level of income	1522.87	816.60	.154	1.85	.070
Level of education	-7661.37	895.95	-.827	-8.55	<.001
Residence	2710.34	757.60	.307	3.57	<.001
Number of family dependant	-5205.54	1579.60	-.308	-3.29	.002
Knowledge of product	-2629.63	1409.30	-.157	-1.86	.070
Fish product	-6062.99	1678.10	-.299	-3.61	<.001
Fish type	-2138.56	1743.60	-.103	-1.22	.228

a. Dependent variable: Consumption rate

Source: Processed data, 2021

Fresh stingrays that are sold can be found in 2 forms, namely small pieces in plate units with prices ranging from IDR 10,000 to IDR 25,000 (Figure 2) weighing between 500 - 750 grams/plate that have been chopped into small pieces. The consumer can also choose in whole form (1 head) which is traded at a price range of IDR 40,000 to IDR 100,000/head (Figure 3).

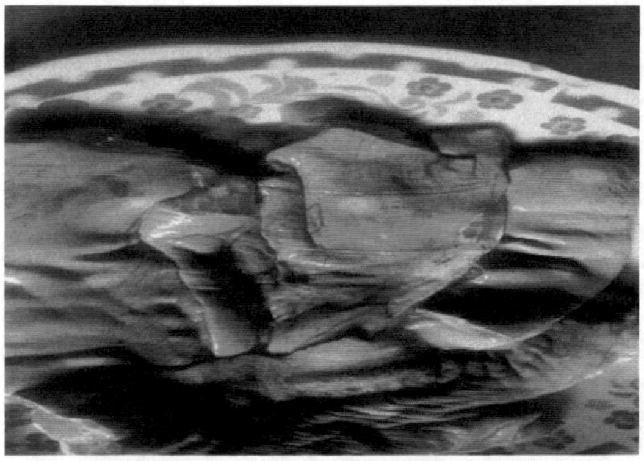

Figure 2. Stingray sold dishes

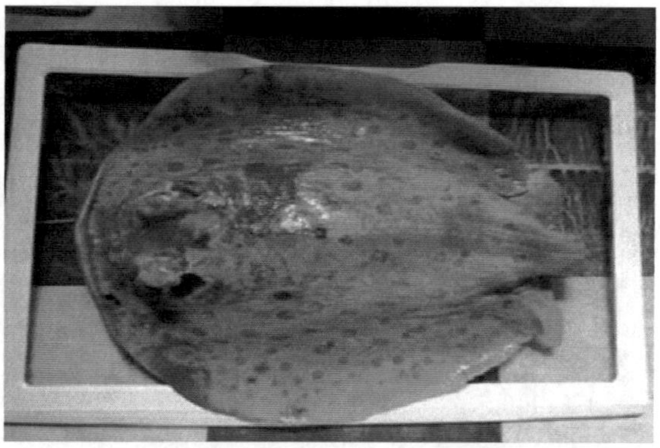

Figure 3. Stingray sold whole

Figure 4. Dried stingray in the form of a bracelet

The price of dried stingray products is around IDR 20,000 to 25,000/bracelet with a weight ranging from 180-240 grams/bracelet (Figure 4) and tied with a string containing 10 fish called welak. Fish are bought during the rainy season, and during the storm season to be stored for a long time. This product can last for a year at room temperature. The significance level for the education level variable (X_2) is 0.001 < 0.1. This shows that the education level variable has a partial effect on the consumption level variable. As many as 2% of consumers have a junior high school education, 13% of consumers have graduated from elementary school, 13% of consumers are high school graduates, and 46% of consumers have a college education. The level of higher education possessed by consumers is a factor that partially influences the consumption rate of stingrays. According to Todaro (2000) in [11], there is a positive correlation between a person's education and the income he will earn. For a person who can complete his secondary education or college, his income will differ between 300 and 800% when compared to those who can only finish a lower-level school. Consumers who have a high level of education have memories of delicious taste memories, chewy meat texture, no real bones, having a distinctive odor, and an increasing appetite for stingrays. The reasons that are most often put forward by consumers are the delicious taste of meat when processed into various types of dishes; it has high nutritional content; can cure diseases. However, this opinion was denied by other respondents who argued that people with asthma should not eat stingray meat for fear of a recurrence of asthma. This was also supported by the opinion of [12], who stated that Chinese medicine-prescribed anti-asthmatic preparations containing 25–107,000 lgG have been linked to acute poisoning in children and adults.

Significance value at residence variable (X_3) is 0,001 < 0,1. This shows that the variable of residence to some degree affects the variable utilization rate. People easily get stingrays because the position of the villages is close to the sea (coastal villages). Stingrays are obtained when the community engages in fishing activities when the moon rises halfway or fish that accidentally enter the fishing nets. Stingrays are very happy to play in surface currents, so they are easily caught by the community either using fishing rods or because they are trapped in nets. They can catch 1-2 stingrays of various sizes and species during the fishing season. This of course will greatly disrupt sustainable fisheries management [13] so it is very important to involve the community and make consensus with fishermen so that they no longer catch stingrays [14].

The variable number of family dependents (X_4) has an importance worth of 0.002 < 0.1 which demonstrates that the variable number of family dependents to some extent affects the utilization level variable (Y). The family's number of dependents and the factors necessary to meet dietary requirements (animal protein), can be a reason for consumers to consume stingrays because the price is affordable and can be cooked in large portions. People in the southern coastal area of Flores Island generally have large family sizes (more than 3 people in one household), so they need lots of side dishes at affordable prices because fish contribute to their food security and nutrition [15].

The product knowledge information variable (X_5) has an importance worth of 0.070 < 0.1 which shows that the information variable to some extent affects the utilization rate. Lack of socialization from the government resulted in a lack of public knowledge of stingray products, causing people to continue to consume stingrays both in fresh and dried form. They consume stingrays without sufficient knowledge about the types of fish that are protected and the impact of their consumption in the future. Product knowledge has a very important role in consumer decision-making [16]. Obviously, this is extremely hindering to customers and the climate so makers must satisfy purchaser needs for trust attributes like wellbeing, nourishment, beginning, and supportability [17].

So far, stingrays caught in nets, regardless of size, are released if they are still alive. If a stingray is found dead because it is entangled in a net, the fish will be consumed and distributed to the family/neighbors, even though based on the directions in the socialization carried out by the government, the fish is buried. Local wisdom says that whatever the sea has given is a fortune. Fishermen also stated that the replacement of nets (which were torn by entangled stingrays) by the local government through the relevant agencies did not match the size of the initial nets, tended to be delayed, were second-hand goods, wasted time and travel costs (to the district capital). This makes fishermen apathetic to report to the relevant agencies when their nets are torn apart by trapped stingrays. The fish will be taken for consumption and sold. The need for food is one of the reasons [18] to continue to consume and sell stingrays. The money from the sale of fish is used to buy thread for sewing broken nets. The contribution of fishers, although small, is very important because it will have an impact on conservation so it is necessary to understand the same perception from various parties to avoid repeated actions in their local areas [19].

According to fishermen, so far, the government has only provided calendars in the form of pictures of protected animals to community monitoring groups without explaining further; so, the general public and consumers, in particular, do not understand the importance of stingrays for ecosystems and the environment. If this overfishing continues there will be permanent damage to the stock and failure to restore original population size [20] so area management should consider other livelihood options for fishermen and traders [21]. Mitigation measures taken so far have not been as effective as expected [22]. Ignorance of consumers about the products they consume is a very big risk for the sustainability of stingray fisheries. if the performance of policy implementation is not carried out optimally, there will be stagnation in the implementation of activities which will certainly affect performance in implementation. Increasing communication support and regional fisheries stakeholder cooperation is an urgent need to create sustainable stingray fisheries [23].

The fish product dummy variable (d_1), both for fresh and dried fish has an importance worth of 0.001 < 0.1, demonstrating that the fish product dummy variable has impacted utilization levels. This is because fish products can be processed into a variety of delicious dishes for families throughout the year. Consumers can choose stingray products according to their taste. Fresh stingrays both cut and whole in the market will always sell out. This is related to consumer tastes for processed stingray products which are quite competitive with other types of processed fish, and the chewy texture that makes stingray meat unique, and of special value.

The dummy variable of protected fish species (d_2) has an importance worth of 0.228 > 0.1 which demonstrates that the variable of secured and unprotected fish species doesn't influence the utilization rate. Reported that there were spotted eagle rays (*Aetobatus narinari*), Chilean devil rays (*Mobula tarapacana*), reef manta rays (*Manta alfredi*), oceanic manta rays (*Manta birostris*), spinetail devil rays (*Mobula japanica*) [24]. Some consumers already know that there are several types of protected species of stingray but they do not know the shape, appearance, or local name of fresh or dried fish products. Government policies are needed that can include improvements to the chain of control and traceability control as well as amendements [25].

3.1.2. Simultaneous test. The F test was utilized to decide the impact of the factors of income level (X_1), an education level (X_2), place of residence (X_3), number of dependents (X_4), product knowledge (X_5), fish product dummy (d_1), and fish type dummy (d_2) together to the variable consumption rate (Y). F table is seen at = 10%, n = 46, k = 10, then the degree of the numerator = k – 1 = 10 – 1 = 9 and the degree of the denominator = n – k = 46 – 10 = 36, so that the F table = 1,81.

Table 2. ANOVA[a]

Model	Sum of Square	df	Mean Square	F	Sig.
1 Regression	2927550925	7	418221560.7	17.6	<.001[b]
Residual	902557770.9	38	23751520.29		
Total	3830108696	45			

a. Dependent variable: Consumption rate
b. Predictors: (Constant). Fish type. Income level. Fish products. Knowledge product. Residence. The number of family dependents. Level of education
Source: Processed data, 2021

The test of the impact of the independent variable on the dependent variable (Y) on the outcomes of data processing yielded the following results. The determined F esteem is 17.608 with an importance worth of 0.001 (Table 2). This shows that the determined F esteem is 17.608 > F table 1.81 so H0 is dismissed and sig < 0.1, and that implies the autonomous variable affects the degree of stingray utilization in individuals in the southern coastal area of Flores Island, East Flores Regency.

3.1.3. Coefficient of determination (R^2). The consequences of the coefficient of assurance (R^2) test results over, the R^2 number is 0.764, and that implies that the greatness of the impact of the autonomous variable on the reliant variable is 76.4% while 23.6% of the Y variable is affected by different factors outside the research model.

3.2. Discussion

The consistent worth of 59316.673 truly intends that assuming the variable worth of income level (X_1), an education level (X_2), place of residence (X_3), number of family dependents (X_4), product knowledge (X_5), fish products (d_1), and the type of fish (d_2) is considered constant or equal to 0, then the consumption rate (Y) has a value of IDR 59,316,673. The worth of the X_1 coefficient of 1522.874 states that on the off chance that the other independent variables have a decent worth and the income level (X_1) has an increase of 1%, the consumption level (Y) of stingrays will increase by IDR 1,522,874. The increase in consumer income results in an increase in the number of stingray products consumed, which of course has an impact on the sustainability of the stingray fishery because it will have implications for higher catching to meet consumer demand. Protection of this species should be increased to avoid further depletion and possible extinction of marine ecosystems [26].

The worth of the X_2 coefficient of - 7661.372 states that if the other independent variables have a decent worth and the degree of education (X_2) has expanded by 1%, then the utilization level (Y) of stingrays will diminish by IDR 7,661,372. In view of the consequences of this examination, the degree of advanced education given by purchasers doesn't ensure that the maintainability of stingray fisheries works out positively. The catching of stingrays that is still happening today is due to the lack of information received by consumers, so it has not had an impact on awareness not to consume these products. Regulations and supervision applied to these water areas [27].

The worth of the X_3 coefficient of 2710.348 states that assuming the other independent variables have a decent worth and the spot of home (X_3) has an extra 1%, the utilization level (Y) of stingrays will increment by IDR 2,710,348 as well as the other way around. If the consumer's residence is outside the research area, the consumption level will decrease by IDR 2,710,348. Living close to the sea (coastal area) allows the surrounding community to easily get stingrays, both during fishing activities during the rising moon with a simple fishing rod or fish that accidentally enter the fishermen's nets as by-catch. Of course, this slowly has an impact on the formation of consumer tastes, which is reinforced by the opinion of Sudarman and Algifari (2006) that one of the factors that influence the demand for an item is consumer taste.

The worth of the X_4 coefficient of - 5205,540 states that assuming the other independent variables have a proper worth and the level of the number of family dependents (X_4) has expanded by 1%, then, at that point, the utilization level (Y) will diminish by IDR 5,205,540. The family can be a justification for shoppers to consume stingrays on the grounds that the costs are reasonable and can be cooked in enormous segments, yet when there is an expansion in the quantity of family, the utilization level of stingrays diminishes. This is because, for a variety of reasons, not all members of a family are able to consume stingray products. As a result, consumers choose to purchase other kinds of fish that are less expensive, easier to obtain, and commonly consumed.

The worth of the coefficient of X_5 is - 2629,631 which expresses that assuming that the other dependent variable have a decent worth and product knowledge (X_5) has expanded by 1%, then the utilization rate (Y) will decrease by IDR 2,629,631. The level of knowledge of the products consumed also affects the sustainability of the stingray fishery. Consumers have the right to know that the products they consume do not damage nature and have permission from the relevant agencies that the products they consume are safe for health. The consumption of stingray meat for people on the southern coast of Flores Island is common, although not often done. Consumers generally know very little about this species and its uses [10]. If consumers are provided with sufficient knowledge about the types and benefits of stingrays for marine ecosystems, massive stingray fishing can be reduced or even stopped.

According to the d_1 coefficient's value of – 6062,992, if the other independent variables are dried fish products and fresh (d_1) have increased by 1%, the consumption rate (Y) will decrease by IDR 6,062,992. Dried stingray meat products available in the market are the cheapest part from an economic point of view. This happens because the main product that is taken is the gills so the supply of dried stingray meat is always available in the market. Although the supply of dried stingray meat is always in the market, the consumption rate has decreased because they have to compete with products from other types of dried fish. Dried fish products from species other than stingrays are cheaper; can be consumed by everyone.

The coefficient d_2 has a value of – 2138,561, which indicates that the consumption level will increase by 1% if the other independent variables remain unchanged and the protected fish species (d_2) is increased (Y) will decrease by IDR 2,138,561. The government and related parties should provide more frequent education on the importance of stingrays for marine ecosystems and provide sanctions by applicable regulations. If counseling to consumers can be carried out properly, and they know the types of stingrays that are protected in the local language and their processed products, the level of awareness of the important role of this wildlife in marine ecosystems will begin to form. Stingray catches will decrease if consumers no longer consume stingray meat. That's when the sustainability of stingray fisheries will be realized because sustainable fisheries resources are an important component of the economic well-being of fishermen as well [28].

4. Conclusion
Intensive outreach to consumers, especially those who have a high level of education regarding the types of stingrays in local languages that are protected and may not be consumed is urgently needed. Knowledge gaps that if not addressed immediately will interfere with the management of conservation areas. Fishery management must be legally binding, and fishermen are also required to report the type of fish caught because good information will be very useful in controlling the trade in stingray meat in this area. As many as 74% of consumers do not know whether the stingrays they have consumed so far are animals protected by the government or not and consumers in the Flores Island region are reckless consumers because they continue to consume these products only based on information about the valuable benefits for the body. Highly educated consumers must be involved because they can influence the community to change their consumption habits of stingray meat so that they become smart consumers who understand product quality and the impact of the food they consume.

Acknowledgment
We thank the Ministry of Research, Technology and Higher Education Scholarships for Domestic Postgraduate Education Scholarships and the respondents who provided valuable information to the authors.

References
[1] Stevens G, Fernando D, Dando M and di Sciara G N 2019 An Introduction to Manta and Devil Rays: in Guide to the Manta and Devil Rays of the World Princeton University Press pp 12–13 doi: 10.2307/j.ctvs32s7t.9 https://www.jstor.org/stable/j.ctvs32s7t

[2] Dulvy N K, Fowler S L, Musick J A, Cavanagh R D, Kyne P M, Harrison L R, Carlson J K, Davidson L N K, Fordham S V, Francis M P, Pollock C M, Simpfendorfer C A, Burger G H, Carpenter K E, Compagno L J V, Ebert D A, Gibson C, Heupel, M R, Livingstone S R, Sanciangco J C, Stevens J D, Valenti S and White W T 2014 Extinction risk and conservation of the world's sharks and rays *eLife Res. Artic.* pp 1–34 doi: 10.7554/eLife.00590.001 https://elifesciences.org/articles/00590

[3] Croll D A, Dewar H, Dulvy N K, Fernando D, Francis M P, Galván-Magaña F, Hall M, Heinrichs S, Marshall A, Mccauley D, Newton K M, Notarbartolo-di-Sciara D, O'Malley M, O'Sullivan

J, Poortvliet M, Roman M, Stevens G, Tershy B R and White W T 2016 Vulnerabilities and fisheries impacts: the uncertain future of manta and devil rays *Aquat. Conserv. Mar. Freshw. Ecosyst.* **26**(3): 562–75 doi: 10.1002/aqc.2591 https://onlinelibrary.wiley.com/doi/full/10.1002/aqc.2591

[4] Pardo S A, Kindsvater H K, Cuevas-Zimbrón E, Sosa-Nishizaki O, Pérez-Jiménez J C and Dulvy N K 2016 Growth, productivity, and relative extinction risk of a data-sparse devil ray *Sci. Rep.* **6**: 1–10 doi: 10.1038/srep33745 https://www.nature.com/articles/srep33745

[5] Dent F and Clarke 2015 State of the global market for shark products Rome, Italy: Food and Agriculture Organization of the United Nations [Online] Available: https://www.fao.org/publications/card/en/c/a3e097e2-960f-41a0-b957-602563de5c7c/

[6] Satu Data KKP Produksi Perikanan Pari Provinsi Nusa Tenggara Timur tahun 2010-2020 (2022) [Online] Available: https://statistik.kkp.go.id/home.php?m=prod_ikan_prov&i=2#panel-footer

[7] Prabuning D, Ningtias P, Harvey A, Setiasih N and Yahya Y 2015 Shark and Ray Supply Chain in NTB (West of Nusa Tenggara) and NTT (East of Nusa Tenggara) in *Simposium Hiu dan Pari di Indonesia 2015*. Kementerian Kelautan dan Perikanan dan WWF-Indonesia pp 127–34 [Online] Available: https://www.researchgate.net/publication/313053442_Shark_and_Ray_Supply_Chain_in_NTB_West_of_Nusa_Tenggara_and_NTT_East_of_Nusa_Tenggara/link/588eea7b92851cef136316e9/download

[8] RPJMD II Kabupaten Flores Timur Gambaran umum kondisi daerah Flores Timur Nusa Tenggara Timur [Online] Available: http://florestimurkab.go.id

[9] Tiffany A E, Mudzakir A K and Wibowo B A 2020 Analisis tingkat konsumsi ikan laut dan faktor-faktor yang mempengaruhi pada masyarakat Semarang *J. Fish. Resour. Util. Manag. Technol.* **9**: 25–34 [Online] Available: https://ejournal3.undip.ac.id/index.php/jfrumt/article/view/29361/24770

[10] Ali L, Grey E, Singh D, Mohammed A, Tripathi V, Gobin J and Ramnarine I 2020 An evaluation of the public's knowledge, attitudes and practices (KAP) in Trinidad and Tobago regarding sharks and shark consumption *PLoS One*. **15**(6): 1–24 doi: 10.1371/journal.pone.0234499 https://journals.plos.org/plosone/article?id=10.1371/journal.pone.0234499

[11] Adiana P P E and Karmini N 2012 Pengaruh pendapatan, jumlah anggota keluarga, dan pendidikan terhadap pola konsumsi rumah tangga miskin di kecamatan Gianyar *E-Jurnal Ekon. Pembang. Univ. Udayana.* **1**(1): 39–48 [Online] Available: https://media.neliti.com/media/publications/44411-ID-pengaruh-pendapatan-jumlah-anggota-keluarga-dan-pendidikan-terhadap-pola-konsums.pdf

[12] Ooi M S M, Townsend K A, Bennett M B, Richardson A J, Fernando D, Villa C A and Gaus C 2015 Levels of arsenic, cadmium, lead and mercury in the branchial plate and muscle tissue of mobulid rays. *Mar. Pollut. Bull.* **94**(1): 251–59 doi: 10.1016/j.marpolbul.2015.02.005. https://www.sciencedirect.com/science/article/pii/S0025326X15000843

[13] Spaet J L Y and Berumen M L 2015 Fish market surveys indicate unsustainable elasmobranch fisheries in the Saudi Arabian Red Sea *Fish. Res.* **161**: 356–64 doi: 10.1016/j.fishres.2014.08.022 https://www.sciencedirect.com/science/article/abs/pii/S0165783614002689

[14] Falla P A M, Castro E R, Ballesteros C A, Hooker H B, Caldas J P, Rojas A and Navia A F 2019 Effect of a precautionary management measure on the vulnerability and ecological risk of elasmobranchs captured as target fisheries *Reg. Stud. Mar. Sci.* **31**: 100779 doi: 10.1016/j.rsma.2019.100779. https://www.sciencedirect.com/science/article/abs/pii/S2352485519301963

[15] Chan C Y, Tran N, Pethiyagoda S, Crissman C C, Sulse T B and Phillips M J 2019 Prospects and challenges of fish for food security in Africa *Glob. Food Sec.* **20**: 17–25 doi: 10.1016/j.gfs.2018.12.002

https://www.sciencedirect.com/science/article/pii/S2211912418300439
[16] McEachern M G and Warnaby G 2018 Exploring the relationship between consumer knowledge and purchase behaviour of value-based labels *Int. J. Consum. Stud.* **32**(5): 414–26 doi: 10.1111/j.1470-6431.2008.00712.x https://onlinelibrary.wiley.com/doi/abs/10.1111/j.1470-6431.2008.00712.x
[17] Alfnes F, Chen X and Rickertsen K 2018 Labeling farmed seafood: A review. *Aquac. Econ. Manag.* **22**(1): 1–26 doi: 10.1080/13657305.2017.1356398 https://www.tandfonline.com/doi/abs/10.1080/13657305.2017.1356398?journalCode=uaqm20
[18] Glaus K B J, Adrian-Kalchhauser I, Piovano S, Appleyard S A, Brunnschweiler J M and Rico C 2019 Fishing for profit or food? Socio-economic drivers and fishers' attitudes towards sharks in Fiji. *Mar. Policy.* **100** (November 2018): 249–57 doi: 10.1016/j.marpol.2018.11.037 https://www.sciencedirect.com/science/article/abs/pii/S0308597X18303270
[19] Satama M and Iglesias E 2020 Fuzzy cognitive map clustering to assess local knowledge of ecosystem conservation in Ecuador. *Sustainability.* **12**(6) doi: 10.3390/su12062550 https://www.mdpi.com/2071-1050/12/6/2550
[20] Bernardo C, Corrêa de Lima Adachi A M, Paes da Cruz V, Foresti F, Loose R H and Bornatowski H 2020 The label 'Cação' is a shark or a ray and can be a threatened species! Elasmobranch trade in Southern Brazil unveiled by DNA barcoding *Mar. Policy.* **116**(December 2019) doi: 10.1016/j.marpol.2020.103920 https://www.sciencedirect.com/science/article/abs/pii/S0308597X19309261
[21] Seidu I, Brobbey L K, Danquah E, Oppong S K, van Beuningen D, Seidu M and Dulvy N K 2022 Fishing for survival: Importance of shark fisheries for the livelihoods of coastal communities in Western Ghana *Fish. Res.* **246**(October 2021): 106157 doi: 10.1016/j.fishres.2021.106157 https://www.sciencedirect.com/science/article/abs/pii/S016578362100285X
[22] de Chelotti L and dos Santos P R S 2020 Biodiversity and conservation of marine elasmobranchs in the extreme south of the Brazil, Southwestern Atlantic *Reg. Stud. Mar. Sci.* **38**: 101371 doi: 10.1016/j.rsma.2020.101371 https://www.sciencedirect.com/science/article/abs/pii/S2352485520304990
[23] Friedman K, Gabriel S, Abe O, Nuruddin A A, Hassan A A R B R, Cadrin S X, Cornish A, De Meulenaer T, Dharmadi, Fahmi, Tuan Anh L H, Kachelriess D, Kissol Jr L, Krajangdara T, Wahab A R, Tanoue W, Tharith C, Torres Jr F, Wanchana W, Win S, Yokawa K and Ye Y 2018 Examining the impact of CITES listing of sharks and rays in Southeast Asian fisheries *Fish Fish..* **19**(4): 662–76 doi: 10.1111/faf.12281 https://onlinelibrary.wiley.com/doi/10.1111/faf.12281
[24] Putra M I H 2017 Megafauna Laut di Perairan Solor, Flores Timur - Indonesia 2016 - 2017 (Versi Bahasa) Nusa Tenggara Timur, Indonesia [Online] Available: website: https://www.misoolfoundation.org/savu-sea-alliance%0AEmpowering
[25] Pramod G, Nakamura K, Pitcher T J and Delagran L 2014 Estimates of illegal and unreported fish in seafood imports to the USA *Mar. Policy.* **48**: 102–13 doi: 10.1016/j.marpol.2014.03.019 https://www.sciencedirect.com/science/article/pii/S0308597X14000918
[26] Worm B, Davis B, Kettemer L, Ward-Paige, C A, Chapman D, Heithaus M R, Kessel S T and Gruber S H 2013 Global catches, exploitation rates, and rebuilding options for sharks *Mar. Policy.* **40**(1): 194–204 doi: 10.1016/j.marpol.2012.12.034 https://www.sciencedirect.com/science/article/abs/pii/S0308597X13000055
[27] Furlong-estrada E, Galván-magaña F and Tovar-ávila J 2017 Use of the productivity and susceptibility analysis and a rapid management-risk assessment to evaluate the vulnerability of sharks caught off the west coast of Baja California Sur, Mexico *Fish. Res.* **194**: 197–208 doi: 10.1016/j.fishres.2017.06.008 https://www.sciencedirect.com/science/article/abs/pii/S0165783617301637

[28] Jensen F, Nielsen M and Ellefsen H 2019 Defining economic welfare in fisheries *Fish. Res.* **218**: 138–54 doi: 10.1016/j.fishres.2019.05.011 https://www.sciencedirect.com/science/article/abs/pii/S0165783619301274

IOP Conf. Series: Earth and Environmental Science 1289 (2023) 012002 doi:10.1088/1755-1315/1289/1/012002

Optimising Empowerment of Fostered Villages for Increased Income of Tambak Farmers in Bulu Cindea Village

R Rahmi[1*], A Syarif[2], J Ibrahim[3], F Indah Yani[4], A Chadijah[1], A R Nanda[5], A S Syamsuri[6], F Fitriyani[7], S Syamsia[8]

[1]Aquaculture Study Program, Faculty of Agriculture, Muhammadiyah University of Makassar, Indonesia.
[2]Agribusiness Study Program, Faculty of Agriculture, Muhammadiyah University of Makassar, Indonesia.
[3]Medical Study Program, Faculty of Medicine and Health Sciences, Muhammadiyah University of Makassar, Indonesia.
[4]Aquaculture Study Program, Faculty of Agriculture, Muhammadiyah University of Pare pare, Indonesia.
[5]Irrigation Engineering Study Program, Faculty of Engineering, Muhammadiyah University of Makassar, Indonesia.
[6]Library Science Study Program, Faculty of Adab and Humanities, Alauddin State Islamic University Makassar, Indonesia
[7]Fisheries Departement, Faculty of Marine and Fisheries, Hasanuddin University, Indonesia.
[8]Agrotechnology Study Program, Faculty of Agriculture, Muhammadiyah University of Makassar, Indonesia.

*Corresponding author mail: rahmiperikanan@unismuh.ac.id

Abstract. The increase in salt imports by the Indonesian government should prompt coastal communities endowed with extensive marine resources to improve their salt production, subsequently catering to the demands of the industrial sector. The primary objective of the target village empowerment program is to assess the impact of implementing geomembrane technology on the augmentation of revenue among salt producers residing in Bulu Cindea Village, located in the Pangkep Regency. The present study employed descriptive analysis and income analysis methodologies. Descriptive analysis offers a more accurate depiction of the indigenous knowledge and practices used by salt pond producers. The examination of income assesses the earnings of salt pond farmers about implementing the PDB program. The present study examined the potential enhancement of pond farmers' income through implementing the PDB program. This was accomplished by analyzing variance, followed by a t-test, to assess the impact of salt farmers' income on adopting geomembrane technology. The results revealed a statistically significant difference ($p < 0.05$) between salt farmers' income before and after the utilization of geomembrane technology, indicating a positive effect on their overall income. The findings suggest that implementing the PDB program, specifically through geomembrane technology, can enhance the socioeconomic conditions of salt producers residing in Bulu Cindea Village.

1. Introduction

Salt is an essential food element indispensable in domestic settings and a fundamental component in the food manufacturing sector. The current salt production still needs to satisfy the existing market demand. Salt production capacity varies across different regions, affecting the overall supply and demand dynamics [1]. The substantial need for salt necessitates alternative means of meeting the demand beyond relying solely on domestic salt production [2]. The demand for salt by the population surpasses the existing manufacturing capability [1]. Upon examining the state of salt production in Indonesia, it can be deduced that the production of domestic salt, particularly table salt, only coincides with domestic salt production [3]. Simultaneously, the sector continues to rely on imported goods. To solve these concerns, it is essential for the government to prioritize domestic production for the manufacturing of salt of superior quality and to incentivize domestic salt manufacturers. This is particularly crucial considering the higher demand for industrial salt than table salt [4]. The current productivity levels exhibit a moderate degree of elevation due to the absence of substantial investments in the effective management of natural resources, such as ponds. To effectively tackle these concerns, the government must prioritize domestic production in the manufacturing of premium salt and incentivize domestic salt producers. This is particularly crucial considering the higher demand for industrial salt than table salt [4]. The current productivity levels exhibit a moderate degree due to a lack of substantial investment in effectively managing natural resources, such as ponds.

Given these concerns, the government must prioritize domestic production to foster the creation of premium salt that will incentivize domestic salt manufacturers to utilize it. This is particularly crucial considering the higher demand for industrial salt compared to table salt [5]. As per the Regulation of the Minister of Maritime Economy and Fisheries of the Republic of Indonesia, specifically No. PER.41/MEN/2011 [6], the Indonesian government has implemented various measures to promote domestic salt production for local consumption. These initiatives involve enhancing the People's Salt Business (PUGAR) and exploring the untapped potential of existing salt deposits. PUGAR is a component of the Independent National Program for Fisheries and Marine Communities (PNPM Mandiri-KP/National Program for Independent Marine and Fisheries Communities) [7]. The primary objective of PUGAR is to attain national self-sufficiency in salt production by enhancing both the quality and quantity of salt production. This is intended to meet the domestic need for salt, which has traditionally been fulfilled through imports from foreign nations.

The enhancement of salt quality depends upon the concentration of sodium chloride inside the salt [8]. The sodium chloride level in salt is contingent upon the concentration of seawater utilized throughout the salt production process and the specific geographical source from which the seawater is collected. In addition to this, the quality of salt is significantly impacted by the geographical location in which it undergoes crystallization [9][10]. The conventional method including the utilization of a designated area of land, commonly referred to as a salt table, represents a crystallization environment whereby salt crystal grains are generated [11][12]. The production of salt via this technology is expected to yield substandard quality due to the potential mixing of salt crystals with soil particles, resulting in a deviation from the desired pure white colouration [13]. Hence, there is an urgent need to incorporate novel technological advancements into the salt crystallization process in order to enhance the quality and quantity of salt production. The implementation of geomembrane technology is anticipated to enhance both the quality and quantity of production [14][15]. A geomembrane is a synthetic layer characterized by a low level of permeability, rendering it impervious even to acidic liquids [16]. The utilization of this substance is prevalent in artificial ponds, sanitary landfills, and other similar contexts [17][18]. Implementing geomembrane technology holds significant advantages for salt farmers since it enables the salt crystallization process to be conducted without direct soil contact. The geomembrane has a higher rate of solar absorption compared to conventional technology, hence enhancing crop quality and reducing the duration of the harvest chain [19][20][21]. Bulu Cindea, a village located in Pangkep Regency, is predominantly inhabited by salt pond farmers. The implementation of geomembrane technology in this region is anticipated to transform it into a prominent salt production hub. This technological advancement can directly enhance salt pond farmers' financial prospects by augmenting their income.

2. Research Methods

The study was carried out between the months of January and June in the year 2023. The study was carried out at Bulu Cindea Village, located in the Bungoro area of the Pangkajene and Islands (Pangkep) Regency in South Sulawesi, Indonesia. The town of Bulu Cindea is situated in the western region of the Pangkep Regency, exhibiting a highly advantageous geographical location due to its proximity to the provincial capital of South Sulawesi. Bulu Cindea Village is situated inside the

Bungoro Sub-district of Pangkep Regency from an administrative standpoint. The primary emphasis of the study is centered around a salt facility, hence the selection of the study's location was deliberately made in close proximity to a salt pond situated near the shoreline. The research locations are depicted in the following map (Fig. 1).

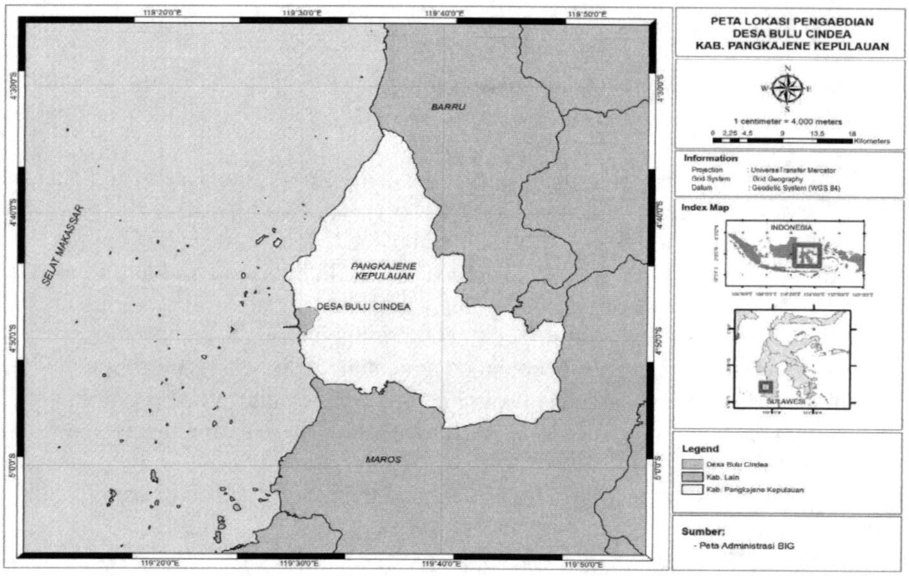

Figure 1. The map of location research at Bulu Cindea Village

The data was gathered using several methods, including observation, interviews, and a literature review focusing on salt farmers operating in salt ponds [22]. The study's sample consisted of 827 farmers, who were categorized into two groups: 65% (538 individuals) were engaged in salt pond farming, while the remainder farmers were involved in rice field cultivation. The sample was picked in a random manner using a simple random technique using the Slovin formula [23], resulting in a final count of 127 respondents. The employed data analysis approaches encompass descriptive analysis and income analysis. The descriptive approach offers a more accurate depiction of the indigenous knowledge held by salt pond producers. The utilization of income analysis serves as a means to ascertain the earnings of those engaged in salt pond farming. The calculation of income is determined by employing the formula as stated in reference [24].

$$TR = P\ X\ Q; TC = TFC + TVC;\ \pi = TR - TC$$

Notes:
TR = Total Revenue (IDR), TC = Total Cost (IDR), P = Price (IDR/Kg), Q = Quantity (Kg), TFC = Total Fixed Cost (IDR), VC = Total Variabel Cost (IDR), dan π = Income (IDR)

The study is conducted by analyzing the pre- and post-implementation revenue levels in relation to the aided village empowerment activities (PDB). The analysis of income disparities between the two time periods was conducted using the paired sample t-test in the SPSS version 16 software. Paired samples refer to a type of sampling design in which the same subjects are subjected to two measurements, one before and one after a specific treatment.

The permissible level of inaccuracy is set at 5% (0.05), corresponding to the alpha value of 5%. The test findings may possess statistical significance if the p-value is less than 0.05. However, the test findings are considered statistically insignificant if the p-value is more than 0.05.

3. Results and discussion

3.1. Characteristics of Research Locations

Bulu Cindea Village is situated within the Bungoro Sub-district of Pangkep District, sharing its borders with the village of Bulu Cindea. The northern boundary of the area is demarcated by the Labakkang Sub-district, characterized by its extensive shrimp farming ponds and salt ponds. To the south, it shares a border with Bori Appaka Village. The western boundary is defined by the Liukang Tuppabiring Sub-district, encompassed by the vast expanse of the ocean. Finally, the eastern boundary is formed by the village of Bowong Cindea, renowned for its salt production activities. The village of Bulu Cindea exhibits considerable potential due to its abundant fisheries resources, particularly in the realm of salt production. The total size of the pond is measured to be 429.5258 hectares.

Bulu Cindea settlement is situated inside the Bungoro Sub-district and has four distinct hamlets. These four hamlets, numbered from left to right: 1) Majannang Hamlet, 2) Bujung Tangaya Hamlet, 3) Jollo Hamlet, and 4) Biringkassi Hamlet, are dispersed over a 7 km^2 region and are situated 7 km from Bungoro District [25]. This community is situated along the seacoast and has diverse species, showcasing significant biodiversity within its boundaries.

According to projections, it is anticipated that the population of Bulu Cindea Village would reach 5119 individuals in the year 2022. This population will be further divided into two distinct groups: males, 2540 individuals, and females, 2579 individuals. The majority of the village community comprises individuals engaged in pond farming and fishing activities. The table presented below (Table 1) provides an overview of the various aspects of the lives of the village community.

Table 1. The population of Bulu Cindea Village, Bungoro Sub-district, Pangkep Regency in 2022

No	Type of Livelihood	Total (Head of Family)	Percentage (%)
1.	Farmer	827	55,47
2.	Trader	200	13,41
3.	Fisherman	385	25,82
4.	Breeder	65	4,36
5.	Government employees	14	0,94
	Total	1.491	100

Source: Population data of Bulu Cindea Village, August 2022

According to the data presented in Table 1, it can be observed that around 55.47% of the populace residing in Bulu Cindea Village depend on agricultural activities as their primary source of income, with a majority of them being involved in salt farming. Furthermore, it is worth noting that a significant proportion of the population, specifically 18.42%, is engaged in salt trading as a livelihood. The majority of the villagers possess a level of education that extends to high school completion. The population data is displayed in Table 2 as presented below:

Table 2. The population of Bulu Cindea Village, Bungoro Sub-district, Pangkep Regency, based on education level in 2022.

No	Level of Education	Total (People)	Percentage (%)
1.	Collage Graduate	67	4,57
2.	Senior HS Graduate	132	9,01
3.	Junior HS Graduate	352	24,03
4.	Elementary School Graduate	423	28,87
5.	Not finished Elementary School	491	33,52
	Total	1.465	100

Source: Population data of Bulu Cindea Village, August 2022

According to the data presented in Table 2, a significant proportion of villagers, approximately 28.87%, possess a primary school education. Additionally, a considerable portion, up to 33.52%, have yet to complete their elementary school education. On the other hand, only 4.57% of villagers have successfully attained an undergraduate level of education. According to the findings from interviews conducted with local community members, a significant proportion of individuals who have attained educational qualifications up to the undergraduate level are offspring of individuals engaged in salt pond farming activities. This study demonstrates that the revenue generated by salt pond farmers positively contributes to the overall well-being of rural communities.

3.2. Description of Research Respondents

The present study encompassed a sample population consisting of individuals engaged in salt pond farming inside the Bulu Cindea community. There are 425 residents that decide to pursue farming, including 225 salt farmers. The sample population of individuals engaged in salt pond farming, who were selected as respondents by a random sampling method, consists of 127 individuals, each possessing the following sample identities:

Table 3. The condition of the sample population of villagers based on age level, education level, and experience as salt pond farmers in 2023.

No	Category	Respondents (People)	Percentage
(%) Age (year)			
1.	15-30	39	30,71
2.	31-46	58	45,67
3.	47-62	27	21,26
5.	>62	3	2,36
	Total	127	100
Education Level			
1.	Collage Graduate	2	1,57
2.	Senior High School	13	10,24
3.	Junior High School	45	35,43
4.	Elementary School	44	34,65
5.	Not finished Elementary School	23	18,11
	Total	127	100
Experience as a salt farmer (year)			
1.	<10	23	18,12
2.	10-15	48	37,79
3.	>15	56	44,09
	Total	127	100

Source: Population data of Bulu Cindea Village, August 2022

Table 3 reveals that 58 respondents, or roughly 45.67% of the total, were still in the 31-46 year age range. The data suggests that most participants between the ages of 15 and 30 (30.71%) and 47 and 62 (21.26%) fall within the productive age range, possessing the necessary physical and psychological capabilities to effectively engage in salt farming activities. According to the educational attainment of the inhabitants of Bulu Cindea Village, it was found that 18.11% of the participants still needed to complete their elementary education. Additionally, 34.65% successfully completed elementary school, while 35.43% attained education up to the junior high school level.

The correlation between this condition and the educational attainment of the village community, as indicated by the respondents in this study, is notable, with a predominant representation of individuals with primary education. The educational attainment of the residents in Bulu Cindea Village is consistent with the observations made among the salt farmers who participated in the research. Specifically, a significant majority of the respondents, around 44.09%, have been engaged in salt farming for over fifteen years. This phenomenon exhibits a strong correlation with acquiring knowledge within official educational establishments [26], hence augmenting one another through the integration of diverse information derived from our practical experiences as pond farmers in the village of Bulu Cindea.

3.3. Salt Management Mechanism in Bulu Cindea Village

A technological advancement facilitating the augmentation of salt production has been discovered. The continued reliance on traditional salt production methods in various places suggests that not all salt-producing areas in central Indonesia have adopted modern technological advancements in this industry. In the past, the salt pond farmers residing in the village of Bulu Cindea employed a traditional approach to regulate salt application within their vicinity. The utilization of traditional management approaches may yield more favourable outcomes. The salt management system in Bulu Cindea Village is characterized by a semi-traditional approach, which incorporates a blend of technologies that have been enhanced by implementing geomembrane technology. The geomembrane system has been specifically engineered to facilitate salt production, employing a foundation composed of black tarpaulin material (see Figure 2.). The primary objective of technological advancements in salt production is to enhance the quality of salt by creating cleaner environmental conditions and increasing yields. This is achieved by using methods that prevent water used in the salt-making process from seeping into the ground, prolonging its usability (see Figure 3). Before entering the salt table, the seawater undergoes a filtration procedure. Additionally, a tarpaulin is utilized to ensure the preservation of salt quality and expedite the production process. The process of seawater evaporation is facilitated through the utilization of solar energy. The process involves the collection of seawater into a designated holding pond, followed by the subsequent transfer of the water to the plots where it is exposed to direct sunlight for seven days to facilitate evaporation. The crystallization rate from saltwater to aged seawater is accelerated due to the excellent heat absorption of black plastic under sun exposure [27]. Upon completing a ten-day storage period, the subsequent batch of salt will be gathered following the development of salt crystals.

Figure 2. Use of geomembrane technology in salt ponds in Bulu Cindea Village, Bungoro Sub-district, Pangkep Regency

Figure 3. Salt production using geomembrane technology in Bulu Cindea village, Bungoro Sub-district, Pangkep Regency

Since 2019, the people of Bulu Cindea Village, especially traditional salt farmers, have switched to using geomembrane technology to produce community salt. Geomembrane technology is versatile in various areas of the Pangkep Regency. Before pond farmers used this trend as the basis for salt production in ponds, geomembranes were also widely used in fishponds, sewage systems, and waterways. Geomembrane is made of HDPE (High-Density Polyethylene) polymer for good strength [28][29]. The various advantages of using this technology include more intensive absorption of solar heat resulting from the black colour effect of the geomembrane material[30]; the raw material for making salt in the form of seawater is not easily absorbed due to the shrinkage of the bottom of the salt pond, which is mostly just compacted soil [31]; the crystallization process takes more time[13]; and it is more efficient in terms of production time [32]. The salt produced is whiter and cleaner because it is not contaminated with impurities from the pond's bottom soil [33]. Salt farmers in the village of Bulu Cindea carry out salt production activities during the dry season (September–December).

3.4. *Differences in Income of Salt Farmers*

The present study examines the income of salt pond farmers subsequent to the adoption of geomembrane technology, as well as their income before introducing this technological innovation. Research on the financial aspects of installing geomembrane technology was conducted for one season, specifically from December 2022 to June 2023. The income analysis on geomembrane technology was conducted by computation of earnings within salt farming communities. The income of the salt pond farmer can be determined by calculating the difference between the total revenue (TR) and the total cost (TC). The revenue salt pond farmers generate in a given season is determined by multiplying the total salt production by the prevailing price per unit (Rp). The table presented below displays the production quantities according to the number of salt pond farmers in Bulu Cindea Village.

Table 4. Salt production in Bulu Cindea Village in 2023

No	Category of Production (Kg)	Number of Salt farmers (People)			
		Before	%	After	%
1.	High (≥ 4800)	0	0	25	19,68
2.	Medium (4001-4750)	23	18,11	88	69,29
3.	Low (≤4000)	104	81,89	14	11,02
	Total	127	100	127	100

According to the data presented in Table 4, the population of salt pond farmers before adopting the geomembrane system was distributed across the medium and low categories. Within the medium group, around 18.11% of the total respondents, or 23 individuals, were salt pond farmers who reported a production range of 400-4750 kg. Around 81.89% of the respondents, or 104 salt pond farmers, reported achieving production levels below 5000 kg within the low group. When implementing the geomembrane system technology, a majority of the participants (69.29%) reported achieving production levels within the specified category, while 25 respondents (19.68%) reported achieving output levels in the high category (≥ 9800). Based on the analysis, it can be inferred that implementing the geomembrane system technology will lead to ideal production outcomes. The utilization of this technology is expected to yield superior production outcomes compared to the conventional system. Specifically, the resulting sodium chloride (NaCl) content is measured at 99.96%, with a moisture content of 0.04%. Additionally, implementing this technology is anticipated to reduce the time required for harvesting in the Jeneponto regency [13]. Based on the findings of a study conducted by the cited source [34], the use of geomembrane technology in Jeneponto regency has the potential to enhance salt production. Table 5 below presents the income of salt pond farmers, which is derived from the quantity of output and the price of salt (Rp/kg), based on the production data provided.

Table 5. Salt farmer income in Bulu Cindea Village in 2023.

No	Category of Revenue (IDR)	Number of Salt farmers (People)			
		Before	%	After	%
1.	High (≥ 21.120.000)	0	0	34	26,77
2.	Medium (17.600.000-20.900.000)	28	22,05	69	54,34
3.	Low (≤17.500.000)	99	77,95	24	18,89
	Total	127	100	127	100

According to Table 5, the total revenue of salt pond farmers prior to the adoption of the geomembrane technology fell within the medium and low-income brackets. The majority of salt pond farmers exhibit moderate (22.05%) to low (77.95%) levels of revenue, ranging from IDR 17,600,000 to IDR 20,900,000 per season, equivalent to a monthly basis. Subsequently, following the implementation of geomembrane technology, most salt pond farmers (54.34%) were classified within the high category (26.77%). Following the implementation of the geomembrane technology, there was a notable increase in the revenue of salt producers. However, the expenses related to salt production in the pond continued to impact this revenue.

The analysis of variance using the t-test revealed a significance value of less than 0.05 ($0.002 < 0.05$), indicating a statistically significant difference in the revenue of salt pond farmers before and after implementing geomembrane technology. The disparity in revenue observed in the analysis findings can be attributed to the utilization of geomembrane technology. This technology offers enhanced salt quality by preventing its mixture with the soil and facilitating a faster production process. In the context of salt pond farming, it is observed that farmers employing geomembranes typically allocate a limited period of 5 to 10 days for maintenance activities prior to the harvest [35][36]. Hence, it is imperative to ensure the continuous advancement and refinement of geomembrane technology to achieve optimal and sustainable economic outcomes for individuals.

4. Conclusions

Enhancing the well-being of the inhabitants of Bulu Cindea Village, with a particular focus on salt pond farmers, will play a crucial role in advancing the adoption and implementation of geomembrane technology within the salt pond farming sector. The production value can attain a maximum of 69.29% on a medium size and a maximum of 19.68% on a high scale. This method can significantly enhance the revenue of salt pond farmers in the medium category by a maximum of 54.34% and in the high category by a maximum of 26.77%. The analysis of variance using the t-test revealed a statistically significant difference ($p < 0.05$) in the revenue of salt producers before and after implementing geomembrane technology. It is imperative to sustain and advance the progress of geomembrane technology to provide economic and socio-cultural advantages for the inhabitants of Bulu Cindea Village.

5. Acknowledgement

The author's thanks go to the Ministry of Education, Culture, Research and Technology Directorate General of Higher Education, Research and Technology for Funding the 2023 Research and Community Service Program (No. 0557/E5.5/AL.04/2023), the University of Muhammadiyah Makassar, and Muhammadiyah University of Pare-Pare and students for their cooperation in carrying out this service. The author also thanks the Head of Bulu Cindea village and all parties involved in this community service program.

6. References

[1] Qadir M, Quillérou E, Nangia V, Murtaza G, Singh M, Thomas R J, ... and Noble A D

2014
Natural Resources Forum **38** (4) 282-295
[2] Aris T, Mamahit D A and Ras A R 2022 *Jurnal Pamator: Jurnal Ilmiah Universitas Trunojoyo* **15**(1) 1-13
[3] Haendra A, Maarif M S, Affandi J and Sukmawati A 2021 *Jurnal Manajemen & Agribisnis*
18(2) 193-193
[4] Lindkvist K B and Sanchez J L 2008 *Regional Studies* **42**(3) 343-354
[5] Jadhav M H and Mannar M G V 2021 *Global Health: Science and Practice* **9**(4) 832-845
[6] Sintaningrum S and Felfina L 2016 *IOP Conf. Ser.: Inter. Multi. Confr.* **1**1
[7] Aditomo A B, Sulardiono B and Muskananfola M R 2020 *J. of Asian Multi. Res. for Eco. and Man. Study* **1**(2) 19-24
[8] Nasri N, Beno N, Septier C, Salles C and Thomas-Danguin T 2011 *Food Quality and Preference* **22**(7) 678-682
[9] McPherson A 2004 Introduction to protein crystallization *Methods* **34**(3) 254-265
[10] Ruedrich J and Siegesmund S 2007 *Env. Geology* **52** 225-249
[11] Goodall T M, North C P and Glennie K W 2000 *Sedimentology* **47**(1) 99-118
[12] Jackson M P and Hudec M R 2017 *Salt tectonics: Principles and practice* (Cambridge: Cambrigde University Press)
[13] Ramly Z A, Ahmad N and Juhaseng N A 2022 *Indo. J. of Env. Man. and Sust.* **6**(3) 76-81
[14] Asbar Y, Iswadi I, Mahdaliana M, Ula M and Razif R 2022 *Irpitage Journal* **2**(2) 79-84.
[15] Gozan M *et al* 2021 *IOP Conf. Ser.: Earth Env. Sci.* **940** 012024
[16] Sulistyaningsih T and Alighiri D 2018 *IOP Conf. Ser.: J. of Phys.* **983** 012166
[17] Shen L, Luo S, Zeng X and Wang H 2011 *Procedia Engineering* **26** 1803-1809
[18] Lugli F and Mahler C F 2016 *Waste Management & Research* **34**(5) 482-486
[19] Tansuchat R 2023 *Agriculture* **13**(4) 802
[20] Okcavia S C, Rudiyanto R, Suwarno P, Widodo P and Saragih H J R 2022 *Qistina: Jurnal Multidisiplin Indonesia* **1**(2) 173-184
[21] Aris T, Mamahit D A and Ras A R 2022 *Pamator: J. Ilmiah Univ. Trunojoyo* **15**(1) 1-13
[22] Gani E and Gitayuda M B S 2020 *Media Trend* **15**(2) 263-274
[23] Tejada J J and Punzalan J R B 2012 *The Philippine Statistician* **61**(1) 129-136
[24] Soekartawi 2003 *Teori Ekonomi Produksi* (Jakarta: Rajawali Press)
[25] Parawangsa R and Lestari I 2020 *Jurnal Ilmiah MEA* **4**(3) 1766-1783
[26] Kim M and Tan H T 2013 *Int. J. of Sci. Edu.* **35**(3) 357-387
[27] Hidayah F F *et al* 2017 *IOP Conf. Ser.* **1** 2
[28] Morsy M S and Rowe R K 2020 *Canadian Geotechnical Journal* **57**(1) 61-72
[29] Rowe R K, Abdelaal F B, Zafari M, Morsy M S and Priyanto D G 2020 *Canadian Geotechnical Journal* **57**(10) 1550-1565
[30] Dolez P I, Beaumier D, Taghizadeh A and Blond E 2017 *Geotechnical Frontiers* pp 259-266
[31] Haug M D and Pauls G 2002 *Mend Report* **2** 142
[32] Dwiyitno D, Sturm M T, Januar H I and Schuhen K 2021 *Env. Sci. and Poll. Res.* **28**(23) 30409-30413
[33] Sidik R F and Efendy M 2019 *Jurnal Pena Sains* **6**(1)
[34] Lestari W *et al* 2019 *IOP Conf. Ser.: J. of Phys* **1424** 012050
[35] Helmi A and Sasaoka M 2018 *Journal of Rural Studies* **59** 88-97
[36] Muhandhis I, Susanto H and Asfari U 2019 Procedia *Computer Science* **161** 867-875

A case study of anchovy diurnal migration patterns in Teluk Bakau, Bintan Regency, Riau Island, Indonesia

Try Febrianto[1], Asep Ma'mun[1*], Arif Pratomo[1], Esty Kurniawati[1], Mario Putra Suhana[1], Donny Apdillah[1,2], Ahmad Zahid[3]

[1] Department of Marine Science, Marine Science and Fisheries Faculty, Raja Ali Haji Maritime University, Indonesia

[2] Magister of Environmental Science, Maritime Raja Ali Haji University

[3] Department Of Water Resources Management, University of Maritime Raja Ali Haji, Jl. Raya Dompak, Dompak, District. Bukit Bestari, Tanjung Pinang, Riau Island Indonesia

*Corresponding author mail: asepmamun@umrah.ac.id

Abstract. Teluk Bakau is an area of water on the shore of Bintan Regency where the people make their livelihoods through fishing. There is now a lot of data showing a decline in catches, but it is unclear why this is happening. Changes in migratory patterns are regarded to be one of the primary reasons. The daily diurnal migratory pattern of anchovies is assumed to correspond with the movement of plankton. Anchovies are a type of tiny fish larvae whose life cycle is tied to plankton. It is possible to see microorganisms by converting the mean volume backscattering strength (MVBS) value into a measure of anchovy quantity and presence. Anchovy diurnal vertical migration was discovered using Diel Vertical Migration (DVM) measurements, which also revealed that the fish were in the MVBS range of -40 dB to -60 dB (e.g. FL: 5–10 cm). The migration of anchovies seen by ADCP shows they typically move towards the surface in the afternoon and that their numbers generally expand until night. Along with this an increase of anchovies, a lot of plankton has been rising to the surface. The presence of an anchovies moves into deeper waters in the early morning, and until daytime, they are decreasingly abundant. An indication that anchovies are increasingly plentiful and more present is in the afternoon

1. Introduction

Tropical sea waters are habitat to an extensive diversity of biological resources, particularly an abundance of aquatic resources including coral reefs, sea grass beds, and other aquatic organisms. One of the water areas on the coast of Bintan Regency that has these three ecosystems in order to maintain the waters' stability despite the effects of land and sea is Teluk Bakau. Anchovies are one of the organisms that the local community uses extensively as a business or as a source of support for economic existence today. Anchovies are frequently captured using traditional methods and fishing techniques that have been carried down from the previous generation to the last. There is a lot of knowledge available currently, but catches have a tendency to decline. It is unclear why this is the case. One of the main explanations of this is due to changing migratory or movement patterns that are out of step with traditional fishing trends. Anchovies belong to the classification of small fish larvae, and owing to its life cycle's association to plankton, it is believed that their daily migratory pattern follows that of the plankton. Plankton can often be used as a sign of potential fishing areas, such as with the coastal ecosytem [14],[19],[20]. Observation of micro-organisms can be carried out by processing the backscatter values and vertical velocity profiles from the ADCP instrument [4].

2. Material and Method

This study was carried out in the waters at Teluk Bakau, Bintan Regency, Riau Island-Indonesia (Figure 1).The following assumptions about the division of time were established in the research of anchovy migration: morning (04:00-09:00), daylight (09:00-14:00), evening (14:00-19:00), and night (19:00-04:00). This divide of time is based on the extent that the sun penetrates the study site (Figure 1). Using the instruments and setting specified in Table1 and Table 2, measurements of migration patterns, environmental factors, and the existence of plankton as a food source for anchovies were carried out. Sampling adapts to its concept of a fixed time period. The amplitude data produced by ADCP in the form of backscattering strength values may be used to visualize the movement pattern of anchovies. After excluding acoustic signals received as echo intensity data, acoustic backscattering values in the form of mean volume backscattering strength (MVBS), which is used as anchovy presence indicator. Observation of the behavior of nekton and zooplankton in the form of diel vertical migration (DVM) was carried out by utilizing the vertical velocity profile of the ADCP [10]. Validation of the measurement results is of course carried out by looking at the catches at the same time span during the study.

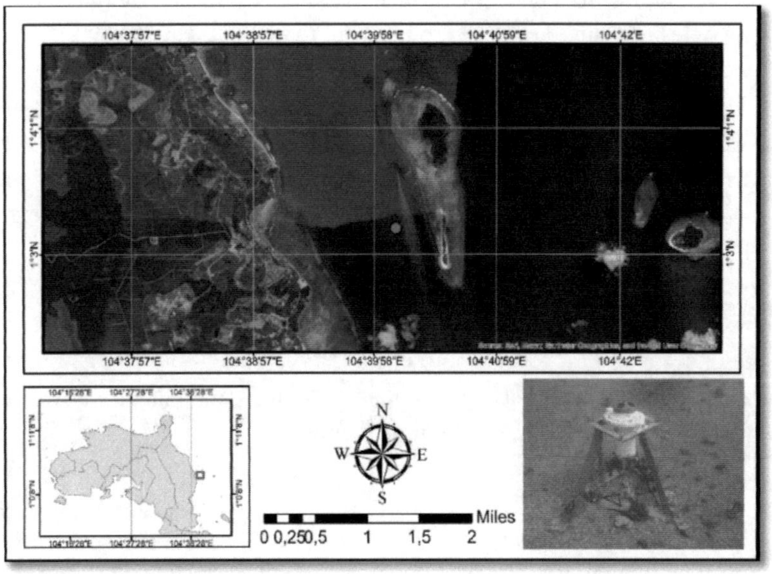

Figure 1. Study Area in Teluk Bakau, Bintan

Table 1. ADCP Setting

Parameter	Setting
Frekuensi	500 kHz
Max.Distance	300 m
Pulse duration	1 s
Beam width	25 deg
Mode Running	*upward-looking*

Table 2. Research Instrument

No.	Name of instrument
1	ADCP Sentinel V50 SN:23979
2	Housing ADCP
3	CTD
4	DO meter
5	pH meter
6	Global Posititioning System (GPS)
7	SCUBA Diving
8	Fishing Instrument
9	Planktonet
10	PC / Laptop Postprocessing

The PCA (Principal Component Analysis) statistical analysis approach is used to correlate the important variables that influence the density of anchovies. This approach is better equipped to give a broad picture of the degree to which variables and environmental variation values are correlated [6],[8]. Using such an approach, it can be accomplished to identify the important variables that actually affect anchovy existence as well as the relationships between various environmental conditions. A hypothesis offered looking forward is: H_0: There is no significant connection between the variables; H_1: There is at least one significant correlation between the variables. In generally, PCA measures the ecluidean distance (angle), which is the sum of the squares of the differences between individuals (rows) for the relevant variable (column). The similarities between the two variables increase with decreasing distance and vice versa. Teritorial maps were produced using the PCA analysis findings to locate anchovy abundance data in relation to environmental parameters. A new hypothesis or starting hypothesis is generated for multiple linear regression analysis utilizing the t-test and F-test after learning some of the key factors that determine the occurrence of anchovies. Impact the presence of anchovies. The following is the hypothesis: H_0 indicates that the main components have no significant impact on the dependent variable, H_1 indicates that one or more main components have a significant impact, and H_2 suggests that the main components as a whole have a considerable impact. By assessing the determination coefficient, one may determine the extent to which the primary component has an impact on the dependent variable.

2.1 *Pre Processing*

The acoustic signal amplitude was sent to the ADCP instrument in counts of echo intensity (E). Er is the noise level of all beams by calculating the minimum echo intensity value. The Er value used in this study was 15 counts for ADCP (307.2 kHz). After knowing the values of two-way transmission loss (2 TLs) and echo level detected on the transducer (RL), the value of acoustic backscatter (RB) can be calculated by use the two values. The RB value may be determined using the following equation, according to previous studies [3],[7]:

$$RL = Kc\ (E - Er) \tag{1}$$

$$Kc = 127.3 / \ + 273 \tag{2}$$

$$TL = 20 \log 10\ R + 2\ \alpha R \tag{3}$$

$$RB = RL + 2\ TL \tag{4}$$

$$SSC = 10^{(A + B \cdot RB)} \tag{5}$$

To convert the echo intensity value to mean volume backscattering strength (MVBS) (dB 1 m^{-3}), use the formula [3],[12] :

$$MVBS = C + 10) \log((Tx + 273.16)R^2 - L_{DBM} + P_{DBW} + 2\alpha R + 10\log (10^{kc(E-Er)/10} - 1 \quad (6)$$

Where is:
C : Constant
Tx : Temperature measured at the transducer (°C)
L_{DBM} : Logarithmic transmitted wavelength
P_{DBW} : Logarithmic transmit power (log(W))
Kc : Conversion factor
E : Amplitude that returns to the measured instrument (count)
Er : The relative amplitude constant of the returned signal strength indicator (RSSI) which is determined as the noise level on all transducers by calculating the minimum value of EI in the time series data (count).
b : Blanking zone (m)
P : The wavelength emitted from the transducer to the water column (m),
D : Bin size (m)
n : Number of bins
θ : Transducer tilt angle (°)
c : Average sound speed of the instrument (ms-1)
co : Speed of sound in water used by ADCP and set at 1475.1 (ms-1).
α : Absorption coefficient (dB m-1) calculated by considering the environmental parameters and the frequency of the instruments used in the study

Table 3. ADCP Configuration for Post Processing

Setting	unit	Moored ADCP
C (Constant of system)	dB	-160,5
Tx (Temp of transduser)	°C	28,06
P (panjang gelombang)	mm	2
LDBM (logaritmik pulsa)	dB	-6,98
PDBW (logaritmik power)	dB	7,95
Kc (faktor skala	dB $count^{-1}$	0,72
Er (nilai minimum RSSI)		43
α (koefisien absorpsi)	dB m^{-1}	0,2444
c (kecepatan suara)	ms^{-1}	1519,23

Source : RDI Istrument, Field Survey

The theoretical TS value is used to compute the plankton abundance. Based on the proposed equations the MVBS value for zooplankton analysis is used to assess the abundance of zooplankton. The following formula is used [2],[9],[17]:

$$(f) = n\, Ts\ (\mathrm{f}) \quad (7)$$
$$MVBS_f\,(\mathrm{dB}) = 10 \log \rho \cdot TS_f\,(\mathrm{dB}) \quad (8)$$
$$TS_f = 10 \log ((f_0, a_0)/4\pi \quad (9)$$
$$(z) = <(f0,z)> / \sigma(f0,a0) \quad (10)$$

The theoretical TS produced from the DWBA model [4] was used in this study as the TS of zooplankton (dB) values. N(z) is the zooplankton abundance at an offered depth; σ_v value is the ensemble average

IOP Conf. Series: Earth and Environmental Science 1289 (2023) 012003 doi:10.1088/1755-1315/1289/1/012003

of individual pings on the volume backscatter strength; σ is the acoustic scattering cross-section with the equipment and zooplankton size; f_0 is the frequency used in ADCP; and a_0 is the zooplankton size.

3. Result and Discussion

The acoustic backscattering values of mean volume backscattering strength (MVBS) during 4 periods, namely morning, daytime, afternoon, and night, could be used to observe the Diel Vertical Migration (DVM) of anchovies. The ADCP data recording results show that anchovies have a Target Strength range or equivalent to length of fish (FL cm), meaning TS -40 dB to -60 dB (e.g. FL: 5-10 cm), with an average abundance of 13266 individual (afternoon), 55922 (night), 15436 (morning), and 2058 (daytime). Figure 2(a) shows the percentage at each time, while Figure 2(b) fishing catch is used for confirmation.

The DVM's movement is seen via the Deep Scattering Layer (DSL) layer's peak and valley patterns. According to [11], DSL is comprised of up of live things such phytoplankton, zooplankton, and small fishes. The migration of anchovies seen by ADCP shows they typically move towards the surface in the afternoon and that their numbers generally expand until night. Along with this an increase of anchovies, a lot of plankton has been rising to the surface. The presence of an anchovies moves into deeper waters in the early morning, and until daytime, they are decreasingly abundant. An indication that anchovies are increasingly plentiful and more present is in the afternoon [14],[20].

Plankton provides the primary food supply for anchovies in a process known as the food chain, which causes this movement. Thus, it can be concluded that the mobility and presence of plankton are closely related to the existence of anchovies [14],[19],[20]. This migratory pattern moves like a twilight migration pattern, migrating to the surface in the evening and to deeper depths in the morning [10],[17],[18]. [21] found that there are not many differences in the rate of plankton migration between morning and evening.

Both internal and external variables contribute to the vertical migration trend. Internal considerations include the presence of a biological clock that naturally controls the rhythm of each person's daily activities. Environmental parameters such as light, temperature, salinity, oxygen concentration, hydrostatic pressure, and feed availability affect external factors [5]. The anchovies swim upward to leave the deeper habitat around or after nightfall. Following the movement of the plankton, anchovies fish travel up and down over a period of 4-5 hours [10]. Anchovies fish migrate to the surface or swim into the water following the pattern of plankton movement, having a vertical speed calculated from the ADCP measurement used in this research, which has a swimming speed between (1-4 cm/s). this is in accordance with previous research conducted on zooplankton which has a strong relationship with anchovies [14], [19],[20].

According to [11], anchovies have a distinct vertical migration pattern, spending the day at the bottom of the ocean and migrating to shallow regions before spending the night on the top. The nighttime vertical movement of anchovies is thought to be motivated by a need to find food. Because of the charts lighting, the water is bright, which helps anchovies catch their food. Accordingly, the interaction depicted in the illustration shows zooplankton being consumed by anchovies while the anchovies themselves are being eaten by larger fish.

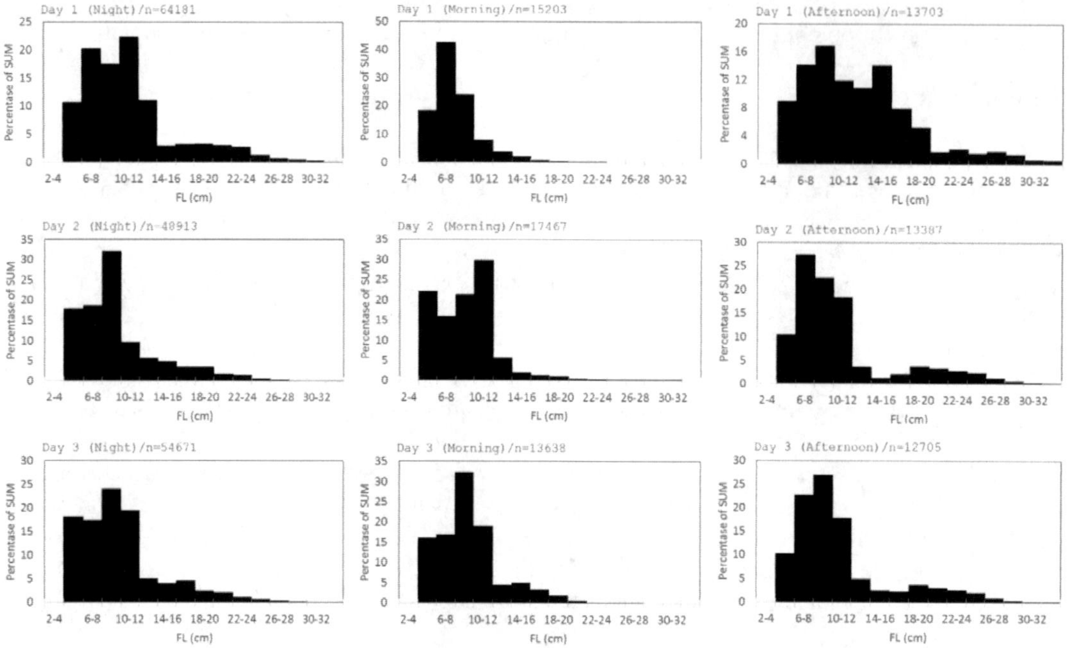

(a)

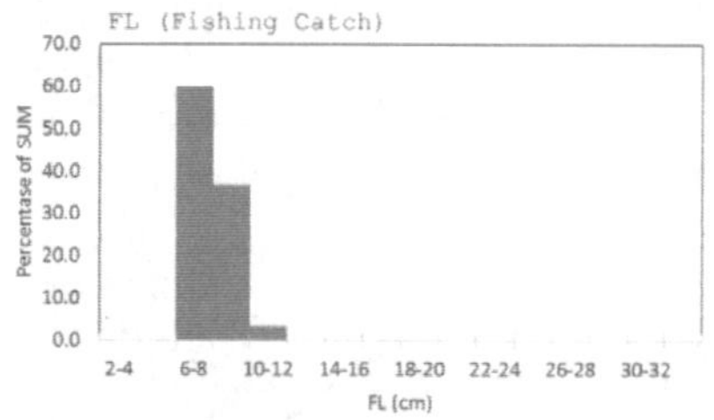

(b)

Figure 2. (a) Fish Length (TLcm) distribution of MVBS
(b) Fish Length (TLcm) distribution of Fishing Catch

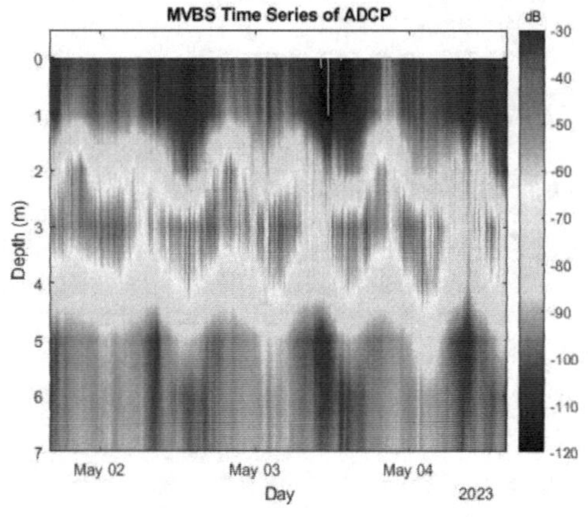

(a)

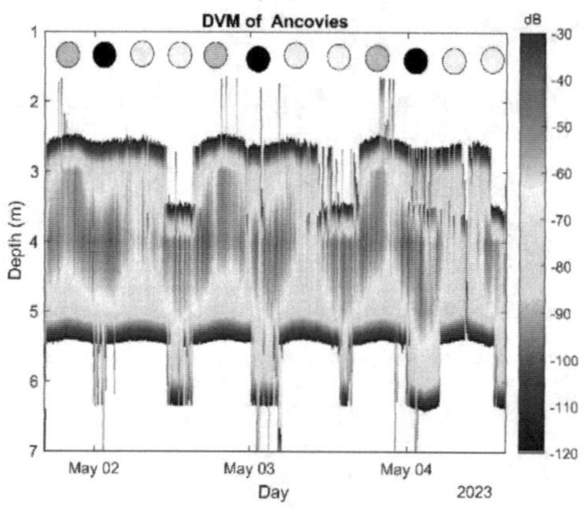

(b)

IOP Conf. Series: Earth and Environmental Science 1289 (2023) 012003 doi:10.1088/1755-1315/1289/1/012003

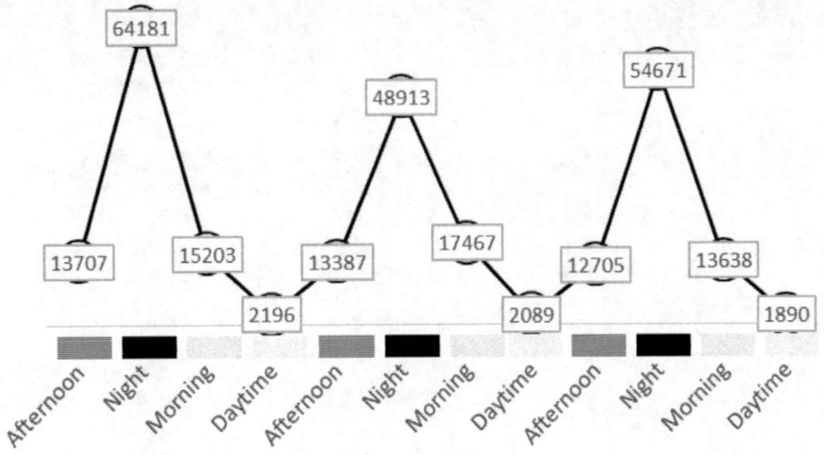

(c)

Figure 3. (a) MVBS Time Series of ADCP, (b) Diel Vertical Migration (DVM) of anchovies, (c) Migration pattern of N-Individual

A series of statistical tests are used to investigate relations between the abundance of anchovies, which is strongly relations to the MVBS value, and environmental parameters such as temperature, salinity, Do, PH, plankton abundance, current direction, and speed. According to the PCA analysis results, the correlation matrix may be used to determine the relationship between the variables have become.

Table 4. Correlation matrix between variables

Variables	Speed	Direction	Temp	Salinity	PH	Do	Plankton	MVBS
Speed	**1**							
Direction	**1.0000**	**1**						
Temp	**-0.4805**	**-0.4805**	**1**					
Salinity	**0.6010**	**0.6010**	0.0048	**1**				
PH	**-0.4833**	**-0.4833**	**0.6616**	**-0.3919**	**1**			
Do	**0.1704**	**0.1704**	**-0.1088**	**0.2673**	**0.0320**	**1**		
Plankton	**0.2333**	**0.2333**	**-0.1008**	**0.3346**	0.0198	**0.9793**	**1**	
MVBS	**-0.5210**	**-0.5210**	**0.3965**	**-0.1297**	**0.2937**	**0.5236**	**0.5524**	**1**

Values in bold are different from 0 with a significance level alpha=0.95

Table 5. Root values and percentage diversity of each component

	F1	F2	F3	F4	F5	F6	F7
Eigenvalue	3.4868	2.4756	1.0632	0.6694	0.2067	0.0917	0.0066
Variability (%)	43.5848	30.9452	13.2897	8.3672	2.5837	1.1465	0.0828
Cumulative %	43.5848	74.5300	87.8197	96.1870	98.7707	99.9172	100.0000

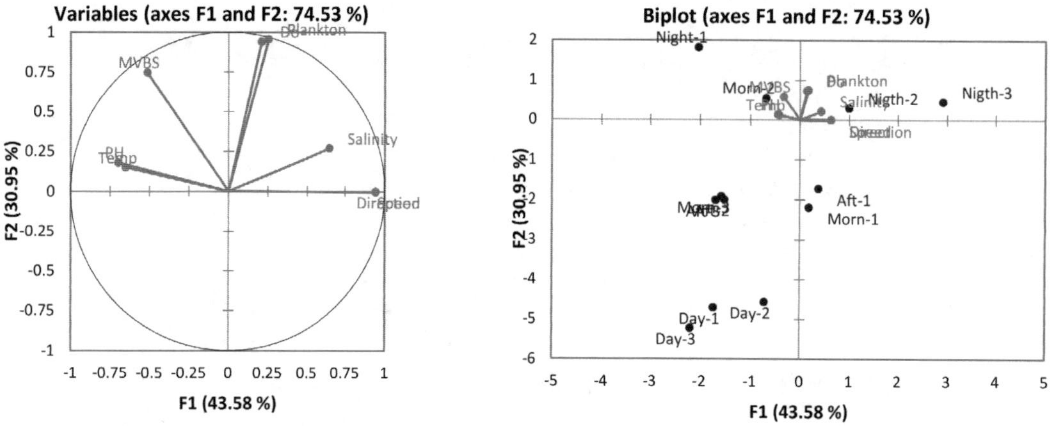

(a) (b)

Figure 4. (a). Correlation of components factor F1 and F2 towards ecosystem variables
(b). Territorial maps of environmental variables and the density of Anchovies

Table 6. Results of t-test dependent variable to main environment

Coefficient [a]								
Model		Unstandardized Coefficients		Standardized Coefficients	t	Sig.	95.0% Confidence Interval for B	
		B	Std. Error	Beta			Lower Bound	Upper Bound
1	(Constant)	-104.700	46.843		-2.235	0.061	-215.467	6.066
	Temp	3.163	1.604	0.498	1.972	0.089	-0.629	6.955
	PH	-2.200	2.199	-0.243	-1.000	0.351	-7.400	3.000
	Do	-11.651	4.759	-2.531	-2.448	0.044	-22.905	-0.397
	Plankton	0.083	0.028	3.055	2.929	0.022	0.016	0.150

a. Dependent Variable: MVBS

Table 7. F-Test results from dominant main environment to dependent variable

ANOVA [a]						
Model		Sum of Squares	df	Mean Square	F	Sig.
1	Regression	20.154	2	10.177	4.469	.0035[b]
	Residual	24.096	9	2.277		
	Total	44.250	11			

a. Dependent Variable: MVBS
b. Predictors: (Constant), Plankton, Do

The KMO value (Kaiser-Meyer-Olkin test) is reached at 0.76 with a significance of <0.05 so that all measured parameters are practical for component analysis despite knowing that the majority of correlation coefficients between variables involved in anchovy are <0.5 (Table 4). The two greatest common factors, F1 and F2, with a cumulative fraction of diversity of 74.53%, are produced based on the eigenvalues for components > 1 (Table 5), with 5 other shared factors contributing a total of 25.47%.

Due to the cumulative percentage of the variance getting the biggest and being totally satisfied (Table 5), only the F1 and F2 components was used to determine the correlation.

The environmental components that are directly related to each other are (ph-temp), (do-plankton) and (current-speed) with smallest angle (ecluidean) to the MVBS (closely with anchovies abundance). Both very closely compared to the other components (Figure 6a). While other environmental factors partially correlated with the presence of anchovies (Figure 6b).The abundance of anchovies is spatially and significantly connected with the environmental variables temperature, pH, Do, and plankton, according to the results of the spatial correlation between environmental parameters and the presence of anchovies. This order is determined by their distance from and proximity to the line established between the correlation center point and the environmental variables.

The statistical test results showed that of the four factors tested using the t-test with a 95% confidence interval, two main environmental factors, Do (t-hit=2.448) and plankton (t-hit=2.929), with a sig value 0.05 and t-hit> t-table (2.228), had a significant influence on abundance anchovies (Table 6).An F-test with a 95% confidence interval was conducted to determine the simultaneous impact of the two key factors impacting the presence and abundance of anchovies (Table 7). These results indicate that the two main variables concurrently and jointly influence the presence and abundance of anchovies. The value of the proximity of the relationship between the components (R-square), which is equal to 0.807 or may be stated with a percentage range of 81%, expresses of the effect components. Photosynthesis, respiration, decomposition, and diffusion and aeration all have an ongoing impact on the concentrations of dissolved oxygen. Plankton is an organism that participates in photosynthesis in order to meet the requirements of its life cycle. While different elements, such the existence of fish, have a big impact, especially in thriving coastal areas [1], [15]

4. Conclusion

Anchovy diurnal vertical migration was discovered using Diel Vertical Migration (DVM) measurements, which also revealed that the fish were in the MVBS range of -40 dB to -60 dB (e.g. FL: 5–10 cm). The migration of anchovies move towards the surface in the afternoon and that their numbers generally expand until night. Along with this an increase of anchovies, a lot of plankton has been rising to the surface. The presence of an anchovies moves into deeper waters in the early morning, and until daytime, they are decreasingly abundant. An indication that anchovies are increasingly plentiful and more present is in the afternoon. This migratory pattern moves like a twilight migration pattern, migrating to the surface in the evening and to deeper depths in the morning. A series of statistical tests are used to investigate the abundance of anchovies, which is strongly connected to the MVBS value, and environmental parameters such as temperature, salinity, Do, PH, plankton abundance, current direction, and speed. The statistical test results showed that two main environmental factors, Do (t-hit=2.448) and plankton (t-hit=2.929), with t-hit> t-table (2.228), had a significant influence on abundance anchovies

Acknowledgments

This research was funded by LPPM, Maritime Raja Ali Haji University, Riau Island-Indonesia, Republic of Indonesia.

References

[1] Aronica S, Fontana I, Giacalone G, Lo Bosco G, Rizzo R, Mazzola S, Basilone S, Ferreri R, and Bonanno A.2019. Identifying small pelagic Mediterranean fish schools from acoustic and environmental data using optimized artificial neural networks. *Ecol. Inform.* **50** 149–161

[2] Brierley A S, Saunders R A, Bone D G, Murphy E J, Enderlein P, Conti SG and Demer DA. 2006. Use of moored acoustic instruments to measure short term variability in abundance of Antarctic krill. *Limnol. Oceanogr. Methods* **4** 18–29

[3] Deines K L.1999.Backscatter estimation using broadband acoustic Doppler current profilers. *In*

Proceedings of the IEEE Sixth Working Conference on Current Measurement (Cat. No.99CH36331) pp 249–253

[4] Dwinovantyo, A., Manik, H. M., Prartono, T., Susilohasi, S., Mukai, T. 2019. Variation of ZooplanktonMean Volume Backscattering Strength from Moored and Mobile ADCP Instruments for Diel Vertical Migration Observation. *Applied Science*. 2: 1-21.

[5] Dypvik, E., Kaartvedt, S., 2013. Vertical migration and diel feeding periodicity of the skinnycheek lanternfish (*Benthosema pterotum*) in the Red Sea. Deep-Sea Res. I 72, 9–16.

[6] Faucon, M. P., Parmentier, I., Colinet, G., Mahy, G., Luhembwe, M. N., & Meerts, P. 2011. May rare metallophytes benefit from disturbed soils following mining activity? The Case of the Crepidorhopalon tenuis in Katanga (D. R. Congo). *Restoration Ecology*, *19*, 333–343

[7] Gartner J W.2004.Estimating suspended solids concentrations from backscatter intensity measured by acoustic Doppler current profiler in San Francisco Bay, California. *Mar. Geol.* **211** 169-187

[8] Grenouillet, G., Buisson, L., Casajus, N., & Lek, S. 2011. Ensemble modelling of species distribution: the effects of geographical and environmental ranges. *Ecography*, *43*, 9−17

[9] Holliday D V and Pieper R E 1995. Bioacoustical oceanography at high frequencies. *ICES J. Mar. Sci.* **52** (3-4) 279-296

[10] Kaartvedt, S, Anders R, Svenja C, Thor A. K. 2020. Diel vertical migration and individual behavior of nekton beyond the ocean's twilight zone. *Deep Sea Research Part I: Oceanographic Research Papers*. Volume (160). https://doi.org/10.1016/j.dsr.2020.103280.

[11] Laevastu, T., Murray L.H.1981. Fisheries Oceanography and Ecology. England: Fishing News Book Ltd. Farnham-Surrey. 199p.

[12] Lurton, X. 2002. *An Introduction to Underwater Acoustics: Principles and Applications*. Hal. 83. Praxis Publishing Ltd.. Chichester, UK. xxv + 341 h.

[13] Mullison J.2017.Backscatter estimation using broadband acoustic Doppler current profilers - updated. *In Proceedings of the ASCE Hydraulic Measurements & Experimental Methods Conference*, Durham, NH. pp 1–5

[14] Ochoa, J., Maske, H., Sheinbaum, J., Canedla, J., 2013. Diel and lunar cycles of vertical migration extending to below 1000 m in the ocean and the vertical connectivity of depth-tiered populations. Limnol. Oceanogr. 58, 1207–1214, 2013.

[15] Otero J, Álvarez-Salgado XA and Bode A.2020. Phytoplankton Diversity Effect on Ecosystem Functioning in a Coastal Upwelling System. Front. Mar. Sci. 7:592255. doi: 10.3389/fmars.2020.592255

[16] Ridha, U., M.R. Muskananfola dan A. Hartoko. 2013. Analysis of Distribution of Catches of Lemuru Fish (*Sardinella lemuru*) Based on Satellite Data of Sea Surface Temperature and Chlorophyll-A in the Bali Strait Waters. Diponegoro Journal Of Maquares *Management Of Aquatic Resources* 2(4): 53-60.

[17] Roman M Holliday D and Sanford L.2001.Temporal and spatial patterns of zooplankton in the Chesapeake Bay turbidity maximum. *Mar. Ecol. Prog. Ser.* **213** 215-227.

[18] Sawada K, Mukai T, Fukuda Y, and Matsuura T 2016 Comparison of zooplankton density estimated by acoustic inversion method and net sampling. *J. Acoust. Soc. Am.* **140** (4) 3243-3243.

[19] Van Haren, H., 2007. Monthly periodicity in acoustic reflections and vertical motions in the deep ocean. Geophys. Res. Lett. 34, 1–5. https://doi.org/10.1029/ 2007GL029947. L12603.

[20] Van Haren, H., Compton, T.J., 2013. Diel vertical migration in deep sea plankton is finely tuned to latitudinal and seasonal day length. PLoS One 8 (5), e64435. https://doi. org/10.1371/journal.pone.0064435.

[21] Wade, I. P., dan K. J. Heywood. 2001. Acoustic Backscatterer Observation of Zooplankton Abundance and Behaviour and the Influence of Oceanic Fronts in the Norteast Atlantic. Deep Sea Research II 48. P 899-924.

IOP Conf. Series: Earth and Environmental Science 1289 (2023) 012004 doi:10.1088/1755-1315/1289/1/012004

Business Analysis of Processed Baby Squid Chili Products (Micro, Small and Medium Enterprises MSMEs "Dapoer Ikan Diana"): Case Study in Samarinda City, East Kalimantan, Indonesia

Eko Sugiharto[1*], Elly Purnamasari[1], Komsanah Sukarti[2]

[1]Department Social Economic Fisheries, Faculty Fisheries and Marine Science, Mulawarman University, Jl Gn Tabur No 1, Kampus Gn Kelua, Samarinda City, East Kalimantan, Indonesia

[2]Department Aquaculture, Faculty Fisheries and Marine Science, Mulawarman University, Jl Gn Tabur No 1, Kampus Gn Kelua, Samarinda City, East Kalimantan, Indonesia

*Corresponding Author e-mail: eko.sugiharto@fpik.unmul.ac.id

Abstract. Micro, Small, and Medium Enterprises (MSMEs) are the most essential pillar in the Indonesian economy. One of them is Dapoer Ikan Diana, which is a processing business using fish-based ingredients. In running a business, it is inseparable from the costs incurred. This research aims to calculate the business profits of MSMEs Dapoer Ikan Diana, conducted from December 2021 to November 2022. This study used purposive sampling, obtaining the sample from one respondent, who is a business owner The result showed that there were two types of investment costs incurred, namely production investment costs of IDR 10.680.714 with depreciation costs of IDR 90,767/month and marketing investment costs of IDR 10,565,925 with depreciation costs of IDR 132,262/month. Operational costs are divided into two types, namely production operational costs incurred in the amount of IDR 2,069,950 with a total cost (TC) of IDR 8.626.707/month and marketing operational costs incurred in the amount of IDR 4,271,675 with a total cost (TC) of IDR 13,121,411/month. The revenue obtained is IDR 14,400,000/month. The business profit obtained is IDR 1,278,589/month.

1. INTRODUCTION

At the end of 2019, the world was hit by the COVID-19 pandemic. Coronavirus, also known as COVID-19, is an outbreak that originated in Wuhan, China. In less than two months, this coronavirus has caused 80 thousand cases and 3,000 deaths [2]. The COVID-19 pandemic has impacted all aspects of life, including the economic sector. One of the most significant economic supporters in Indonesia is MSMEs. Based on the Ministry of Cooperatives and Small and Medium Enterprises (SMEs) data, the current number of MSMEs reaches 64,2 million with a contribution to GDP of 61,07%, or IDR 8,573,89 trillion [3]. According to the Office of Cooperatives and SMEs in Samarinda City, there are 39,955 MSMEs in Samarinda City [3].

MSMEs are businesses operated by individuals, households, or small business entities. MSMEs are classified based on limits on annual income turnover, total assets, and number of employees. Other businesses not categorized as MSMEs, such as big businesses, are productive economic businesses run by business entities with a total net worth or annual sales result greater than medium-sized businesses [4]. MSMEs contributed 60.3% of Indonesia's total gross domestic product in 2017, absorbing 97% of the total workforce and 99% of total employment. MSME development is also one of the efforts put forward in the ASEAN framework for Equitable Economic Development, in reducing development disparities, strengthening the quality of human resources, increasing social welfare, and widening participation in the ASEAN integration process [1].

An MSMEs under the auspices of the Maritime Affairs and Fisheries Service of East Kalimantan Province is Dapoer Ikan Diana. Opened in December 2017, it sells processed products with fish as the main ingredient. Dapoer Ikan Diana. One of their products is Baby Squid Sambal, which is not purchasable by those who do not like spicy foods.

As businesses are inseparable from the costs incurred from their inception until operation, Dapoer Ikan Diana requires money to run, but the profit received since it was incepted is not yet known. Based on this background, the author is interested in conducting research entitled "Business Analysis of Baby Squid Sambal: A Case Study of MSMEs Dapoer Ikan Diana in Samarinda City". This study aims to determine the business profits.

2. Research Methodology

The research took place at MSMEs Dapoer Ikan Diana, KS Tubun Street, Samarinda City from December 2021 to November 2022. The data types used in this research consisted of primary data, obtained from interviews with respondents using a prepared questionnaire, and secondary data, obtained from the literature. The research uses a purposive sampling method, which according to [7], is a technique with certain considerations. Therefore, MSMEs Dapoer Ikan Diana owner is the only respondent.

Data Analysis

A. Descriptive Analysis

According to [6], descriptive analysis method is statistics used to analyze data by describing the data collected as it is without intending to make general conclusions or generalizations.

B. Business Profit Analysis

1. Total Business Cost

Total cost (Total Fixed Cost/TFC) are marketing cost incurred by the company to fund short-term marketing activities. According to [5], the following formula can be used to calculate the total cost.

$$TC = TFC + TVC$$

2. Total Revenue

$$TR = P \times Q$$

3. Profit

$$\pi = TR - TC$$

3. Result and Discussion

A. Business Cost for Baby Squid Sambal

1. Investment cost

a.Production Investment Cost

Table 1. Production Investment Cost of Baby Squid Sambal from MSMEs Dapoer Ikan Diana

No	Description	Investment Cost (Rp)	Cost of Depreciation (Rp/month)
1	Sub-amount of production equipment	1.680.714	15.767
2	Sub-amount supports equipment	9.000.000	75.000
Subtotal 1 and 2		10.680.714	90.767

It can be seen that the total production investment cost incurred by MSMEs Dapoer Ikan Diana amounts to IDR 1,680,714 with a depreciation fee of IDR 15,767/month. Then, the sub-total cost of equipment supporting production investment costs incurred is IDR 9,000,000 with a depreciation fee of IDR 75,000/month. The cost incurred at the inception of the business included the equipment used for production activities and the cost of supporting equipment needed to produce Baby Squid Sambal. The investment cost used to analysis the business is the cost of equipment, divided into several products using the same equipment for the processing of MSMEs Dapoer Ikan Diana, both production investment costs and marketing investment costs.

b. Marketing Investment Cost

Table 2. Marketing Investment Cost of Baby Squid Sambal from MSMEs Dapoer Ikan Diana

No	Description	Investment Cost (IDR)	Cost of Depreciation (IDR/month)
1	Sub-amount marketing equipment	315,952	40,595
2	Sub-amount marketing support equipment	10,250,000	91,667
	Subtotal 1 and 2	10,565,925	132,262

The table shows that the total cost of marketing investment equipment incurred by MSMEs Dapoer Ikan Diana is IDR 315,925 with a depreciation fee of IDR 40,595/month. Then, the sub-total cost of marketing support equipment issued is IDR 10,250,000 with a depreciation fee of IDR 91,667/month. Costs incurred at the inception of the business are the cost of marketing equipment, including vacuum sealers, presses (sealer), bottle cap presses, scissors and scales, and the cost of marketing support equipment including motorbikes, and cell phones.

2. Operational Cost
   a.Production Operational Cost

Table 3. Production Operational Cost of Baby Squid Sambal from MSMEs Dapoer Ikan Diana

No	Operational Cost	Total Cost (Rp/Product)	Total Cost (TC) Rp/month
1	Sub-amount production variable cost	2,069,950	8,310,940
2	Sub-amount production fixed cost	-	315,767
	Subtotal 1 and 2	2,069,950	8,626,707

It can be seen that there are two types of production operational cost, namely variable cost incurred by MSMEs Dapoer Ikan Diana in the form of a total cost of IDR 2,069,950 per production, and total cost (TC) of IDR 8,310,940/month. The sub-amount of fixed cost incurred is IDR 315,767/month. The sub-total TC, namely variable cost and fixed cost for production, amounted to IDR 2,069,950 per production and the sub-total TC, namely variable cost and production fixed cost, amounted to IDR 8,626,707/month. The cost used to carry out production activities, both variable costs and fixed costs, including the cost of purchasing grow materials and the wages of the labour force in the production section of Baby Squid Sambal. Electricity and after cost are not included due to the privacy of the respondents.

b. Operational Marketing Cost

Table 4. Operational Cost Marketing of Baby Squid from MSMEs Sambal Dapoer Ikan Diana

No	Operational Cost	Total Cost (Rp/Product)	Total Cost (TC) Rp/month
1	Sub-amount marketing variable costs	902,473	3,671,675
2	Sub-amount marketing fixed cost	-	600,000
	Subtotal 1 and 2	902,473	4,271,675

It can be seen that there are two types of marketing operational costs, namely variable costs and marketing fixed costs. Marketing variable costs in the form of a total cost of IDR 902,473 per production, and TC of IDR 3,671,675/month. Then, the sub-amount of fixed marketing costs incurred is IDR 600,000/month. Sub-total total costs, namely variable costs and marketing fixed costs, amounting to IDR 902,473 per production and sub-total TC, namely variable costs and fixed costs for marketing, amounting to IDR .271.675/month. Variable costs include the cost of purchasing packaging bottles, printing stickers, motor oil, and gasoline, and fixed costs include the cost of marketing labor wages.

B. Total Revenue of Sambal Baby Squid Processed Products from MSMEs *Dapoer Ikan Diana*

Total Revenue can be calculated as follows:

$$TR = P \times Q$$

= Rp 30,000 x Rp 480
= Rp 14,400.000/month

The revenue Dapoer Ikan Diana earns is IDR 14,400,000, obtained from the selling price of the product multiplied by the number of products produced for a month.

Business Profit

Profits Dapoer Ikan Diana earned can be calculated as follows:

$$\pi = TR - TC$$

= Rp 14,400,000 – Rp 13,121,411
= Rp 1,278,589/month

This profit is the total profit received per month. Based on the calculation, the profit earned is IDR 1,278,589/month with revenues of IDR 14,400,000 minus the total cost (TC) of IDR 13,121,411. The TC is obtained from the total production cost of IDR 8,626,707 obtained from the total investment costs of production depreciation plus the total production operational costs divided by the number of products marketed, namely 480 bottles, and the total marketing costs of IDR 4,271,675 obtained from the total marketing depreciation investment costs plus the total marketing operational costs divided by the total marketed products. These sales include sales from homes, participating in major activities, events and online sales.

4. Conclusion and Suggestion

Based on the result and discussion above, it can be concluded that the income received is IDR 14,400,000/month and the business profit earned is IDR 1,278,589/month. Dapoer Ikan Diana should put more pressure on the marketing costs incurred to make the profits from marketing institutions more profitable. MSMEs Dapoer Ikan Diana should open job vacancies, especially in the marketing section so that they can assist in marketing products. For the government, it is best to evaluate all MSMEs in Samarinda so that they can further advance existing MSMEs.

REFERENCES

[1]. Gloria, 2020. MSMEs Important Actors in the ASEAN Economic Community. Universitas Gajah Mada. https://www.ugm.ac.id./id/berita/19821-umkm-aktor-important-actor-in-asean-economic-society (Accessed 23 February 2022).

[2]. Junaedi Dedi and Salistia Faisal, 2020. The Impact of the Covid-19 Pandemic on the Economic Growth of Affected Countries. [Journal] National Symposium on State Finance.National Islamic Institute (IAI-N) Laa Roiba. Bogor. Page 995-1115.

[3]. Coordinating Ministry for the economy of the republic of Indonesia, 2021.UMKM is an Important Pillar in the Indonesian Economy. Jakarta.

[4]. Rafada, 2022.Micro small medium enterprises.Jakarta

[5]. Soekartiwi. 1991. Basic Principles of Marketing Management of Agricultural Products. RajawaliPress. Jakarta [Journal] Agroland

[6]. Sugiyono, 2014. Quantitative, Qualitative, and R&B Educational Research Methods. Bandung: Alfabeta.

[7]. Sugiyono, 2016. Quantitative, Qualitative and R&D Research Methods. Bandung: PT Alfabet.

IOP Conf. Series: Earth and Environmental Science 1289 (2023) 012005 doi:10.1088/1755-1315/1289/1/012005

Nutritional value of sea cucumber *Acaudina molpadiodes* from Delta Wulan Estuary, Demak Regency, Central of Java, Indonesia

W Widianingsih*, and R Hartati

Department of Marine Sciences, Faculty of Fisheries and Marine Sciences, Diponegoro University, Jl. Prof. Soedharto, SH. Tembalang, Semarang, Indonesia

*Corresponding author mail: widia2506@gmail.com

Abstract. Sea cucumber *Acaudina molpadiodes* has a nutritional value that can be utilized as a functional food for humans. Therefore, this study aims to examine the composition of proximate, fatty acid, and amino acids in *A. molpadiodes*. The samples were collected from Delta Wulan Estuary, Demak Regency, Central Java, Indonesia. Based on the result of this study, the average value of moisture, protein, carbohydrate, total lipid, and ash contents in *A. molpadiodes* ranged between 90.44 – 91.23%, 4.44 – 4.97%, 4.12-4.35%, 0.16-0.18%, and 0.05-0.06% respectively. According to the result of amino acid analysis, the profile of essential amino acids comprised phenylalanine, isoleucine, valine, lysine, leucine, threonine, histidine, cysteine, methionine, and tryptophan. The average value of leucine content was 2273.95 ± 291.61 $mg.kg^{-1}$, and the threonine content was 2732 ± 276.93 $mg.kg^{-1}$. The non-essential amino acid profile consisted of glutamic acid, alanine, glycine, aspartic acid, tyrosine, proline, and taurine. The glycine and glutamic acid content in *A. molpadiodes* are 6843.67 ± 393.71 $mg.kg^{-1}$ and 5975.06 ± 687.85 $mg.kg^{-1}$. Fatty acid profiles in *A. molpadiodes* are DHA (docosahexaenoic), EPA (Eicosapentaenoic acid), and AA (Arachidonic acid with an average value of 2.03 ± 0.17 $mg.100g^{-1}$, 2.42 ± 0.83 $mg.100g^{-1}$, and 6.20 ± 1.56 $mg.100g^{-1}$.

1. Introduction

One species abundant in the Delta Wulan estuary is *Acaudina molpadiodes,* a translucent, pink, smooth-body sea cucumber. This species can easily be distinguished from other sea cucumbers because of its fusiform shape and its more tapering posterior than the anterior end. Its body lacks tube feet, secretes mucus, and has 15 tentacles in its mouth to collect the food [1]. It belongs to Caudinidae, order Molpadida, class Holothuroidea, and phylum Echinodermata [2].

Many studies investigated Caudinid's bioecology, habitat, population, and reproduction [1–4]. Some works also have been conducted on the bioactive compounds in them [5,6]. While research on nutrition, such as proximate analyses, amino acid, and fatty acid, is mainly on *Paracaudina australis, Stichopus vastus, Holothuria polii, H. tubulosa, H. mammata, H. atra, Actinopyga echinites, H. scabra* also have been conducted [7–10]. Studies on the collagen of *A. molpadiodes* have been done by researchers [6,11], but not in their nutrition, such as protein, lipid, carbohydrate, amino acid, and fatty acid. Researchers stated that those nutrition values vary with the sampling location [7,8], which is possibly due to different food resources. Hence, this research examines the proximate composition, amino acid, and fatty acid profile in *A. molpadiodes* caught from Delta Wulan Estuary, Demak Regency, Central Java, Indonesia.

2. Material and Methods

The samples of *A. molpadiodes* were taken from Delta Wulan Estuary, Demak Regency, Central Java, Indonesia. The geography position of sampling stations located at latitude and longitude ranging from 6°42'59.17'' to 6°45'13.46'' and 110°31'18.53'' to 110°33'12.07''. Samples of *A. molpadiodes* were collected monthly in April, May, and June 2021. Their body wall was separated from their visceral organ and coelomic fluid, washed with fresh water, and cut into small pieces (1-1.5cm). After that, the body wall samples were stored at -4°C until further analysis.

For analysis, the body walls were homogenized using the food processor. The body wall sample was dried in the oven at 105°C for moisture content until it obtained a constant weight. Ash and total lipid were determined by AOAC method analysis [12,13]. The total lipid was extracted using the method of [14]. Fatty acid was hydrolyzed and derived from total lipids using the procedure [15]. Amino acid analysis was carried out by the method described in AOAC [12].

3. Results and Discussion

3.1. Proximate in A. molpadiodes

According to research results, the moisture content in *A. molpadiodes* is the highest (90.85±0.40%) (Table 1) compared to *P. australis* (74.93±0.25%), *H. polii* (81.24%), *H. mammata* (85.24%), and *H. tubulosa* (84.30%) [7,9]. The average value of protein was 4.72 ± 0.27% (Table 1) and lower than *P. australis, H. polii, H. tubulosa*, and *H. mammata,* i.e., 20.22, 8.66, 8.82, 7.88% respectively [7,9]. Based on the analysis, the carbohydrate content in *A. molpadiodes* is 4.24 ±0.12 % (Table 1). Most sea cucumbers, such as *H. polii, H. tubulosa, H. mammata,* and *S. vastus,* do not have carbohydrate content [7,8].

Table 1. Proximate Content in Sea Cucumber *A. molpadiodes*

No.	Parameter	April	May	June	Average ± SDev
1	Protein (%)	4.97	4.76	4.44	4.72 ± 0.27
2	Total lipid (%)	0.18	0.17	0.16	0.17 ± 0.01
3	Carbohydrate (%)	4.35	4.25	4.12	4.24 ± 0.12
4	Ash (%)	0.06	0.06	0.05	0.06 ± 0.01
5	Moisture (%)	90.44	90.89	91.23	90.85 ± 0.40

3.2 Amino Acid in A. molpadiodes

Based on the result of amino acid analysis, *A molpadiodes* has nine essential amino acids, i.e., phenylalanine, valine, lysine, leucine, threonine, histidine, cystine, methionine, tryptophan, and eight non-essential amino acids, i.e. glutamic acid, alanine, glycine, aspartic acid, tyrosine, proline, serine, and taurine. Threonine was the highest essential amino acid with an average value of 2732.19 ± 276.93 mg.kg^{-1} (Table 2), and lower than *S. vastus* (16963.45 mg.kg^{-1}) [10]. Threonine is known as α – amino-β- hydroxybutyric acid and is a significant bioactive molecule that has vital medicine effects on protein synthesis, energy metabolism, and nutrient absorption [16]. In the body of marine organisms, threonine is required for nutritional regulation and metabolism [16–18].

The second-highest amino acid in *A. molpadiodes* was leucine (2273.95±291.61 mg.kg^{-1}). These results were in line with the research by [12]. In *A. molpadiodes* taken from Zhejiang Province in China also contained threonine, leucine, glycine, etc. The value of threonine, leucine, and glycine content indicates that *A. molpadiodes* can be used as traditional medicine and functional food for humans and livestock nutrition.

The third major essential amino acid found in *A, molpadiodes* was valine (1890 ± 199.53 mg.kg^{-1}), followed by phenylalanine (1678.41 ± 146.46 mg.kg^{-1}), isoleucine (1397.35 ± 177.35 mg.kg^{-1}), lysine (1329.14 ± 258.17 mg.kg^{-1}), histidine (763.80 ± 91.98 mg.kg^{-1}), cysteine (282.11 ± 9.07 mg.kg^{-1}), and tryptophan (58.85 ± 5.06 mg.kg^{-1}). The methionine has the lowest average value (48.18 ± 0.02 mg.kg^{-1}) among its essential amino acids (Table 2). Other essential amino acids in *A. molpadiodes* were histidine, cystine, methionine, and tryptophan. In this research, their values were as low as those found in *H. scabra, S. vastus,* and *P. australis* [9,10,19]. Even though their value was low, this essential amino acid is important in neurotransmitter and immune response [10].

The profile of non-essential amino acids in *A. molpadiodes* consists of glutamic acid, alanine, glycine, aspartic acid, tyrosine, proline, serine, and taurine (Table 2). The highest value of non-essential amino acids was glycine, i.e., 6843.67 ± 393.71 mg.kg^{-1} or 0.68 ± 0.039 g.100g^{-1}, which is similar to that in *H. scabra, H. leucospilota, A. mauritania* [20]. But it is lower than in *P. australis, and S. vastus* [9,10]. Glycine is one of the amino acids needed for producing collagen and muscle tissue and is important in maintaining the central nervous and digestive systems. Glycine also protects against some types of cancer [21] and can reduce cholesterol levels in the human blood [20].

Table 2. The composition of the amino acid (mg.kg^{-1}) of *A. molpadiodes*

No.	Parameter	April	May	June	Average ± SDev (mg.kg^{-1})
Essential Amino Acid					
1.	Phenylalanine	1565.48	1625.86	1843.89	1678.41 ± 146.46
2.	Isoleucine	1296.11	1293.14	1602.80	1397.35 ± 177.35
3.	Valine	1779.25	1771.17	2120.73	1890.38 ± 199.53
4.	Lysine	1193.82	1166.76	1626.84	1329.14 ± 258.17
5.	Leucine	2113.08	2098.21	2610.56	2273.95 ± 291.61
6.	Threonine	2590.21	2555.04	3051.31	2732.19 ± 276.93
7.	Histidine	721.29	700.76	869.34	763.80 ± 91.98
8.	Cysteine	292.42	275.40	278.50	282.11 ± 9.07
9.	Methionine	48.20	48.18	48.16	48.18 ± 0.02
10.	Tryptophan	64.44	54.58	57.53	58.85 ± 5.06
Non-Essential Amino Acid					
1.	Glutamic Acid	5612.72	5544.12	6768.33	5975.06 ± 687.85
2.	Alanine	3615.24	3687.17	3972.48	3758.30 ± 188.94
3.	Glycine	6623.10	6609.69	7298.23	6843. 67 ± 393.71
4.	Aspartic Acid	3465.55	3408.94	4192.45	3688.98 ± 436.94
5.	Tyrosine	1194.99	1192.76	1346.26	1244.67 ± 87.99
6.	Proline	3035.27	3008.49	3531.55	3191.77 ± 294.56
7.	Serine	2162.81	2211.00	2569.70	2314.50 ± 181,52
8.	Taurine	75.53	73.02	74.05	74.20 ± 1.26

3.3. Fatty Acid in Acaudina molpadiodes

The composition of fatty acids in *A. molpadiodes* is Docosahexaenoic Acid (DHA), Eicosapentaenoic (EPA), and Arachidonic Acid (AA) (Table 3). The highest fatty acid content was arachidonic acid (C20: n4-6), i.e., 6.20 ± 1.56 mg.100g^{-1} That value was lower than the result of the study conducted by [7] in which the arachidonic acid content in *H. tubulosa, H. polii*, and *H. mammata* ranged from 10.55-23.06%. Even though the Arachidonic Acid content in *A. molpadiodes* is the smallest among other sea cucumbers, it has a vital role as a neurotransmitter and should be added to food [24]. In addition, Arachidonic Acid is a nutritional element that has a significant role in the growth and development of nerves in the brain, forming brain fat tissue, and maintaining the interconnection of brain nerve cells, especially in influencing brain development [22].

IOP Conf. Series: Earth and Environmental Science 1289 (2023) 012005 doi:10.1088/1755-1315/1289/1/012005

The second major fatty acid in *A. molpadiodes* was EPA (Eicosapentaenoic Acid) with a value of 2.42 ± 0.83 mg.100g^{-1} (Table 3.), while the DHA content was 2.03 ± 0.17 mg.100g^{-1}. These EPA and DHA values were higher than in *S. horrens* [15]. DHA is needed as a plate-forming element for the container of rhodopsin, a vital compound that senses and sends the signals the eyes receive back to the brain [15].

Table 3. The composition of fatty acid in *A. molpadiodes*

No.	Parameter	Unit	April	May	June	Average ± SDev
1	Poly-unsaturated Fatty Acid	%	0.06	0.06	0.02	0.5 ± 0.02
2.	Unsaturated Fatty Acid	%	0.17	0.16	0.08	0.14 ± 0.04
3.	C22: 6n-3: DHA (Omega-3) (Docosahexaenoic Acid)	mg.100g^{-1}	2.1	2.2	1.8	2.03 ± 0.17
4.	C20: 5n-3: EPA (Eicosapentaenoic Acid)	mg.100g^{-1}	2.9	3.1	1.25	2.42 ± 0.83
5.	C20: 4n-6: AA (Omega-6) (Arachidonic Acid)	mg.100g^{-1}	7.3	7.3	4.0	6.20 ± 1.56
6.	Mono-unsaturated Fatty Acid	%	0.10	0.10	0.05	0.08 ± 0.02
7.	Saturated Fatty Acid	%	0.14	0.14	0.24	0.17± 0.05

4. Conclusion

The nutritional value of *A. molpadiodes* varies and consists of protein, lipid, carbohydrate, amino, and fatty acids. *A. molpadiodes* have a high moisture content of 90.85 ± 0.40%, 4.72 ± 0.27% protein, 4.24 ± 0.12% carbohydrate, and 0.17 ± 0,01% lipid. The protein content in *A. molpadiodes* is low. Still, the profile amino acids show many essential amino acids, such as threonine, leucine, valine, isoleucine, phenylalanine, histidine, lysine, cysteine, methionine, and tryptophan. Threonine is the highest essential amino acid with an average value of 2732.19 ± 276.93 mg.kg^{-1}. Although the values of Arachidonic Acid (AA), Eicosapentaenoic (EPA), and Docosahexaenoic Acid (DHA) were not high, they still could give an excellent potential for functional foods.

Acknowledgments

The authors thank the Research Institute, Diponegoro University, for supporting this research through Contract No. 185-30/UN7.6.1/PP/2021.

References

[1] Hartati R, Redjeki S, Riniatsih I, Widianingsih W, Nuraini R A T, Endrawati H and Mahendrajaya R T 2021 *IOP Conference Series: Earth and Environmental Science* vol 944 (IOP Publishing) p 012023

[2] O'Loughlin P M and Ong J Y 2015. *Raffles Bulletin of Zoology* **31** 292–302

[3] Choo P S 2012 *SPC Beche-de-mer Information Bulletin* **32** 43–8

[4] Hartati R, Zainuri M, Ambariyanto A, Redjeki S, Riniatsih I, Azizah R T N and Endrawati H 2019 *Ilmu Kelautan: Indonesian J Marine Sciences* **24** 121–6

[5] Abdul M, Mondol M, Shin H J, Rahman M A and Islam M T 2017 *Mar Drugs* **15** 317

[6] Li J, Li Y, Li Y, Yang Z and Jin H 2020 *Mar Drugs* **17** 370

[7] Aydın M, Sevgili H, Tufan B, Emre Y and Köse S 2011 Turkey *Int J Food Sci Technol* **46** 500–8

[8] Ibrahim M Y, Elamin S M, Gideiri Y B A and Ali S M 2015 *Food Science and Quality Management* **41** 11–6

[9] Widianingsih Z, Zaenuri M, Anggoro S and Kusumaningrum H P S 2016 *Aquat Procedia* **7**

271–6
[10] Rasyid A 2018 *Squalen Bulletin of Marine and Fisheries Postharvest and Biotechnology* **13** 9–15
[11] Jin H-X, Xu H-P, Li Y, Zhang Q-W and Xie H 2019 *Mar Drugs* **17** 169
[12] Horwitz W and Latimer G W 2005 Amino Acids in feed, AOAC official method 994.12, Official methods of analysis of AOAC international, 18th edition 12p
[13] Horwitz W and Latimer G W 2005 Fat (crude) in seafood, AOAC International method 964.12, official methods of analysis of AOAC International, 18th edition *AOAC International press* **35** 11
[14] Bligh E G and Dyer W J 1959 *Can J Biochem Physiol* **37** 911–7
[15] Ridzwan B H, Hanita M H, Nurzafirah M, Norshuhadaa M P S and Hanis Z F 2014 *Int J Biosci Biochem Bioinforma* **4** 204
[16] Chen Y P, Cheng Y F, Li X H, Yang W L, Wen C, Zhuang S and Zhou Y M 2017 *Poult Sci* **96** 405–13
[17] Estalkhzir F M, Khojasteh S and Jafari M 2013 *The Journal of Novel Applied Sciences* **2** 382–6
[18] Karau A and Grayson I 2014 *Biotechnology of food and feed additives* 189–228
[19] Sroyraya M, Hanna P J, Siangcham T, Tinikul R, Jattujan P, Poomtong T and Sobhon P 2017 *Functional Foods in Health and Disease* **7** 168–81
[20] Wen J, Hu C and Fan S 2010 *J Sci Food Agric* **90** 2469–74
[21] Yahyavi M, Afkhami M, Javadi A, Ehsanpour M, Khazaali A and Mokhlesi A 2012 *Afr J Biotechnol* **11** 2862
[22] Hanna V S and Hafez E A A 2018 *J Adv Res* **11** 23–32

Echinoderm diversity in the soft bottom of Wulan Estuary, Demak, Indonesia

Retno Hartati[1]*, Widianingsih Widianingsih[1], Muhammad Zainuri[2], Edy Supriyo[3]

[1]Department of Marine Science, Faculty of Fisheries and Marine Science, Diponegoro University. Jl. Prof. Soedharto SH. Tembalang, Semarang
[2]Department of Oceanography, Faculty of Fisheries and Marine Science, Diponegoro University. Jl. Prof. Soedharto SH. Tembalang, Semarang
[3]Department of Industrial Technology, Vocational School of Diponegoro University Jl. Prof. Soedharto SH., Kampus Tembalang, Semarang, Indonesia

*Corresponding author mail: retnohartati.undip@yahoo.com

Abstract. The majority of research on echinoderms in the tropics has involved coral reefs. There are relatively few studies on echinoderms connected to soft bottoms and estuaries. This work describes echinoderm species found in the soft-bottom sediments of the Wulan estuary in Demak, Indonesia, are described. Twelve species from Holothuroidea, Ophiuroidea, Asteroidea, and Echinoidea were identified. The three most prevalent species identified were *Astropecten* sp., *P. chilensis*, and *Paracaudina* sp. The species found differ depending on the sampling sites. Still, this knowledge expands our understanding of the richness of soft-bottom megabenthic fauna, especially the echinoderm. It could be used as a baseline to track the consequences of potential changes on this heavily sedimented river.

Keywords: biodiversity; echinodermata; holothuroidea, asteroidea; *Paracaudina chilensis*

1. Introduction

The Echinodermata consists of five classes, namely Holothuroidea, Crinoidea, Asteroidea, Echinoidea, and Ophiuroidea[1]. These species inhabit every ocean, in all climates, depths, and zone ranges [2][3]. Echinoderms are crucial in marine ecosystems, particularly within diverse food chains, and could be herbivores, carnivores, detritivores, and omnivores [4–6]. They may compete with demersal fish for food resources, prey on commercially essential bivalves, and serve as a significant food source for fish [7]. Additionally, Echinoderms interact with their environment in various ways and act as bioturbators, modifying seafloor sediments through activities like ingesting sediment, extracting organic matter during digestion, and expelling the remaining sediment, similar to the behavior of sea cucumbers. For example, large populations of sea cucumbers can result in a substantial turnover of surface sediments and lead to chemical and physical component changes in the sediment [8]. Sea cucumbers and sand dollars bury themselves in the sand, increasing the depth of oxygen and allowing for more species [9]. Echinoderms also contribute to the preservation of coral reef health by limiting

IOP Conf. Series: Earth and Environmental Science 1289 (2023) 012006 doi:10.1088/1755-1315/1289/1/012006

algae growth. Starfish and sea urchins have been observed to curb algae growth on coral reefs [10]. Moreover, these organisms have a global impact on the marine inorganic and organic carbon cycle[11]. Notably, echinoderms, particularly sea cucumbers, have significance beyond ecological roles. They have been utilized as food sources for humans[12–16] and have been explored for their medicinal properties [17–20]

Echinoderms within the tropical region, such as in Indonesia, have mainly been studied concerning coral reefs [21–23], seagrass beds [24–26], or mixing both ecosystems [27,28]. Studies focusing on echinoderms found in soft-bottom environments and estuaries have been limited [29], explicitly examining only sea cucumbers. This research aims to fill this gap by reporting on the various echinoderm species that inhabit the soft-bottom sediments in the Wulan estuary in Demak, Indonesia.

2. Materials and methods

In June 2022, a research study was conducted at the Estuary of Wulan River in Demak, Central Java, Indonesia. Echinoderm specimens were collected from the seabed at 11 different sites within the Wulan estuary, ranging in depths from 2 to 20 meters. The sampling locations were the same as those used in the study by [29]. The samples were collected with a 1,3-meter dredge, anesthetized in a 7.5% magnesium chloride solution, fixed, and kept in 70% denatured ethanol. Whenever possible, the live color of the specimens was observed and recorded.

The collected samples were sorted into their respective classes: Holothuroidea, Echinoidea, Asteriopde, and Ophiuroidea. A total of 256 samples were analyzed and classified according to their morphological character. The process to identify Holothuroidea followed the guidelines provided [30–39]. Echinoids were identified using the method outlined [40], Asteoidea using [41], and Ophiouroid identification followed the references[36,42,43]. The diversity of species was described and compared. Young individuals (small size) were also recorded in each sampling site.

3. Results and Discussion

The echinoderm samples were collected using a dredge, and the typical samples are presented in Figure 1. The present work found twelve species of echinoderm belonging to four classes, namely Holothuroidea (9 species), and one species of Asteroidea, Ophiouroidea, and Echinoidea (Table 1). They are *Acaudina sp, A. molpadioides, Paracudina sp. P. australis, P chilensis, P spiculata, Phyrella sp Colochirus quadrangularis, Leptosynapta sp Astropecten cf, spinolosus, Amphiodia sp., Spatangus sp.* The sea cucumber species found in this study was more than in previous work by [29] in 2021 and found in every sampling site. Asteroidea and Ophiuroidea were only found in sampling sites 1, 2, and 7; while Echinoidea exsisted at sampling sites 1, 2, and 6 (Table 2).

Figure 1. The echinoderm samples collected by a dredge

Table 1. The Echinoderm species found in the Wulan Estuary and their classification

Class	Ordo	Family	Genus	Species
Holothuroidea	Malpodiida	Caudinidae	Acaudina	*Acaudina sp.*
				Acaudina molpadioides
			Paracaudina	*Paracudina sp*
				Paracaudina chilensis
				Paracaudina australis
	Dendrochirotida	Phylloporidae	Phyllophorus	*Phyllophorus spiculata*
				Phyrella sp
		Cucumariidae	Colochirus	*Colochirus quadrangularis*
	Apodida	Synapridae	Leptosynapta	*Leptosynapta sp.*
Asteroidea	Paxillosida	Astropectinidae	Astropecten	*Astropecten cf, spinolosus*
Ophiuroidea	Ophiurida	Amphiuridae	Amphiodia	*Amphiodia sp.*
Echinoidea	Spatangoida	Spatangidae	Spatangus	*Spatangus sp.*

Table 2. The Echinoderm species found in the sampling site of the Wulan estuary

No.	Species	Sampling site										
		1	2	3	4	5	6	7	8	9	10	11
	HOLOTUROIDEA											
1	*Phyrella sp.*	+	+	+	+	+	+	-	+	-	-	-
2	*Acaudina malpodida*	+	-	+	+	+	+	+	+	+	+	+
3	*Acaudina sp.*	+	+	+	+	-	+	+	+	+	+	+
4	*Paracaudina chilensis*	+	+	+	+	+	+	+	+	+	+	+
5	*Paracaudina australis*	+	+	+	-	+	+	+	+	+	+	+
6	*Paracaudina* sp.	+	+	+	+	+	+	+	+	+	+	+
7	*Leptosynapta* sp	+	-	-	-	-	+	+	+	-	-	-
	ASTEROIDEA											
8	*Astropecten* sp.	+	+	-	-	-	-	+	-	-	-	-
	OPHIUROIDEA											
9	*Amphiodia* sp.	+	+	-	-	-	-	+	-	-	-	-
	ECHINOIDEA											
10	*Spatangus* sp.	+	+	-	-	-	+	-	-	-	-	-

Seven species of sea cucumber revealed in this study, *Acaudina sp., A. molpadioides, Paracudina sp, P. chilensis,* and *P. australis* [29], and during present work they were found in almost all sampling sites (Table 2). Therefore, this paper only described the two species, i,e, *Phyrella* sp., and *Leptosynapta* sp. Phyrella is commonly found in shallow waters (less than 10 meters) and is distributed in the Indo-West Pacific, including Indonesia. It belongs to the Family Phyllophoridae Östergren, 1907 [44]. The species in Figure 2 has 14–20 tentacles of various diameters and is tiny to medium in size. These tentacles can be found in pairs or alone, and are often arranged in an outer and an inner circular. Numerous cylindrical podia with terminal suckers with a diameter size from 100 to 300 μm and 1-3 mm in length can be found on the body surface of Phyrella. Five sparsely calcified podia eventually grow into anal teeth and surround the conical posterior end, which gradually tapers to a little cloaca. The calcareous ring consists of both posterior prolongation radials and inter-radials [35] During our research, *Phyrella* sp. was discovered in Sampling sites 1, 2, 3, 4, 5, 6, and 8.

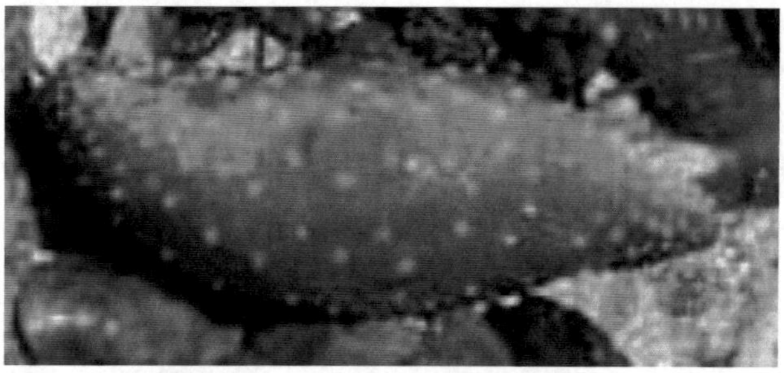

Figure 2. *Phyrella* sp of Phylloporidea

Leptosynapta sp. (Figure 3) was discovered in Sampling sites 1, 6, 7, and 8, belonging to the Family Synaptidae. Synaptidae is unique among Holothuroidea members because they lack podia or tube feet and no respiratory trees on the body's surface. But they have anchor-shaped spicules in their body wall that support their ability to move. While Synaptidae are commonly found in seagrass areas, there have been no specific observations of their existence in soft-bottom sediment until now. Identified two species of Synaptidae, at the intertidal zone of Sanur Beach, Bali., i.e., *Synapta maculata* and *Opheodesoma grisea*[25].

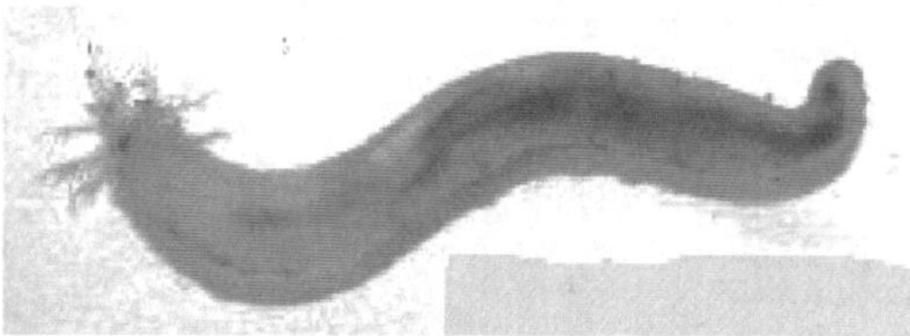

Figure 3. *Leptosynapta* sp of Holothuroidea

The Synaptidae's adaptability to extreme environments is evident through their osmoregulation to salinity fluctuations, wide tolerance of temperature changes, defense mechanisms against wave motion, possession of sticky skin, and autotomy mechanism[25]. Unique species of Synaptidae, including *O. grisea, S. maculata*, and *S. recta*, in the Seagrass bed of Elnusa, Timor Island, Nusa Tenggara Timur[24]. *Leptosynapta* sp. of Wulan Estuary exhibited distinct characteristics, with adults typically ranging from small to moderate sizes, measuring 12-30 cm in length and mostly around 12-15 cm in diameter at the midbody. The longitudinal muscles were easily visible, and the body wall was thin and transparent (Figure 3). Well-developed, numerous, and uniformly spaced papillae were seen from the anterior to the posterior end. It has 12 well-extended tentacles. *Leptosynapta* sp. Inhabits gravel, coarse to fine sand, and mud, typically in the intertidal zone to a depth of about 50 meters [32], and they are known for their burrowing habits [44]. Although they are usually found in dense populations [32], this was not the case in the Wulan estuary.

The member of class Asteroidae in the Wulan estuary was *Astropecten* sp (Figure 4). It belongs to Ordo Paxillosida and Family Astropectinidae. Astropecten starfish typically inhabit sandy, muddy, or gravelly seafloors. Commonly, they spend the day mostly hidden beneath silt [45]. It favors sandy seafloors between 1 and 50 meters deep that are also relatively abundant in algae [46]. It is a carnivore that eats mollusks, which it captures with its arms before putting them in its mouth and keeping them there with its long, moving prickles [45]. The long, narrow arms of *Astropecten* sp. taper quickly to a minor point. The paxillar region is constrained. The disk's paxillae have a few central granules surrounded by slender rod-shaped spinelets [47]..

Figure 4. *Astropecten cf. spinulosus* of Asteroidea

The Echinoid discovered in the bottom of Wulan Estuary was *Spatangus sp.* (Figure 5). It belongs to Ordo Spatangoida, Family Spatangidae. It includes irregular sea urchins or Irregularia subgroups, consisting of Neognathostomata (cassiduloids, sea biscuits, and sand dollars) and Atelostomata (heart urchins and allies). This subgroup has bilaterally symmetrical forms covered by small, specialized spines and lives as infaunal organism. The clade of echinoids known as atelostomata is considered the most varied [48]. Besides Spatangus, there are Plethotaenia, and Granopatagus as members of the family Spatangidae[48–50]. Spatangus is the most widespread worldwide[50]. A test oval outline with a heart shape characterizes the family Spatangidae. The apical system is ethmolytic. It has well-developed petals. Ventral portions of poriferous zones are often simple [51].

Figure 5. *Spatangus sp.* of Echinoidea

A genus of Spatangus has a stiff, broad, oval shape test with a frontal ambulacral groove that is moderately deep. When ambulacra are paired, the petals are pretty distinct, virtually closed distally, and not depressed. It has four genital holes and a subcentral apical system with ethmolytic function. Labrum is noticeable. Most sternums are narrow. *Spatangus* sp is a considerable partially flattened heart urchin. The dorsal surface petals of the test are not depressed, but the test's front has a deep cleft corresponding to the frontal ambulacrum. They use their spines to move and bury [52].

Because they are non-selective deposit-feeding, able to move and burrow beneath the sediment surface, Spatangoid urchins play a significant role in sediment bioturbation [53]. Their bioturbation activities have important ramifications, including enhancing the diversity of infaunal and microbial communities [54], facilitating solute exchange across the sediment-water interface[55], removing contaminants that are bound to the sediment, and enhancing nutrient and particle mixing [56]. Bioturbation also increases oxygen penetration into the sediment, affecting its oxygenation depth[57, 58].

The Ophiuroid inhabited in Wulan Estuary was *Amphiodia* sp, (Figure 6) a Family Amphiuridae and Order of Ophiurida member. With 2,096 known species, the Class of Ophiuroidea is the biggest group within the phylum Echinodermata [59]. Morphologically, the bodies of ophiuroids have a five-fold symmetry. The standard body structure comprises a pentagonal to spherical core disk and five sharply separated arms. On the other hand, some species have six, seven, or even ten arms that can branch once or multiple times [60]. Species with branching arms are frequently called basket stars, while species without branched arms are called brittle stars. The maximum diameter ranges between 3 and 50 mm, and the arms' length is sometimes described as n times the disk diameter. Ophiuroids live in marine benthic habitats, while certain species can survive in brackish settings [59].

Amphiodia sp. possesses a rounded disk covered with numerous small, overlapping scales. The radial shields are convex externally and have a straight inner margin, with most of them connected, except for one or two small triangular and elongated scales at the proximal ends. The ventral interradius scales are slightly smaller than the dorsal scales. Narrow bursal slits are present and have an arrow-shaped oral shield. The madreporite is positioned close to the distal border and contains one or two pores. The aboral shields are thin and laterally expanded. Each side of the jaw angle has two oral papillae, with the distal papilla being longer and broader. Small infra-dental papillae are also present. The dorsal arm plate is rectangular and wider than long, while the ventral arm plate is pentagonal and broader. This particular species has been found in various habitats such as sand, mud, gravel, and algae-covered substrates [61], similar to its discovery in the Wulan Estuary where it was found buried in the silt.

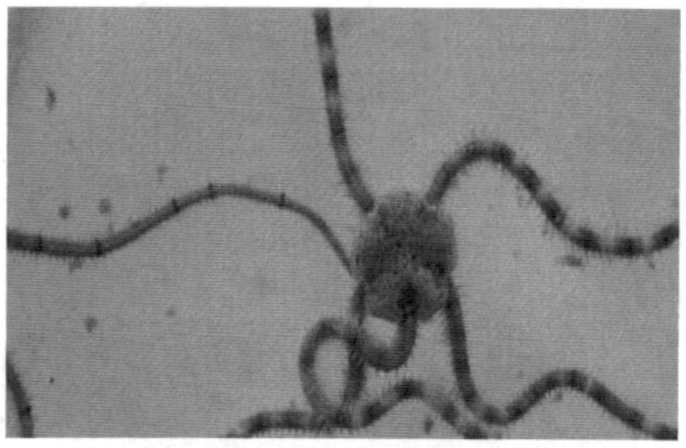

Figure 6. *Amphiodia* sp,. discovered in the Wulan Estuary's soft bottom sediment

Ophiuroids comprise a sizeable component of both epi- and infaunal populations with various lifestyles. The distribution and quantity of ophiuroids in benthic habitats can be used to confirm their ecological importance. While burrowing species may number in the thousands in infaunal groups, epifaunal ophiuroids may form dense carpets over the seafloor [60,61].

4. Conclusion

Twelve species were recorded, belonging to the family of Holothuroidea, Ophiuroidea, Asteroidea, and Echinoidea. *Paracaudina* sp., *P. chilensis*, and *Astropecten* sp were the most common species. The asteroid, echinoid, and ophiuroid found were *Astropecten cf. spinulosa*, *Spatangus* sp., and *Amphiodia* sp. This contribution could be viewed as a baseline for assessing the impact of potential changes on the heavily silted-up Wulan River.

Acknowledgment

The authors would like to thank the Rector of Diponegoro University who funded this work with Funding of "Other than APBN" in the scheme of International Publication Research 2023 No185-72/UN7.6.1/PP/2022-2023

References

[1] Zulliger D E, and Lessios H A 2010. *Zootaxa* **2504** 1-19.

[2] Brusca R C and Brusca G J 1990 *Invertebrates* (Sunderland, Massachusetts: Sinauer. Associates, Inc)

[3] Uthicke S, Schaffelke B, and Byrne M 2009 *Ecol Monogr* **79** 3–24

[4] Whitmore E 2017 *Echinoderms: Ecology, Habitats and Reproductive Biology* (Nova Publisher)

[5] Harrold C 1987 *Echinoderm studies 2*
[6] Luna Salguero B M and Reyes Bonilla H 2010 *Hidrobiológica* **20** 127–34
[7] Packer D B, Watling L and Langton R W 1994 *J Exp Mar Biol Ecol* **179** 207–22
[8] Hermosillo-Núñez B B 2020 *Hydrobiologia* **847** 739–56
[9] Vardaro M F, Ruhl H A and Smith K J L 2009 *Limnol Oceanogr* **54** 2081–8
[10] Iken K, Konar B, Benedetti-Cecchi L, Cruz-Motta J J, Knowlton A, Pohle G, Mead A, Miloslavich P, Wong M, and Trott T 2010 *PLoS One* **5** e13845
[11] Lebrato M, Iglesias-Rodríguez D, Feely R A, Greeley D, Jones D O B, Suarez-Bosche N, Lampitt R S, Cartes J E, Green D R H and Alker B 2010 *Ecol Monogr* **80** 441–67
[12] Jangoux M and Lawrence J M 1982 *Echinoderm nutrition* (London: CRC Press) 700p. DOI https://doi.org/10.1201/9781003078920
[13] Fernandez C and Boudouresque C-F 2000 *Mar Ecol Prog Ser* **204** 131–41
[14] Rahman M A, Ismail M, Parvez M S, Asadujjaman M, Ashik A-A and Molla M H R 2022 *Journal of Biological Studies* **5** 413–43
[15] Wulandari D A and Warsito M F 2022 *Omni-Akuatika* **18** 101–11
[16] Furlam E 2023 *Novel Techniques in Nutrition & Food Science* **7** 672–4
[17] Kelly M S 2005 *Prog Mol Subcell Biol* **39** 139–65
[18] Marmouzi I, Tamsouri N, El Hamdani M, Attar A, Kharbach M, Alami R, El Jemli M, Cherrah Y, Ebada S S and Faouzi M E A 2018 *Revista Brasileira de Farmacognosia* **28** 575–81
[19] Stabili L, Acquaviva M I, Cavallo R A, Gerardi C, Narracci M and Pagliara P 2018 *Evidence-Based complementary and alternative medicine* **2018** 8p https://doi.org/10.1155/2018/7891748
[20] Zhukova N V 2022 *Mar Drugs* **21** 1 21 https://doi.org/10.3390/md21010021
[21] Handayani T, Sabariah V and Hambuako R R 2017 *Jurnal Perikanan Universitas Gadjah Mada* **19** 45–51
[22] Hartati R, Meirawati E, Redjeki S, Riniatsih I and Mahendrajaya R T 2018 *Jurnal Kelautan Tropis* **21** 41–8
[23] Tarigan R S, Hartati R and Widowati I 2020 *IOP Conference Series: Earth and Environmental Science* **530** (IOP Publishing) p 012042
[24] Purwati P and Wirawati I 2008 *Oseanologi dan Limnologi di Indonesia* **34** 371–84
[25] Setyastuti A 2014 *Jurnal Oceana* **XXXIX** 1 1-9
[26] Muzaki F K, Setiawan E, Insany G F A, Dewi N K and Subagio I B 2019 *Biodiversitas* **20** 7 1787-1793 DOI: 10.13057/biodiv/d200701
[27] Setyastuti A, Purbiantoro W and Hadiyanto H 2018 *Biodiversitas* **19** 5 1919–25 DOI: 10.13057/biodiv/d190544
[28] Jailani A K, Sahami F M and Kadim M K 2020 *The NIKe Journal* **8** 2 33-36
[29] Hartati R, Redjeki S, Riniatsih I, Widianingsih W, Nuraini R A T, Endrawati H and Mahendrajaya R T 2021 *IOP Conference Series: Earth and Environmental Science* **944** 012023
[30] Clark A M and Rowe FWE. 1971. *British Museum (Natural History) Publications* **690** 234
[31] Thandar A S 1990 *Afr Zool* **25** 207–23
[32] Sewell M A, Chia F-S and Thandar A S 1995 *Can J Zool* **73** 469–85
[33] Massin C 1996 *Bulletin de l'Institut royal des Sciences naturelles de Belgique, Biologie* **66** 151–78
[34] Massin C 1999 *Zoologische verhandelingen* **329** 1–144
[35] Michonneau F and Paulay G 2014 *Zootaxa* **3760** 101–40
[36] Olbers J M, Griffiths C L, O'Hara T D and Samyn Y 2019 *Field guide to the brittle and basket stars (Echinodermata: Ophiuroidea) of South Africa* (CEBios, Royal Belgian Institute of Natural Sciences)
[37] O'Loughlin M, Barmos S and Spiegel V D 2011 *Mem Mus Vic* **68** 37–65
[38] Purcell S W, Lovatelli A, González-Wangüemert M, Solís-Marín F A, Samyn Y and Conand C

2023 *Commercially important sea cucumbers of the world* (Rome: FAO Species Catalogue for Fishery Purposes)

[39] Samyn Y 2003 Towards an understanding of the shallow-water holothuroid fauna (Echinodermata: Holothuroidea) of the western Indian Ocean (Universiteit Brussel)

[40] Stara P, Melis R, Bellodi A, Follesa M C, Corradini C, Carugati L, Mulas A, Sibiriu M and Cannas R 2023 *Front Mar Sci* **10** 1033710

[41] Lawrence J M 2013 *Starfish: biology and ecology of the Asteroidea* (Baltimore, MD, USA: JHU Press)

[42] Pomory C M 2007 *Caribb J Sci* **10** 1–42

[43] Gondim A I, Alonso C, Dias T L P, Manso C L C and Christoffersen M L 2013 *Zookeys* 45

[44] McEuen F S 1986 The reproductive biology and development of twelve species of holothuroids from the San Juan Islands, Washington

[45] Coll M, Piroddi C, Steenbeek J, Kaschner K, Ben Rais Lasram F, Aguzzi J, Ballesteros E, Bianchi C N, Corbera J and Dailianis T 2010 *PLoS One* **5** e11842

[46] Costello M 2001 *European register of marine species: a check-list of the marine species in Europe and a bibliography of guides to their identification* (Paris: Muséum national d'histoire naturelle)

[47] Thompson B A W and Riddle M J 2005 *Mar Ecol Prog Ser* **290** 135–43

[48] Kroh A 2020 Chapter 1 - Phylogeny and classification of echinoids *Developments in Aquaculture and Fisheries Science* vol 43, ed J M Lawrence (Elsevier) pp 1–17

[49] Smith Ab and Kroh A 2011 The echinoid directory. WorldWide Web electronic publication

[50] Kroh A and Mooi R 2022 World echinoidea database. Spatangus Gray

[51] Shin S 2013 *Anim Syst Evol Diversity* **29** 308–11

[52] Kanazawa K 1992 *Palaeontology* **35** 733

[53] Hollertz K and Duchêne J-C 2001 *Mar Biol* **139** 951–7

[54] Widdicombe S, Austen M C, Kendall M A, Warwick R M and Jones M B 2000 *Island, Ocean and Deep-Sea Biology: Proceedings of the 34th European Marine Biology Symposium, held in Ponta Delgada (Azores), Portugal, 13–17 September 1999* (Springer) pp 369–77

[55] Ford P W, Bird F L and Hancock G J 1999 *Mar Freshw Res* **50** 523–32

[56] Sandnes J, Forbes T, Hansen R and Sandnes B 2000 *Mar Ecol Prog Ser* **197** 151–67

[57] Osinga R, Lewis W E, Wopereis J L M, Vriezen C and van Duyl F C *Ophelia* **41** 221–36

[58] Widdicombe S and Austen M C 1998 *J Exp Mar Biol Ecol* **228** 241–55

[59] Stöhr S, O'Hara T D and Thuy B 2012 *PLoS One* **7** e31940

[60] Turon X, Codina M, Tarjuelo I, Uriz M J and Becerro M A 2000 *Mar Ecol Prog Ser* **200** 201–12

[61] Metaxas A and Giffin B 2004 *Deep Sea Research Part I: Oceanographic Research Papers* **51** 1307–17

IOP Conf. Series: Earth and Environmental Science 1289 (2023) 012007 doi:10.1088/1755-1315/1289/1/012007

Variation in Eastern Little Tuna *(Euthynnus affinis)* catches related to El Niño Southern Oscillation (ENSO) events in the Makassar Strait

M L Syamsuddin[1*], A R Puspita[2], F Syamsudin[1,3], Y N Ihsan[1], Sunarto[1] and M Zainuddin[4]

[1]Department of Marine Sciences, Faculty of Fisheries and Marine Sciences, Universitas Padjadjaran, Jatinangor 45363 Indonesia
[2]Master of Marine Conservation Study Program, Faculty of Fisheries and Marine Sciences, Universitas Padjadjaran, Jatinangor 45363 Indonesia
[3]National Research and Innovation Agency (BRIN), Jalan M. H. Thamrin No. 8, Jakarta 10340 Indonesia
[4]Department of Fisheries, Faculty of Marine and Fisheries Sciences, Hasanuddin University, Makassar 90245 Indonesia

*Corresponding author e-mail: mega.syamsuddin@unpad.ac.id

Abstract. In order to determine how the El Niño Southern Oscillation (ENSO) affects Eastern Little Tuna catch rates in the Makassar Strait (MS), satellite-derived oceanographic variables, including sea surface temperature (SST), chlorophyll-a, salinity, sea surface height (SSH), ocean current, Niño index 3.4, and Eastern Little Tuna *(Euthynnus affinis)* catch data were used. Analyses were conducted for a seven-year period (January 2015-December 2021), which included the El Niño (2015) and La Niña (2021) events. A higher CPUE value is obtained in 2021 when La Niña occurs with a CPUE value of 3973 kg/trip, compared to the El Niño event. Based on seasonal variations, the highest catch was recorded during the northwest monsoon (706.02 kg/trip) with 2789 fishing points, while the lowest occurred during the southeast monsoon (700 kg/trip) with 2942 fishing points. The CPUE was 16.25 kg/trip with 1133 fishing points during El Niño and 23.56 kg/trip with 1027 fishing points during La Niña, respectively. The fishing grounds area was wider during the La Niña period, at 2 - 8°S and 115 - 120°E. This study revealed a strong impact of the La Niña events, significantly increasing Eastern Little Tuna catch in the MS.

1. Introduction

The Fisheries Management Area (FMA) 713, which includes the Makassar Strait (MS), is responsible for Indonesia's second-largest fish production [1]. The oceanographic variations in MS are primarily influenced by monsoon systems and Indonesian Throughflow (ITF) [2]. Thus, the regions are very dynamic and favorable for pelagic fish migration. According to information obtained from FMA 713, the Indonesian Statistical Data Agency reported that the fishery yield for 2019 was 558,000 tons. Various environmental and ecological factors are related to Eastern Little Tuna behavior. Examining the Eastern Little Tuna catch rates necessitates an examination of multiple oceanographic factors.

Eastern Little Tuna is found in regions between 40°N and 40°S, they typically prefer water temperatures ranging from 29 to 33°C, with a lower Chlorophyll-a concentration of 0.3 mg/m^3 [3]. Tuna species usually prefer waters with salinity ranging from 32 – 37 psu [4, 5]. In general, Eastern Little Tuna lives on the surface layer down to 40 meters [6]. Regional climate related to the El Niño Southern Oscillation (ENSO) have an effect on global tuna catches [7, 8]. The ENSO is a widespread pattern of climate change that has a considerable impact on a significant portion of the world. It has been established that the ENSO cycle causes warm (El Niño) and cool (La Niña) phases that have an impact on the Pacific Ocean's tuna fishing yields [9, 10]. Previous research investigated how the ENSO

IOP Conf. Series: Earth and Environmental Science 1289 (2023) 012007 doi:10.1088/1755-1315/1289/1/012007

variability affects environmental variability and their relationship with Eastern Little Tuna in the Java Sea [11].

Production from fishing activities is highly dependent on the fishing location itself [12, 13], and the determining factor for fishing ground affects the size of the catch. In principle, estimating fishing grounds consists of examining relationships between oceanic parameters and fish schools [14]. The existence of fish schools can be estimated by understanding the ocean factors in a particular area and can be employed for fishing intentions [15, 16]. As a result, comprehending the influence of changes in ocean climate variation on the distribution of catches is crucial for the responsible control of Eastern Little Tuna resources management in the MS. The correlation between oceanographic parameters and migration patterns of pelagic fish has been conducted [17, 18].

The application of remote sensing satellite data is crucial in addressing the issue of establishing the connection between the marine environment and the quantity of fish stock [19, 20]. The objective of this research is to analyze how the climate variability of ENSO affects Eastern Little Tuna catch rates in the MS.

2. Materials and methods

2.1 Study area

The research location was within the MS region, between 1°N - 5°S and 115 - 121°E (Figure 1). According to FMA 713, the area was chosen based on the ocean dynamics and fishing potential of the MS.

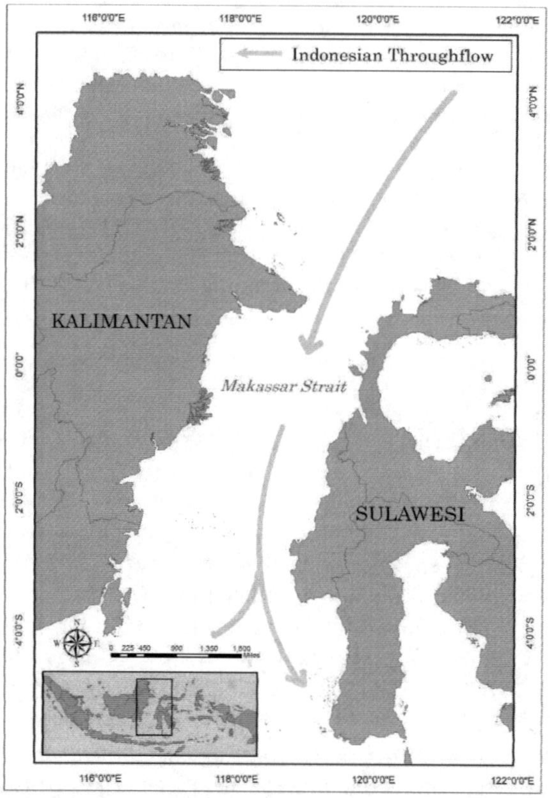

Figure 1. Map of the Indonesian Sea with the inset box designating the Makassar Strait research region. The blue line with the arrow shows the Indonesian Troughflow (ITF).

2.2 Data

The period of data processing is seven years, starting from January 2015 to December 2021. The monthly basis of spatial data is cropped based on the research area in the MS (Table 1).

Table 1. Overview of specification for Eastern Little Tuna and oceanographic parameter data

	Data	Unit	Spatial Resolution	Sources
1.	Sea Surface Temperature	°C	4 km	https://oceancolor.gsfc.nasa.gov
2.	Chlorophyll-a	mg/m^3	4 km	https://oceancolor.gsfc.nasa.gov
3.	Salinity	ppt	40 km	https://poaac.jpl.nasa.gov/SMAP
4.	Ocean Currents	m/s	8 km	https://marinecopernicus.eu
5.	Sea Surface Height	m	8 km	https://marinecopernicus.eu
6.	Niño Index (ONI)	-	-	https://origin.cpc.ncep.noaa.gov
7.	Eastern Little Tuna catches	-		Ministry of Marine Affairs and Fisheries, Republic of Indonesia

2.2.1 Eastern Little Tuna catch data processing

Catch data is represented by catch per unit effort (CPUE). Data consists of fishing location, number of fishing days, fish weight (in kg), vessel identification, and fishing gear. CPUE is a technique used to assess the fishery output fluctuations, which are averaged on an annual basis and influenced by the spatial yield. The CPUE calculation is derived from the formula:

$$CPUE_I = \frac{Catch_i}{Effort_i} \quad (1)$$

Where:
$CPUE_i$ = absolute catch of species per fishing effort (kg/trip)
$Catch_i$ = total species taken in year t (kg)
$Effort_i$ = total number of boat days in year t (trip)

Fishing gear standardization is carried out prior to calculating CPUE when multiple types of fishing gear are used. Gill nets and purse seines are used to catch a large number of Eastern Little Tuna in the MS. The use of the following formula allows the standardization of fishing gear:

$$FPI = \frac{CPUE_i}{CPUE_s} \quad (2)$$

where:
FPI = The fishing gear type's catch-effort factor
$CPUE_i$ = Catch per yearly effort of other fishing gear (kg/trip)
$CPUE_s$ = Catch per yearly effort of standard gear (kg/trip)

Utilizing R programming, a histogram is created utilizing these CPUE values to determine fish production from environmental factors. This histogram is then used to examine the preferred oceanographic variables for Eastern Little Tuna.

2.2.2 Oceanographic data processing

This study utilized satellite-derived oceanographic parameters. The monthly data on oceanographic parameters were analyzed and processed in terms of both temporally and spatially distributions. Spatial data processing encompasses cropping, extraction, and visualization. The temporal distribution patterns of oceanographic parameters corresponding to fishing points were extracted and averaged for each month to explore their seasonal and interannual variability. All oceanographic parameters data were resampled to 9 km spatial resolution. To more accurately capture changes in oceanographic characteristics for Eastern Little Tuna catch rates, the study overlaid the fishing points with the oceanographic variables at a 9 km spatial resolution.

2.2.3 Climatic index data processing

The Niño 3.4 index is utilized as an approximation of the ENSO conditions. The Niño 3.4 index measures the average sea surface temperature anomaly in the area between 120° - 170° W and 5° N - 5° S. When the 5-month running average of the Niño index is greater than +0.5°C for El Niño and -0.5°C for La Niña for at least five consecutive months, the El Niño and La Niña event is said to have occurred. In order to determine the variations in climate conditions during a strong El Niño event (2015) or La Niña event (2021), the Eastern Little Tuna catch and satellite remotely sensed data were evaluated throughout 7 years datasets from January 2015 to December 2021 (Table 2).

Table 2. Climate condition during 2015 - 2021

Year	Climate Condition
Jan 2015 – Apr 2016	El Niño
Oct 2017 – Apr 2018	La Niña
Sep 2018 – Jun 2019	El Niño
Aug 2020 – Dec 2021	La Niña

3. Results and discussion

3.1 Temporal variability of Eastern Little Tuna catch rates

The following results will show how El Niño and La Niña phases affect Eastern Little Tuna catches in the MS. Figure 2 displays the Eastern Little Tuna annual CPUE and Niño 3.4 index from 2015 to 2021. This graph displays the annual variation in catch rates from 2015 to 2021 (blue bars) and sea surface temperature anomaly from the Niño 3.4 index (black line). CPUE increases during the negative Niño index (La Niña) and decreases during positive Nino index (El Niño). Catches had significant increment during La Niña (2017, 2020, and 2021) compared to during El Niño events (2015, and 2018).

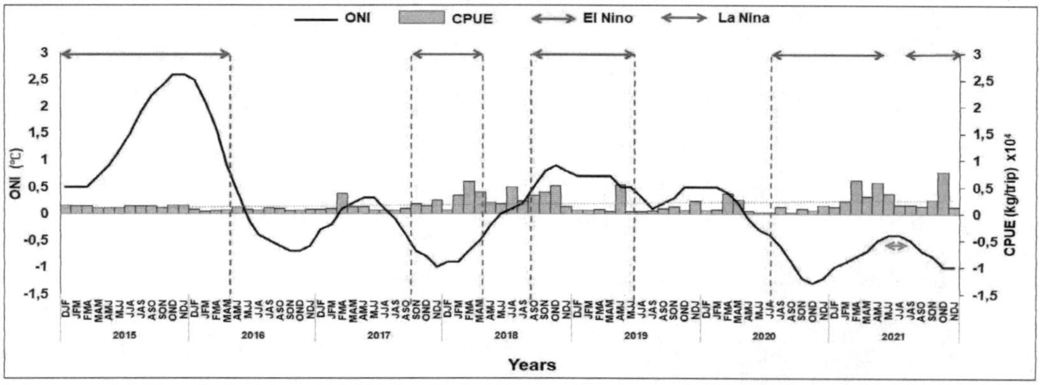

Figure 2. Changes in the catch rates of Eastern Little Tuna (blue bars) and SST anomalies from Niño 3.4 index (black line) from 2015-2021. The horizontal lines with arrows above this graph showed the period of the El Niño event (red line), and the La Niña event (blue line).

From August to December 2015, there was a strong El Niño phase with an average CPUE value of 11.29 – 15 kg/trip. A weak La Niña phase occurred from October 2017 to April 2018 with average CPUE ranging from 15.35 – 59.55 kg/trip. In 2018, a weak El Niño phase occurred from September to December with average CPUE values between 13.42 – 52 kg/trip. With the exception of June and July, a strong La Niña phase lasts from September 2020 to December 2021, with a 3.53 – 34.4 kg/trip average CPUE value. Increases in Eastern Little Tuna CPUE are often greater during the La Niña event than the El Niño event. Small pelagic catches in the MS were less favorable during the El Niño event.

3.2 Mean monthly climatological oceanographic parameters
The dynamics of the environmental variability are influenced by climate variations and thus affect the distribution of Eastern Little Tuna [11]. Figure 3 shows the mean monthly climatological of the five-oceanographic parameters and CPUE to give a clear understanding of the effect of ocean variability, and Eastern Little Tuna catches during 2015-2021.

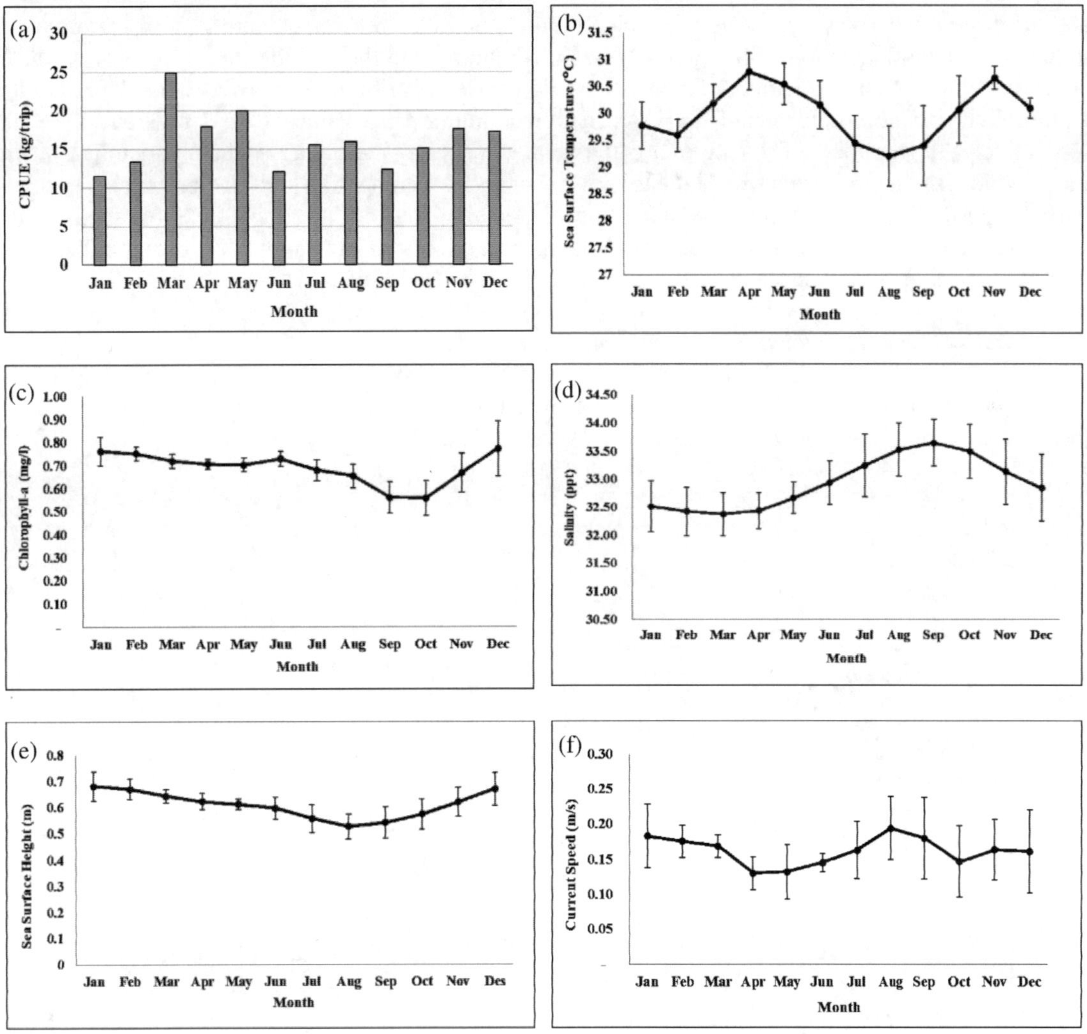

Figure 3. The time series plots of the monthly mean climatology of CPUE and oceanographic parameters during 2015 – 2021 (a) CPUE in kg/trip, (b) SST in °C, (c) chlorophyll-a in mg/m^3, (d) salinity in ppt, (e) SSH, in m, and (f) current speed in m/s. The vertical line indicates the standard deviation of the parameter's oceanographic values.

Figure 3a shows the monthly average CPUE for Eastern Little Tuna in the MS. The highest average CPUE reached 17.8 – 24.8 kg/trip during the transitional season I (March-May). With a CPUE of 24.8 kg/trip in March, CPUE achieved its highest peak. On the other hand, the southeast monsoon (June-Aug) had the lowest average CPUE with 12.1 – 15.9 kg/trip. High probabilities of catches corresponded with an SST value of 30.5° C, Chlorophyll-a concentration of 0.7 mg/m^3, Salinity of 32.5 ppt, SSHA of 0.65 m, and current speed of 0.17 m/s, respectively. Based on the time series plots, most oceanographic parameters showed strong seasonality. Seasonal features dominate all selected ocean parameters, with SST, Chlorophyll-a, salinity, SSH, and current speed exhibiting annual and semi-annual variability, respectively. A distinct semi-annual pattern is portrayed in SST distribution, where high SST characterizes the Eastern Little Tuna habitat during the first transitional season (March to May) and transitional season II (September to November). This is consistent with other studies [21] stating that SST in the transitional season is higher than in other seasons. The highest SST values were in April and November (30.5°C).

On the other hand, the SST distribution is lower during the northwest monsoon (December-January) and the eastern monsoon (June-August). This is consistent with [22] who stated that temperature tends to have the same fluctuations. SST typically begins to increase to the maximum in March or April. In contrast, Chlorophyll-a (Figure 3c) and SSH (Figure 3e) distributions showed annual variability. Chlorophyll-a concentrations and SSH reached maximum peak values (0.77 mg/l and 0.68 m, respectively) in December. The temporal patterns of salinity showed evident annual variability, where high and low salinity occurred from June to November and December to May. The temporal patterns of current speed (Figure 3f) show similar patterns, indicating that both variables exhibit semi-annual variations. The current speed reached the maximum value of 0.19 m/s in August. On the other hand, the minimum peak of current speed occurred in April and May at 0.13 m/s.

3.3 Variation in the spatial distribution of Eastern Little Tuna

Based on their spatial catch distribution, Eastern Little Tuna can be found throughout the area of the MS, especially in the south area of the MS. The spatial distribution of Eastern Little Tuna, CPUE and total fishing points based on different seasons showed in Figure 4.

(a)

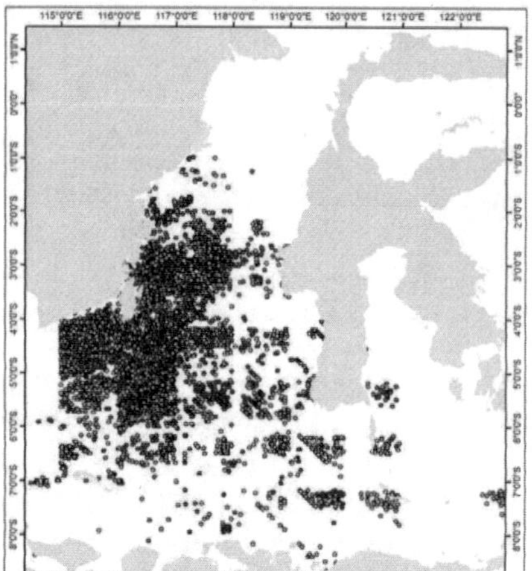

(b)

Figure 4. (a) Spatial distribution of Eastern Little Tuna based on seasonal variations, and (b) CPUE value (blue bars) and total fishing locations (green bars) during the northwest and southeast monsoon from 2015 – 2021.

Most fishing activity located in the area with coordinates 2-6°S and 115-118°E. The seasonal spatial pattern over the 7-year datasets showed that fishing hotspots for Eastern Little Tuna were frequently occurred in this location. During northwest monsoon (blue dots) fishing locations are found more frequent than during southeast monsoon (red dots) in the MS. Furthermore, based on the temporal distribution during the northwest monsoon period, there were 2.789 of total fishing locations with CPUE 706.2 kg/trip. In the southeast monsoon, there were 2.942 fishing locations with a CPUE value of 700 kg/trip. In order to give more detail understanding of ENSO impacts on spatial distribution, we differentiated the spatial distribution map during El Niño and La Niña events, as shown in Figure 5.

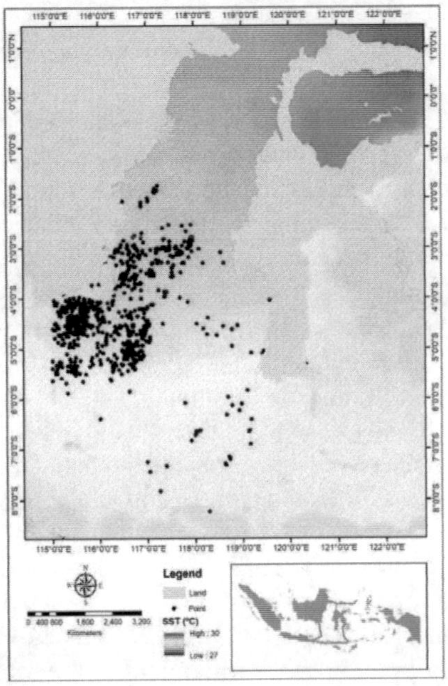

(a)

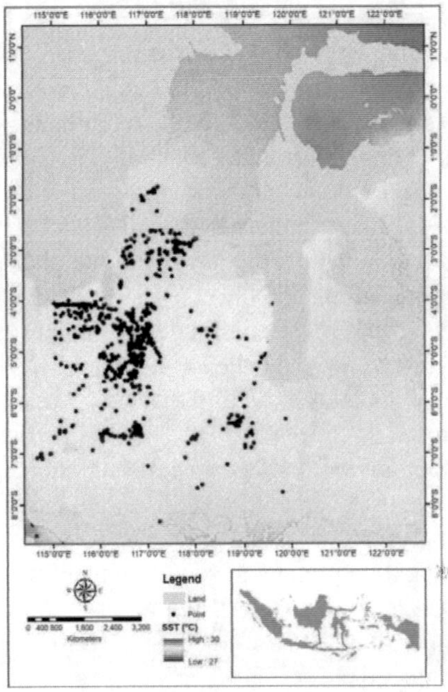

(b)

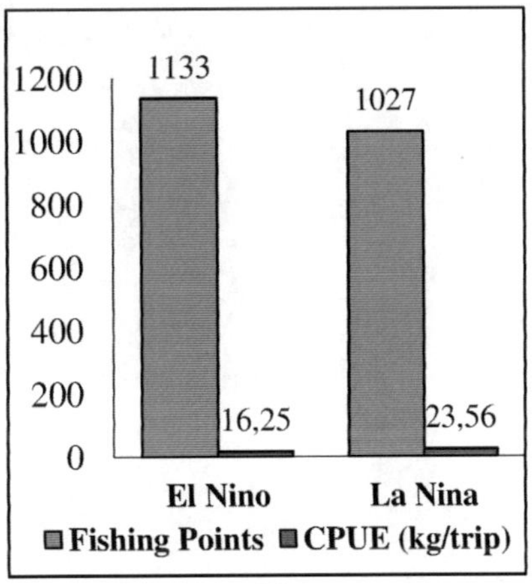

(c)

Figure 5. Spatial distribution of fishing locations of Eastern Little Tuna overlaid with SST during (a) El Niño phase, (b) La Niña phase, and (c) CPUE and fishing points during ENSO events 2015-2021

During El Niño, fishing locations have more conducted area as sparse with coordinates 3.22 – 6.59 °S and 108.20 – 109.67 °E. On the other side, the fishing grounds were coordinated between 5.39 – 6.60 °S and 108.64 – 109.22 °E during La Niña. SST values in the Makassar Strait are higher throughout the El Niño event (28.5 to 29.5 °C) than during the La Niña event (28 to 29°C). There were 1.133 fishing locations during the El Niño phase with an average CPUE of 16.25 kg/trip. Meanwhile, 1.027 fishing activities were carried out during La Niña with a mean CPUE of 23.56 kg/trip. Eastern Little Tuna catches varied significantly during the ENSO event due to changes in the environment [11]. The La Niña events caused an increased in Eastern Little Tuna catch rates and decrease during El Niño. This revealed that the environmental condition during La Niña more conducive to Eastern Little Tuna and resulted in higher catch rates.

3.4 Features of the spatial distribution of Eastern Little Tuna overlaid on monthly mean images of oceanographic variables

The results below show how different stages of ENSO affected the Makassar Strait water conditions. Images of five ocean parameters (SST, Chlorophyll-a, salinity, SSH, and current speed) overlaid with fishing locations can be shown in Figure 6 and Figure 7. Each map represents the values for four different months during different periods of El Niño (Figure 6) and La Niña (Figure 7). We put only the monthly peak of the climatic index during September-December as the representative months with a strong Niño index representing El Niño (2015) and La Niña (2021) periods.

According to the spatial distribution of Eastern Little Tuna spatial distribution during the strong El Niño phase (Figure 6), the suitable areas for Eastern Little Tuna off the Makassar Strait corresponded to areas with intense fishing activity. The fishing location during the El Niño phase of 2015 between coordinates 1 – 5 °S and 115 – 118 °E. A relatively diverse habitat zone was present from September to December in the southern Makassar Strait. The CPUE obtained is between 1 – 15 kg/trip during September - December 2015. The range of SST was 28 – 29°C from September to October and 29 – 31°C in November and December, respectively. These results are similar to the previous studies that found that tuna prefer SST conditions in the range of 29°C to 30°C [23]. In addition, Chlorophyll-a and salinity values were 0.5 – 1 mg/m^3 and 32 – 34 ppt, respectively. The range of SSH was 0.4 – 0.55 m in September, 0.5 – 0.55 in October, 0.55 – 0.6 m in November, and 0.6 – 0.65 m in December. Current speeds ranged from 0.1 – 1 m/s, with the maximum current speed in the northern part of the Makassar Strait as the Indonesian Throughflow entrance with a current speed value of 0.8 – 1 m/s. In December, the current velocity decreased by 0 – 0.5 m/s. In addition to oceanographic parameters like SST and chlorophyll-a, the current speed in a specific area affects the distribution of Eastern Little Tuna [24].

IOP Conf. Series: Earth and Environmental Science 1289 (2023) 012007 doi:10.1088/1755-1315/1289/1/012007

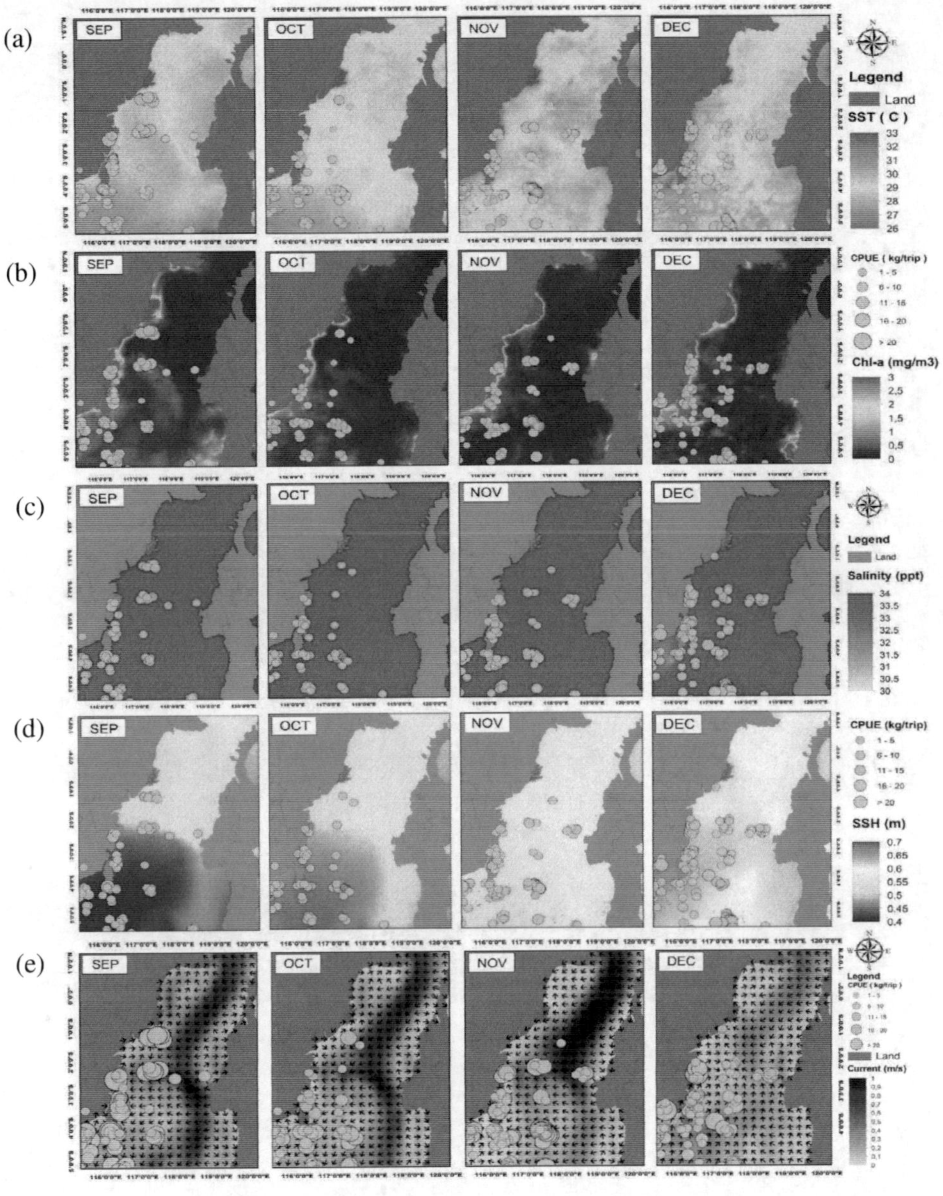

Figure 6. A map of Eastern Little Tuna spatial distribution overlaid on oceanographic parameters monthly mean images in 2015 (strong El Niño) for (a) SST, (b) Chlorophyll-a, (c) Salinity, (d) SSH, and (e) Current speed. Fishing locations with CPUE values are shown as pink dots.

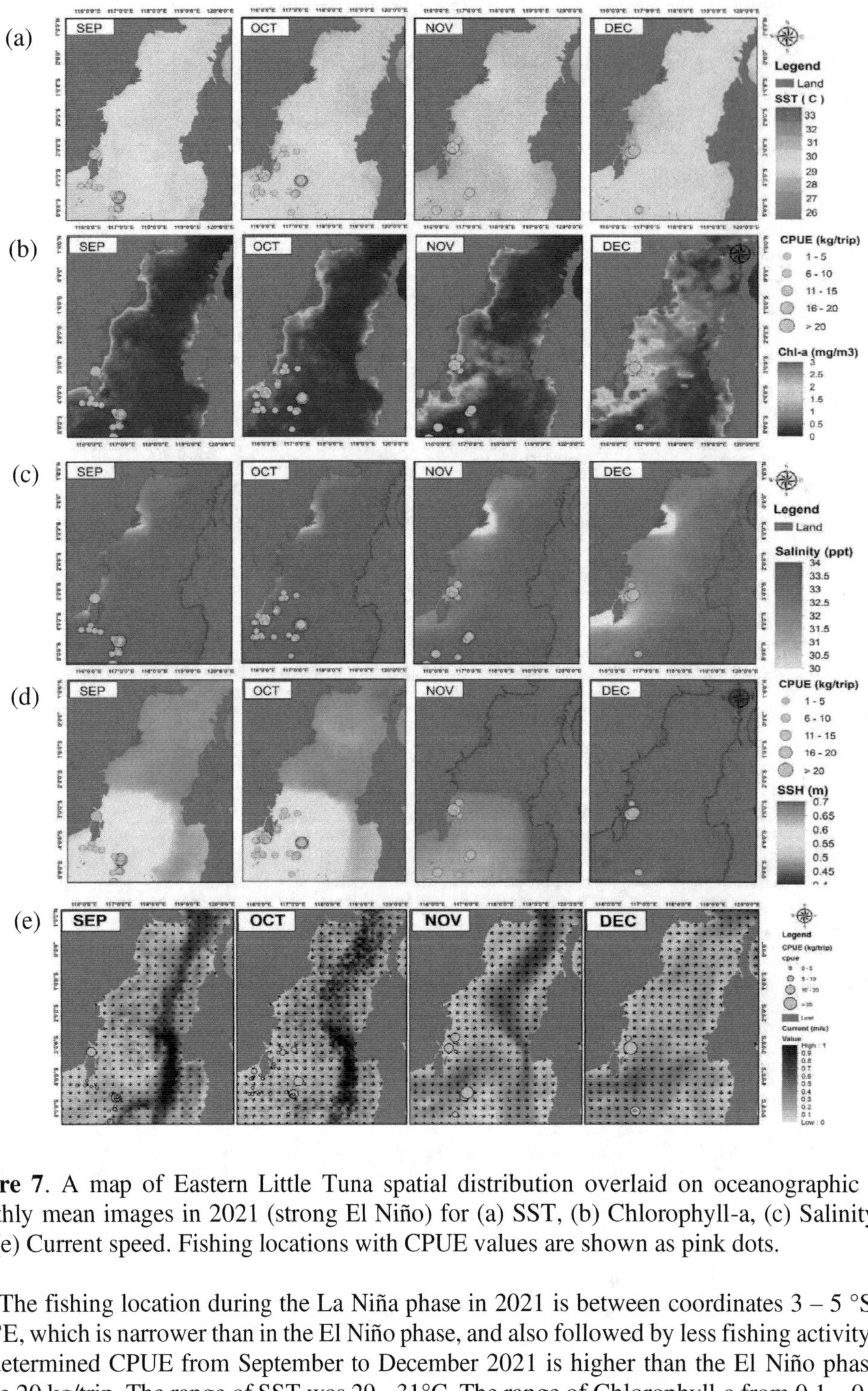

Figure 7. A map of Eastern Little Tuna spatial distribution overlaid on oceanographic parameters monthly mean images in 2021 (strong El Niño) for (a) SST, (b) Chlorophyll-a, (c) Salinity, (d) SSH, and (e) Current speed. Fishing locations with CPUE values are shown as pink dots.

The fishing location during the La Niña phase in 2021 is between coordinates 3 – 5 °S and 115 – 117 °E, which is narrower than in the El Niño phase, and also followed by less fishing activity. However, the determined CPUE from September to December 2021 is higher than the El Niño phase, which is above 20 kg/trip. The range of SST was 29 - 31°C. The range of Chlorophyll-a from 0.1 – 0.5 mg/m^3 in September and October, then increased in November and December (0.5 – 2.5 mg/m^3). The salinity values were 32.5 – 34 ppt from September to October and 30.5 – 33 ppt in November and December, respectively. The range of SSH was 0.55 – 0.65 m in September and October, 0.65 – 0.7 m in November, and 0.7 m in December. Current speeds range from 0 – 1 m/s, with the maximum current speed was in the northern part of the Makassar Strait as the Indonesian Throughflow entrance with the value of 0.8 – 1 m/s. The current velocity slows by 0 – 0.6 m/s in November and December. In general, there are contrasting oceanographic conditions between 2015 El Niño and 2021 La Niña. When compared to the La Niña event, El Niño has higher SST, salinity, and current speed. On the other hand, El Niño events have lower Chlorophyll-a and SSHA than La Niña events.

A histogram is used to show the correlation between the CPUE value of Eastern Little Tuna and five oceanographic parameters from 2015 to 2021 (Figure 8). Values of oceanographic parameters extracted at each position catch rate location are plotted against CPUE.

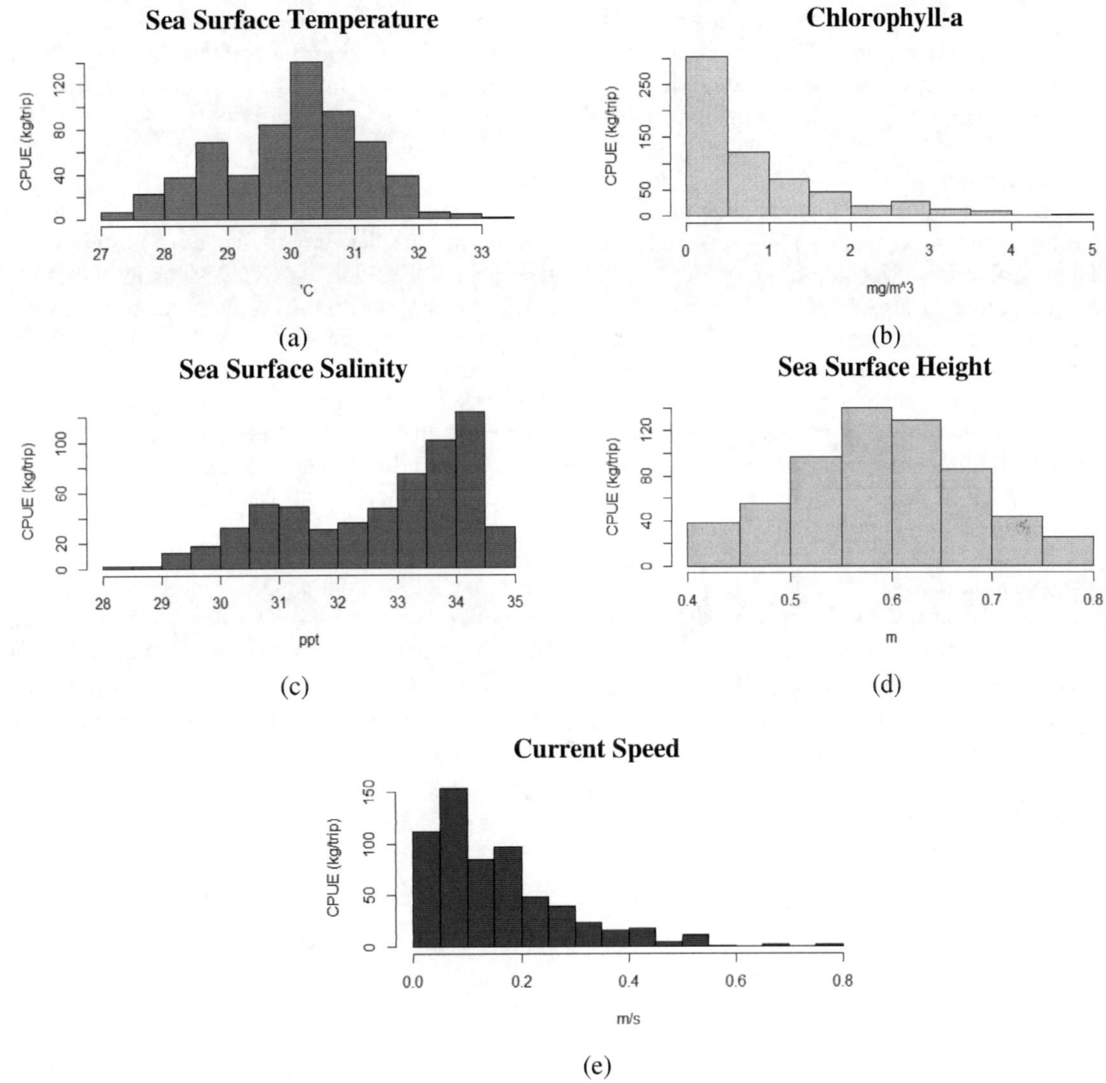

Figure 8. Histogram of CPUE in relation to oceanographic parameters: (a). SST in °C, (b) Chlorophyll-a in mg/m^3, (c) Salinity in ppt, d) SSH in m, and e) Current speed in m/s. Oceanographic factors are displayed on the x-axis, while the CPUE value is displayed on the y-axis.

With a CPUE value of 120 kg/trip, which is consistent with another study by Puspita et al. [21], the maximum CPUE values for Eastern Little Tuna were recorded at SST values between 30 – 30.5°C. This indicates that Eastern Little Tuna prefers warmer waters. Therefore, a Chlorophyll-a concentration range of 0.1 – 0.5 mg/m^3 is related to a high Eastern Little Tuna CPUE of about 250 kg/trip. Additionally, the maximum CPUE was observed within the oceanographic factors ranges of salinity value between 34 – 34.5 ppt, 0.55 – 0.6 m of SSH, and 0.05 – 0.1 m/s of current speed. This histogram analysis seemed consistent with favorable oceanographic conditions as demonstrated by the habitat preferences of Eastern Little Tuna during the La Niña event.

Fishing grounds are affected by SST, Chlorophyll-a, salinity, SSHA, and current speed. Fish productivity and distribution are influenced by environmental changes such as SST, salinity, and current [25]. This implies that various climatic phenomena may result in distinct oceanographic circumstances that are advantageous for the catches of Eastern Little Tuna. The outcomes of the data's spatial and temporal distribution indicate that, in terms of catch rates and oceanographic factors, the La Niña period shows more suitable area. This is supported by a histogram fit visualizing the preferred habitat (Figure 8). Variations in oceanographic conditions during the ENSO event influence the quantity and fish distribution in relation to environmental change.

4. Conclusion

The results further highlight catches of Eastern Little Tuna tend to be higher during La Niña and northwest monsoon, while El Niño and southeastern monsoon cause less conducive to the Eastern Little Tuna so that it will decrease the amount of the total catch. The CPUE was 16.25 kg/trip with 1133 fishing points during El Niño and 23.56 kg/trip with 1027 fishing points during La Niña, respectively. The fact that the fishing grounds area was larger during the La Niña period, at 2 - 8°S and 115 - 120°E, suggests that different climatic events may result in diverse oceanographic circumstances that are favorable for the catch of Eastern Little Tuna. These findings further emphasize the strong impact of the La Niña events, which significantly increased the MS's catch of Eastern Little Tuna. Further research of the ITF influence on catch rates needs to be addressed in the future. Our findings would support the conservation of pelagic fish diversity in order to mitigate the impacts of climate variations, as well as assist policymakers in effectively overseeing a sustainable catch in the MS.

Acknowledgments

The authors wish to thank GSFC/DAAC for Chlorophyll-a and SST data, AVISO for SSHA data, Copernicus Marine Service for Salinity and Current Velocity, and the NOAA Climate Prediction Center for the use of the Niño 3.4 Index. We also thank the Ministry of Marine Affairs and Fisheries Republic of Indonesia, Universitas Padjadjaran, Minister of Education, Culture, Research, and Technology and also the ISMFR 2023 committee member.

References

[1] Koeshendrajana S, Rusastra I W, and Martosubroto P 2019 *Potential of Marine Fisheries Resources FMA 713* (Jakarta:AMAFRAD) p 713

[2] Potemra, J T, and Schneider N 2007. Interannual Variations of the Indonesian Throughflow. *J. Geoph. Res.* **112** 1–13

[3] Hidayat T, Febrianti E, Restianingsih, Y H 2016 Pattern and Spawning Season of Kawakawa (*Euthynnus affinis*, Cantor 1850). Bawal, **8** 101-108

[4] Arrizabalaga H, Dufour F, Kell L, Merino G, Ibaibarriaga L, Chust G, and Iriogen X 2015 Global habitat preferences of commercially valuable tuna. *Deep. Sea Res. Part II Top. Stud. Oceanogr.* **113** 102-112

[5] Song L M, Zhang Y U, Xu L X, Jiang W X, and Wang J Q 2008 Environmental preferences of longlining for yellowfin tuna (*Thunnus albacares*) in the tropical high seas of the Indian Ocean. *Fish. Oceanogr.* **17** 239–253

[6] Laevastu T, and Hayes M L 1981 *Fisheries Oceanography and Ecology* (New York: Fishering News Book Ltd)

[7] Howell E A and Kobayashi D R 2006 El Niño effects in the Palmyra Atoll region: oceanographic changes and bigeye tuna (*Thunnus obesus*) each rate variability. *Fish. Oceanogr.* **15** 477-489

[8] Lehodey P, SeNiña J S, Bopp L, Calmettes B, Hampton V, and Murtugudde R 2010 Preliminary forecast of Pacific bigeye tuna population trends under the A2 IPCC scenario. *Prog. Oceanogr.* **86** 302-315

[9] Lehodey P, Bertignac M, Hampton J, Lewis A, and Pieaut J 1997 El Niño Southern Oscillation and tuna in the western Pacific. *Nature* **389** 715-718

[10] Lehodey P 2001 The pelagic ecosystem of the tropical Pacific: dynamic spatial modeling and biological consequences of ENSO. *Prog. Oceanogr.* **49** 439-468

[11] Syamsuddin M L, Sunarto, and Yuliadi L 2016 Environmental Variability Associated with Eastern Little Tuna [*Euthynnus affinis* (Cantor, 1849)] Catches: A Case for The North Indramayu Waters, Java Sea. *J Tek,* **78** 139-143

[12] Mujib Z, Boesono H and Fitri A D P 2013 Mapping the Distribution of Tuna (*Euthynnus* sp.) using Modis Image Chlorophyll-a Data on Danish Seine in Pelabuhanratu Bay, Sukabumi, West Java. *J. Fish Res Util Man and Tech* **2** 150-160

[13] Gaol J L and Sadhotomo B 2007 Characteristics and variability of oceanographic parameters in the Java Sea and its relationship with the distribution of fish catch. *J. Pen. Perik Indonesia* **13** 201-211

[14] Zuur A F 2007 *Analyzing Ecological Data.* In *Text*

[15] Tangke U, Karuwal J C, Zainuddin M, and Mallawa A 2015 Distribution of Sea Surface Temperature and Chlorophyll-a and its Effects on Catches of Yellowfin Tuna (*Thunnus albacares*) in the Southern Halmahera Sea. *J. IPT. PSP,* **2** 248-260

16] Tussadiah A, Pranowo W S, Syamsuddin M L, Riyantini I, Nugraha B, Novianto D 2018 Characteristic of eddies kinetic energy associated with yellowfin tuna in southern Java Indian Ocean. *IOP Conf. Ser.: Earth Environ. Sci.* **176** 1-10

[17] Ranintyari M, Sunarto, Syamsuddin M L, Astuty S 2018 Effects of oceanographic factors on the spatial distribution of Whale sharks in Cendrawasih Bay National Park, West Papua. *IOP Conf. Ser.: Earth and Environ. Sci.* **149** 1-10

[18] Ahmad A. L., Syamsuddin M. L., Purba N. P., Sunarto, 2019 Thermal front condition through El Niño and Indonesian throughflow phase in southern sea of East Java and Bali on the east monsoon. IOP Conference Series: Earth and Environmental Science **303** 1-8

[19] Zainuddin M, Safruddin S, Farhum A, Budimawan B, Hidayat R, Selamat M.B, Wiyono E S, Ridwan M, Syamsuddin M, and Ihsan Y N 2023 Satellite-Based Ocean Color and Thermal Signatures Defining Habitat Hotspots and the Movement Pattern for Commercial Skipjack Tuna in Indonesia Fisheries Management Area 713, Western Tropical Pacific. *Rem. Sens.* **15** 1-22.

[20] Harahap S A, Syamsuddin M L, and Purba N P 2020 Range of sea surface temperature and chlorophyll-α values based on mackerel catches in the northern waters of West Java, Indonesia. *AACL Bioflux* **13** 2265-2272

[21] Puspita A R., Syamsuddin M L, Subiyanto, Syamsudin F, and Purba N P 2023 Predictive Modeling of Eastern Little Tuna (*Euthynnus affinis*) Catches in the Makassar Strait by Using Generalized Additive Model. *J. Mar. Sci. Eng.* **11** 165

[22] Abidin A F, Amron, and Marzuki M I 2020 Relationship between chlorophyll-a and sea surface temperature to tuna catch in the Southern Water of Java. *J. Pen. Sains* **22** 55-68

[23] Syamsuddin M, Sunarto, and Yuliadi L 2018 Oceanographic factors related to Eastern Little Tuna (*Euthynnus affinis*) catches in the west Java Sea. *IOP Conf. Ser.: Earth Environ. Sci.* **162** 012044

[24] Ikhwan M, Haditiar Y, Wafdan R., Setiawan I, Marwan, Muchlisin Z A, & Rizal S 2019 Wind driven circulation in Makassar Strait during monsoon 2017. *IOP Conf. Ser.: Earth and Environ. Sci.* **348** 1

IOP Conf. Series: Earth and Environmental Science 1289 (2023) 012007 doi:10.1088/1755-1315/1289/1/012007

[25] Syah A F, Gaol J L, Zainuddin M, Apriliya N R, Berlianty D, Mahabror D 2020 Detection of potential fishing zones of Bigeye tuna (*Thunnus obesus*) at the profoundity of 155 m in the eastern Indian Ocean. *Indonesian J. Geogr* **52** 29-35

IOP Conf. Series: Earth and Environmental Science 1289 (2023) 012008 doi:10.1088/1755-1315/1289/1/012008

Estimation of carbon storage of tropical mangroves in the North Maluku Islands region

Najamuddin*, I Tahir, M I Pujiati, Z A Harahap

Marine Science Study Program, Fisheries and Marine Science Faculty, Khairun University, Indonesia.

*Corresponding author e-mail: najamuddin313@gmail.com

Abstract. The mangroves possess a remarkable capacity to absorb carbon dioxide (CO2), surpassing tropical rainforests by fourfold. This underscores the crucial role played by mangroves in global efforts to combat climate change. This study aimed to assess the biomass, carbon content, and CO2 absorption capabilities of mangrove ecosystems. Data collection followed the line transect method, utilizing the Mangrove Watch guide, which involved recording the diameter of mangroves at breast height (DBH), typically around 1.3 meters above the ground, as well as identifying the mangrove species and quantifying the number of stands for each tree species. The analysis of mangrove biomass, carbon content, and CO2 absorption capacity was conducted through an allometric modeling approach. The findings revealed the presence of seven distinct mangrove tree species at the study site, namely *Rhizophora mucronata, R. apiculata, R. stylosa, Sonneratia alba, Bruguiera gymnorrhiza, Xylocarpus granatum*, and *Ceriops decandra*. The mangrove density in this area was recorded at 1406 trees per hectare, with a corresponding biomass of 630,508 tons per hectare, a carbon content of 246,035 tons of carbon per hectare, and an impressive CO2 absorption capacity of 902,127 tons of CO2 per hectare. Notably, the mangrove vegetation in this research site exhibited a substantial capacity for CO2 absorption and biomass storage, surpassing many other locations in different regions. These findings underscore the significant role that island mangrove ecosystems can play in mitigating the effects of climate change.

1. Introduction

The primary driver of global warming and climate change is the increasing presence of CO_2 in the atmosphere. To combat this issue, preserving mangrove ecosystems along coastlines has emerged as a key mitigation strategy. Mangroves have the unique ability to absorb CO_2 from the atmosphere through photosynthesis and store it within their tissues and sediment.

Research indicates that mangroves excel in carbon absorption, surpassing tropical rainforests by a factor of four, with the potential to store up to 1200 Mg C/ha [1,2,3,4]. For instance, more detailed studies, such as [5], report that mangroves exhibit above-ground biomass of 433.69 t/ha, an estimated carbon stock of 303.83 t C/ha, and a carbon dioxide (CO_2) absorption rate of 748.07 tCO_2/ha. Furthermore, [6] highlights an above-ground biomass of 666,967 t/ha, an estimated carbon stock of 313.52 t C/ha, and a CO_2 absorption of 1149.56 tons CO_2/ha. These findings underscore the crucial role of mangroves in mitigating global climate change.

Beyond their climate mitigation role, mangroves offer a myriad of ecosystem services, including supporting fisheries, controlling sediment, absorbing pollutants, regulating seawater intrusion, providing coastal protection against storms and tsunamis, maintaining nutrient cycles, and serving as habitat for diverse fauna. Unfortunately, mangrove areas have been dwindling due to deforestation for activities like pond cultivation and infrastructure development, with a 30–50% decline over the past half-century [7,8,9,10]. Consequently, conservation and restoration efforts are imperative, as mangrove degradation not only reduces CO_2 absorption but also increases CO_2 emissions.

International climate agreements, such as REDD+ (Reducing Emissions from Deforestation and Forest Degradation), emphasize the significance of curbing emissions from deforestation and degradation as a cost-effective climate change mitigation strategy. REDD+ and similar initiatives demand accurate monitoring of carbon storage and emissions, underscoring the need for precise estimates, particularly in forests with high carbon stocks and those facing uncontrolled conversion, such as mangrove forests [11].

Despite their recognized carbon storage capacity, comprehensive data on carbon storage for entire mangrove ecosystems remains limited [12,51], including in North Maluku Province. Reports on carbon storage, especially standing biomass, are still scarce [13,14]. Similarly, data regarding the total organic matter content of mangrove soil is limited [15,52], even in the North Maluku region.

North Maluku boasts coastal mangrove ecosystems, yet many of them lack proper protection and sustainable management. While research has explored aspects of mangrove community condition and structure in the region, studies on mangrove carbon storage are lacking. Therefore, research on the carbon storage capabilities of tropical mangroves in North Maluku Islands is essential. Such research can establish a valuable database for assessing the role and contribution of mangroves in CO2 emissions reduction and serve as a basis for policymaking related to Indonesian mangrove forest management, particularly concerning climate change mitigation efforts. This study aims to analyze mangrove biomass, carbon content, and their ability to absorb CO2 from the environment in the North Maluku Islands region.

2. Materials and methods

Data collection was performed in the Guraping mangrove ecosystem conservation area, North Maluku Province. The zone of the mangrove conservation is 151.3 ha established in 2021. The mangrove data collection method was carried out by line transects adopting [16]. The observation sampling points consisted of 12 plots divided into 4 transects located perpendicular to the shoreline in the mangrove conservation area. The dimensions of the plot area used is 10 m x 10 m which was placed parallel to the transect line which represents the mangrove forest near the sea, the middle part, and the area near the land. The mangrove data collected consists of mangrove tree species, number of mangrove tree stands in each plot area, and mangrove tree diameter at breast height or 1.3 meters high. The identification of mangrove species was undertaken using Global Mangrove Watch manual and [16,17]. The picture of the research place is provided in Figure 1.

Analysis of the mangrove community structure to determine the condition of the mangrove ecosystem was undertaken by calculating the mangrove density using below formula [18]:

$$D = \frac{Ni}{Ai} \quad \text{......(1)}$$

Where D = density of mangrove tress (trees/m^2), Σni = number of mangrove trees, and Ai = the area of sampling unit or plot areas (m^2).

The biomass of mangrove vegetation was determined applying an allometric model provided by [13]:

$$B = 0{,}251\ \rho\ DBH^{2{,}46} \quad \text{......(2)}$$

Where B = biomass of mangrove species (kg), ρ = mangrove wood density (g/cm^3), and DBH = diameter at breast height (cm).

Carbon content was counted by using formula provide by [11]:

$$C = \text{Biomass x } 0.47 \quad \text{......(3)}$$

Where C = mangrove carbon content (kg), biomass = biomass of mangrove species (kg), and 0.47 = constant carbon content in living things.

The mangrove carbon absorption capacity was calculated using formula provide by [19]:

WCO_2 = Cn (Mr. CO_2/Ar. C) ....(4)

Where WCO_2 = mangrove absorption capacity, Cn = mangrove carbon content (kg), Mr. CO_2 = relative molecule mass of CO_2, and Ar. C = relative atomic mass of carbon.

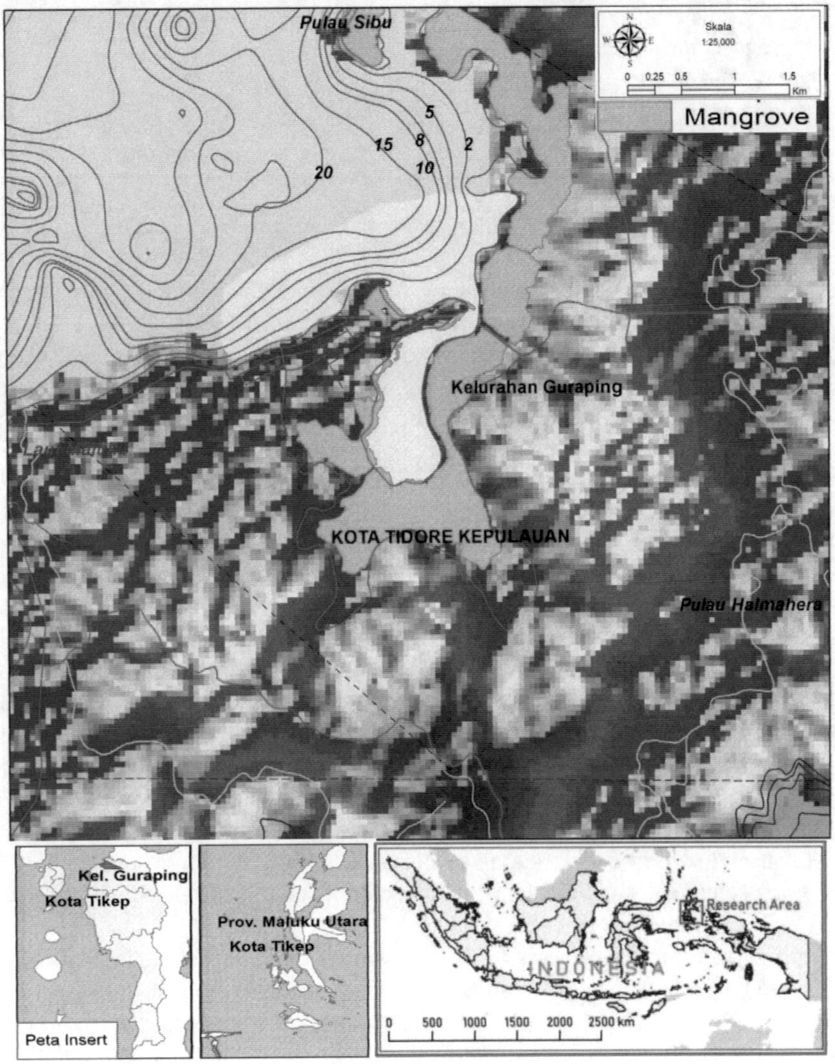

Figure 1. Map of research location.

3. Results and discussion

3.1. Mangrove species composition and density

There were 7 species of mangroves found at the study site, namely *Rhizophora mucronata, R. apiculata, R. stylosa, Sonneratia alba, Bruguiera gynorrhiza, Xylocarpus granatum* and *Ceriops decandra.* The seven species of mangroves are true mangroves comes from the *Rhizophoraceae, Sonneratiaceae*, and *Meliaceae* families [20,17]. Distribution of those seven mangrove species at observation stations provided in Table 1.

Table 1. Composition of mangrove species in the study location

No	Mangrove Species	Station 1	Station 2	Station 3	Station 4
1	*Rhizophora mucronata*	√	√	√	√
2	*Rhizophora apiculata*	√	√	√	√
3	*Rhizophora stylosa*	√	√	√	√
4	*Sonneratia alba*	√	√	-	-
5	*Bruguiera gynorrhiza*	√	-	-	-
6	*Xylocarpus granatum*	√	-	-	-
7	*Ceriops decandra*	-	√	-	-

Note: √ = exist,
- = none,

The greatest species diversity was observed at station 1 and station 2, primarily because these areas are more shielded from anthropogenic activities, owing to their distance from residential zones. Conversely, stations 3 and 4 exhibited lower species composition due to their proximity to local settlements, making them more accessible for activities like logging. The escalating population and rapid development along coastal regions for various purposes such as habitation, fishing, ports, tourism, and land reclamation are exerting ecological stress on coastal ecosystems, particularly mangrove environments [21]. This heightened pressure is causing a decline in quality and damage to mangrove ecosystems, both directly through logging and land conversion and indirectly through pollution stemming from diverse human activities. Similar concerns were articulated by [22], emphasizing that the most significant contributors to mangrove destruction in Indonesia are the felling of mangrove trees for domestic and commercial purposes, along with mangrove conversion for ponds, agriculture, and settlements.

The composition of mangrove types observed at the research site mirrors findings from several other locations in North Maluku. For instance, Bacan Island in South Halmahera hosts six species [23], Morotai Island boasts fourteen species [24], the Minaluli Coast in North Mangoli, Sula Islands, features seven species [25], and South Jailolo houses nine species [26]. Further studies have demonstrated varying species compositions among Indonesian mangroves; for instance, the Bandar Bakau Dumai area in Riau has ten species [27], Perancak in Jembrana-Bali has nine species [28], South Minahasa in North Sulawesi has seven species [29], Wonorejo on the East Coast of Surabaya features six species [30], Indragiri Hilir in Riau comprises five species [31], and the mangrove areas of Kema in North Sulawesi contain eight species [32].

Regarding mangrove tree density, *Rhizophora mucronata* exhibited the highest density at Station 1, with 570 trees/ha, while *Bruguiera granatum* and *Xylocarpus granatum* had the lowest density, each with 70 trees/ha. At Station 2, *Rhizophora mucronata* displayed the highest tree density (1000 trees/ha), whereas *Sonneratia alba* exhibited the lowest density (67 trees/ha). Station 3 featured *Rhizophora apiculata* with the highest density (1400 trees/ha), while *Rhizophora stylosa* had the lowest density (100 trees/ha). Lastly, at Station 4, *Rhizophora apiculata* had the highest mangrove tree density at 600 trees/ha, with *Rhizophora mucronata* having the lowest at 30 trees/ha. Figure 2 illustrates the distribution of mangrove density across the study locations.

The dominance in the distribution pattern of mangrove types at the research site was attributed to *Rhizophora apiculata, Rhizophora stylosa*, and *Rhizophora mucronata*. Among these, *Rhizophora sp.* displayed superior adaptability compared to other species, making it a common dominant species. *Rhizophora sp.* has a remarkable ability to thrive in high salinity environments and on hard substrates [33,34]. Additionally, various research findings have consistently shown that mangroves in North Maluku are predominantly characterized by *Rhizophora apiculata* [35,36].

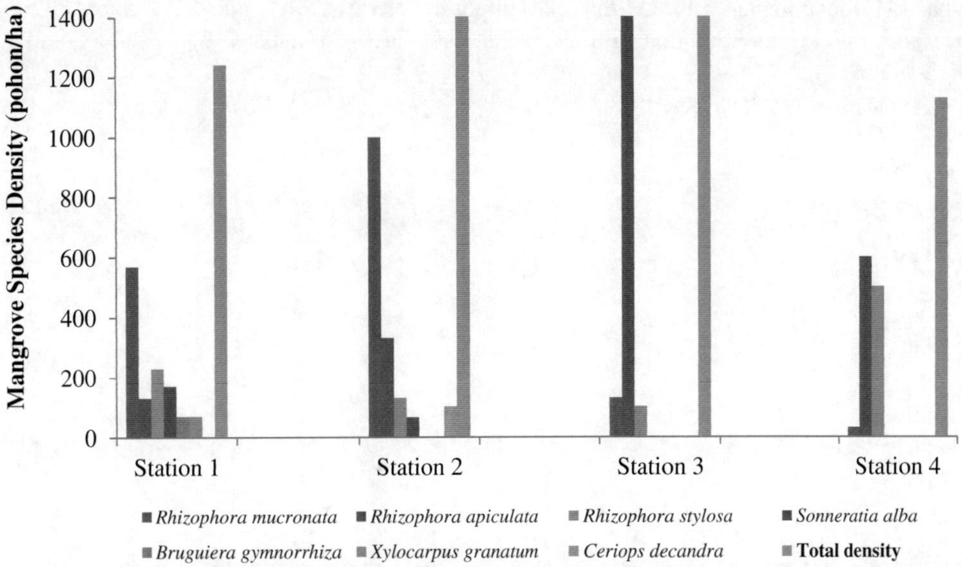

Figure 2. Mangrove types density at the study location

Variations in mangrove density at the study site are notably influenced by the utilization patterns of the local community. Stations 1 and 2, situated within the core conservation area, benefit from better maintenance and protection efforts. In contrast, stations 3 and 4 lie outside the core zone, making them more accessible to local communities. Additionally, geographical characteristics distinguish these four research stations. Stations 3 and 4, positioned in the lower watershed area, experience fluctuating salinity levels that impact mangrove development. Meanwhile, stations 1 and 2 are situated in coastal regions characterized by stable salinity and favorable substrate conditions for mangrove growth. Consequently, both stations exhibit a higher mangrove species composition and tree density.

The average mangrove density in the research location stands at 1406.75 trees per hectare. According to the Minister of Environment Decree number 201 of 2004, this density value falls within the medium density range, defined as a density value ranging from 1000 to 1500 trees per hectare. Notably, the mangrove density in this research location is lower than that on Morotai Island, where it reaches 2920 trees per hectare, classifying it as a dense mangrove category [24].

3.2. Mangrove species biomass

The mangrove species with the highest biomass at station 1 was *Rhizophora mucronata*, totaling 133.09 tons per hectare (t/ha), while the lowest biomass was observed in *Xylocarpus granatum*, amounting to 0.43 t/ha. Moving to station 2, *Rhizophora mucronata* exhibited the highest biomass at 351.00 t/ha, whereas *Ceriops decandra* had the lowest at 2.14 t/ha. Station 3 showcased *Rhizophora apiculata* with the highest biomass at 1190.66 t/ha, while *Rhizophora stylosa* had the lowest at 4.82 t/ha. Finally, at station 4, *Rhizophora apiculata* stood out with the highest biomass at 203.80 t/ha, whereas *Rhizophora mucronata* recorded the lowest at 0.06 t/ha. Figure 3 illustrates the distribution of mangrove species biomass at the study site. Overall, *Rhizophora sp*. emerged as the species with the highest biomass at the study site, aligning with findings from previous studies [37,38,28,30].

Across stations 1, 2, 3, and 4, the total biomass of mangrove vegetation reached 161.66 t/ha, 497.06 t/ha, 1207.13 t/ha, and 252.38 t/ha, respectively. Consequently, the research site exhibited an average mangrove vegetation biomass of 529.56 t/ha. Notably, this biomass exceeds the values reported in some earlier studies: [39] recorded a range of 62.9-398.8 t/ha, [27] reported 115.85 t/ha, [40] found

IOP Conf. Series: Earth and Environmental Science 1289 (2023) 012008 doi:10.1088/1755-1315/1289/1/012008

139.11 t/ha, [41] documented 115.85 t/ha, [28] reported 187.212 t/ha, and [41] noted 83.51 t/ha. However, it's essential to mention that [6] reported an even larger biomass of 666.97 t/ha compared to the research site.

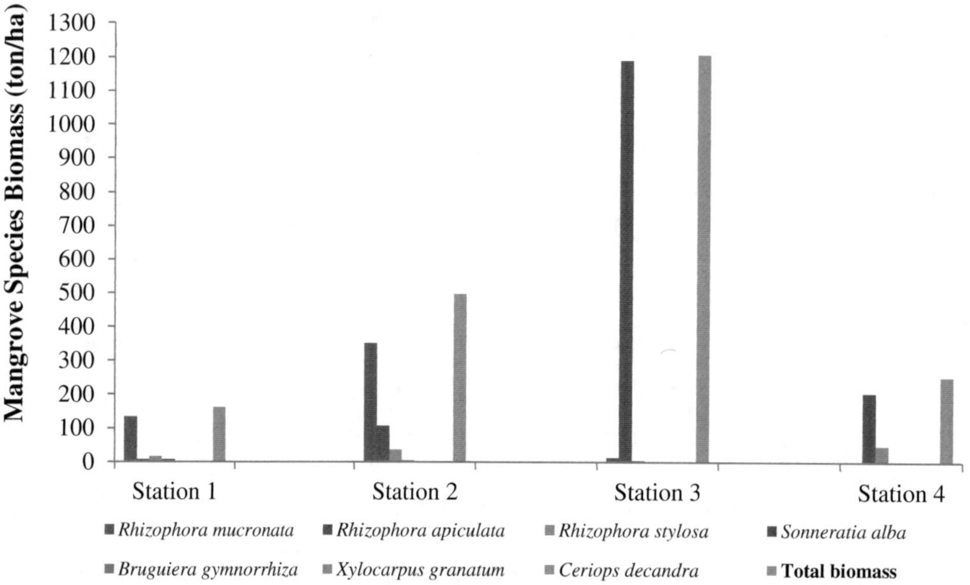

Figure 3. Biomass of mangrove species at the study site.

The biomass and carbon storage levels within ecosystems exhibit variations dependent on multiple factors. These factors include the diversity and density of plant species, tree characteristics, soil composition, and various environmental factors such as solar radiation, water availability, temperature, soil fertility, and the management practices employed within the ecosystem [43]. It's worth noting that aside from tree density, biomass is also influenced by tree diameter and height. In essence, larger trees, characterized by greater diameter and height, tend to store more biomass. As highlighted by [44], the impact of tree diameter and height on biomass surpasses that of tree density. This connection is attributed to the trees' capacity to absorb CO2 from the atmosphere through the process of photosynthesis, resulting in the accumulation of biomass in various tree components, including leaves, branches, stems, and roots.

Furthermore, the biomass of mangrove ecosystems is shaped by additional factors, such as the age of the trees, soil nutrient content, and the type of substrate in which mangroves thrive. Soil fertility levels and tree density significantly influence whether a mangrove ecosystem exhibits high or low biomass levels [41]. Likewise, [45] suggests that biomass increases with tree age due to the radial growth of the tree's diameter, which leads to the development of new cells.

It's crucial to recognize that biomass holds a direct relationship with the stored carbon content within a stand of trees. Approximately 50% of a plant's biomass consists of carbon [46]. Consequently, as biomass levels increase, so does the stored carbon content. When forested areas undergo transformation into agricultural, plantation, or industrial land uses, the result is a continuous accumulation of CO2 in the atmosphere. This occurs because the trees, which serve as natural carbon sinks, are removed. Moreover, the stored carbon within the trees is released into the atmosphere, further contributing to the rise in atmospheric CO2 levels and exacerbating the issue of global warming.

3.3. Mangrove species carbon content

The analysis results pertaining to the carbon content of mangrove vegetation, categorized by species and the overall carbon storage levels at each observation station, are depicted in Figure 4. Notably, at station 1, the species *Rhizophora mucronata* exhibited the highest carbon content at 62.55 t C/ha, whereas *Xylocarpus granatum* demonstrated the lowest carbon content at 0.203 t C/ha. Transitioning to station 2, *Rhizophora apiculata* displayed the highest carbon content, reaching 156.90 t C/ha, while *Ceriops decandra* had the lowest at 0.955 t C/ha. Moving on to station 3, *Rhizophora apiculata* maintained its position as the species with the highest carbon content, registering 559.61 t C/ha, whereas *Rhizophora stylosa* recorded the lowest carbon content. Station 4 featured *Rhizophora apiculata* with the maximum carbon content at 95.784 t C/ha, while *Rhizophora mucronata* exhibited the lowest at 0.026 t C/ha. It's worth noting that according to [45], *Rhizophora stylosa* species possessed a carbon content of 196.855 t C/ha, while [37] reported that *Rhizophora mucronata* exhibited a carbon content of 217.22 t C/ha. In general, the study found that *Rhizophora sp.* consistently exhibited the highest carbon content across the study sites, aligning with the findings of prior research [37, 38, 28, 30].

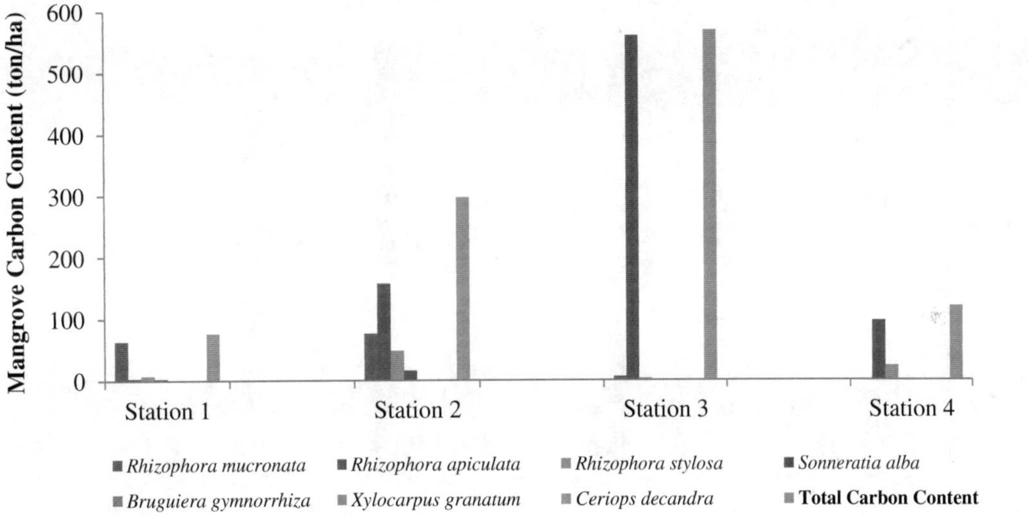

Figure 4. The carbon content of mangrove vegetation at the study site

The total carbon content within the mangrove vegetation across the study stations displayed variation: station 1 recorded 75.891 t C/ha, station 2 showed 296.839 t C/ha, station 3 exhibited 567.351 t C/ha, and station 4 registered 118.621 t C/ha. On average, the carbon content of mangrove vegetation at the study site amounted to 264.698 t C/ha. Comparatively, the carbon storage within the mangrove vegetation at this research location surpassed the findings of previous studies. Specifically, it exceeded the carbon content reported in [41] (57.91 t C/ha), [19] (49.13 t C/ha), [27] (24.56 t C/ha), [40] (69.56 t C/ha), [47] (141.64 t C/ha), [48] (198.61 t C/ha), [28] (86.117 t C/ha), and [42] (39.25 t C/ha). However, it fell short of the results presented in [6], which documented a mangrove vegetation carbon content of 313.52 t C/ha.

The variations in carbon content are contingent upon multiple factors, encompassing biomass, mangrove species density, age of the mangroves, overall mangrove density, nutrient composition, and the type of substrate in which the mangroves thrive. Notably, carbon content within mangrove biomass is positively correlated with the quantity of mangrove biomass, with each mangrove biomass component storing approximately 46-50% carbon [44]. Furthermore, according to [49], the carbon storage capacity of mangroves can reach levels between 800 and 1,200 tons per hectare.

3.4. Mangrove species absorption capacity

The analysis of CO2 absorption capacity among mangrove species at the study site yielded the following results, as illustrated in Figure 5. Among the mangrove species, *Rhizophora apiculata* and *Rhizophora mucronata* demonstrated the highest ability to absorb CO_2 from the environment. Specifically, *Rhizophora apiculata* exhibited a range of CO_2 absorption capabilities between 11.865 and 2051.903 tons of CO_2 per hectare, while *Rhizophora mucronata* displayed a capacity ranging from 0.1 to 575.284 tons of CO_2 per hectare.

The total CO_2 absorption capacity of mangrove vegetation varied across the study stations. Station 3 recorded the highest capacity at 2080.29 tons of CO_2 per hectare, whereas station 1 exhibited the lowest capacity at 278.597 tons of CO_2 per hectare. On average, the CO_2 absorption capacity across the study sites amounted to 902.127 tons of CO_2 per hectare. Notably, the CO_2 absorption capacity at this research site exceeded the values reported in [19] (90.16 t CO_2/ha), [47] (519.73 t CO2/ha), and [37] (805.68 t CO_2/ha). However, it fell short of the results documented in [6], which reported a CO_2 absorption capacity of 1149.56 t CO_2/ha.

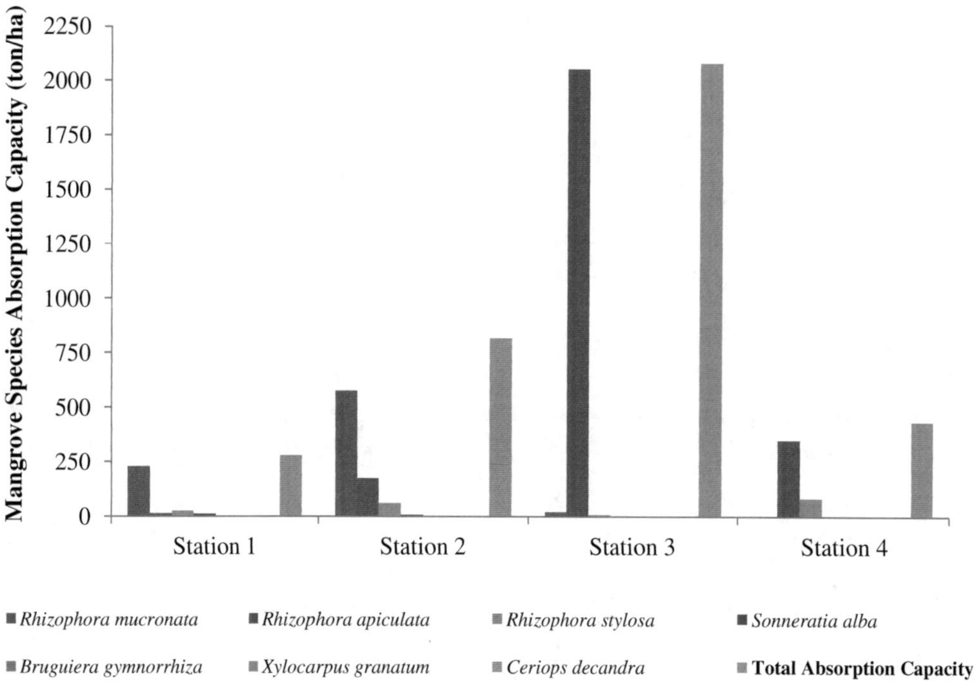

Figure 5. Absorption capacity of mangrove species

Variations in the CO_2 absorption capacity of mangroves are influenced by the same factors that impact mangrove biomass and carbon storage. These factors encompass mangrove species density, mangrove age, tree diameter, overall mangrove density, nutrient content in the substrate, the type of substrate where mangroves grow, the predominant tree species, and environmental variables such as sunlight, temperature, and the management approach applied to the mangrove ecosystem.

As highlighted in [37], the term "carbon value" for plants signifies their ability to capture CO_2 from the atmosphere. Plants take in CO_2 from the air and transform it into organic matter through the process of photosynthesis, utilizing it for their growth. According to [50], the growth phase of mangroves also plays a role in their capacity to absorb CO_2 from the atmosphere. During the net growth phase, they tend to absorb more CO_2 from the atmosphere compared to mature trees. Figure 6 visually depicts the relationships between density, biomass, carbon content, and the absorption capacity of mangrove vegetation.

Figure 6 illustrates a direct correlation between the density of mangrove vegetation, its biomass, carbon content, and its capability to absorb CO_2. The amount of CO_2 absorbed by mangroves from the atmosphere is contingent upon their density, biomass, and the carbon content stored within various vegetative tissues, including leaves, twigs, stems, and roots, as described in [2,4,6].

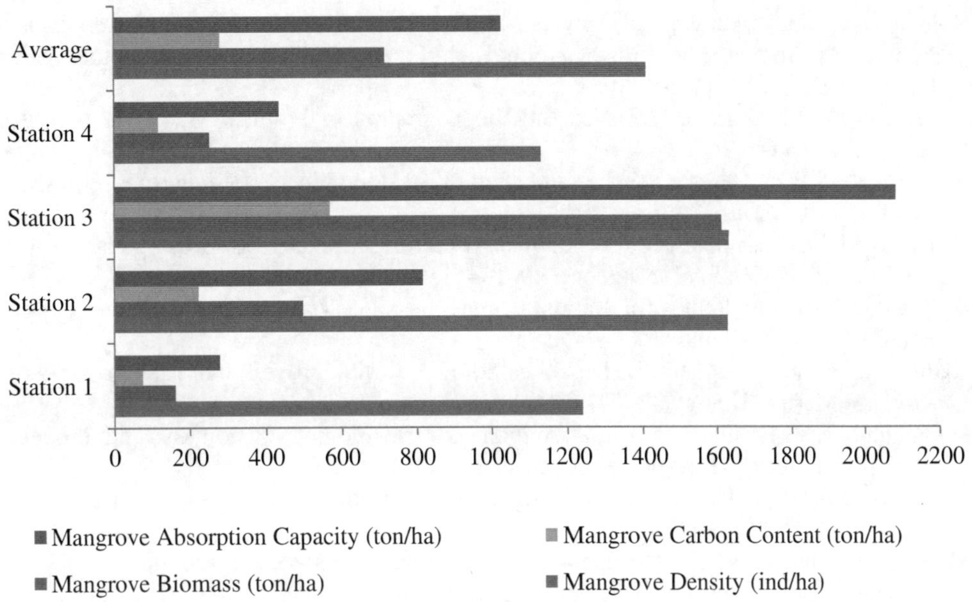

Figure 6. The relationship between density, biomass, carbon content, and the absorption capacity of mangrove

4. Conclusion

Addressing the issue of rising CO_2 emissions into the atmosphere, which contributes to global warming, can be partially alleviated by strengthening the capacity of mangroves to absorb CO_2 from the air. The findings of this study, along with numerous other research endeavours, underscore the remarkable ability of mangrove vegetation to capture substantial amounts of CO_2 and sequester carbon. Recognizing the strategic significance of mangroves in mitigating climate change and combatting global warming, it is imperative to advocate for and bolster initiatives aimed at safeguarding, conserving, and restoring mangrove ecosystems.

References

[1] Kauffman J B and Donato D C 2012 Protocols for the Measurement, Monitoring and Reporting of Structure, Biomass and Carbon Stocks in Mangrove Forests. Working Paper 86. CIFOR. Bogor.

[2] Donato D C, Kauffman J B, Murdiyarso D, Kurnianto S, Stidham M, Kanninen M 2012 Mangroves are one of the richest carbon forests in the tropics. *Center for International Forestry Research* (CIFOR), Bogor, Indonesia.

[3] Purnobasuki H 2012 Utilization of Mangrove Forests as Carbon Storage. Bulletin of environmental management of Surabaya University 28 : 3-5.

[4] Ati R N A, Rustam A, Kepel T L, Sudirman N, Astrid M, Daulat A, Suryono D D, Puspitaningsih Y, Mangindaan P, Hutahaean A 2014 Carbon Stock and Mangrove Community Structure as Blue Carbon in Tanjung Lesung, Banten. Jurnal Segara, 10 (2):98-171.

[5] Bachmid F, Sondak C F A, Kusen J D 2018 Estimation of Carbon Absorption in the Bahowo Mangrove Forest, Tongkaina Village, Bunaken District. Thesis, Department of Marine Science. Faculty of Fisheries and Marine Science. Sam Ratulangi University, Manado.
[6] Amanda Y, Mulyadi A, Siregar Y I 2021 Estimation of Carbon Reserved in Mangrove Forest at the Estuary of the Batang Apar River, North Pariaman District, Pariaman City, West Sumatra Province. 9(1): 38-48.
[7] Donato D C, Kauffman J B, Murdiyarso D, Kurnianto S, Stidham M, and Kanninen M 2011 Mangroves Among The Most Carbon Rich Forests in The Tropics. Nature Geoscience. DOI:10.1038/NGEO1123.
[8] Duke N C 1992 Mangrove Floristics and Biogeography. In Tropical Mangrove Ecosystems 41: 63-100.
[9] Alongi D M 2002 Present State and Future of the World's Mangrove Forests. Journal Environmental Conservation. 29: 331-349.
[10] Giri C 2011 Status and distribution of mangrove forests of the world using Earth Observation Satellite data. Glob. Ecol. Biogeogr. 20, 154–159 (2011).
[11] IPCC 2006 IPCC Guidelines for National Greenhouse Gas Inventories—Volume 4: Agriculture, Forestry and Other Land Use (AFOLU).
[12] Kathiresan K and Bingham B L 2001 Biology of Mangroves and Mangrove Ecosystems. Advance Marine Biology, 40:81-251.
[13] Komiyama, A., J.E. Ong, dan S. Poungparn. 2008. Allometry, Biomass, and Productivity of Mangrove Forest: A Review. Aquatic Botany, 89: 128-137.
[14] Tomlinson P B 1986 The Botany of Mangroves. Cambridge University Press, Cambridge, U.K., p.419.
[15] Matsui, N. Estimated stocks of organic carbon in mangrove roots and sediments in Hinchinbrook Channel, Australia. Mangr. Salt Marsh. 2, 199–204 (1998).
[16] Dharmawan I W S, Suyarso, Ulumuddin Y I, Prayudha B, Pramudji 2020 Mangrove community structure monitoring guide in Indonesia. PT Media Sains Nasional.
[17] Noor Y R, Khazali M, Suryadiputra I N N 2012 Introduction guide mangroves in Indonesia. Wetlands International dan Ditjen PHKA.
[18] Badan Standardisasi Nasional 2011 SNI7724 Ground Based Forest Carbon Accounting.
[19] Bismark M, Heriyanto N M, and Iskandar S 2008 Biomass and Content in Production Forest in the Siberut Island Biosphere Reserve, West Sumatra. Jurnal Penelitian Hutan dan Konservasi Alam, 5(5): 397-407.
[20] Onrizal, Rugayah, and Suhardjono 2005 Tree-habitus mangrove flora in the Angke-Kapuk Protected Forest. Biodiversitas 6 (1): 34-39.
[21] Bengen D G 2004 Synopsis of Coastal and Marine Ecosystems and Natural Resources and the Principles of Their Management. Bogor: Center for Coastal and Marine Resources Studies, IPB.
[22] Saparinto 2007 Mangrove ecosystem utilization. Dahara Press. Semarang
[23] Tolangara and Ahmad 2017 Mangrove Density and Its Conservation in Bacan, Halmahera Regency South of North Maluku Province. Techno: Research journal. 06 (02): 22-29
[24] Ahmad F 2014 Mangrove Species in Daruba and Wayabula, Morotai Island, North of Molucca. Berita Biologi 13(3) - Desember 2014.
[25] Akbar N, Baksir A, Mutmainnah, Ismail F, Tahir I, Marus I, Nurdiansah D and Wahab I 2019 Assessment of Community and Environmental Linkages to Mangrove on the Coast of Minaluli Village, North Mangoli District, Sula Archipelago Regency, North Maluku Province. Environmental Science Journal. Jurnal Ilmu Lingkungan 17 (3): 507-514.
[26] Taher M M, Kamal M, and Zairion 2012 Potential and management of mangrove ecosystem in South Jailolo Sub-district, West Halmahera, North Moluccas. Bonorowo Wetlands 2 (1): 19-30, June 2012.

[27] Mandari D Z, Gunawan H, Isda M N 2016 Assessment of Biomass and Carbon Stored in Forest Ecosystems Mangroves in the Dumai Mangrove City Area. Jurnal Riau Biologia 1 (3): 17-23, Januari 2016.
[28] Suryono N, Soenardjo, Wibowo E, Ario R, Rozy E F 2018 Perancak Mangrove Forest, Jembrana Regency, Bali Province. Thesis, Department of Marine Science, Faculty of Fisheries and Marine Sciences, Diponegoro University.
[29] Schaduw, J. N. 2015. Mangrove bioecology of a community-based marine protected area in Blongko Village, Sinonsayang District, South Minahasa Regency, North Sulawesi Province. Jurnal LPPM Bidang Sains dan Teknologi, 2(1), 89-102.
[30] Nedhisa P I and Tjahjaningrum I T 2019 Biomass Estimation, Carbon Stock and Sequestration Mangrove Carbon on Rhizophora mucronata in Wonorejo Surabaya with Allometric Equations. Jurnal Sains dan Seni. 8(2) :2337-3520.
[31] Hilmi E and Siregar A S 2006 Mangrove Vegetation Biomass Estimation Model in Indragiri Hilir Regency, Riau. Biosfera 23 (2) Mei 2006.
[32] Kepel T L, Suryono D D, Ati R N A, Salim H L, and Hutahaean A A 2017 Important value and Economic Value Estimation of Carbon Storage of Mangrove Vegetation in Kema, North Sulawesi. Jurnal Kelautan Nasional. 12(1), 19-26.
[33] Kolinug K H, Langi M A, Ratag S P, and Nurmawan W 2014 Zoning of the main plants that make up mangroves based on the level of seawater salinity in Teling Village, Tombariri District. COCOS. 5(4).
[34] Muzaki F, Dian S, Dwianita S, and Aries 2012 Exploring the Mangrove Surabaya.Pusat Studi Kelautan LPPM,Surabaya.
[35] Prawiroatmodjo S, Pramudji, Sapulate D and Budiman A 1987 Mangrove forest ecology in Kao Bay, Halmahera. 3rd Seminar Proceedings Mangrove Ecosystems. Denpasar, Bali 5 – 8 August, 1986, 93-97.
[36] Pramudji 1988 Preliminary Study on Mangrove Forests on Several Islands in the Aru Archipelago, Southeast Maluku. 3rd Seminar Proceedings Mangrove Ecosystems. Denpasar, Bali 5 – 8 August, 1986, 74-79.
[37] Heriyanto N M and Subandono E 2012 Composition and Structure, Biomass, and Potential Carbon Content of Mangrove Forests in Alas Purwo National Park. Jurnal Penelitian Hutan dan Konservasi Alam. 9 (1): 023-032.
[38] Rachmawati D, Setyobudiandi I, and Hilmi E 2014 Potential estimation of carbon stored in mangrove vegetation in the coastal area of Muara Gembong, Bekasi Regency.Omni-Akuatika, 10(2)85-91. DOI : 10.20884/1.oa.2014. 10.2.22.
[39] Kusmana C 2002 Management of mangrove ecosystems in a sustainable and community-based manner. National workshop on mangrove ecosystem management. Jakarta.
[40] Rianti A P 2012 Carbon Potential Stored in Mangrove Forest Stands at Marine Station, West Dumai District. Thesis, Faculty of Mathematics and Natural Sciences, Riau University.
[41] Dharmawan I W S and Siregar C A 2008 Soil Carbon and *Avicennia marina* Carbon Estimation (Forsk.) Vierh di Ciasem, Purwakarta. Jurnal Penelitian Hutan dan Konservasi Alam 5, 317-328.
[42] Lumbu T, Rumengan A P, Paruntu C P, Darwisito S, Ompi M, Mandagi S 2022 Study of Carbon Storage in Mangrove Biomass at The Coastal Village of Tatengesan, Pusomaen District Southeast Minahasa Regency North Sulawesi Province. Jurnal Pesisir dan Laut Tropis. 10(1): 63-71.
[43] Hairiah K and Rahayu S 2007 Measurement of Carbon Stored in Various Land Uses. World Agroforestry Centre. Bogor.
[44] Adinugroho C W and Kade S 2001 Model for estimating the above-ground biomass of mahogany trees (*Swietenia macrophylla* King). Jurnal penelitian Hutan dan Konservasi Alam. 3(1): 103 – 117.

[45] Imiliyana A, Muryono M, Purnobasuki H 2012 Carbon Stock Estimation in *Rhizopora stylosa* at Camplong Beach, Sampang-Madura. 10 November Institute of Technology. Surabaya.
[46]Brown S 1997 Estimating Biomass and Biomass Change of Tropical Forests: a Primer. (FAO Forestry Paper - 134). FAO. Rome.
[47] Irsadi A, Martuti N K R, and Nugraha S B 2017 Estimation of Mangrove Carbon Stock in Tapak Hamlet, Tugurejo Village, Semarang City. Jurnal Sains dan Teknologi, 15(2): 122.
[48] Cahyaning W, Setiawan A, and Rusita 2016 Estimation of Carbon Stored in Mangrove Forests in Margasari Village, Labuhan Maringgai District, East Lampung Regency. Jurnal Sylva Lestari, 6(1): 66-74.
[49] Fajar R, Hairul B, and Sufardi 2015 The Potential Carbon Stored on Mangrove and Fishpond Area InThe Coastal Areas of Banda Aceh. J. Manaj. Sumberd. Lahan. 4(1) :527–534.
[50]Dharmawan I W S and Samsoedin I 2012 Dynamics of Carbon Biomass Potential in Logged-over Forest Landscapes in the Malinau Research Forest. Jurnal Penelitian Sosial dan Ekonomi Kehutanan, 9(1) : 12-20.
[51] Eong O J 1993 Mangroves a carbon source and sink. Chemosphere 27, 1097–1107.
[52] Bouillon S, Rivera-Monroy V H, Twilley R R, Kairo J G 2009 Mangroves in the management of natural coastal carbon sinks (eds Laffoley, D. & Grimsditch). IUCN.

Tropical cyclone Saudel impact on ocean dynamic over the South China Sea

S R Arum[1], A Wicaksono[1,2*] and A D Siswanto[1,2]

[1]Department of Marine Science, Universitas Trunojoyo Madura
[2]Laboratory of Oceanography, Department of Marine Science, Universitas Trunojoyo Madura
PO.Box.2 Kamal, Bangkalan-Madura, 69162 Phone : (031) 3013234, Fax. (031) 3011506

*Corresponding author mail: ashari.wicaksono@trunojoyo.ac.id

Abstract. Tropical cyclones (TC) are atmospheric phenomena categorized as extreme weather that is perilous and destructive. The parameters used include sea surface temperature (SST), wind speed (WS), and air pressure (AP), which are factors in the formation of TC. The emergence of TC occurred in October 2020, when it grew in Philippine waters and died out in Vietnamese waters. The characteristics of the waters of the South China Sea (SCS) in particular and the Pacific Ocean in general make them one of the waters with the highest storm potential in the world; therefore, this research was conducted. The purpose of this research is to understand the characteristics of Saudel Cyclone (SC) based on the parameters SST, WS, and AP, which are obtained from the ECMWF website, to be able to describe the process of growth until the extinction of SC. Data from the ECMWF website has limitations because it is model data (re-analysis). The results of this study can determine the process of starting the extinction of SC as well as the characteristics of SST, AP, and WS in the conditions of starting SC, during the peak of SC, and starting the extinction of SC.

1. Introduction

A tropical cyclone (TC) is an atmospheric phenomenon categorized as extreme weather that is perilous destructive [11][20]. The formation of TC is initiated by the presence of low AP (Table 2), so high-speed winds are formed (Tables 1 and 2), and it develops into a storm [23]. TC has a devastating impact because it has a huge strength with an average radius of 150 to 200 km, so it can cause highly impacts on a large scale, such as heavy rain, strong winds, landslides, and high waves [1][2][7][20]. One of the important parameters of TC formation is a SST of at least 26.5°C [35][36], accompanied by the formation of comulonimbus clouds, or thunder clouds, and low AP [27]. The general characteristics of TC over the ocean, in addition to a minimum SST of 26.5°C, are also characterized by low AP around TC, so that it will affect the increase in SST and ultimately form a cyclonic storm [14]. SC storm as one of the TC over the ocean (western Pacific Ocean, especially the Philippine waters) and across the SCS until it reached the mainland of Vietnam during the period from October 18 to 25, 2020 [33]. The SC occurred over a period of more than 162 hours [17], with the intensity increasing slowly since pass into the waters of the SCS on October 21, 2020, then moving faster and getting stronger on October 22, 2020. The SC continued to move towards the northwest and then weakened on October 24, 2020, at 08:00 until it finally completely faded on October 25, 2020 [25]. Therefore, it is important to understand the characteristics of SC based on SST, AP, and WS parameters using the ECMWF data as an initial investigation in understanding the characteristics of the SCS in particular and the Pacific Ocean in general as one of the waters with the highest storm potential in the world [12].

Table 1. Saffir-Simpson scale of TC [3]

Category	Wind speed threshold	
	1-minute average sustained	10-minute average sustained
Category 1	64-82 knots 32.9-42.2 m/s	29.0-37.1 m/s
Category 2	83-95 knots 42.7-48.9 m/s	37.6-43.0 m/s
Category 3	96-112 knots 49.3-57.6 m/s	43.4-50.7 m/s
Category 4	113-136 knots 58.1-70 m/s	51.1-61.6 m/s
Category 5	≥ 137 knots ≥ 70 m/s	≥61.1 m/s

Table 2. Saffir-Simpson scale of TC [27]

Saffir-Simpson Categories	WS max (ms^{-1}, knot)	AP Minimum (mb)	Storm surge (m, ft)
1 (Minimum)	33-42 m/s [64-83 knots]	>=980 mb	1.0-1.7 m [3-5 ft]
2 (Moderat)	43-49 [84-96 knots]	979-965	1.8-2.6 [6-8 ft]
3 (Extensive)	50-58 [97-113 knots]	964-945	2.7-3.8 [9-12 ft]
4 (Extrime)	59-69 [114-135 knots]	944-920	3.9-5.6 [13-18 ft]
5 (Catastrophic)	>69 [>135 knots]	<920 mb	>5.6 [>18 ft]

2. Methods

We used met-ocean data that include SST, AP, and WS from https://cds.climate.copernicus.eu/cdsapp#!/search?type=dataset,. The data is analyzed to obtain the pattern of SST, AP, and WS during the period and pathways of SC movement. Detail information regarding SC can be found at http://agora.ex.nii.ac.jp/digital-typhoon/summary/wnp/s/202017.html.en. This page provides information regarding the development of various TC in the SCS as time passed (Table 3). The reference data matches the location and periodization of the SC (Figure 1).

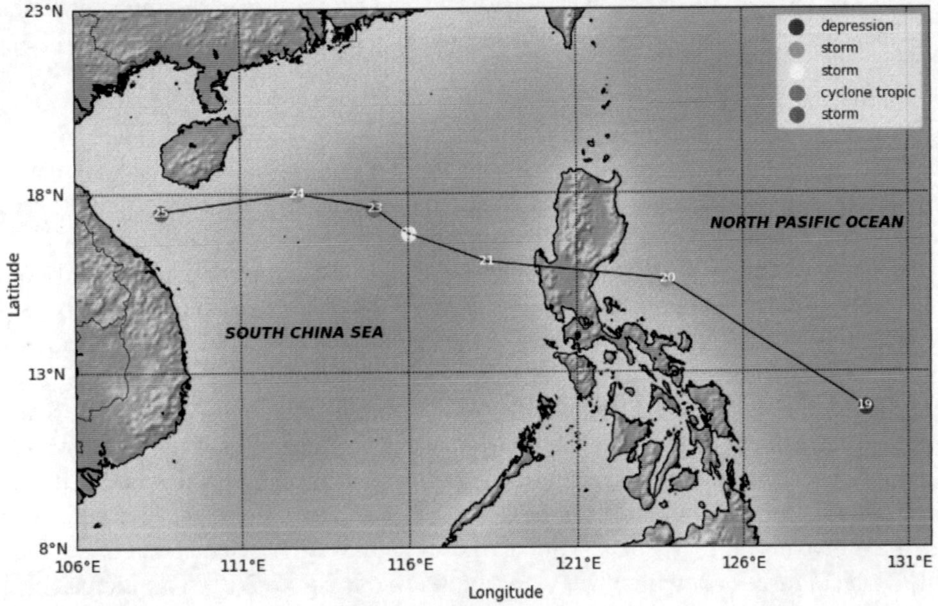

Figure 1. Trajectory of the SC
(Reproduce from https://www.hko.gov.hk/en/informtc/saudel20/saudel.html)

Table 3. Periodization of TC formation in the South China Sea in October 2020

No	Number	Name	Basin	Birth (UTC)	Death (UTC)	Duration (days, hours)	Min. res.
1	202014	CHAN-HOM	W. N. Pacific	05/10/2020 00:00	11/10/2020 18:00	6, 18	965
2	202015	LINFA	W. N. Pacific	10/10/2020 18:00	11/10/2020 12:00	0, 18	994
3	202016	NANGKA	W. N. Pacific	12/10/2020 06:00	14/10/2020 12:00	2, 6	990
4	202017	SAUDEL	W. N. Pacific	20/10/2020 00:00	25/10/2020 12:00	5, 12	975
5	202018	MOLAVE	W. N. Pacific	24/10/2020 06:00	28/10/2020 18:00	4, 12	940
6	202019	GONI	W. N. Pacific	28/10/2020 18:00	05/11/2020 12:00	7, 18	905

3. Results

TC occurs due to several factors, which are often used as indicators. When the SST increases, heat energy will be absorbed, which will also increase air humidity, triggering TC [31]. When the sea surface warms, it pushes up evaporation, leading to a rise in the water vapor content of the atmosphere. The increase in absolute humidity causes the atmosphere to radiate more strongly to the surface and heat it up, so the SST will growth. In addition to SST, WS is also an important indicator in TC studies [24]. WS can change significantly due to the influence of AP [29]. WS is affected by AP differences, which are called the air pressure gradient (APG). The greater the difference in AP, the tighter the pressure gradient and the higher the WS generated in the area between the APG, and vice versa [26]. Based on the research by [13], each T-Number can be used to estimate the maximum WS, lowest AP, and its equivalence to the TC category on the Saffir-Simpson scale. In his research, T-Number Cat 1 has a WS value of 65 knots and an AP of 976 millibars. So, the greater the difference in AP, the higher the WS. Low AP will form a vortex until it becomes the center of the TC [25]. The SC path starts from Philippine waters to Vietnamese waters from October 18 to 25, 2020. The 18th is a condition where SC has not yet formed, so it is used as a comparison between before and the start of SC formation. On the 19th, a tropical depression began to occur, then on the 20th there was an increase, which triggered the formation of the SC vortex, and continued to the 21th stronger, which formed the SC vortex, then moved towards the waters of the SCS on the 22nd and continued to increase and move until the 23th, which was the peak of the SC vortex. On the 24th, the SC began to weaken and became a tropical storm due to several influencing factors, and on the 25th, it continued to weaken until the extinction of the SC [32][33].

3.1. Wind Direction and Speed Analysis

The pattern of wind direction and speed at H±7 (Figure 2) shows a typical ST-like trend [21]. The SC WS range is 35–65 knots and illustrates the SC formation process. Changes in WS can occur during the TC formation process due to high AP differences at the centre [29]. Increased WS due to pressure differences can result in strong pressure gradients [16][18]. The Coriolis effect can cause the wind in TC to rotate clockwise in the northern hemisphere and counter clockwise in the southern hemisphere [16][18] (Figure 2) as part of the process of SC on October 18–25, 2020. On the 18th (Figure 2(a)), the initial process of SC growth begins with various atmospheric disturbances that trigger winds centered on locations that have low AP. The 19th (Figure 2(b)) experienced a tropical depression that formed in the Philippine Sea. The 20th (Figure 2(c)) developed and intensified into a tropical storm, which moved towards the SCS. The 21th (Figure 2(d)) moved faster and stronger towards the SCS, which had a SC vortex in Philippine waters, and continued on the next day (Figure 2(e)), characterized by a vortex, and continued to move closer to Vietnam with an increase in SC grade. The peak of the SC was shown on the 23rd (Figure 2(f)), which was a continuation of (Figure 2(e)) with increasing WS, ranging from 30 to 40 m/s or 64 kt. Furthermore, on October 24th, the SC moved towards Vietnam waters and began to weaken into a tropical storm influenced by high AP, resulting in weakening winds (Figure 2(g)), and further weakened on the 25th (Figure 2(h)), accompanied by the growth of a new TC in Philippine waters. According to the categorization of the Saffir-Simpson Scale [3], SC is in category 1 with speeds of 64–82 knots.

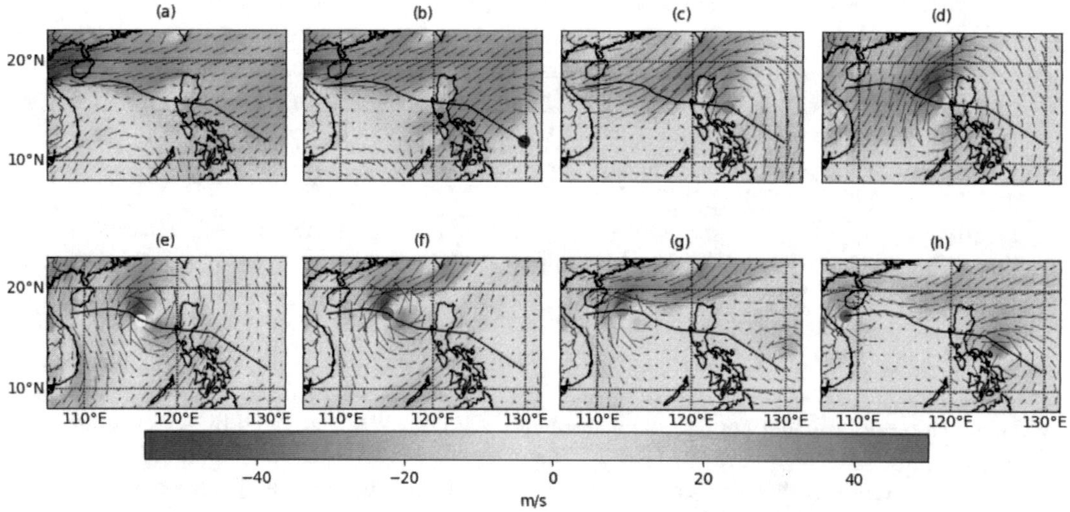

Figure 2. H±7 wind direction and speed pattern of SC during October 2020. (a) 18th, b) 19th, c) 20th, d) 21st, e) 22nd, f) 23rd, g) 24th, h) 25th. Wind data has two components, u and v, and are combined in the processing. Meaning of positive (+) and negative (-) wind speed: u positive (+) indicates eastward, u minus (-) indicates westward, v positive (+) indicates northward, v minus (-) indicates southward.

3.2. Air Pressure Analysis

Low AP is an important trigger in the process of TC formation. The maximum WS reached 18 m/s or more, and the minimum AP was 998 hPa. Environmental conditions are formed because high surface temperatures can cause the surrounding air to be more effective and move at a high speed, thus reducing AP [31]. Figure 3 shows the process of growth until the extinction of SC; the centre of the cyclone is shown by the turquoise color resembling a circle. AP when SC occurs decreases significantly so that it becomes < 1000 mb [28] and the cyclone centre is formed, which is influenced by several factors characterized by temperature conditions that tend to be constant and stable (or increasing) so that it will affect other factors that will ultimately reduce AP [8][19]. Figure 3 shows the stages of SC formation. Before the peak of SC, the region around the cyclonic vortex had a high pressure above the 1000 mb value. Letter (a) shows the process of entering the beginning of SC emergence (1000 mb pressure), then the SC seedlings begin to appear accompanied by the formation of the cyclone centre and move towards Philippine waters (letters (b) and (c)). The cyclone centre begins to form and is marked by the green color at the centre of the vortex (letter (d)), which indicates an area of air pressure < 1000 mb. It then begins to move (letter (e)) as a result of the influence of lower pressure at the centre of the cyclone on the wind and will continue to move towards areas with low AP (letter (f)). The peak of the SC occurred on October 23, 2020 (letter (f)) with very low-pressure values (below 980) and moved towards Vietnam, then gradually weakened (letter (g)) and finally became extinct (letter (h)). The SC centre area, which is dominated by low-pressure (< 980 mb) (letter (e)) and reaches its lowest peak on October 23, 2020 (letter (f)), is influenced by SST. The SC centre has low AP due to high SST, thus causing the air above it to also be warm, and eventually the area with low AP becomes gradually hotter [37]. The increase in SST also triggers a change in WS (letters (e) to (f)) at the centre of the low-pressure cyclone, resulting in the mixing of sea surface water in the cooling process. Strong winds can also affect surface currents and lift cold water from the deep ocean layers to the surface, resulting in a process of decreasing SST [30].

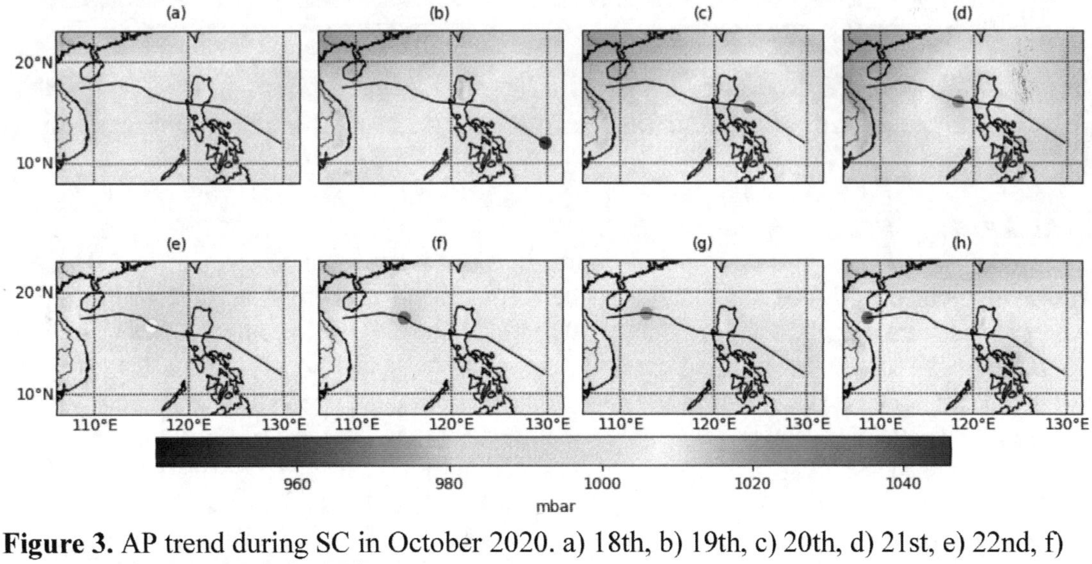

Figure 3. AP trend during SC in October 2020. a) 18th, b) 19th, c) 20th, d) 21st, e) 22nd, f) 23rd, g) 24th, h) 25th.

3.3. Analysis Sea Surface Temperature (SST)

The SST around the SC ranged from ± 28–30 °C (Figure 4). Water conditions do not yet show the potential for ST formation, especially SC, because the temperature is still in the range of 28°C to 29°C (a). SC began to form (Figure 4(b)), characterized by an increase in SST. Based on the research of [38], the entire SST variability track looks the same when SST changes, and it becomes a larger track dispersion when SST increases. In general, the temperature heights (Figure 4(a-d)) look almost the same (28–30 °C). SC forms when the SST is warm (above 26.5°C, approximately 27–28 °C, with an average annual maximum of 27.3°C) [10][34]. Low AP in a high-SST (= warm) environment can trigger the formation of a storm vortex [31], and especially SC in the form of high-SST, dense temperature color triggers even lower AP, thus affecting WS [30] (Figure 4 (e)). Figure 4 (f) shows the highest peak of SC (October 23, 2020) due to the influence of stable high temperature conditions in the previous period, resulting in high AP. High WS causes a decrease in temperature (slightly faded color) so that it experiences extinction (October 24, 2020) (Figure 4 (g)) and SC moves towards Vietnam [30]. SST can trigger the formation of tropical storms, as the minimum SST is 26–27 °C [25]. An increase in SST during a tropical storm will trigger low pressure and result in an increase in WS. Temperatures that can trigger tropical storms range from 26–27 °C [10][25][31][34].

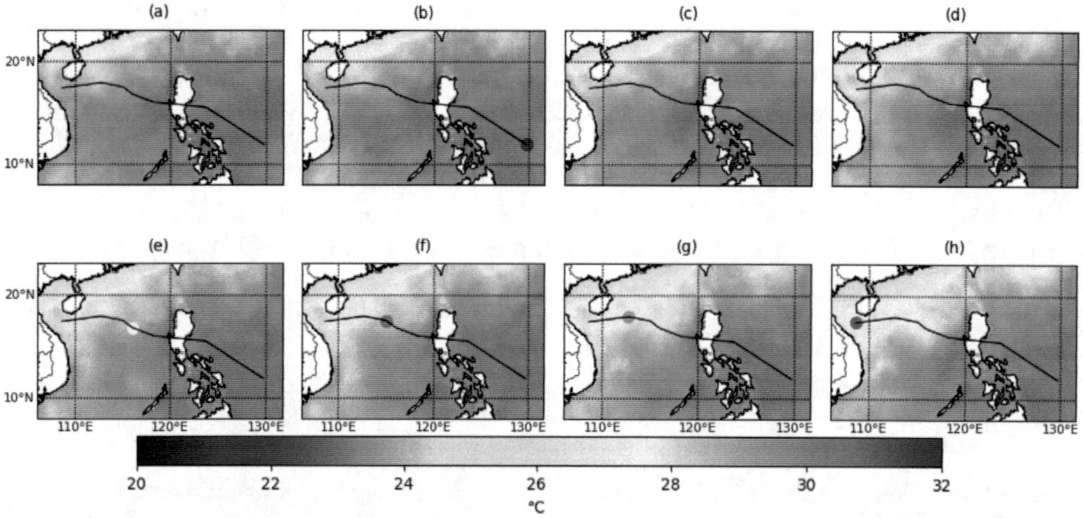

Figure 4. SST profile during the SC period (data processed from the ECMWF website). a) 18th, b) 19th, c) 20th, d) 21st, e) 22nd, f) 23rd, g) 24th, h) 25th.

3.4. The Process of SC Formation

Figures 2, 3, and 4 show the development of the SC since it appeared on October 19, 2020, and was characterized by a disturbance due to the formation of a depression zone in Philippine waters in the form of an area of low AP (<1,000 hPa). The SST showed a significant increase on October 20, 2020. Slowly, the depression zone strengthened and formed a low-pressure system accompanied by a cyclonic wind pattern (as the initial stage of SC development and the formation of a cyclone centre) with low AP below 1,000 on October 21, 2020. The SC fully developed over the SCS on October 22 and persisted until the 23rd, which was the peak of the SC. It was characterized by WS ranging from 40 to 60 m/s, AP less than 1,000 hPa at the eye of the cyclone, and a stable SST. On October 24, 2020, the SC moved towards Vietnam, characterized by WS that began to decrease, followed by AP that began to increase and temperatures that began to decrease. October 24, 2020, is also the extinction date of SC until October 25, 2020. The formation of low pressure cannot be separated from SC generation in the waters of the SCS. Observational evidence of the formation of a low-pressure system is illustrated in the AP on October 21, when the AP began to drop to 990 and even lower. The decrease in AP before SC can be related to the occurrence of strong winds that affect the decrease in SST, so that AP will rise. Thus, decreasing AP, stabilizing SST, and increasing WS can play an important role in developing SC, and vice versa.

4. Discussion

SST changes in the TC near the cyclone eye are related to changes in heat flux from the ocean to the atmosphere and are linear. A decrease in surface flux values below the eye of the cyclone has a greater potential to affect storm intensity compared to changes in surface flux at a certain distance from the TC center [4][5][6]. At the beginning of the SC formation on October 19, 2020, the SST showed high values (reaching 30 °C), and in general, the waters of the SCS were dominantly warm, triggering the cyclone (Figure 5 (a)). The dominant pattern of SST at the beginning of SC formation tends to be stable over a period of time. On the same date, the AP was 1000 mbar (Figure 5(b)) and the WS was 15 to 20 m/s (Figure 5(c)). The average SST at the onset of the cyclone was around 28.04 °C and continued to increase to 29.49°C, almost close to the SST value at the onset of SC (30 °C) [14]. Research states by [28], that there is a decrease in AP becomes < 1000 mbar when TC begins to form, including the formation of SC. The WS at the start of the TC formation was ± 28 m/s and the AP was 1000 mbar [15].

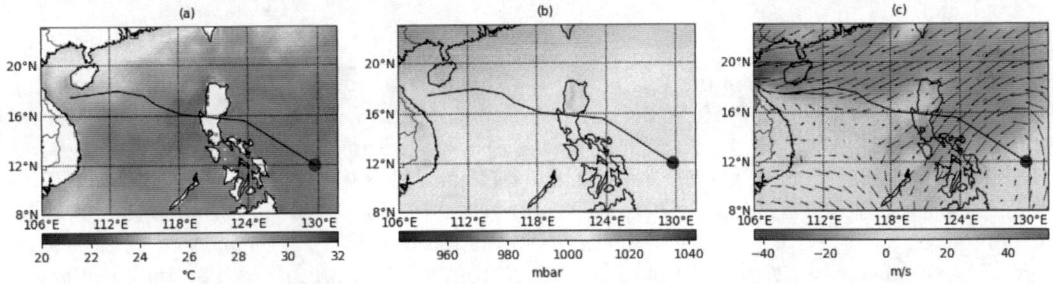

Figure 5. Environmental conditions at the beginning of SC formation on October 19, 2020. (a) SST, (b) AP, and (c) WS.

Figure 6 shows the peak of SC on October 23, 2020. Environmental conditions show the SST in the warm category (± 28°C) (Figure (a)), low AP (± 980 mBar) (Figure (b)), and WS ± 40 m/s (Figure (c)). Slightly different results were shown by [15], the AP decreased (930 mBar) and the WS was ± 56 m/s at the peak of TC. Several cyclones grew in the SCS, i.e. TC Molave (formed east of Luzon Island) with a maximum WS of 38 m/s and an AP of 965 hPa near the center of the cyclone [9]. The TC SST forms over a warm ocean that reaches more than 26.5°C [34]. AP at the peak of TC decreased (from 992.6 mBar to 969.3 mBar), and WS reached 10 m/h and increased (from 65 km/h to 94 km/h) [22].

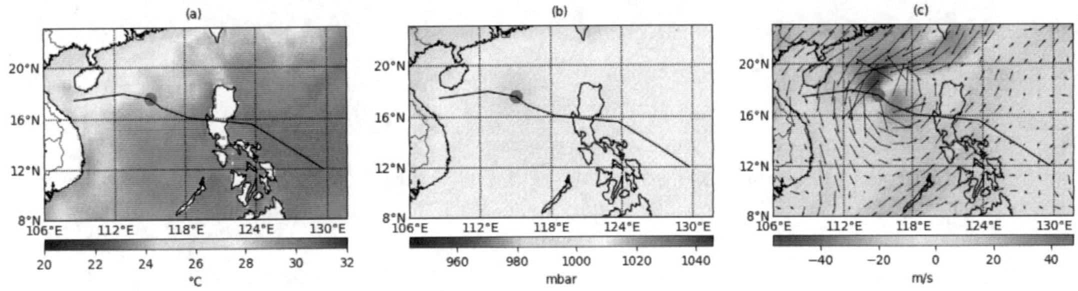

Figure 6. Environmental conditions during the SC peak on October 23, 2020. (a) SST, (b) AP, and (c) WS

Environmental conditions during the weakening of SC intensity on October 24, 2020 (Figure 7) show a decrease in SST due to strong winds at the peak of the cyclone, followed by an increase in AP and a decrease in WS [30]. Figure 7(a) shows an SST of 27 °C with an air pressure of 1000 (Figure 7(b)) and a WS of 30 m/s (Figure 7(c)). An important and highly distinctive characteristic is that TC weakening is characterized by a decrease in air pressure (to 980 mBar) and wind speed (± 26 m/s) weakening (to ± 16 m/s) [15].

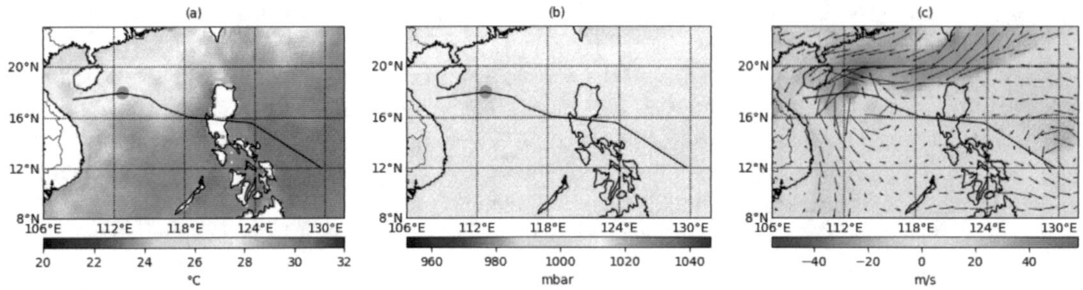

Figure 7. Environmental conditions when the SC began to weaken on October 24, 2020. (a) SST, (b) AP, and (c) WS

5. Conclusion

The SC grew from extinction in the Philippines to extinction in Vietnam from October 18 to 25, 2020. Starting the SC on October 19, 2020, the SST was 30°C, the AP was 1000 mBar, and the WS was 15-20 m/s. At the peak of SC on October 23, 2020, the SST was warm at 28°C. AP reaches a low of ± 980 mBar and a WS of ± 40 m/s.SC extinction began on October 24, 2020, with an SST of ± 27 °C, an AP of ± 1,000, and a WS of ± 30 m/s. There were changes in the values of several important parameters, including the SST by ± 3°C (the beginning of SC formation was 30°C, at the peak of SC ± 28°C, at the start of SC weakening ± 27°C), AP by ± 20 mBar (from ± 1,000 mBar to ± 980 mBar; although AP again reached around ± 1,000 mBar on October 24, 2020), and an increase in WS by 20–25 m/s (changing from 15-20 m/s on October 19 to ± 40 m/s on October 23 at the peak of SC).

6. Acknowledgment

This paper was funded by the DIPA Institute for Research and Community Service (LPPM) Universitas Trunojoyo Madura 2023 with scheme research by a novice lecturer (PDP). We appreciate the time and effort of the editors and anonymous reviewers whose comments and suggestions helped improve the manuscripts.

References

[1] Bagtasa, G. (2021). Analog forecasting of tropical cyclone rainfall in the Philippines. Weather and Climate Extremes, 32(February), 100323. https://doi.org/10.1016/j.wace.2021.100323

[2] Bakker, T. M., Antolínez, A. A., Leijnse, T. W. B., Pearson, S. G., & Giardino, A. (2022). Estimating tropical cyclone-induced wind , waves , and surge : A general methodology based on representative tracks. 176(August 2021). https://doi.org/10.1016/j.coastaleng.2022.104154

[3] Bloemendaal, N., de Moel, H., Muis, S., Haigh, I. D., & Aerts, J. C. J. H. (2020). Estimation of global tropical cyclone wind speed probabilities using the STORM dataset. Scientific Data, 7(1), 1–11. https://doi.org/10.1038/s41597-020-00720-x

[4] Cione, J. J., and E. W. Uhlhorn, 2003: Sea surface temperature variability in hurricanes: Implications with respect to intensity change. Mon. Wea. Rev., 131, 1783–1796.

[5] D'Asaro, E. A., T. B. Sanford, P. P. Niiler, and E. J. Terrill, 2007: Cold wake of Hurricane Frances. Geophys. Res. Lett., 34, L15609, doi:10.1029/2007GL030160.

[6] Dare, R. A., & Mcbride, J. L. (2011). Sea surface temperature response to tropical cyclones. Monthly Weather Review, 139(12), 3798–3808. https://doi.org/10.1175/MWR-D-10-05019.1

[7] Dela, G., & Santos, C. (2021). ScienceDirect 2020 tropical cyclones in the Philippines : A review. Tropical Cyclone Research and Review, 10(3), 191–199. https://doi.org/10.1016/j.tcrr.2021.09.003

[8] Fang, R., Chen, S., Zhou, M., Li, W., Xiao, H., Zhan, T., Wu, Y., Liu, H., & Tu, C. (2021). Periodic Cycles of Eyewall Convection Limit the Rapid Intensification of Typhoon Hato (2017). Advances in Meteorology, 2021. https://doi.org/10.1155/2021/5557448

[9] Gao, S. Z., & Lu, X. Y. (2012). Diagnostic analysis on the distribution of rainfall associated with typhoon "molave" (0906). Journal of Tropical Meteorology, 18(2), 220–227. https://doi.org/10.3969/j.issn.1006-8775.2012.02.011

[10] Guan, S., Zhao, W., Sun, L., Zhou, C., Liu, Z., Hong, X., Zhang, Y., Tian, J., & Hou, Y. (2021). Tropical cyclone-induced sea surface cooling over the Yellow Sea and Bohai Sea in the 2019 Pacific typhoon season. Journal of Marine Systems, 217(January), 103509. https://doi.org/10.1016/j.jmarsys.2021.103509

[11] Hoque, M. A. A., Phinn, S., Roelfsema, C., & Childs, I. (2018). Modelling tropical cyclone risks for present and future climate change scenarios using geospatial techniques. International Journal of Digital Earth, 11(3), 246–263. https://doi.org/10.1080/17538947.2017.1320595

[12] Huang, H., Yuan, J., Wen, G., Bi, X., Huang, L., & Zhou, M. (2021). Identifying the development of a tropical depression into a tropical storm over the south china sea. Weather and Forecasting, 36(4), 1299–1328. https://doi.org/10.1175/WAF-D-20-0186.1

[13] Ismail, P., Hidayat, N. M., & Siadari, E. L. (2019). Analisis Siklon Tropis Nock-Ten Berbasis Data Satelit Himawari. Jurnal Meteorologi Klimatologi Dan Geofisika, 4(3), 16–23. https://doi.org/10.36754/jmkg.v4i3.49

[14] Koropitan, A. F., Khaldun, M. H. I., & Naulita, Y. (2022). Impact of tropical Cyclone Marcus on ocean subsurface and surface layers. Global Journal of Environmental Science and Management, 8(3), 353–368. https://doi.org/10.22034/gjesm.2022.03.05

[15] Lavender, S. L., Hoeke, R. K., & Abbs, D. J. (2018). The influence of sea surface temperature on the intensity and associated storm surge of tropical cyclone Yasi: A sensitivity study. Natural Hazards and Earth System Sciences, 18(3), 795–805. https://doi.org/10.5194/nhess-18-795-2018

[16] Li, L., Li, Z., He, Z., Yu, Z., & Ren, Y. (2022). Investigation of Storm Tides Induced by Super Typhoon in Macro-Tidal Hangzhou Bay. Frontiers in Marine Science, 9(June), 1–24. https://doi.org/10.3389/fmars.2022.890285

[17] Lin, K., Chung, K., Wang, S., Chen, L., Liou, Y., Lin, P., Chang, W., Chiu, H., & Chang, Y. (2023). Evaluation of hygroscopic cloud seeding in warm rain process by a hybrid microphysics scheme on WRF model : a real case study. April.

[18] Liu, P., Zhao, L., Fang, G., & Ge, Y. (2021). Explicit polynomial regression models of wind characteristics and structural effects on a long-span bridge utilizing onsite monitoring data. Structural Control and Health Monitoring, 28(5), 1–24. https://doi.org/10.1002/stc.2705

[19] Liu, W. C., & Huang, W. C. (2020). Investigating typhoon-induced storm surge and waves in the coast of Taiwan using an integrally-coupled tide-surge-wave model. Ocean Engineering, 212(April), 107571. https://doi.org/10.1016/j.oceaneng.2020.107571

[20] Lu, X., Wong, W. K., Au-Yeung, K. C., Choy, C. W., & Yu, H. (2022). Verification of tropical cyclones (TC) wind structure forecasts from global NWP models and ensemble prediction systems (EPSs). Tropical Cyclone Research and Review, 11(2), 88–102. https://doi.org/10.1016/j.tcrr.2022.07.002

[21] Lu, L. N., Scussolini, P., Kew, S., Philip, S., Hariadi, M. H., Vautard, R., Van Mai, K., Van Vu, T., Truong, K. B., Otto, F., van der Schrier, G., van Aalst, M. K., & van Oldenborgh, G. J. (2021). Attribution of typhoon-induced torrential precipitation in Central Vietnam, October 2020. Climatic Change, 169(3–4), 1–22. https://doi.org/10.1007/s10584-021-03261-3

[22] Medvedev, I. P., Rabinovich, A. B., & Šepić, J. (2022). Destructive coastal sea level oscillations generated by Typhoon Maysak in the Sea of Japan in September 2020. Scientific Reports, 12(1), 1–12. https://doi.org/10.1038/s41598-022-12189-2

[23] Prasetya, R., As'ari, A., & Dayantolis, W. (2014). Analisis Dampak Siklon Tropis Nangka, Parma dan Nida pada Distribusi Curah Hujan di Sulawesi Utara. Jurnal Fisika Dan Aplikasinya, 10(1), 1. https://doi.org/10.12962/j24604682.v10i1.816

[24] Randall, D. A., Cess, R. D., Blanchet, J. P., Boer, G. J., Dazlich, D. A., Del Genio, A. D., Deque, M., Dymnikov, V., Galin, V., Ghan, S. J., Lacis, A. A., Le Treut, H., Li, Z. X., Liang, X. Z., Mc Avaney, B. J., Meleshko, V. P., Mitchell, J. F. B., Morcrette, J. J., Potter, G. L., … Zhang, M. H. (1992). Intercomparison and interpretation of surface energy fluxes in atmospheric general circulation models. Journal of Geophysical Research, 97(D4), 3711–3724. https://doi.org/10.1029/91JD03120

[25] Shi, C., Shi, C., Zhang, T., Du, J., & Zheng, H. (2022). Preliminary Analysis of Multi-Typhoon Influence on Hainan Province and Its Circulation Background in Autumn 2020. 1, 24–37. https://doi.org/10.23977/envcp.2022.010103

[26] Sumaja, K. (2021). Mitigating The Impacts Of Heavy Rain And Strong Wind On Aviation Due To The Effect Of Tropical Cyclones In I Gusti Ngurah. December.

[27] Surinati, D., & Kusuma, D. A. (2018). Karakteristik Dan Dampak Siklon Tropis Yang Tumbuh Di Sekitar Wilayah Indonesia. Oseana, 43(2), 1–12. https://doi.org/10.14203/oseana.2018.vol.43no.2.16

[28] Tanaka, H. K. M., Gluyas, J., Holma, M., Joutsenvaara, J., Kuusiniemi, P., Leone, G., Lo Presti, D., Matsushima, J., Oláh, L., Steigerwald, S., Thompson, L. F., Usoskin, I., Poluianov, S., Varga, D., & Yokota, Y. (2022). Atmospheric muography for imaging and monitoring tropic cyclones. Scientific Reports, 12(1), 1–14. https://doi.org/10.1038/s41598-022-20039-4

[29] Wang, C., Li, Z., Hu, L., Zhao, Z., Luo, Q., Hu, J., & Zhang, X. (2019). Field research on thewind-induced response of a super high-rise building under Typhoon. Applied Sciences (Switzerland), 9(11). https://doi.org/10.3390/app9112180

[30] Wang, Y. (2020). Composite of Typhoon-Induced Sea Surface Temperature and Chlorophyll-a Responses in the South China Sea. Journal of Geophysical Research: Oceans, 125(10). https://doi.org/10.1029/2020JC016243

[31] Wu, Z., Jiang, C., Chen, J., Long, Y., Deng, B., & Liu, X. (2019). Three-dimensional temperature field change in the South China Sea during Typhoon Kai-Tak (1213) based on a fully coupled atmosphere-wave-oceanmodel. Water (Switzerland), 11(1). https://doi.org/10.3390/w11010140

[32] You, S., Kang, Y. H., Kim, B. U., Kim, H. C., & Kim, S. (2021). The role of a distant typhoon in extending a high PM2.5 episode over Northeast Asia. Atmospheric Environment, 257, 118480. https://doi.org/10.1016/j.atmosenv.2021.118480

[33] Zhang, W., Liu, B., Zhang, W., Li, S., Cao, X., & Xing, X. (2022). Effect of Flow-Dependent Unbalanced Background Error Variances on Tropical Cyclone Forecasting. Journal of Marine Science and Engineering, 10(11). https://doi.org/10.3390/jmse10111653

[34] Zhu, W., Subrahmanyam, M. V., Wang, L., & Guo, B. (2023). Dynamic Response of Atmospheric and Ocean Parameters and Their Relation to Typhoon Haikui (2012) Using Satellite Data. 2012.

[35] Kotal, S. D., Bhattacharya, S. K., & Roy Bhowmik, S. K. (2019). Estimation of tropical cyclone intensity and location over the north Indian Ocean–a challenge. *Meteorological Applications*, *26*(2), 245-252.

[36] Emanuel, K. (2003). A similarity hypothesis for air–sea exchange at extreme wind speeds. *Journal of the atmospheric sciences*, *60*(11), 1420-1428.

[37] Seager, R., Cane, M., Henderson, N., Lee, D. E., Abernathey, R., & Zhang, H. (2019). Strengthening tropical Pacific zonal sea surface temperature gradient consistent with rising greenhouse gases. *Nature Climate Change*, *9*(7), 517-522.

[38] Noyelle, R., Ulbrich, U., Becker, N., & Meredith, E. P. (2019). Assessing the impact of sea surface temperatures on a simulated medicane using ensemble simulations. *Natural Hazards and Earth System Sciences*, *19*(4), 941-955.

First record and phylogenetic relationship of the endangered species *Betta foerschi* (Vierke, 1979) (Anabantiformes : Osphronemidae) from Belitung Island, Indonesia

AF Syarif[1]*, FS Valen[1], M Herjayanto[2], S Aisyah[3]

[1] Department of Aquaculture, Faculty of Agricultural Fisheries and Biology, University of Bangka Belitung, Indonesia
[2] Department of Fisheries Science, Faculty of Agriculture, University of Sultan Ageng Tirtayasa, Indonesia
[3]Departement of Aquatic Resource Management, Faculty of Agricultural Fisheries and Biology, University of Bangka Belitung, Indonesia

*Corresponding author: ahmadfahrulsyarif@gmail.com

Abstract. *Betta foerschi* (Vierke, 1979) is a betta fish group from the Osphronemidae family which is only known from Central and West Borneo Island, Indonesia. This species is classified as an Endangered (EN) by the IUCN Red List, where the population has recently been moderately decreased. The new distribution of *B. foerschi* fish has been found on Belitung Island based on morphological similarities. In this research, we report the first record and phylogenetic relationship of *B. foerschi* based on the DNA Barcoding method using the Cytochrome C Oxidase Subunit I (COI) gene from Belitung Island Indonesia. Additionally, the DNA barcoding for *B. foerschi* not only represents the first DNA record for Belitung Island, but also the first sequence reported in the Genbank database. The result showed that *B. foerschi* forms the same clade within the Osphronemidae family and Betta genus solidifying its position in this taxonomic group with the closest genetic distance that occurs between *B. foerschi* and *B. persephone* of 0.15 using the Maximum Composite Likelihood model analysis.

Keywords: *Betta foerschi*; COI, First Genetic Record; Belitung Island

1. Introduction

Indonesia's Greater Sunda Islands area include Belitung Island. In addition, Up to 700 kilometers separate Belitung Island from the closest town in Peninsular Malaysia, 250 kilometers separate Belitung Island from the closest town in Bangka Island, 600 kilometers separate Belitung Island from the closest town in Sumatra, and 500 kilometers separate Belitung Island from the closest town in Borneo. [1][2][3]. Due to its location in Sundaland and with the presence of multiple once-connected ancient rivers [4]. Belitung Island has a great biodiversity and a diverse range of indigenous species. Several species of wild betta are also inhabit in inland waters of Belitung Islands due to their location is filled with peat swamp forest and highly acidic blackwater, which is the main habitat for *Betta* sp.[5].

In the Osphronemidae family, the Betta fish genus is the most varied. South-East Asia has more than 65 species of Betta, which are split into numerous species groups. [6][7][8]. *Betta* sp. can be able to survive at pH 3.0 or 4.0. [9][10][11] and inhabit peat swamp woodlands in shallow pools of buried leaf litter. Several wild betta species have been recorded in the Belitung Islands, such as *Betta chloropharynx*, *Betta edithae* and *Betta uberis* [12]. In this research we discovered the presence of *Betta foerschi* from Belitung Island Indonesia.

Betta foerschi is an endemic species from Borneo Island. This species usually found in the creeks of the rainforest of Central Kalimantan province, and around the settlement of Kubu in Western Kalimantan provinces, Borneo, Indonesian. Moreover, the IUCN Red List classifies this species as Endangered (EN) and Threatened Species [13], indicating that its population has lately been moderately reduced. In our last record we found *Betta foerschi* in Belitung Island Indonesia. However, as has been reported from Borneo [14], the number of *Betta foerschi* populations in Belitung is also endangered with a declining population. Even though the existence of this species on Belitung Island adds to the diversity of species. For this reason, domestication and conservation efforts are very urgent to be carried out. To ensure the effectiveness of domestication and conservation initiatives, this species must be validated. A genetic study approach is required to determine species certainty.

The aim of this study is providing a genetic identification of the *Betta foerschi* from Belitung Island, Indonesia, based on the COI gene. The COI gene has been selected as a standard tool for molecular taxonomy and identification of animals including fishes [15][16][17]. The COI gene has been effectively used as a species identification tool for freshwater fish identification in Indonesia [18][19][20][21]. DNA barcoding will be added to the Genbank NCBI database as a standard for identifying the Wild *Betta foerschi* species. This sequence will be the first record of Wild *Betta foerschi* in the GenBank database. The sequence will aid in the speedy identification of species, the understanding of biodiversity, the increased knowledge of genetic diversity, and the comprehension of life histories. In addition, studies regarding genetic diversity and population structure will also be supported by this DNA Barcoding, especially for the selection of brood in domestication efforts and to see the number of populations based on genetic diversity to determine the number of species populations needs to be carried out for proof and the first steps of domestication.

2. Materials and Methods

2.1. The Sampling Sites and Fish Collection

The study was conducted from 7 to 27 June 2023 in the swamp forest stream of Limbongan village, Gantong district, East Belitung Regency, Belitung Island, Indonesia (**figure 1**). We collect samples using environmentally acceptable fishing equipment, i.e. Traps [22]. A total of Seven (7) specimens of *Betta foerschi* were caught using the fish trap during a week of traping. The remaining four (4) preserved as livestock at the Fish Reproduction Laboratory, Bangka Belitung University, Indonesia due to the domestication act. Moreover, one wild specimen was killed before being identified. Killing fish is viewed as an ethical method of using live fish for research [23]. The right body's pectoral fin of one specimen was removed for DNA extraction and the left side was used for morphometric data collection and is frequently visible in photographs. The specimen was preserved in a 90% alcohol solution [24]. The other one specimen was preserved in a 7% formalin solution [25] and deposited at the zoology laboratory, at Bangka Belitung University. The specimen's diagnostic morphological traits were examined in accordance with Tan & Ng [10]. The specimens are identified as *Betta foerschi* and range in size from 5-6 cm. The determined meristic characteristics include dorsal, ventral, pectoral, and anal fin rays.

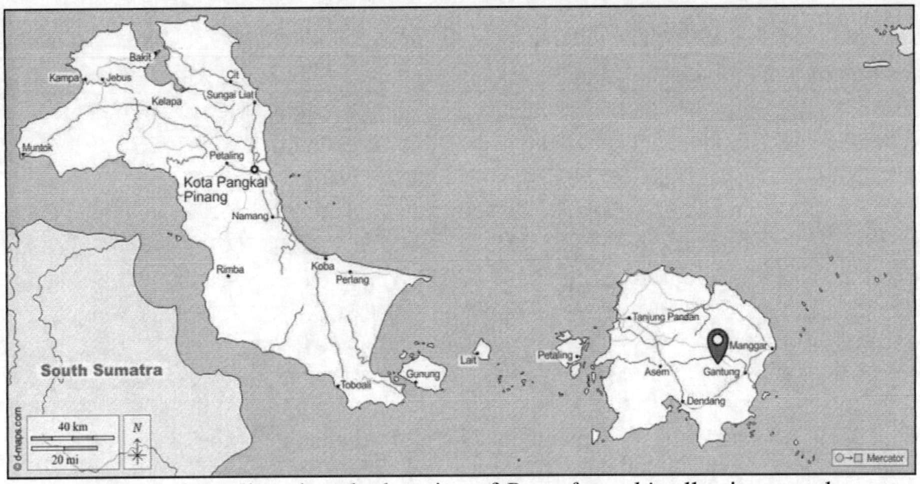

Figure 1. Sampling sites the location of *Betta foerschi* collecting samples

2.2. DNA Extraction and Amplification

DNA separation was carried out utilizing with the NexproTM DNA kit at the Basic Biology Laboratory, University of Bangka Belitung, with the FISH-F2 and FISH-R2 primers [26]. The PCR process refers to the procedure of Aisyah et al [27] by using 2% forward and reverses primers, 50% PCR mix of NexproTM Master mix, 36% ddH2O, and 10% DNA template. The reaction mixture was then amplified using a BioRad type T100TM. PCR cycling parameters included an initial pre-denaturation at 95 °C for 1 minute, denaturation at 95°C for 15 seconds, annealing at 54°C, extension at 72°C for 10 seconds (35 cycles), and final extension at 72°C for 10 minutes. The PCR results were then visualized in 1% agarose gel via electrophoresis by staining Nucleic Acid Gel Stain (GelRed®) [28]. The agarose was a mixture with 5 ml of TAE buffer and 45 ml of distilled water, and 8 µl of red gel stain was added to the solution 5 µl of DNA sample was homogenized with 1 µl of Loading Dye and added to the agarose gel wells. Positive sample (sparkling DNA bands) was then processed for DNA reading (sequencing) using the Sanger dideoxy method at PT. Genetika Science Jakarta.

2.3. Data Analysis

In order to identify the specimens, additional sequences were obtained using BLAST searches of the Genbank database. We register the DNA Barcode of the sequence to the Genbank (National Center for Biotechnology Information) (https://www.ncbi.nlm.nih.gov/genbank/) with the accession number OR192539. We also present the evolutionary history of the species by using the Neighbor-Joining method [29] with the bootstrap test (1000 replicates) [30]. The Maximum Composite Likelihood technique was used to compute the evolutionary distances [31] in MEGA X software [32].

3. Results and Discussions

3.1. Sampling Sites and Fish Collection

The *Betta foerschi*'s DNA-barcode from Belitung Island, which has a base-pair length of 686 bp, was successfully sequenced. (**Table 1**). According to Guo et al [33], the mitochondrial DNA Cytochrome Oxidase Subunit I (COI) gene has the potential to function as a barcode of all animals, and sequences of COI genes with more than 658 base pairs can be used as a benchmark for differentiating animals.

IOP Conf. Series: Earth and Environmental Science 1289 (2023) 012010 doi:10.1088/1755-1315/1289/1/012010

Table 1. DNA Barcoding of *Betta foerschi* from Belitung Island, Indonesia

DNA Barcoding of ***B. foerschi***
GGTGCATGGGCCGGAATAGTCGGCACCGCTCTAAGTCTGCTCATTCGGGCAGAACTGAGCCAACCAGGG GCCCTCCTTGGTGACGACCAGATCTATAATGTAATTGTTACGGCACACGCTTTTGTAATAATTTTCTTT ATGGTAATGCCTGTTATAATCGGGGGTTTCGGAAATTGACTAGTCCCACTAATAATTGGAGCACCAGAC ATGGCCTTTCCCCGGATAAACAATATAAGCTTCTGACTTTTACCCCCATCTTTCCTGCTATTACTAACA TCCTCCGGGGTAGAAGCTGGGGCAGGTACTGGCTGAACCGTCTACCCCCCATTAGCTAGCAATTTAGCA CACGCGGGCGCCTCTGTGGATTTAACAATTTTTTCATTACATCTTGCAGGTGTCTCATCTATTTTAGGG GCTATTAACTTTATTACCACAATTATTAACATAAAACCACCCGCAATCTCCCAATACCAAACACCTCTA TTTGTGTGGGCTGTTTTGGTCACAGCTGTCCTACTTCTCTTATCACTACCCGTCTTAGCTGCCGGAATC ACGATACTCTTAACAGACCGAAACCTAAACACAACCTTCTTTGACCCCGCAGGAGGCGGAGA

DNA barcoding is crucial for advancing science, particularly in the areas of taxonomy and biotechnology expertise [34]. Based on the Cytochrome Oxidase Subunit I (COI) gene, *Betta foerschi*'s DNA Barcode is the first barcode filed in Genbank and the first report from Indonesia. As part of the Barcode of Life (BOL) project, this sequence also intends to catalog the biodiversity of eukaryotes. The genetic samples of species are also useful for understanding the population structure where the populations of several native and endemic fishes are declining on a global scale lately caused by the presence of invasive and non-native species [35][36][37][38][39].

Based on this results of *Betta foerschi* sequence analysis based on the COI gene from Belitung Island, it was found that the DNA Barcode of *Betta foerschi* had high average amounts of adenine and thymine called A-T rich group (A-T rich) around 60% and C-T Group around 40%. The A-T hydrogen bond consists of 2 hydrogen bonds which are weaker than the G-C hydrogen bond which has 3 hydrogen bonds, so the possibility of species mutation is higher [43]. The complete percentage of nucleotide composition of *Betta foerschi* was T (28.6%), C (26.8%), A (25.59%), and G (18.62%).

3.2. Molecular Identification

The DNA Barcoding of the sequence from the Belitung Island was analyzed and compared to the NCBI GenBank (National Center for Biotechnology Information) via BLAST (Basic Local Alignment Search Tool-nucleotide) method (https://blast.ncbi.nlm.nih.gov) and the BOLD SYSTEM via Identification Engine to find out and analyze a sequence homology (**Table 2**).

Table 2. The species identification and similarity *Betta foerschi* from Belitung Island, Indonesia

Species Outcome	Family	Accession ID	Query Coverage (%)	Percent Identity (%)
B. persephone	**Osphronemidae**	**KM485407.1**	**99**	**86.68**
B. hipposideros	Osphronemidae	KM485457.1	100	86.16
B. pulchra	Osphronemidae	KM485449.1	100	86.02
B. anabantoides	Osphronemidae	GQ911723.1	100	85.69
B. waseri	Osphronemidae	KM485458.1	100	85.69
B. prima	Osphronemidae	KM485329.1	100	85.67
B.palida	Osphronemidae	KM485429.1	100	85.50
B. enisae	Osphronemidae	KM485402.1	100	8581

Based on the results of the analysis of BLAST DNA sequence from Belitung Island was in a group to Betta genus with a Quary cover 99-100% and level of similarity of about 86.68% and an E-value 0.0. According to Hebert et al [40], species with 97-100% similarity levels are identical. So that species differences above 3% based on COI genes lead to new species. In fact, our DNA has no species resemblance to the GenBank database. Nonetheless, our species is *Betta foerschi* morphologically, but the GenBank database does not contain the COI *Betta foerschi* gene sequence. Furthermore, in order to facilitate future identification, we have registered our sequences in the GenBank Database under the accession code OR192539.

3.3. Molecular Identification

Additionally, to illustrate the family's evolutionary history, we also create a phylogenetic tree in this work using the Neighbor-Joining method based on the Tamura-Nei model with MEGA X and 1000 replicates (**Figure 2**). We also took the sequences of betta (21 species and 10 groups) from the Genbank to complete the phylogenetic tree based on the COI gene.

As can be seen, the phylogenetic tree of *Betta foerschi*was clustered in a clade in the *Betta*. This supports our morphological analysis, which indicates that our specimen is a member of the genus Betta. However, to clarify the connection of Betta species, we strongly encourage registering the COI gene of each Betta species, especially for the 13 Betta groups with 54 species dispersed across western Indonesia. By using various sequences from the Genbak database, we discovered that *Betta foerschi* is genetically closer to *Betta presephone* than the other sequences, with a genetic distance of 0.1. This number, however, does not represent similarities within a species, but rather similarities within a genus.

Table 3. Betta genus estimates of evolutionary divergence based on the COI gene and the maximum composite likelihood model

		1	2	3	4	5	6	7	8
1	***Betta foerschi***								
2	*Betta persephone*	0,16							
3	*Betta hipposideros*	0,16	0,19						
4	*Betta prima*	0,17	0,21	0,13					
5	*Betta ferox*	0,17	0,20	0,10	0,18				
6	*Betta enisae*	0,17	0,18	0,12	0,16	0,13			
7	*Betta pulchra*	0,16	0,20	0,11	0,16	0,13	0,14		
8	*Betta waseri*	0,17	0,19	0,03	0,12	0,10	0,12	0,11	

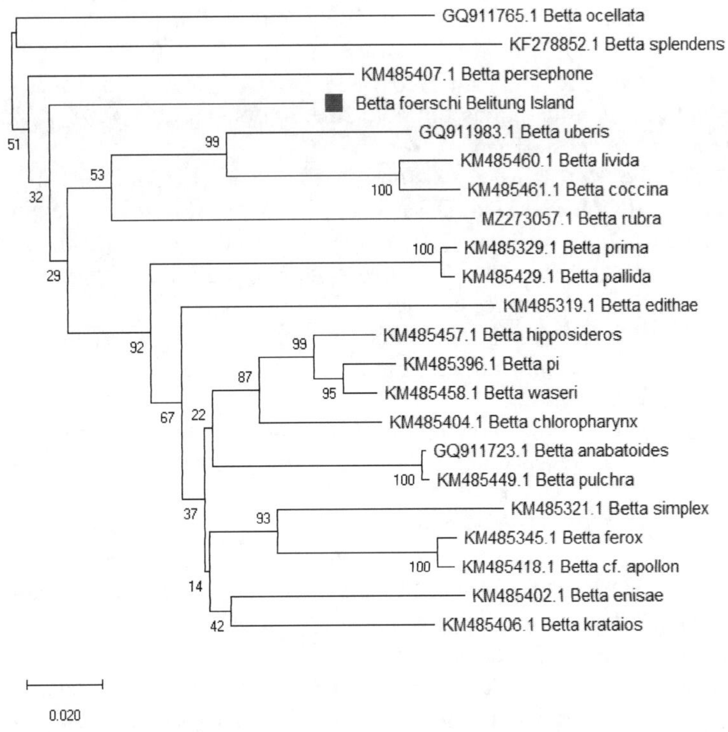

Figure 2. The evolutionary tree of based on COI of *Betta foerschi* from Belitung Island, Indonesia

4. Conclusion

This study recorded the first DNA Barcoding of *Betta foerschi* from the Swamp Forest Stream of Limbongan village, Gantong district, East Belitung Regency, Belitung Island, Indonesia. The DNA barcoding of *Betta foerschi* from Belitung Island yielded a 667 base-pair long sequence of gen COI. The 667 bp COI gene sequence satisfied the requirements for animal differentiation and identification. The study confirmed that the COI gene accurately identifies species, as the sequence belonged to the Betta genus consistent with morphology-based studies. The DNA barcode of *Betta foerschi* from Belitung Island was the first genetic record for Belitung and Indonesia. The genetic information obtained from this research has significant implications for enriching scientific knowledge, especially in the fields of taxonomy and biotechnology. It contributes to the registration of eukaryotic biodiversity and supports the Barcode of Life project. The phylogenetic analysis revealed that *Betta foerschi* from Belitung Island formed a clade with Betta genera. This study supports the morphological identification of species. These findings have important implications for conservation efforts and the management of fish populations, including the selection of prospective parents for domestication programs and determining the number of species populations based on genetic diversity.

References

[1] Hasan V, South J 2023 First record of the Vulnerable Malayan Flat-shelled Turtle, *Notochelys platynota* (Gray 1834) (Reptilia, Testudines, Geoemydidae), in Belitung, Indonesia *Check List* **19** 301-304

[2] Kusumah W, Hasan V, Samitra D 2023 Rediscovery of the Billiton Caecilian, *Ichthyophis billitonensis* Taylor, 1965 on Belitung Island, Indonesia, after more than five decades *Herpetol Notes* **16** 95-97

[3] Kurniawan A, Kurniawan A, Hariati, A M, Kurniawan, N, Nugroho, T W, Bidayani, E, Syarif, A F, Wiadnya D G R 2022 Genetic diversity analysis and phylogeography of *Osteochilus spilurus* (Cyprinidae: Labeoninae) from Bangka, Belitung, and Kalimantan Islands using Cytochrome b gene *Biodiversitas Journal of Biological Diversity* **23** 4738-4746

[4] Bennett R H, Ellender B R, Mäkinen T, Miya T, Pattrick P, Wasserman R J, Woodford D J, Weyl O L F 2016 Ethical considerations for field research on fishes *Koedoe-African Protected Area Conservation and Science* **58** 1353

[5] Syarif A F, Tiandho Y, Robin R, Gustomi A 2020 Karakter Morfometrik Ikan Tepalak (Wild Betta) Asal Pulau Belitung Sebagai Dasar Pengembangan Akuakultur. Seminar Nasional Biologi "Inovasi Penelitian dan Pembelajaran Biologi IV (IP2B IV) **1** 23-27 [In Indonesian]

[6] Schindler I, Linke H 2013 *Betta hendra*-a new species of fighting fish (Teleostei: Osphronemidae) from Kalimantan Tengah (Borneo, Indonesia). *Vertebr Zool* **63** 35-40

[7] Kottelat M 2013 The fishes of the inland waters of southeast Asia: a catalogue and core bibliography of the fishes known to occur in freshwaters, mangroves and estuaries *Raffles Bulletin of Zoology Supplement* **27** 1-663

[8] Britz R 2001 The genus Betta – monophyly and intrarelationships, with remarks on the subfamilies Macropodinae and Luciocephalinae (Teleostei: Osphronemidae) *Ichthyological Exploration of Freshwaters* **12** 305-318

[9] Tan H, Ng P K L 2005a The Fighting Fishes (Teleostei: Osphronemidae: Genus Betta) Of Singapore, Malaysia And Brunei *Raffles Bulletin of Zoology* **13** 43-99

[10] Tan H, Ng P K L 2005b The Labyrinth Fishes (Teleostei: Anabantoidei, Channoidei) Of Sumatra, Indonesia *Raffles Bulletin of Zoology* **13**115-138

[11] Kottelat M, Ng P K L 1994 Diagnoses of five new species of fighting fishes from Banka and Borneo (Teleostei: Belontiidae) *Ichthyological Exploration of Freshwaters* **5** 65- 78

[12] Syarif A F, Tiandho Y, Robin R, Gustomi A 2021 Perbandingan pola rasio morfometrik dan karakteristik habitat dua spesies ikan wild betta asal pulau belitung *Bioscientist : Jurnal Ilmiah Biologi* **9** 20-28 [In Indonesian]

[13] Low, B W 2019 *Betta foerschi. The IUCN Red List of Threatened Species* 2019: e.T91309890A91309895. https://dx.doi.org/10.2305/IUCN.UK.2019-2.RLTS.T91309890A91309895.en. Accessed on 28 June 2023
[14] Vierke J 1979 *Betta anabatoides* und *Betta foerschi* spec. nov., zwei Kampffische aus Borneo *Aquarium Aqua Terra* **13** 386-388
[15] Liu K, Zhao S, Yu Z, Zhou Y, Yang J, Zhao R, Yang C, Ma W, Wang X, Feng M, Tang Y, Li K, Zhou C 2020 Application of DNA barcoding in fish identification of supermarkets in Henan province, China: More and longer COI gene sequences were obtained by designing new primers *Food Res Int* **136**:109-516
[16] Bingpeng X, Heshan L, Zhilan Z, Chunguang W, Yanguo W, Jianjun W 2018 DNA barcoding for identification of fish species in the Taiwan Strait *PLoS One* **13** 198109
[17] Robin R, Valen F S, Ramadhanu D, Nomleni A, Turnip G, Insani L 2023a A new distributional record of flying barb, *Esomus metallicus* (Actinopterygii: Cyprinidae), from Kapalo Banda River, West Sumatra, Indonesia *International Journal of Aquatic Biology* **11**: 59-69
[18] Valen F S, Widodo M S, Islamy R A, Wicaksono K P, Soemarno S, Insani L, Hasan V 2022a Molecular phylogenetic of silver barb *barbonymus gonionotus* (bleeker, 1849) (cypriniformes: Cyprinidae) in Java, Indonesia *IOP Conference Series: Earth and Environmental Science* **1036** 012011
[19] Robin R, Insani L, Swarlanda S, Prananda M, Valen F S 2023b Range extension of Spanner barbs, *Barbodes lateristriga* (Valenciennes, 1842) (Cypriniformes: Cyprinidae) to Bangka Island, Indonesia *Iranian Journal of Ichthyology* **9** 149–157
[20] Insani L, Jatayu D, Valen F S, Widodo M S, Hasan V 2023 Comparing genetic *Mystacoleucus marginatus* and *Mystacoleucus padangensis* (Cypriniformes: Cyprinidae) based on *Cythochrome C Oxidase sub unit I* (COI) gene *Iranian Journal of Ichthyology* **9** 195–203
[21] Hasan V, Valen F S, Islami R A, Widodo M S, Saptadjaja A M, Islam I 2021 Short Communication: Presence of the vulnerable freshwater goby *Sicyopus auxilimentus* (Gobiidae, Sicydiinae) on Sangihe Island, Indonesia *Biodiversitas* **22** 573-581
[22] Hasan V, Andraini N E, Isroni W, Sari L A, Nafisyah A L, Dewi N N, Putri D N A, Prasasti T A B, Ramadhani A A, Daniel K, South J, Vieira L O, Ottoni F P, Maftuch M, Faqih A R, Wirabuana P Y A P, Tamam M B, Valen F S 2023 Fish diversity of the Bengawan Solo River estuary, East Java, Indonesia *Biodiversitas* **24** 2207-2216
[23] Metcalfe J D, Craig J F 2011 Ethical justification for the use and treatment of fishes in research: An update *Journal of Fish Biology* **78** 393-394
[24] Valen F S, Hasan V, Ottoni F P, Nafisyah A L, Erwinda M, Annisa A N 2022b Description of Silver Barb *Barbonymus gonionotus* (Bleeker, 1849) (Cypriniformes: Cyprinidae) from Madura Island, Indonesia *IOP Conference Series: Earth and Environmental Science* **1036** 012066
[25] Valen F S, Soemarno S, Widodo M S, Wiadnya D G R, Hasan V 2020 Contemporary distribution records of yellow finned barb *mystacoleucus marginatus* (Valenciennes, 1842) in brantas basin, Indonesia. *Ecology, Environment and Conservation* **26** S40-S43
[26] Ward R D, Zemlak T S, Innes B H, Last P, Hebert P D N 2005 DNA Barcoding Australia's Fish Species. *Phil Trans R Soc B* **360** 1847-1857
[27] Aisyah S, Thania O S, Syarif A F 2021 Identifikasi molekuler sirip ikan hiu menggunakan gen mitokondria *Cytochrome C Oxydase Subunit I* (COI) yang didaratkan di pesisir kabupaten bangka selatan. *Jurnal Enggano* **6** 99-109 [In Indonesian]
[28] Valen F S, Sambah A B, Wicaksono K P, Widodo M S, Soemarno S, Hasan V 2021 Genetic diversity of Yellow Finnedbarb *Mystacoleucus marginatus* (Valenciennes, 1842) (Teleostei, Cyprinidea) in Brantas basin Upstream, Indonesia *Eco. Env. & Cons* **27** 695-699
[29] Saitou N, Nei M 1987 The neighbor-joining method: A new method for reconstructing phylogenetic trees *Molecular Biology and Evolution* **4** 406-425
[30] Felsenstein J 1985 Confidence limits on phylogenies: An approach using the bootstrap. *Evolution* **39** 783-791

[31] Tamura K, Nei M, Kumar S 2004 Prospects for inferring very large phylogenies by using the neighbor-joining method *Proceedings of the National Academy of Sciences (USA)* **101** 11030-11035

[32] Kumar S, Stecher G, Li M, Knyaz C, Tamura K 2018 MEGA X: Molecular Evolutionary Genetics Analysis across computing platforms *Molecular Biology and Evolution* **35** 1547-1549

[33] Guo M, Yuan C, Tao L, Cai Y, Zhang W 2022 Life barcoded by DNA barcodes *Conservation Genet Resour* **14** 351–365

[34] Tadmor-Levi R, Borovski T, Marcos-Hadad E, Shapiro J, Hulata G, Golani D, David L 2022 Establishing and using a genetic database for resolving the identification of fish species in the Sea of Galilee, Israel *PLoS One* **17** e0267021

[35] Valen FS, Hasan V, Ottoni F P, Nafisyah A L, Erwinda M, Annisa A N, Adis MA 2022c First country record of the bearded gudgeon *Pogoneleotris heterolepis* (Günther, 1869) (Teleostei: Eleotridae) from Indonesia. *IOP Conference Series: Earth and Environmental Science* **1036** 012074

[36] Robin R,Valen F S, Nomleni A, Turnip G, Luhulima M Y, Insani L 2023c Presence of non-native freshwater fish in Indonesia: A review - Risk and ecological impacts *AACL Bioflux* **16** 66-79

[37] Ihwan I, Pratama F S, Yonarta D, Faqih A R, Widodo M S, Valen F S, Tamam M B, Hasan V 2020 Presence of asian catfish *Clarias batrachus* (Siluriformes, Clariidae) in Madura Island, Indonesia *AACL Bioflux* **13** 958-962

[38] Serdiati N, Insani L, Safir M, Rukka A H, Mangitung S F, Valen F S, Tamam M B, Hasan V 2021 Range expansion of the Invasive Nile Tilapia Oreochromis niloticus (Perciformes: Cichlidae) in Sulawesi Sea and first record for Sangihe Island, Tahuna, North Sulawesi, Indonesia *Ecology Environment and Conservation* **27** 168-171

[39] Valen F S, Sambah A B, Wicaksono K P, Widodo M S, Soemarno S, Hasan V 2021 Genetic diversity of Yellow Finnedbarb *Mystacoleucus marginatus* (Valenciennes, 1842) (Teleostei, Cyprinidea) in Brantas basin Upstream, Indonesia *Eco. Env. & Cons* **27** 695-699

[40] Hebert P D N, Cywinska A, Ball S L, deWaard J R 2003 Biological identifications through DNA barcodes *Proceedings of the Royal Society of London* **270** 313–321

Acknowledgments

The authors express their gratitude to the Directorate of Research, Technology and Community Services (DRTPM) of the Ministry of Education, Culture, Research, And Technology, Republic of Indonesia in research grant "Penelitian Fundamental Reguler" financial years in 2023 (Contract No: 116/E5/PG.02.00.PL/2023 & 1275.A/UN50/L/PP/2023) for their support in facilitating the research and publication process.

A new distributional record of spanner barbs, *Striuntius lateristriga* (Valenciennes, 1842) (Cypriniformes: Cyprinidae) in Belitung Island, Indonesia

AF Syarif[1]*, AM Hafidz[1], FS Valen[1], M Herjayanto[2], R Robin[1]

[1] Department of Aquaculture, Faculty of Agricultural Fisheries and Biology, University of Bangka Belitung, Indonesia
[2] Department of Fisheries Science, Faculty of Agriculture, University of Sultan Ageng Tirtayasa, Indonesia

*Corresponding author: ahmadfahrulsyarif@gmail.com

Abstract. The Spanner Barb *Striuntius lateristriga* is a freshwater species from the Cyprinidae family that spread in Southern Thailand; the Malay Peninsula, Singapore and Indonesia. The previous presence of the Spanner barb in Indonesia was recorded in West Kalimantan and Bangka Island. This study was carried out to record for the first time the presence of *S. lateristriga* on Belitung Island, Indonesia. The presence of *S. lateristriga* in the Belitung archipelago adds knowledge about the new distribution and biogeography of freshwater species. Moreover, information on this species' new distribution is critical for supporting suitable conservation policies and measures. The twenty *S. lateristriga* specimens were collected on January 20, 2023 utilizing a fish trap at the Gurok Beraye waterfall, Gunung Tajam, Belitung Island, Indonesia. This specimen was identified based on specific literature. The new record of *S. lateristriga* provided here is the southernmost record for this species, expanding its geographic distributing. In addition, the new record site is 250 km southeast of the nearest locality in Bangka Island, and about 500 km south-west from the nearest locality in Borneo. Futrhermore, understanding biodiversity, range extension, and biogeography of the species in Sundaland will enrich knowledge in science.

Keywords: *Striuntius lateristriga*, Distribution, Freshwater fishes, Sundaland, Belitung Island

1. Introduction

The Spanner Barb *Striuntius lateristriga* Valenciennes, 1842 is a freshwater species native to Southeast Asia such as Southern Thailand; the Malay Peninsula, Singapore and Indonesia [1-6]. In Indonesia, *S. lateristriga* was recorded in West Kalimantan and Bangka Island [7][8]. In Bangka Island, this species is frequently found in the mountain, below waterfalls, with streams strewn, with rocks and boulders, this species lives as a benthopelagic and feeds on worms, crustaceans, insects, and plant matter in the tropical freshwater [9]. After the previous record, we believe that we can find *S. lateristriga* in other Indonesian archipelagos. Thus, In January 2023, we eventually discovered other record for *S. lateristriga* from Belitung Island. This new record will increase the information of extend geographic distribution of *S. lateristriga* to another island of the Indonesian [10-13].

Indonesia has an remote island called Belitung Island that extends 226 kilometers to the east of Bangka Island. In fact, the existence of freshwater species in the Belitung Islands is quite threatened due to the destruction of their natural habitat and environment due to the negative impacts of open-pit tin mining. [14]. Furthermore, the presence of invasive fish such as Nila Tilapia, Red devils and Redclaw crayfish is also a threat to *S. lateristriga* which causes competition for resources, predation, habitat modification, and disease transmission [15-18]. The discovery of a new distribution of *S. lateristriga* in the Belitung Islands provides basic information for scientific progress, especially as a basis for species management, domestication, and conservation [19][20].

2. Materials and Methods

2.1. The Sampling Sites and Samples Collection

The study was conducted on January 2023 in Gurok Beraye Waterfall, Gunung Tajam, Belitung Island, Indonesia. The twenty specimens were caught using a fish traps [22] and scoop net [23] during sampling activity. The six samples were maintained as livestock at the Fish Reproduction Laboratory, Bangka Belitung University, Indonesia. Two specimens were fixed in 7% formalin and deposited at the Aquaculture Laboratory of University of Bangka Belitung, Indonesia. Two specimen was preserved in 90% alcohol for further analysis and the other specimens were released to their natural habitat. The fish caught do not require special permits because they are in the Least Concern (LC) status based on the list on the IUCN Redlist [30].

2.2. Morphological Identification

The morphological analysis consists of meristic and morphological characters of *Striuntius lateristriga* following Lim et al [5], Sutin et al [6], Robin et al. [7] and Alfred [21].

3. Results and Discussions

3.1. New Record of Striuntus lateristriga

Last records: Indonesia: Bangka Belitung Islands Province; **New Record**: Indonesia: Belitung Island: Gurok Beraye Waterfall, Gunung Tajam; 20th January 2023, twenty specimens, TL 7,5 cm and 8.5 cm, and the photograph (**Figure 1**).

Figure 1. Specimen of *Striuntius lateristriga,* from Gurok Beraye Waterfall, Gunung Tajam, Belitung Island, Indonesia (Photograph by AM Hafidz)

3.2. Morphological Profile

The morphological analysis consists of meristic and morphological characters. In addition, the meristic characters of *S. lateristriga* are given in **Table 1**. Other specific morphological characters of *S.*

lateristriga are as follows: The body is compressed; mouth terminal; teeth villiform; scale ctenoid; tail forked; stomach J-shape; maxillary barbels present; eye moderately large; rostral barbels present; 24-25 scales on lateral line row on the body; the general body color is yellowish olive, with a faint reddish tint. Vertical stripe starting below the dorsal fin; black patch above the anal fin; average total length 7-8 cm. In this study, we found *S. lateristriga* from Belitung Island is smaller than the specimen from Bangka Island, whereas about 7.5-8.5 cm [7]. however, the specimens from Belitung is bigger than the specimen from Singapore which is around ± 3.2 [6].

Table 1. Meristics of *S. lateristriga* from Gurok Beraye Waterfall, Gunung Tajam, Belitung Island, Indonesia (n=20 specimens)

Meristics (counts)	Present Study	Robin et al., 2023	Alfred, 1966
Pored Lateral line scales	24	24-25	22-24
Dorsal fin rays	IV.8	II.8	IV.8
Anal fin rays	iii.5	iii.6-7	iii.5-6
Pectoral fin rays	i.12	i.12-13	i.12-13
Ventral fin rays	i.7	i.7-8	i.8

In general, the meristic calculations for *S. lateristriga* between Belitung Island and Alfred 1966 are nearly same, except for a tiny discrepancy in the number of ventral fin rays, which is 7 in the Belitung specimen and 8 based on Alfred 1966. A remarkable disparity in the number of Dorsal fin rays and Anal fin Rays is observed in Robin et al 2023, which is likely owing to changes that occur between species due to various environmental stressors and geographic isolation. Belitung Island is influenced by open-pit tin mining, so the environmental stress that occurs may play a role in changing the shape and number of meristics in fish.

3.3. Presence Record and Their Habitats

The discovery of *Striuntius lateristriga* in Gurok Beraye Waterfall, Gunung Tajam, Belitung Island, Indonesia is a new record that represents a range extension of about 226 km to the east from Bangka Island (**Figure 2**). The publication of new fish records improves the extent of the species' recorded distribution, which enriches the study of species diversity and biogeography [26]. The previous presence of the Spanner barb in Indonesia was recorded in West Kalimantan [27][28] and Bangka Island [7] Indonesia. Moreover, this species was also previously found in the Malay Peninsula, Singapore and Southern Thailand [1-6]

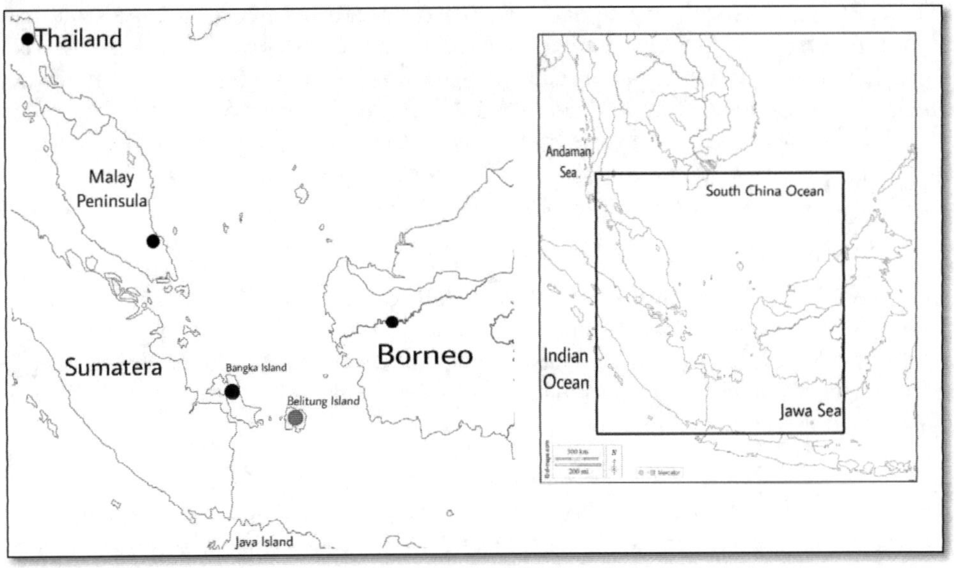

Figure 2. Distribution of *S. lateristriga* in South East Asian Regions. Black circles are the previous record. Red circle is the recent record from Belitung Island, Indonesia

The Malay Peninsula, Singapore, Southeast Thailand, West Kalimantan, and the Bangka Islands also consist of *S. lateristriga*, indicating that these regions were formerly connected to Sundaland, an ancient river that existed thousands of years ago. The Malay Peninsula, Sumatra, Java, and the Borneo Islands are all part of Sundaland, which has a total area of over 1.800.000 km^2 [29].

Figure 3. The location of Gurok Beraye Waterfall, Belitung Island, Indonesia

S. lateristriga was discovered on Belitung Island by Gurok Beraye Waterfall of Gunung Tajam, Belitung Island, Indonesia (**Figure 3**). The water was clear and the bottom of the waterfall hill was rocky. This ecosystem was still pristine and far from residential areas. He had the same habitat as *Barbodes sellifer.* Habitats of *S. lateristriga* have also been recorded in tropical freshwater, under waterfalls, and in clear mountain streams, usually strewn with rocks and debris [6,7]. Some species there are *Channa lucius, Channa striata, Sphaerichthys osphromenoides, Silurichthys indragiriensis, Rasbora chepalottenia, Hamibagrus nemurus, Cryptopterus macrocephalus, Pristolepis fasciata, Hamirhampodon pogonatus, Macrobrachium sp, Parathelphusa maculate, Anemaththys apogon,* and *Osteochilus haseltii* were also found in the same habitat of *Striuntius lateristriga* during the sampling.

4. Conclusion

Specimens were collected from Gurok Beraye Waterfall, Gunung Tajam, Belitung Island, Indonesia valid as *Striuntius lateristriga* based on morphological approach. After previously being discovered in Singapore, Borneo, southern Thailand, the Malay Peninsula, and Bangka Island, *S. lateristriga* has now been discovered in Belitung Island. Information on a new record of species is needed to increase knowledge of the distribution of a species.

References

[1] Q. Ren, L. Yang, C. H. Chang, R. L. Mayden, Zoologica Scripta. **49**(6), 697-709 (2020).
[2] C. K. C. Ng, T. Y. Lim, A. B. Ahmad, MdZ. Khaironizam, Zootaxa. **4567**(3), 515-545 (2019).

[3] M. R. A. Halim, M. I. N. Lokman, D. M. Belabut, R. Ramli, M. S. Azirun, K. M. Zain, Malayan Nature Journal. **70**(4), 409-420 (2018).
[4] M. Aqmal-Naser, B.A. Amirrudin, Journal of Wildlife and Parks. **33**, 95-106 (2018).
[5] T. Lim, C. K. Chay, W. H. Lim, Y. Cai, Singapore. 56-84 (2016).
[6] S. Sutin, M. Pollar, M. Jaroensutasinee, Proceedings Of World Academy Of Science. Engineering And Technology. **26**, 1307-6884 (2007).
[7] R. Robin, L. Insani, Swarlanda, M. Prananda, F. S. Valen, Iranian Journal of Ichthyology, **9**(3), 149–157 (2023).
[8] M. Kottelat, E. Widjanarti, Raffles Bulletin of Zoology Suppl. **13**, 139-173 (2005).
[9] R. Robin, F. S. Valen, A. Nomleni, G. Turnip, M. Y. Lahulima, L. Insani, AACL Bioflux. **16**(1), 66-79 (2023).
[10] R. Robin, F.S. Valen, D. Ramadhanu, A. Nomleni, G. Turnip, L. Insani, International Journal of Aquatic Biology. **11**(1) (2023).
[11] F. S. Valen, V. Hasan, F. P. Ottoni, A. L. Nafisyah, M. Erwinda, A. N. Annisa, M. A. Adis, IOP Conference Series: Earth and Environmental Science. **1036**(1), 012-074 (2022).
[12] V. Hasan, F. S. Valen, R. A. Islami, M. S. Widodo, A. M. Saptadjaja, I. Islam, Biodiversitas. **22**(2), 573-581 (2021).
[13] I. Ihwan, F. S. Pratama, D. Yonarta, A. R. Faqih, M. S. Widodo, F. S. Valen, M. B. Tamam, V. Hasan, AACL Bioflux. **13**(2), 958-962 (2020).
[14] W. Kusumah, V. Hasan, D. Samitra, Herpetology Notes. **16**, 95-97 (2023).
[15] L. Insani, V. Hasan, F. S. Valen, F. S. Pratama, M. S. Widodo, A. R. Faqih, R. A. Islamy, A. T. Mukti, W. Isroni, Ecology Environment and Conservation. **26**(3), 1115-1118 (2020).
[16] V. Hasan, M. S. Widodo, A. R. Faqih, G. Mahasri, M. Arief, F. S. Valen, M. B. Tamam, D. Yonarta, F. S. Pratama, R. Fitriadi, Ecology, Environment and Conservation, **26**(February Suppl. Issue). S73-S75 (2020).
[17] N. Serdiati, D. Yonarta, F. S. Pratama, A. R. Faqih, F. S. Valen, M. B. Tamam, Y. I. G. Hamzah, V. Hasan, AACL Bioflux. **13**(1), 137-141 (2020).
[18] S. K. Bariyyah, S. M. Saleh, L. Insani, N. Seridati, F. S. Valen, Ecology, Environment and Conservation. **27** (November Suppl. Issue), S272-S275 (2021).
[19] V. Hasan, F. P. Ottoni, J. South, Check List. **19**(1), 51-55 (2023).
[20] F. S. Valen, V. Hasan, Ottoni, F. P. Ottoni, A. L. Nafisyah, M. Erwinda, A. N. Annisa,, IOP Conference Series: Earth and Environmental Science. **1036**(1), 012-066 (2022).
[21] E. R. Alfred, Zoologische verhandelingen. **78**, 1-68 (1966).
[22] F. S. Valen, Soemarno, M. S. Widodo, D. G. R. Wiadnya, V. Hasan (2020), Ecology, Environment and Conservation. **26**, S40-S43 (2020).
[23] N. Serdiati, L. Insani, M. Safir, A. H. Rukka, S. F. Mangitung, F. S. Valen, M.B. Tamam, V. Hasan, Ecology Environment and Conservation. **27** (1), 168-171 (2021).
[24] L. Insani, D. Jatayu, F. S. Valen, M. S. Widodo, V. Hasan, Iranian Journal of Ichthyology. **9**(4), 195–203 (2023).
[25] Lutfiatunnisa, M. S. Widodo, A. R. Faqih, L. D. Sari, F. S. Valen, AACL Bioflux, **13**(5), 2828-2834 (2020).
[26] F. S. Valen, V. Hasan, F. P. Ottoni, A. L. Nafisyah, M. Erwindah, A. N. Annisah, M. A. Adis, IOP Conference Series: Earth and Environmental Science, **1036**(1), 012-074 (2022).
[27] I. Rachmatika, Treubia, Museum Zoologicum Bogoriense, **33**(2), 181-190 (2004).
[28] Roberts T. R., Mem. Calif. Acad. Sci, **14**:210 (1984).
[29] H. H. Ng, M. Kottelat, Zootaxa, **8**(1): 4188 (2016).
[30] Lumbantobing D. 2019. *Barbodes lateristriga*. https://dx.doi.org/10.2305/IUCN.UK.2019-2.RLTS.T180977A89799617.en. Accessed on 28 November 2023.

Acknowledgments
The authors are grateful to the Institute of Research and Community Service (LPPM) University of Bangka Belitung for the fundings of "Penelitian Akselerasi Lektor ke Lektor Kepala" in 2021 and Directorate of Research, Technology and Community Services (DRTPM) of the Ministry of Education, Culture, Research, And Technology, Republic of Indonesia in research grant "Penelitian Fundamental Reguler" financial years in 2023 for their support in facilitating the research and publication process.

Sustainable marine ecotourism governance based on tourism preferences

Atikah Nurhayati[1],Titin Herawati[*], Asep Agus Handaka [*],Wahyuniar Pamungkas*, [*]Nora Akbarsyah [*], Agnes Sudarmono [2]

[1*] Faculty of Fisheries and Marine Science, Universitas Padjadjaran
Jl. Bandung- Sumedang KM.21, Hegarmanah, District. Jatinangor, Sumedang Regency, West Java Province- Indonesia 45363

[2.] Study Program of Master of Fisheries Management, Open University
Jalan Cabe Raya, Pondok Cabe, Pamulang, Tangerang Selatan 15437, Banten – Indonesia

Corresponding author e-mail: atikah.nurhayati@unpad.ac.id

Abstract. The utilization of coastal areas and small islands in Indonesia is for marine ecotourism. Marine tourism areas need to be managed and regulate marine tourism activities that form interconnectivity between sectors including the conservation of fisheries and marine resources. Based on landscape conditions (bio-ecoregion) and tourist preferences, choosing marine tourism destinations. This research aims to analyze sustainable marine ecotourism governance based on tourism preferences. the research was conducted from August 2022 - to February 2023 using primary and secondary data. The technique for taking respondents used a purposive sampling of 100 respondents. The analytical tool used to analyze the factors influencing tourists to visit marine tourism destinations uses the Partial Least Square – Structural Equation Modelling (PLS-SEM) method. Based on the results of research on factors that influence tourists' decision-making in choosing marine tourism destinations, namely: income (X_1), level of education (X_2), distance traveled to tourist destinations (X_3), accessibility (X_4), accommodation (X_5), infrastructure (X_6), information from social media (X_7), security and safety (X_8). Governance is carried out by considering social-ecological system aspects based on coastal area utilization zones, but not all of them are implemented properly, so collaboration between stakeholders is needed to support the sustainability of marine tourism.

1. Introduction

Indonesia is the largest archipelago in the world and has the potential to develop natural resource-based marine tourism areas. Marine tourism is one segment of tourism that is growing worldwide, including in West Java Province, Indonesia. Marine tourism has great potential in Indonesia, as an archipelago country, as well as the potential for marine tourism attractions to attract tourists. The United Nations World Tourism Organization (UNWTO), addressed the development of marine tourism related to SDG-14, which ensures benefits for host countries from tourism activities through the sustainable use of marine tourism resources[1]. Marine tourism is part of the management of coastal areas. The coastal area is a transitional area between affected terrestrial and coastal ecosystems by the tides [2,3,4,5]. Coastal areas can make a large contribution to the community's economy through fisheries, forestry, industry, tourism, mining, and other sectors. Marine tourism plans usually focus on zoning marine spatial use, fisheries and marine resource conservation zones, types of tourist attractions, infrastructure, and the number of tourist visits, but recently practitioners have included environmental considerations, local wisdom, community empowerment, and socio-economics. [5,6,7]. The marine tourism sector also offers many opportunities to use nature-based solutions.

Besides marine tourism, nature tours provide education on the conservation of natural resources, usually called ecotourism. Ecotourism is a tourist trip to the natural and artificial environment and existing culture that is informative and participatory which aims to ensure the conservation of natural resources and local wisdom [5]. Ecotourism pays attention to the following aspects: (1) able to contribute to conservation activities and maintain biodiversity; (2) able to improve the welfare of the local population; (3) tourists who come to gain experience and knowledge; (4) the participation of the local community to play an active role in the tourism activities being developed [8,9,10]. Marine ecotourism is a special interest tourism activity whose activities are related to marine activities, both carried out under the sea and above sea level, under the sea, and in coastal areas, including sports and cultural attractions. [5,11]. Marine ecotourism is tourism based on marine attractions in coastal areas that are dominated by waters or the sea.

The marine ecotourism sector is a complex network of various interests and activities that operate at different ranges of space, time, scale, and intensity in coastal areas, such as in West Java Province, Indonesia [5,12]. The potential for marine ecotourism in West Java province is located in the southern coastal area of West Jawa Province. Marine Ecotourism, provides natural resource products for marine ecosystems in the form of mangrove forests, marine parks, as well as various fauna, both in the sea and around the coast. For example, mangroves can reduce the occurrence of erosion due to sea tides, a place for a diversity of aquatic and non-aquatic biota. The functions of the mangrove ecosystem are ecological and economic functions. The ecological function of mangroves is to protect the coastline and prevent seawater intrusion, while the economic function of mangrove forests is to produce household needs, industrial needs, and ecotourism.

The marine ecotourism sector can help local communities improve their livelihoods through diversification work. It can create employment opportunities through small-scale medium enterprises, improve the capacity of soft-skill and hard-skill local communities, develop local supply chains and products, including promote local cultural heritages or local wisdom [13,14]. Marine tourism is a recreational activity that involves traveling away from one's place of residence and which has its host or focusing on the fisheries and marine resources [5,15,16]. The coastal area is a growth area for the marine tourism industry. Marine tourism is an area-based activity, which is carried out along the coastline of the archipelago while diving, snorkeling, and community cultural heritage are place-based activities. Marine tourism activities include viewing the sunrise and sunset, fishing tours, surfing, boating, diving, watching dolphins, cruising, tracking, and many more activities. Marine tourism areas need to be managed to manage and regulate marine tourism activities that form inter-connectivity between sectors including the conservation of fisheries and marine resources.

The scope of marine ecotourism marketing includes market research, tourism product planning, promotion of tourist destinations, and providing information about where tourists can get marine tourism products or services. Tourist behavior is a key support of all marine ecotourism marketing activities carried out for the development, promotion, and sale of tourism products and processes to learn why people buy the products they buy and how to make decisions choosing marine ecotourism. [5,17]. Tourist preferences show consumer preferences from a wide selection of marine tourism products including marine ecotourism

Marine ecotourism can contribute to the social economic and ecological system of coastal communities. It creates awareness and support for conversations about fisheries resources. The main challenge for marine tourism is to preserve fisheries and marine resources by preserving marine ecosystems, reducing marine pollution, using marine resources in a sustainable manner, and being able to increase income for coastal communities. it is necessary to manage visitor marine ecotourism characteristics and their preferences in comparison with the availability of tourist attractions, activities, and facilities. Seeing the potential of coastal resources that are still abundant along with the prospect of developing development needs and the introduction of coastal resource-based tourist destinations, this study aims to analyze sustainable marine ecotourism governance based on tourism preferences

2. Material and Methods

This research uses a descriptive method, which collects information about the status of the existing indications. This research was conducted in West Java Province. The method used in this research is the survey method. This research was conducted from August 2022 - February 2023 using primary and secondary data. The technique for taking respondents used a purposive sampling of 100 respondents. The analytical tool used to analyze the factors influencing tourists to visit marine tourism destinations uses the Partial Least Square – Structural Equation Modelling (PLS-SEM) method

Partial least squares (PLS) analysis is an alternative to ordinal least square regression, canonical correlation, or covariance-based structural equation modelling (SEM) of systems of independent and response variables. These systems are commonly called composite-based SEM which is the usual type. [18]. The advantages of PLS include the ability to model multiple, the ability to handle multicollinearity among the independents, robustness in the face of data noise and data loss and the creation of independent latent variables directly based on the involving cross-product response variable, making predictions more robust.

The relationship between latent factors is shown by the regression coefficient or SEM path. Exogenous latent variables are all causal variables that cannot be directly observed. Observation of these variables is carried out through exogenous manifest variables, namely measurable indicators that can be observed directly to measure exogenous latent variables. Construction of observable endogenous and latent exogenous variables, as follows:

$$X_{ij} = \lambda_j X_i + \gamma_j \qquad \text{...(1)}$$

Information:
λ: Loading component
γ: Measurement error
X_n: Latent exogenous variable

$$Y_{ij} = \lambda_j Y_i + \gamma_j \qquad \text{...(2)}$$

Information:
λ: Loading component
γ: Measurement error
Y_n: Latent edogenous variable

Factors influence tourists' decision-making in choosing marine tourism destinations using a measurement scale as a reference to produce quantitative data.

Table 1 Range to scale for factors that influence tourists' decision-making in choosing marine tourism destinations

Scoring	Information
1	Strongly disagree
2	Disagree
3	Hesitant
4	Agree
5	Strongly agree

The Likert scale is designed to measure 'attitude' in order to be scientifically accepted and validated. Variables included in the dependent or endogenous variables are, factors that influence tourists' decision-making in choosing marine tourism destinations (Y), while the exogenous or

IOP Conf. Series: Earth and Environmental Science 1289 (2023) 012012 doi:10.1088/1755-1315/1289/1/012012

independent variable is internal factors, external factors, and marine ecotourism: such as income (X_1), level of education (X_2), distance traveled to tourist destinations (X_3), accessibility (X_4), accommodation (X_5), infrastructure (X_6), information from social media (X_7), security and safety (X_8)

Table 2 Measurement *Goodness of fit* (GOF)

	Measurement *Goodness of fit* (GOF)	Stability Level Target
1	*Incremental Fit Indeks* (IFI)	IFI ≥ 0,90
2	*Root Mean Square Error of Approximation* (RMSEA)	RMSEA ≤ 0,08
3	*Non- Normed Fit Indeks* (NNFI)	NNFI ≥ 0,90
4	*Comparative Fit Indeks* (CFI)	CFI ≥ 0,90
5	*Goodness of Fit Indeks* (GFI)	GFI ≥ 0,90

Partial Least Square (PLS) analysis is an alternative to ordinary least square regression, covariance-based structural equation modelling (SEM) of systems of independent and response variables. PLS is known as a predictive technique and not as an interpretive technique, except for exploratory analysis as a precursor to interpretive techniques such as multiple linear regression or covariance-based structural equation modelling (SEM). Structural Equation Modelling Analysis (SEM) is a set of approaches aimed at expressing hypotheses about the mean, variance and covariance of the observed data in terms of reducing the number of "structural" parameters determined by the underlying model [21].

3. Result

3.1 Research Location

The development of marine tourism in West Java Province is planned and developed in an environmentally friendly, conservation-based manner by minimizing damage to natural resources and social degradation, but is maintained for sustainable use. Sustainable tourism development is a development that is ecologically supported in the long term, while being economically viable, ethically and socially just. The development of sustainable marine tourism in West Java is based on ensuring the sustainability of supporting resources for tourism development that are integrated with the natural, cultural, and socio-economic environment and local wisdom.

West Java Province has a variety of marine tourism attractions spread across the southern part of West Java. These tourist attractions have different levels of development with different segments of the tourist market. Grouping adjacent marine tourism attractions in one cluster can strengthen the competitiveness of marine tourism in order to attract visits from various segments of the tourist market. The factors considered in determining marine tourism cluster, namely: bio-ecoregion factors, and geographical proximity is an important factors that must be considered in determining the grouping of marine tourism areas

Figure 1 Research Location

Marine ecotourism areas must be supported with good accessibility. Even though the location of tourist objects and attractions are close together, if they are not supported by easy accessibility, then the grouping will be carried out with other tourist areas with better accessibility. Characteristics of superior marine tourism products that are the same complementary. A coastal area should have superior marine tourism products that can be used as development themes so that they can emerge the identity of marine ecotourism areas.

3.2. Characteristic Respondents

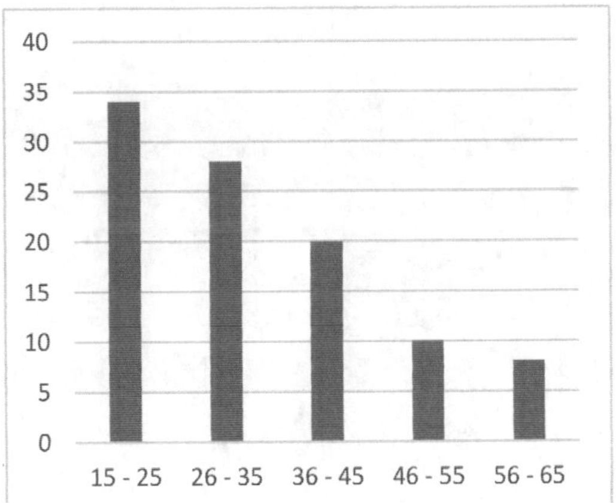

Figure 2 Characteristics Respondent Based On Age

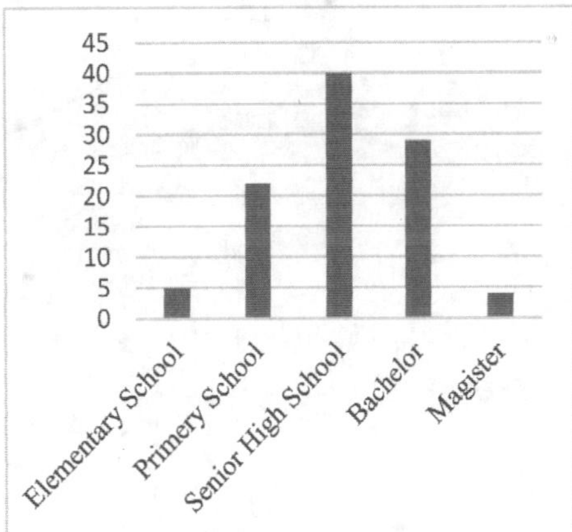

Figure 3 Characteristics Respondent Based On Education

Tourist respondents at each location amounted to 100 respondents. Human age is grouped into several age ranges that describe the stages of human productivity growth. This grouping was used in this research showed the highest number of tourists with an age range of 15-25 years old, at the research site amounting to 34 %. In Figure 3 the number of tourists based on education showed that tourists with a high school education are the highest in research locations. The characteristics of respondents who visited the research locations based on their work showed differences, over the highest percentage level was student tourists (60%). The difference in the amount of income from tourists is reflected in the resulting data, the highest percentage for this research tourist site is the income of Rp. 0.00 – Rp. 3,000,000.,- This is related to the level of education and type of work of tourists for decision-making to choose destination tourism.

Based on the characteristics of tourists, will relate to tourism needs, vacation desires, the amount of money to be used during the tour, as well as other factors in the development and sustainability of marine ecotourism activities in research sites. Understanding the characteristics of local communities and tourists needs to be done for the development and management of marine ecotourism areas starting from planning, monitoring, and stages of implementation of sustainable tourism. Tourists tend to choose marine tourism attractions over conservation tourism, it is necessary to pay attention to safety, natural conditions, waves, and currents so as not to endanger the safety of tourists. Marine ecotourism on the south coast of West Java is dominated by mangrove ecosystem ecotourism, sea turtle ecosystem ecotourism, and capture fisheries ecosystem ecotourism.

3.3. Factors Influence Tourists' Decision-Making in Choosing Marine Tourism Destinations

Factors influence tourists' decision-making in choosing marine tourism destinations, consists of internal and external factors of marine ecotourism. Existing marine tourism attractions and activities, which is as follows in Figure 4

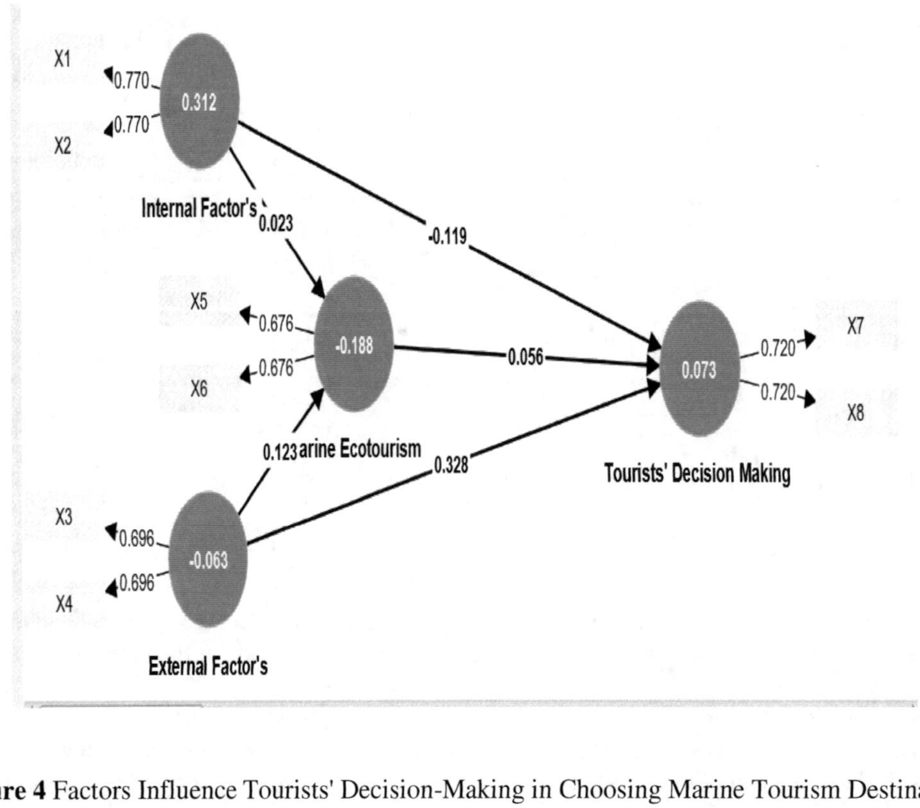

Figure 4 Factors Influence Tourists' Decision-Making in Choosing Marine Tourism Destinations

Tourist preferences can be interpreted as a description of the best values that tourists consider in determining travel choices and choosing marine tourism destinations. Several statements on the questionnaire were submitted to find out the preferences of tourists choosing marine ecotourism. Factors influence tourists' decision-making in choosing marine tourism destinations, such as income (X_1), level of education (X_2), distance traveled to tourist destinations (X_3), accessibility (X_4), accommodation (X_5), infrastructure (X_6), information from social media (X_7), security and safety (X_8). Tourist preferences in making decisions consider the following aspects something to see, to do, to buy, and to learn. Infrastructure is an important component for tourists to visit a marine ecotourism area location and is included in the components of a tourist attraction, providing experiences and growing tourist attraction through e-commerce.

Table 3 Measurement Goodness of Fit Factors Influence Tourists' Decision-Making in Choosing Marine Tourism Destinations

No	Measurement *Goodness of fit* (GOF)	Stability Level Target	Estimation Results	Goodness Fit
1	*Incremental Fit Indeks* (IFI)	IFI ≥ 0.90	0.94	Good Fit
2	*Root Mean Square Error Of Approximation* (RMSEA)	RMSEA ≤ 0.08	0.060	Good Fit
3	*Non Normed Fit Indeks* (NNFI)	NNFI ≥ 0.90	0.91	Good Fit
4	*Comparative Fit Indeks* (CFI)	CFI ≥ 0.90	0.94	Good Fit
5	*Goodness of Fit Indeks* (GFI)	GFI ≥ 0.90	0.92	Good Fit

Table 4 of Exogenous Latent Sub-Variables Indirect Influence of Internal Factors (X_1) to Tourism Decision Making (Y).

No	Internal Factor Exogenous Latent Sub Variable	Estimation	T-Value	R^2	Result
1.	Income (X_1)	0.312	0.770	0.84	Significant
2.	Level Education (X_2)	0.312	0.770	0.64	Significant

Table 5 Exogenous Latent Sub-Variables Indirect Influence of External Factors (X_2) to Tourism Decision Making (Y).

No	External Factor Exogenous Latent Sub Variable	Estimation	T-Value	R^2	Result
1.	Distance traveled to tourist destinations (X_3)	0.188	0.676	0,71	Significant
2.	Accessibility (X_4)	0.188	0.696	0.61	Significant
3.	Accommodation (X_5)	0.188	0.676	0.65	

Table 6 of Exogenous Latent Sub-Variables Indirect Influence of Marine Ecotourism Factors (X_3) to Tourism Decision Making (Y).

No	Ecotourism Factor Exogenous Latent Sub Variable	Estimation	T-Value	R^2	Result
1.	Infrastructure (X_6)	0.063	0.696	0.71	Significant
2.	Information from social media (X_7)	0.063	0.720	0.61	Significant
3.	Security and safety (X_8	0.063	0.720	0.81	Significant

4. Discussion

In Table 1 it can be seen that the overall model suitability test is seen from the parameters Incremental Fit Index The estimated result is 0.94 with the degree of suitability good fit, *Root Mean Square Error Of Approximation* estimation results 0.060 with the degree of suitability good fit. *Non Normed Fit Index* the estimated result is 0.91 with the degree of suitability good fit. *Comparative Fit Index estimation* results in 0.94 with a good fit level. Based on the results of the goodness-of-fit test, the proposed model is able to estimate a population covariation matrix that is not different from the sample data covariation matrix. This indicates that the estimation results obtained from sample data can be used as a basis for making generalizations about the phenomenon under research.

Based on the results of calculations using the SEM (Structural Equations Modelling) model in Figure 4 can be seen the indirect influence of internal, external, ecotourism factors to tourism decision making can be seen in the table 2,3, and 4. All latent variables analyzed, both exogenous latent sub variables Internal Factors (X_1), External Factors (X_2). Marine Ecotourism Factors (X_3) and endogenous latent variables Tourism Decision Making (Y) make a direct impact. where is the $t_{value} > t_{table.}$ Internal factors exogenous latent sub-variables such as Income (X_1) with a coefficient of determination value of 0.84 (84 %) and Level of Education (X_2) with a coefficient of determination of 0.64 (64 %) influence tourists' decision-making in choosing tourist destinations. Based on the results of research in the field, the level of education has an influence on people's income levels, so this variable has an influence on tourists' decision-making in determining tourist destinations. Apart from this, it will also determine the length of time of visit at the tourist destination.

Exogenous latent sub-variables indirect influence of external factors (X_2) to Tourism decision-making (Y) such as distance traveled to tourist destinations (X_3) with a coefficient of determination value of 0.71 (0.71%) and accessibility (X_4) with a coefficient of determination value of 0.61 (61%), accommodation (X_5) with a coefficient of determination value of 0.65 (65%) influence tourists' decision-making in choosing tourist destinations. Accessibility is an important aspect that supports marine ecotourism, involving cross-sectoral development. However, in terms of vehicle availability, it is included in the unfavorable category. This means that tourists are motivated to visit because of the natural scenery so the development of tourist sites can optimize the natural scenery and there are no decisions that are contrary to this.

Ecotourism factors exogenous latent sub-variables such as Infrastructure (X_6) with a coefficient of determination value of 0.71 (71 %) and information from social media (X_7) with a coefficient of determination value of 0.61 (61 %), security and safety (X_8) with a coefficient of determination value of 0.81 (81 %) influence tourists' decision-making in choosing tourist destinations. The development of marine tourism activities in West Java province, especially on the southern coast, has experienced a rapid increase. This is supported by increasingly sophisticated technology and the ease with which people can access the internet so that marine tourism objects will be easily exposed on social media to become the most sought-after destinations by tourists. Important aspects that must be developed in new tourism are safety and health, accessibility, attractions, and amenities.

Maritime tourism accessibility is all types of facilities and transportation infrastructure, especially sea transportation which supports the movement of tourists from the tourist's area of origin to the tourism destination as well as movement within the destination area tourism in relation to motivation for tourist visits. Attractions or marine tourism attractions are all activities found at marine tourism objects that become an attraction so that tourists visit the place and have new experiences and are willing to visit again.

Internal, external and marine ecotourism exogenous latent variables influence tourist decision-making. Marine tourism management is carried out by (1) providing spatial zoning regulations to manage the growth of tourism facilities and activities without hindering the environmental quality and economic growth of other users; (2) allocating human uses, including cultural and social uses, as well as place value and tourism value chains, optimizing the use of space that is in high demand from multiple users while considering environmental and cultural needs.

The regional government of West Java Province should be improve the quality of marine tourism products in the form of attractions and amenities which include developing marine tourism objects, developing regional specialties in coastal areas, improving the availability of clean water, electricity, telecommunications (Wi-Fi) and public facilities at marine tourism objects including toilets, trash cans, and photo spots that are easily accessible to all target tourist audiences. Tourists hope to improve the quality of roads and transportation in the form of public transport to marine tourism objects on the southern coast of West Java Province. Next, they need to involve the community in developing tourist attractions and collaborate with various parties to promote marine tourism villages.

5. Conclusion

The condition of the unplanned marine tourism business shows that it is running well even though it has not run optimally and is accepted by tourists and the local community. So the role of stakeholders with their capacity and capability as part of a support system for local communities as initiators is necessary to accelerate an unplanned tourism business to establish an ecotourism business. Based on the results of research on factors that influence tourists' decision-making in choosing marine tourism destinations, namely: income (X_1), level of education (X_2), distance traveled to tourist destinations (X_3), accessibility (X_4), accommodation (X_5), infrastructure (X_6), information from social media (X_7), security and safety (X_8). governance is carried out by considering social-ecological system aspects based on coastal area utilization zones, but not all of them are implemented properly, so collaboration between stakeholders is needed to support the sustainability of marine tourism.

Acknowledgments. We are grateful to the Faculty of Fisheries and Marine Science, Universitas Padjadjaran for providing facilities and support. This research was funded through Unpad internal resources with contract number: 1549/ UN6.3.1/ PT.00/2023. We also thank the anonymous referees who have given valuable comments and critiques for the earlier version of the manuscript.

References

[1]. UNWTO. 2019. International tourist arrivals reached 1.4 billion two years ahead of forecasts. United World Tourism Organization (UNWTO). http://www2.unwto.org/press-release/2019-01-21/internationaltourist-arrivalsreach-14-billion-two-years-ahead-forecasts.

[2]. Nababan, E.J.K., Qurniati, R. dan Kustanti, A. 2016. Social capital in the management and preservation of mangrove forests in Labuhan Maringgai District, East Lampung Regency [Modal sosial pada pengelolaan dan pelestarian hutan mangrove di Kecamatan Labuhan Maringgai Kabupaten Lampung Timur]. Jurnal Sylva Lestari 4 (2): 89-100

[3]. Suwarsih. 2018. Ecological and economic utilization of the mangrove rehabilitation program in the coastal area of Jenu Village, Jenu District, Tuban Regency [Pemanfaatan ekologi dan ekonomi dari program rehabilitasi mangrove di kawasan pesisir pantai Desa Jenu Kecamatan Jenu Kabupaten Tuban]. Jurnal Techno-fish 2(1): 12-18

[4]. Iswahyudi, Kusmana, C., dan Hidayat, A. 2019. Evaluation of land suitability for mangrove forest rehabilitation in Langsa City, Aceh [Evaluasi kesesuaian lahan untuk rehabilitasi hutan mangrove Kota Langsa Aceh]. Jurnal Matematika, Sains, dan Teknologi 20(1): 45-56.

[5]. Nurhayati.A., et. al. 2019. Model Development of A Synergistic Sustainable Marine Ecotourism—A Case Study in Pangandaran Region, West Java Province, Indonesia. Jornal Sustainability Vo; 11 (12), 3418; https://doi.org/10.3390/su11123418

[6]. Hall, C.M. 2001. Trends in Ocean and Coastal Tourism: the end of the last frontier?. Ocean & Coastal Management, 44: 601-618

[7]. Lachs, L. and Oñate-Casado, J. 2020. Fisheries and tourism: Social, economic, and ecological trade-offs in coral reef systems. In S. Jungblut et al. (Eds.). YOUMARES 9 – The oceans: Our

research, our future: Proceedings of the 2018 conference for Young Marine Researchers in Oldenburg, Germany, Springer International Publishing, 243–260

[8]. Ketjulan R. 2010. Carrying Capacity of Hari Island Waters as a Marine Ecotourism Object [Daya Dukung Perairan Pulau Hari Sebagai Objek Ekowisata Bahari Paradigma. 14(2): 195-204.

[9]. International Ecotourism Society.2023. What is Ecotourism? Available online: https://ecotourism.org/ (accessed on 4 June 2023)

[10]. Higham J 2007 Ecotourism: Competing and Conflicting Schools of Thought. In Critical Issues in Ecotourism: Understanding a Complex Tourism Phenomenon (UK: Oxord Elsevier) p14

[11]. Gier, L., Christie, P. and Amolo, R. 2017. Community perceptions of scuba drive tourism development in Bien Unido, Bohol Island, Philippines. Journal of Coastal Conservation, 21:153–166.

[12]. Munro, Jennifer, Halina Kobryn, David Palmer, Sam Bayley, and Susan A. Moore. 2019. "Charting the Coast: Spatial Planning for Tourism Using Public Participation GIS." Current Issues in Tourism 22 (4): 486–504. https://doi.org/10.1080/13683500.2017.1346589.

[13] Papageorgiou, M. 2016. Coastal and marine tourism: A challenging factor in marine spatial planning. Ocean and Coast Manage, 129: 44–48.

[14]. Ross, S.; Wall, G. Evaluating ecotourism. 1999. The case of North Sulawesi, Indonesia. Tour. Manag. 20, 673–682. [CrossRef]

[15]. Orams.M. 1999. Marine Tourism: Development, Impacts, and Management.Psychology Press. p 155

[16]. Chul Jeong, Chul. (2014.) 'Marine Tourist Motivations Comparing Push and Pull Factors',, Journal of Quality Assurance in Hospitality & Tourism, 15:3, 294-309

[17]. Swarbrooke, John & Horner, Susan. 2007. Consumer Behavior in Tourism second edition. USA p 36-45

[18]. Davies, Anthony M. C. 2001. Uncertainty testing in PLS regression. Spectroscopy Europe, 13(2).Fennel, D.; Dowling, R. Ecotourism policy and planning: Stakeholders, management and governance. Ecotourism Policy Plan. 2003, 1, 331–344

[19]. Joshi, A., S. Kale, S. Chandel dan D. K. Pal. 2015. *Likert Scale: Explored and Explained. British Journal of Applied Science & Technology,* 7(4): 396-403

[20]. Garson, G. D. 2016. *Partial Least Squares: Regression & Structural Equation Models.* Statistical Publishing Associates. United States of America.262 hlm.

[21]. Hair, J. F., G. T. M. Hult, C. M. Ringle, M. Sarstedt, N. P. Danks dan S. Ray. 2021. *Partial Least Squares Structural Equation Modeling (PLS-SEM) Using R a Workbook.* Springer. Switzerland. 197 hlm

COVID-19 pandemic and the momentum of seaweed industry development in Indonesia towards sustainable food system: An overview

B Wardono[1], A S Prabakusuma[2*], A Zulham[1], R Yusuf[3], E S Luhur[3], I Muliawan[3], R Deswati[3], F Y Arthatiani[3]

[1] Research Center for Cooperative, Corporation, and People's Economy, National Research and Innovation Agency, Jakarta 12710, Indonesia
[2] Vocational School of Foodservice Industry, Food Biotechnology Research Group, Universitas Ahmad Dahlan, Yogyakarta 55166, Indonesia
[3] Ministry of Maritime Affairs and Fisheries, Jakarta 10110, Indonesia

*Corresponding author mail: adhita.sriprabakusuma@tp.uad.ac.id

Abstract. The COVID-19 pandemic has had an impact on the seaweed industry due to changes in supply and demand, cost increases, market volatility, and the implementation of lockdown measures. This study aimed to determine the effect of the COVID-19 pandemic on the seaweed industry to support a sustainable food system. This study used both primary data from the seaweed industry, especially the upstream sector, and secondary data to determine the progress of the national and international seaweed industries. The data was analyzed descriptively to obtain information regarding the role of the seaweed industry in accelerating post-pandemic coastal economic recovery. The results indicated that seaweed cultivation can support coastal economic recovery in the post-COVID-19 pandemic. After demand recovers, production can be enlarged in a short time without excessive investment and has the potential to generate income within 30–45 days. It was found that total seaweed exports have dropped by 30.54% and 19.90% by volume and value, respectively, in only the early stages of the COVID-19 outbreak. However, the effect of the pandemic has encouraged accelerative industrialization by increasing the capacity of domestic seaweed factories from 50% to 72%. The pandemic has also triggered the domestic industry to produce seaweed derivative products independently, such as carrageenan, alginate, and agar, to reduce dependence on imports. It is imperative to recommend to the Indonesian government that the export tariff on dried seaweed be increased to boost the competitiveness of the domestic seaweed processing industry and invest in the diversification of processed seaweed derivative products.

1. Introduction

Indonesia, with an expansive ocean area of 6,400,000 km^2 and a coastline of 110,000 km, is endowed with favorable conditions, including a tropical climate, that make it a conducive region for the cultivation of several species of seaweed [1]–[4]. Indonesia, being situated inside the coral triangle, has a diverse array of seaweed species, numbering at least 550, that possess significant commercial value [5]. Naturally, Indonesia provides an appropriate environment to thrive optimally for around 8,000 algae species worldwide, generally referred to as seaweed [1]. Seaweed is a nutritionally nutrient-dense food that contains carbohydrates, protein, dietary fiber, salts, ash, mostly composed of sodium and potassium, and a little fat. Seaweed is rich in several vitamins, including vitamin A, B_1, B_2, B_6, B_9, B_{12}, C, D, E, K, and beta-carotene [6]. Additionally, it is a source of essential minerals such as potassium, calcium, phosphorus, salt, iron, and iodine. Certain types of seaweed exhibit a higher concentration of essential vitamins and minerals, such as iron and calcium, in comparison to fruits and vegetables. The health-promoting attributes often associated with seaweed are mostly linked to the presence of polyphenols, polysaccharides, sterols, and several other bioactive compounds [7], [8]. These compounds have been reported to possess antioxidant, anti-inflammatory, anti-cancer, and anti-diabetic activities [8]. When comparing seaweed to terrestrial plants, it is seen that seaweeds contain a greater amount of essential fatty acids, specifically eicosapentaenoic (EPA) and docosahexaenoic (DHA) fatty acids [9]. Furthermore, seaweed has the capacity to synthesis several secondary metabolites, including terpenoids, oxylipins, phlorotannin, volatile hydrocarbons, and compounds derived from mixed biogenetic pathways [10]. Hence, seaweed has natural characteristics that make it a promising subject of significant scientific attention. They harbor a diverse array of chemicals that exhibit a multitude of biological activities, making them suitable as functional ingredients in various technological applications aimed at producing functional food towards supporting sustainable food system in Indonesia.

In 2022, Indonesian seaweed exports and other algae amounted to 231,829.70 tons, valued at US$397.16 million. In this international trade, China emerged as the primary importing country for Indonesian seaweed exports, receiving a significant amount of 194,115.41 tons, valued at US$335.83 million. The quantity of seaweed exports originating from Indonesia had a massive growth of 12.44% in comparison to the preceding year. These exports amounted to 206,185.10 metric tons, with a corresponding value of US$222.61 million [11]. This, of course, supports Indonesia's important role in the global seaweed trade due to its considerable contribution to both production and exportation, accounting for up to 40% of the world's total seaweed production and over 30% of global seaweed exports [12]. Previously, seaweeds were used as a direct food source in many regions of Asia, sections of South America, and the Pacific Islands [3], [13]. Nowadays, there is a growing interest in scientifically explorating the potential of seaweeds as nutraceuticals or functional foods for promoting human health, particularly during the Coronavirus disease (COVID-19) pandemic [14]. It is worth noting that about 99% of cultivated seaweed is primarily allocated for the manufacturing of thickening and gelling agents used in the pharmaceutical and food sectors. Indonesia pays special attention to the red seaweeds *Kappaphycus alvarezii* and *Eucheuma denticulatum* for the manufacturing of carrageenan, while *Gracilaria* species are used to produce agar and alginate for making various gelling agents in a broad range of applications [3], [15]. So far, Indonesia contributes 66% of the world's supply of these hydrocolloid seaweeds, with the sector expanding fast since 2000.

At present, the production of seaweed in Indonesia is mostly carried out by smallholder farmers with low incomes, especially those located in Eastern Indonesia [3], [16]. These farmers possess the ability to adapt to seasonal and tidal variations and face limited alternative sources of revenue generation. The Indonesian government has placed significant emphasis on the development of the seaweed industry, acknowledging its growing significance for coastal livelihoods. This commitment is exemplified by the recent enactment of Presidential Decree 33-2019, which presents a strategic plan for the growth and advancement of the seaweed sector between 2018 and 2021 [17]. It highlights potential avenues for expanding seaweed production and enhancing local value-added processing. Efforts have been made to enhance the value-added aspects of many goods, including food products, animal feed products, fertilizers, cosmetics, and bioethanol. Seaweed cultivation serves as a significant economic resource for more than 267,000 rural families in Indonesia, as stated in Presidential Decree 33-2019 (p. 17) [15], [17]. These households generate an approximate yearly income of US$2,000, as shown in Presidential Decree 33-2019 (p. 16) [15], [17]. Local farmers often have the perception that seaweed farming is a more financially rewarding endeavor compared to well-established agricultural, fishery, and marine businesses like cocoa, copra, tilapia, catfish, crabs, and other marine products [4], [16], [18]–[20]. Furthermore, several other advantages associated with seaweed farming have been reported, including an enhanced capacity to effectively handle financial resources and accumulate savings for significant expenditures [4], [18]. These benefits coexist with a range of other environmental and social changes linked to the sector, as well as diverse advantages experienced by various members of communities and households [15]–[17], [20].

When Presidential Decree 33-2019 was initially implemented, the COVID-19 pandemic brought significant adverse effects to the Indonesian seaweed industry, particularly in the first quarter of 2020 [21]. Those effects were mainly related to the decrease of export volume and value of dried seaweeds. The Indonesian government's and its trading partners' implementation of lockdown measures has adversely affected the seaweed export performance. Consequently, the export volume and value of dried seaweed had a reduction during the first semester [22]. One contention is that exporters of raw and dried seaweed experienced various issues and limitations due to the COVID-19 pandemic measures. One aspect to consider was the diminished operational efficiency of Indonesian seaweed processing facilities, which stands at around 50% and may be attributed, in part, to the insufficient provision of raw materials [21], [22]. However, there are limited studies on the contribution of the seaweed industry, especially the upstream sector, to accelerating the post-COVID-19 pandemic coastal economic recovery in Indonesia. Therefore, this present study aimed to determine the effect of the COVID-19 pandemic on the seaweed industry to support sustainable food system. This research used primary data collected from the seaweed industry, with a particular focus on the upstream sector, as well as publicly available secondary data to assess the progress achieved within both the domestic and global seaweed sectors.

2. Materials and Methods

2.1. Research sites and data collection

This qualitative study, conducted between December 2020 and November 2021, involved collecting and analyzing non-numerical data, including text, audio, and video, to understand concepts, experiences, and opinions from stakeholders at the upstream level of the seaweed industry. This method was effectively applied to gather in-depth insights into a universal problem and generate beneficial ideas for this research. This study focused on Takalar regency, South Sulawesi province, as one of the biggest seaweed production centers in Indonesia (from S 5°031′ and 5°0381′ and between E 199°0221′ and 199°0391′) and Kupang city, East Nusa Tenggara province (from S 10°36′14″ and 10°39′58″ and between E 123°32′23″ and 123°37′01″). Those selected regions have significant potential for bolstering the food system in the Eastern Indonesia, with a special emphasis on their contribution to the production of seaweed as well as other marine and fishery products.

Primary data was obtained from the combination of in-depth interviews, focus group discussion, and surveys by using questionnaire. According to in-depth interviews with village leaders in the studied locations, seaweed farming was practiced by 90% of the households during the time of the field observation. Additional occupations in those areas include civil servants, teachers, and local entrepreneurs. Beside local seaweed farmers, other respondents included civil servants working at the Department of Marine and Fisheries in Takalar regency and Kupang city, aquaculture extension workers, and managers in the local seaweed processing factories. While the secondary data were retrieved from the Central of Statistics Agency (https://www.bps.go.id/), statistical data reported by Ministry of Marine Affairs and Fisheries (MMAF) (https://statistik.kkp.go.id/home.php), trade statistics for international business development (https://www.trademap.org/Index.aspx), and the United Nations Comtrade database (https://comtradeplus.un.org/).

2.2. Data analysis

Descriptive statistics, including means, standard error, maximum and minimum, were conducted in the different sections. Narrative analysis was performed to understand how research participants construct their condition and narrative from their own personal experience. Besides, content analysis was also conducted to determine the presence of certain information, themes, or concepts within some obtained qualitative data. All statistical data was processed on Microsoft Excell 365 software.

3. Results and Discussion

3.1. Total export value and volume of Indonesian seaweed in the first semester of 2020

Seaweed is one of the main marine commodities that is exported in dry form or processed products such as carrageenan, alginate, and agar [3], [10], [20]. Figure 1 shows the development of the volume and value of Indonesian seaweed exports in the period of October 2019–June 2020. The export volume and value experienced a decline in January and February of 2020 when compared to the corresponding data from 2019 (Fig. 1a). However, there was subsequent growth in the succeeding months. The observed fluctuation may be attributed to the influence of the COVID-19 pandemic measures, particularly in the first stages of the global health crisis [15], [21]. Based on the data, it was found that total seaweed exports have dropped by 30.54% and 19.90% and dried seaweed exports decreased by 32.02% and 30.11 by volume and value in the first semester of 2020, respectively (Fig. 1b; 1c).

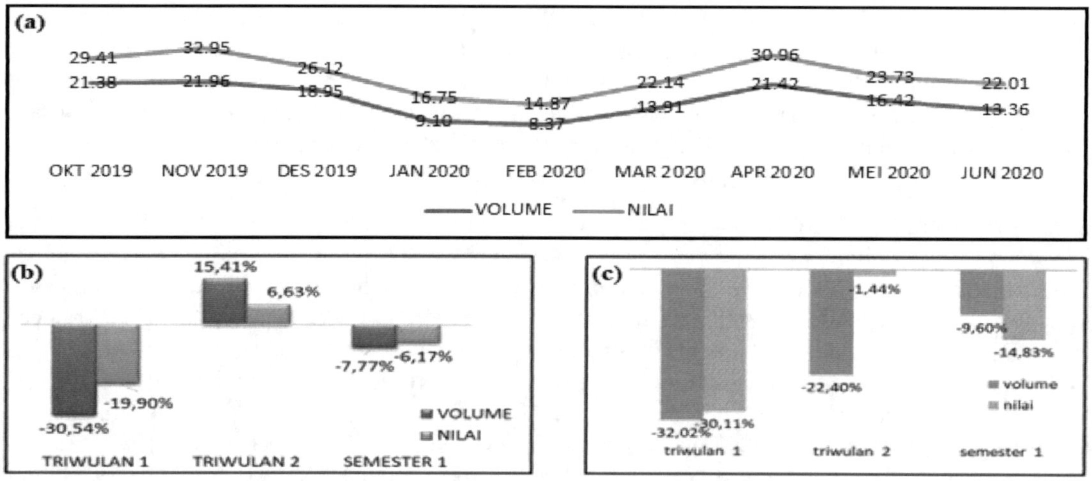

Volume: Total export volume; Nilai: Total export value; Triwulan 1: Quarter 1; Triwulan 2: Quarter 2.

Fig. 1. (a) Total export volume and value of Indonesian seaweed between October 2020 and June 2020; (b) Total export volume and value of Indonesian seaweed within quarters 1 and 2 of 2020; and (c) Total export volume and value of Indonesian dried seaweed products within quarters 1 and 2 of 2020.

The export performance of processed seaweed, particularly carrageenan and agar, showed a growth during Semester I of 2020. In that time period, there was a 16.04% increase in the export volume of carrageenan, accompanied by an 8.61% growth in its export value (Fig. 2a). The export performance of agar products increased in the first and subsequent quarters, demonstrating their significant growth. The volume and value of agar exports experienced a growth of 40.29% and 37.13%, respectively (Fig. 2b).

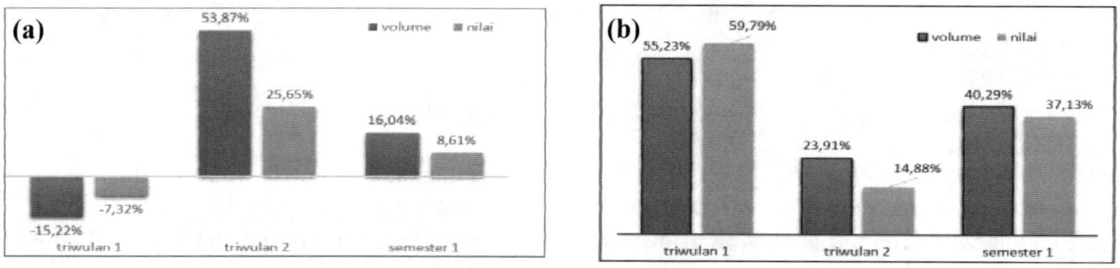

Volume: Total export volume; Nilai: Total export value; Triwulan 1: Quarter 1; Triwulan 2: Quarter 2.

Fig. 2. (a) Total export value and volume of carrageenan within quarters 1 and 2 of 2020; and (b) Total export value and volume of agar within quarters 1 and 2 of 2020.

The hypothesized reason for this situation arises from China's prioritization of controlling the COVID-19 pandemic inside its borders, leading to a halt in the manufacture of processed seaweed products [22]. In contrast to the decline seen in the export performance of dried seaweed during Semester I 2020, which saw a contraction of -15.22% and -7.32% in terms of volume and value, respectively (Fig. 2a). The results indicated that seaweed cultivation can serve as an alternative to coastal economic recovery during COVID-19 pandemic. After seaweed market demand from China and other countries recovers, production can be enlarged in a short time without excessive investment and has the potential to generate income within 30–45 days with its labor-intensive activity. This finding suggests that the national exporters of dried seaweed have had the most significant obstacles, particularly when China closed its country's border and market. The strong reason for the occurrence was the huge proportion of Indonesian dried seaweed exports, with a total of 65%, directed mainly towards China, which serves as the main importing country for Indonesian seaweed export [11]. Main importing countries for Indonesian seaweed exports based on export value in 2020 is presented in Fig. 3 [22].

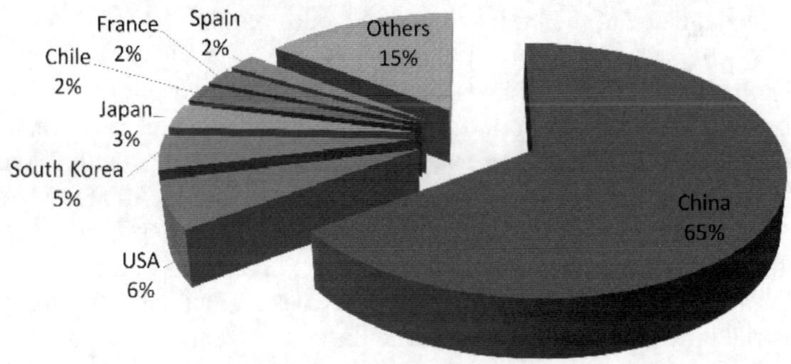

Fig. 3. Main importing countries for Indonesian seaweed exports based on export value in 2020.

Based on the statistical data, the COVID-19 pandemic has negatively affected Indonesian dried seaweed export performance in quarters 1 and 2 of 2020 (Fig. 2a; 2b). The point that may be inferred is that there is a need to decrease the export volume of dried seaweed due to its susceptibility to external disruptions, such as the global disease epidemic, recession, political crisis, or other natural disasters, and the lack of additional value in dried seaweed exports. The COVID-19 pandemic demonstrated a clear indication that Indonesian seaweed is only mainly oriented to the export market, while the domestic market has not been optimally penetrated. The only reliance on the export market will make marine products more susceptible to external disruptions. The orientation of the export market should be balanced with that of the domestic market penetration. In fact, the impact of the pandemic has actually encouraged industrialization by increasing the capacity of seaweed factories from 50% to 72% in accordance with Presidential Decree 33-2019 [17]. This momentum is bolstered by the emergence of new opportunities created by the expansion of domestic industries that use seaweed products as auxiliary materials, particularly in health sectors. Ironically, the processed seaweed product in the form of refined carrageenan is still highly dependent on imports, resulting in yearly volume and value increases of 14.88% and 23.91%, respectively (Fig. 2b). The amount of carrageenan imports increased from 332.7 tons to 703.21 tons equivalent (US$ 8,511,930 in 2019) [11]. These carrageenan products are for the fulfillment of downstream industries such as food, prebiotics, dairy, textiles, toothpaste, pharmaceuticals, and other industries [13]. Therefore, it is necessary to encourage the fulfillment of domestic carrageenan production and reducing imports.

3.2. Effects of COVID-19 pandemic on seaweed processing industry

During the COVID-19 pandemic, an interesting phenomenon was demonstrated by the seaweed processing industry, especially in carrageenan sales. It was observed that the export volume of *Eucheuma cottonii (E. cottonii)* carrageenan increased by 42%, the prices for dried seaweed increased by 3–5%, except for dried *Eucheuma Spinosum (E. Spinosum)*, which decreased by 4%, the price of *E. spinosum* carrageenan increased by 58%, and the frequency of processed seaweed shipments increased by 22% (Tabel 1). In the production aspect, processed seaweed production of refined carrageenan increased by 40%, and dried seaweed prices for *Gracilaria* sp. increased by 8% (Tabel 1). Processed seaweed prices for semi-refined carrageenan increased by 58%. Besides, in the raw material purchasing aspect, the purchase price of *Glacilaria* sp. increased by 10% (Tabel 1). The findings provide a signal for Indonesia that the export of raw material products need to be reduced and processed seaweed products should be increased due to those have received new demand from domestic markets [22]. This fact provides a compelling rationale for the government's ongoing efforts to foster the expansion and advancement of seaweed processing industries. Furthermore, in the context of a sustainable food system, seaweed production and processing have an important role, not only for supplying many kinds of nutritious foods to humans but also for providing high-quality animal feeds and various chemical substances for medicinal and laboratory uses [23]. It can be concluded that, during the COVID-19 pandemic, Indonesian seaweed industry was only affected during the early stages of the pandemic, after which it continued to expand due to the high demand in health sectors.

3.3. Recommended policy options and their impact on the Indonesian seaweed industry

Currently, seaweeds are considered an important component of the blue economy's development in Indonesia and support national economic growth [4]. Indonesian aquaculture receives significant economic benefits from the exportation of seaweed, which has a prominent position as the fifth-ranked export product [22]. The primary outcomes of this study were directed to provide support for initiatives aimed at enhancing the development of the Indonesian seaweed sector, as outlined in Presidential Decree 33-2019 [17]. In this part, recommended policy options were elucidated to strengthen aquaculture production and industrial processing of seaweed, particularly carrageenan and agar, to fulfil the demands of the local industrial sector, decrease reliance on imported carrageenan, and expand the market share of Indonesian processed seaweed products in international trade. Based on the observed issues and primary findings, particularly obtained from field observations and surveys, two potential policy recommendations have been deliberated upon, those were: (1) Increasing the export tariffs on dried seaweed. It is anticipated that the imposition of higher export tariffs on dried seaweed would enhance the competitiveness of the seaweed processing industries. Additional incentives to improve the effectiveness of the seaweed industry encompass the restoration of old machinery, the provision of electricity incentives for nighttime production, the establishment of an institution to supervise the flow of raw materials to the processing industry, the diversification of processed seaweed products through recent technological innovations, and the further processing of by-products and waste from carrageenan, alginate, and agar-agar production [14].

Furthermore, based on the survey results, the cultivation of alternative seaweed types beyond *E. cottonii*, *E. Spinosum*, and *Glacillaria* sp, which have already undergone extensive processing, should also be developed [22]. This cultivation initiative of of alternative seaweed types can serve as a potential alternative employment opportunity [1], [3], [19]. Lastly, the establishment of a novel seaweed processing facility necessitates the streamlining of bureaucratic procedures for investors seeking construction permits, particularly in terms of cost and processing time. Additionally, tax incentives should be provided to stimulate investors to invest new factories in potential regions that provide as hubs for seaweed production [4], [5], [14]. Each of the options presents both advantages and disadvantages were evaluated based on three key criteria: strategic considerations, in alignment with Presidential Decree 39-2019), the significance of addressing the demand for processed products, particularly refined carrageenan products in the domestic market, and the potential for expanding the market share of carrageenan and agar exports in the international market [17]. Table 1 presents a comprehensive overview of the comparison between the two solutions, taking into consideration three specific criteria.

IOP Conf. Series: Earth and Environmental Science 1289 (2023) 012013 doi:10.1088/1755-1315/1289/1/012013

Table 1. Overview of the effects of COVID-19 pandemic on seaweed processing industry.

Raw material purchasing	Warehouse	Manpower	Production	Selling
▪ Purchases of *E. cottonii* decreased by 25–67% ▪ Purchases of *Gracilaria* sp. increased by 27% ▪ Purchases of *E. spinosum* decreased by 50% ▪ Purchase price of *E. Cottonii* and *E. Spinosum* decreased by 10–11% ▪ The purchase price of *Glacilaria* sp. increased by 10% ▪ Purchase frequency decreased 19–50%	▪ Utilized capacity decreased due to the reduction of production ▪ Longer storage time due to inter-regional transportation restrictions	▪ Companies reduced or laid off their workers ▪ Labor wages were reduced, ▪ Companies only retained their permanent employees	▪ Dried seaweed production decreased by 7–47% ▪ Processed seaweed production (non-refined carrageenan) decreased by 50%–100% ▪ Processed seaweed production of refined carrageenan increased by 40% ▪ Dried seaweed prices decreased by 3%–12%, except for gracilaria, which increased by 8% ▪ Processed seaweed prices decreased by 25%–26%, except for carrageenan or semi-refined carrageenan, which increased by 58%	▪ The export volume of dried seaweed decreased by 16–41% ▪ The export volume of *E. cottonii* carrageenan increased by 42% ▪ The export volume of *Glacilaria* sp. carrageenan decreased by 39% ▪ Prices for dried seaweed increased by 3–5%, except for dried *E. spinosum*, which decreased by 4% ▪ *E. cottonii* carrageenan prices fell 11% ▪ The price of *E. spinosum* carrageenan increased by 58% ▪ Export target markets tend not to change ▪ China is still the main export market for both dried and processed seaweed ▪ The frequency of shipments of dried seaweed decreased by 33–57% ▪ The frequency of processed seaweed shipments increased by 22%

Tabel 2. Recommended policy options and their impact on the Indonesian seaweed industry.

Policy options	Strategic Aspect: Supporting Presidential Decree 33-2019		Interest in supplying the needs of processed seaweed products for the domestic market		Increasing the export share of processed seaweed products in the international market	
	Description	**Score**	**Description**	**Score**	**Description**	**Score**
Option 1	Supporting the implementation of the Presidential Regulation 33-2019 target (Increase in processed seaweed products by 2021). The realization of the processing industry production target by 2021: ▪ Alkali treated cottonii (ATC): 3,850 ton. ▪ Semi-refined carrageenan (SRC): 16,174 ton. ▪ Refined carrageenan (RC): 7.814 ton ▪ Agar: 10.393 ton	+++	The demand for carrageenan in industry is rising, and most of it is still supplied by imports, therefore carrageenan imports must be reduced. Carrageenan imports rose dramatically in 2019 (million USD 8,511,930).	+++	Indonesia can become the world's main exporter of carrageenan and agar products. Currently, Indonesia is the 3rd global exporter of carrageenan products with a share of 8.47%, but its exports are dominated by SRC products; and global exporter rank 6 for agar with a share of 5.23%.	+++
Option 2	Increasing the use of seaweed as a product: food, feed, fertilizer, pharmaceutical products, and cosmetic products.	++	Encouraging the realization of increased production of formulated products made from seaweed for food, feed, fertilizers, cosmetic products, and pharmaceutical products in potential areas	++	Increasing value-added export products from waste products and by-products	++
Option 3	Increasing the quality and quantity of processed seaweed products	+	Addressing the demands for processed seaweed products for food, feed, fertilizers, cosmetic products, and pharmaceutical products	++	Increasing market share of processed seaweed products in the international market	++

The proportion of dried seaweed exports is expected to decline in order to facilitate the processing of seaweed into finished derivative products, such as ATC, SRC, RC, and agar. This strategic shift aims to strengthen the volume and value of exported seaweed products. The optimum use of processing industry capacity may lead to an increase in the production of processed seaweed. This strategies aim to address the critical issue of labor and seaweed production at the farmer level in the vicinity of the plant. It is anticipated that those strategies aforementioned have an imperative contribution to stimulating coastal economic growth, benefiting the local community, the national economy, and also natural sustainability [3]–[5]. Those strategies have the potential to improve the competitiveness of Indonesian seaweed exports. While the implementation of a zero-waste approach in the seaweed industry, as an integrated part of a sustainable food system, brings a cutting edge to boost the advantage and efficiency of seaweed processing industries by adding value to waste and by-products, it can also be used to be marketed domestically and abroad [23].

4. Conclusion

The findings of the study suggest that the cultivation of seaweed has the potential to be incorporated as an effective strategy for promoting economic recovery in coastal regions, especially during COVID-19 pandemic. COVID-19 pandemic only affected the national seaweed industry performance in early stage of pandemic, which is quarters 1 and 2 of 2020. In the period of Semester 2 of 2020, once seaweed demand from China and other importing countries has been restored, it is possible to rapidly expand production without requiring excessive investment. This expansion has the ability to generate revenue within a very short timeframe, owing to the labor-intensive activity. At the same time, it has been observed that there has been a decline in the volume and value of dried seaweed exports.

It is important to note that the pandemic has also stimulated an acceleration by facilitating an increase in global demand and the performance of seaweed industries, as shown by the rise in production capacity, in alignment with the guidelines outlined in Presidential Decree 33-2019. Nevertheless, the ongoing epidemic has served as a catalyst for the local industries to engage in the production of seaweed derivative products, including carrageenan, alginate, and agar. This initiative is potentially supporting a decrease in the reliance on imported seaweed derivative commodities. Besides, it is of utmost importance to propose to the government implementing the augmentation of the export tariff on dried seaweed in order to enhance the competitiveness of the seaweed processing sector and allocate resources towards the expansion of processed seaweed derivative products as one of the potential food commodities to support a sustainable food system in Indonesia.

Acknowledgement

We would like to extend our heartfelt appreciation to colleagues, external partners, local seaweed farmers, local government, the National Research and Innovation Agency, Universitas Ahmad Dahlan, Yogyakarta, and MMAF for supporting this research and publication of this study.

Conflicts of Interest

The authors declare that no conflict of interest, personal relationship, or competing financial interest could influence the work presented in this article.

Funding

This study was financially supported by the Center for Socio-Economic Research on Maritime Affairs and Fisheries, MMAF 2021 (Grant No. 53/BRSDMKP-BBRSEKP/TU.110/V/2020).

References

[1] Srihadi T F 2022 *SEEIJ (Social Econ. Ecol. Int. Journal)* **6** 1 1–11
[2] Parasasti M A, Prakoso L Y, and Sudiarso A 2022 *Strateg. Pertahanan Laut* 159–173
[3] Rimmer M A *et al* 2021 *Sustain.* **13** 19 1–22
[4] Aslan L O M *et al.* 2018 *IOP Conf. Ser.: Earth Environ. Sci.* **175** 1 012221
[5] Waters T *et al* 2019 *Nat. Conserv.* **7** 1–47
[6] Lomartire S, Marques J C, and Gonçalves A M M 2021 *Mar. Drugs* **19** 6 342
[7] Peñalver R *et al* 2020 *Mar. Drugs* **18** 6 301

[8] Michalak I *et al* 2022 *Vet. Q.* **42** 1 99–114
[9] Rocha C P *et al* 2021 *Int. J. Environ. Res. Public Health* **18** 9 4968
[10] Rustad T and Hayes M 2012 *Marine Bioactive Compounds* (Boston: Springer)
[11] Sadya S 2023 *Agribusiness and Forestry* [Online] (accessed August 05, 2023)
[12] Herawati H, Saragih A K, and Burhanuddin B 2022 *J. Integr. Agribus.* **4** 1 77–87
[13] Gomez-Zavaglia A *et al* 2019 *Antioxidants* **8** 9 406
[14] Buschmann A H *et al* 2017 *Eur. J. Phycol.* **52** 4 391–406
[15] Langford A *et al* 2021 *Mar. Policy* **127** 104431
[16] Mariño M *et al* 2019 *Mar. Policy* **107** 103600
[17] The Audit Board of the Republic of Indonesia 2019 Presidential Regulation (PERPRES) Number 33 of 2019 concerning the 2018-2021 National Seaweed Industry Development Road Map [Online] (accessed August 05, 2023)
[18] Arsyad M *et al* 2014 *Int. J. Agric. Syst.* **2** 2 119–131
[19] Safni I 2019 *J. Pengabdi. Al-Ikhlas* **4** 1 30–52
[20] Nuryartono N *et al* 2021 *Commodities Pilot Project Report – The Australia-Indonesia Centre* [Online] (accessed August 05, 2023)
[21] Prasetia H *et al* 2022 *Omni-Akuatika* **18** 96–100
[22] Wardono B *et al* 2020 *Policy Brief Socio-Economic Maritime Affairs and Fisheries* [Online] (accessed August 05, 2023)
[23] Sultana F *et al* 2023 *Aquac. Fish.* **8** 5 463–480

IOP Conf. Series: Earth and Environmental Science 1289 (2023) 012014 doi:10.1088/1755-1315/1289/1/012014

Profile of fisheries capture in Cikidang Fish Landing Base (PPI), Pangandaran, West Java Study 2017-2021 and its potential utilization for fisheries product processing industry

T Yuniarti[1*], M Mulyono[1], M Mardiana[2], I Nurmala[2], I J Puspadewi[1], H Triyono[1], M H Alauddin[1], Y P Handoko[1], M Maulita[1] and A Maulani[1]

[1] Department of Utilization of Fishery Resources, Politeknik Ahli Usaha Perikanan (Jakarta Technical University of Fisheries), Jalan AUP No. 1, Pasar Minggu, Jakarta 12520, Indonesia
[2] Freshwater Aquaculture Research Center and Fisheries Extension, BPSDM KP Jl. Sempur No.1, RT.01/RW.01, Sempur, Kecamatan Bogor Tengah, Kota Bogor, Jawa Barat 16129, Indonesia

Corresponding author mail: tatty.yuni@gmail.com

Abstract : Fishing activities in Pangandaran Regency are centered in Cikidang Fish Landing Base (PPI), Babakan Village, Pangandaran Sub-district, Pangandaran Regency, West Java Province. The research objective is to identify the fishing profile in Panagandaran Sub-district during 2017-2021 and identify its potential utilization for the fishery product processing industry. The research method is descriptive data collection. The research location is in Pangandaran sub-district. Data taken are secondary and primary data from 2017-2022. Respondents are Joint Business Groups (KUB) based on Cikidang PPI, Pangandaran Regency Marine Fisheries and Food Security Office, Head of Cikidang PPI, and Pangandaran District village leaders. The results of the study identified the age profile of respondents aged 55-59 years (55%), basic education level (75%), 31-35 years of fishing experience (25%), no side job (85%), monthly income of 1-2 million (60%). The types of fishing gear used by fishermen are Vessel Seine, Drift Net, Liong Bun, Hand Line and seser. The dominant types of fish catches in a row from the highest in 2017 to 2021 are Rebon Shrimp, Yellowfin Layur, Krosok Shrimp, Dogol Shrimp, White Bawal and fish bycatch. By-catch fish are in the top 4 most abundant fish caught using vessel seine and drift net gear. Most of the by-caught fish from middlemen are made into salted fish. Fishing activity decreased in 2020-2021, but the value of fishing production increased. The utilization of caught fish is mostly salted fish, yet to be utilized for the fishery products industry.

1. Introduction

The Cikidang Port is designated as a PPI and must have met the requirements per Article 5 Letter d of the Minister of Marine Affairs and Fisheries Regulation No. 8 of 2012 concerning Fishing Ports. The PPI is determined based on technical and operational criteria, which include technical criteria, such as 1). able to serve fishing vessels conducting fishing activities in Indonesian waters; 2) have mooring facilities for fishing vessels of at least 5 GT; 3) the length of the pier is at least 50 m, with a pool depth of at least minus 1 m; 4) can accommodate fishing vessels of at least 15 units or a total of at least 75 GT; and 5) utilize and manage land of at least 1 Ha. In addition, PPIs are also required to meet operational criteria, which include fish loading and unloading activities and marketing of fishery products averaging 2 tons per day. However, in reality, PPI Cikidang requires improvements in catch transportation, catch data recording, and administrative management [1].

The captured fisheries data is the basis for planning, development, and decision/policy making. Fish catch data (HTI) needed to support the development of capture fisheries is precise and accurate. However, the fish caught by fishermen is not entirely for sale. Sometimes the catch is consumed during trip activities, but it is also distributed to the crew to take home and cook at home. The spread of fish distribution before the weighing process is an obstacle to obtaining actual fish catch data. The data collection on all fishing gear is necessary so that the data recorded and reported is the actual and accurate data [2].

Based on the type of fishing gear, the fish catch can be divided into two groups: main catch and by-catch. The main catch is the fish that is the main target of specific fishing gear. The type of fish not the primary purpose of fishing is called by-catch and is thrown overboard (discarded). The disposal process of by-catch (discarded). By-catch generally consists of fish smaller than the target species that are part of the by-catch as well as non-target species. There are several concerns regarding the disposal process of by-catch, such as: 1) good disposal of potential food; 2) adversely affecting the bottom environment; 3) reducing stocks of commercial target and non-target species [3].

It is well known that fish is an organic material with a composition of nutrients humans need for growth and for the smooth metabolism of their bodies [4]. Fish is a source of animal protein rich in amino acids lysine, threonine, valine, leucine isoleucine, and non-essential amino acids such as glutamic acid and glycine to provide different umami flavors [5]. A study on the nutritional composition of 20 species of non-targeted by-catch fish from the by-catch of tropical tuna revealed that these by-catch fish have a nutritional composition that is suitable for consumption and is recommended to be managed and processed as raw materials for processed fish in the fish processing industry. The fish were excellent sources of protein (min-max = 14.4-25.2 g/100 g fillet), had low omega-6/omega-3 ratios (<1, except for silky shark), and had mercury content below the safety limits defined by sanitary agencies. Silky sharks and Istiophoridae had the highest mercury contents (min-max = 0.029-0.317 ppm w/w) [6]. By-catch can reach 1.16% [7]; 40% of total fish catch [8] and 60% of total fish catch [9].

Based on preliminary interviews, using by-catches has been processed into salted fish. Whereas the quality of fish is indisputable as a source of excellent nutrition, it is necessary to research the potential of fish by-catch from PPI Cikidang for wider utilization based on data on fish caught by fishermen and fishermen's profiles. This study aims to identify the fishing profile and its by-products and assess the potential utilization of fish by-catch from PPI Cikidang, Pangandaran District, Pangandaran Regency, West Java.

2. Research Methods

The research was conducted from June to December 2022 in Cikidang Subdistrict, Pangandaran Regency.

2.1. Data and data sources

The primary data were obtained from interviews using questionnaires with the main actors/fisheries groups who are members of the Cikidang PPI-based Joint Business Group (KUB) in Pangandaran Subdistrict, Pangandaran Regency fisheries extension officers, Pangandaran Regency Marine Fisheries and Food Security Office, Head of Cikidang PPI and Pangandaran Subdistrict village leaders. Primary

IOP Conf. Series: Earth and Environmental Science 1289 (2023) 012014 doi:10.1088/1755-1315/1289/1/012014

data included respondent characteristics, types of fishing gear, types of catches, and number of catches.

The secondary data were obtained from analyzing reports and documents from the Pangandaran Regency fisheries extension officer, BPS (Central Bureau of Statistics) of Pangandaran Regency, in the form of fishing activities from 2017-2021. The determination of the number of respondents uses the Slovin equation so that the sample of respondents taken can represent the entire population, as follows:

$$n = \frac{N}{1+N.e^2} \tag{1}$$

2.2. Data processing method

The management of data analyzing data carried out using qualitative and quantitative descriptive approaches emphasizes the participatory direct observation aspect of the research. The research stages consisted of three stages, i.e.: (1) identifying the profile of fishermen; (2) identifying the main fishing profile and by-catch; (3) analyzing and recommending strategies for utilizing fishermen's by-catch.

2.3. The research respondent

The Cikidang PPI is included in type D fisheries port, a facility for government and fisheries business systems and includes shipping safety facilities and fisheries support activities (Permen KP No. 8 of 2012). The Pangandaran sub-district has 33 KUBs, with the number of members ranging from 6 to 23 members or Fishery Households (RTP) per KUB. The total number of households is 338. The following is the data of the KUBs in Table 1.

Table 1. Data on KUBs in Pangandaran Sub-district, Pangandaran Regency

No.	KUB Name	Address	Members
1	Baruna Jaya	Babakan Village	12
2	Bahtera Biru	Babakan Village	6
3	Banyuasin 2	Babakan Village	10
4	Baruna Jaya	Babakan Village	12
5	Budi Utomo	Babakan Village	3
6	Cikidang Indah	Babakan Village	6
7	Guripa Jaya	Babakan Village	7
8	Inti Jaya	Babakan Village	10
9	Janggrung	Babakan Village	10
10	Kerapu Jaya	Babakan Village	8
11	Lestari	Babakan Village	14
12	Lobster Mandiri	Babakan Village	10
13	Lobster Sejahtera	Babakan Village	7
14	Mina Bahari	Babakan Village	10
15	Mina Kamurang	Babakan Village	16
16	Mina Kelapa Tiga	Babakan Village	10
17	Mina Tuna	Babakan Village	10
18	Minasari Makmur	Babakan Village	23
19	Putra Cikidang	Babakan Village	10
20	Rengganis	Babakan Village	10
21	Samudra Abadi	Babakan Village	6
22	Tunas Bahari	Babakan Village	10
23	Usaha Mandiri	Babakan Village	10

24	Kebon Carik	Dusun Bojongkarekes Rt. 01 Rw. 16 Babakan Village	10
25	Mina Kerapu	Dusun Kalapatiga Rt. 01 Rw. 08 Babakan Village	12
26	Bandeng Jaya	Dusun Kalapatiga Rt. 02 Rw. 07 Babakan Village	10
27	Bahari Lestari	Dusun Kamurang Rt. 02 Rw. 12 Babakan Village	12
28	Bandeng Kembar	Dusun Kamurang Rt. 02 Rw. 12 Babakan Village	12
29	Mina Karanggedang	Dusun Kamurang Rt. 03 Rw. 12 Babakan Village	11
30	Mugia Berkah	Dusun Kamurang Rt. 04 Rw. 12 Babakan Village	11
31	Remond	Dusun Kamurang Rt.01 Rw. 12 Babakan Village	10
32	Jaya Mandiri	Dusun Kamurang Rt.02 Rw. 12 Babakan Village	10
33	Batu Layar	Dusun Karanggedang Rt. 04 Rw. 06 Babakan Village	10
Total number of fishery households (RTP)			338

3. Results and Discussion

3.1. The profile of PPI Cikidang Pangandaran District

The Pangandaran Regency was formed based on Law Number 21 of 2012 concerning the Establishment of Pangandaran Regency in West Java Province. The area of Pangandaran Regency is 168,509 Ha, with a sea area of 67,340 Ha. Pangandaran Regency has a beach length of 91 Km. The law states that Pangandaran Regency originated from part of Ciamis Regency, which consists of Parigi District, Cijulang District, Cimerak District, Cigugur District, Langkaplancar District, Mangunjaya District, Padaherang District, Kalipucang District, Pangandaran District, and Sidamulih District. The capital of Pangandaran Regency is located in Parigi Subdistrict. Pangandaran Regency has 241 Joint Business Groups (KUB).

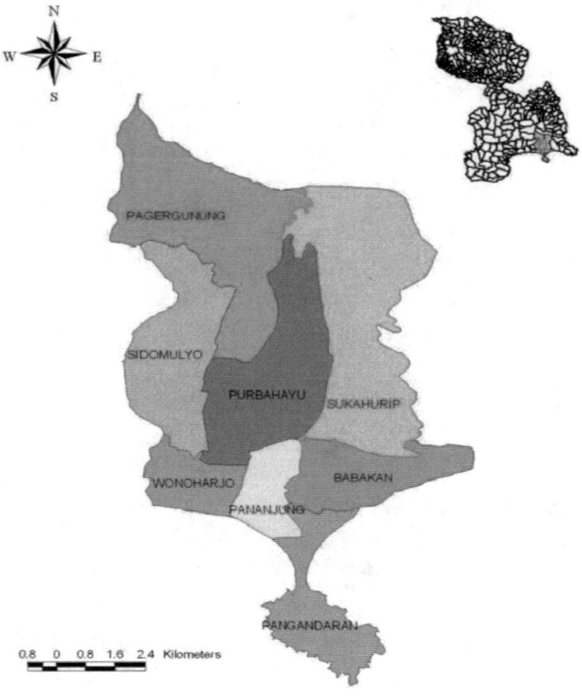

Figure 1. Pangandaran Sub-district Area Map

The sub-district is approximately 23 km from the capital of Pangandaran Regency and 168 km from the capital of West Java Province. The government center of Pangandaran Sub-district is located in Pananjung Village. Geographically, Pangandaran sub-district is located at 6A°19` - 6A°47` South latitude and 106A°21` - 10A°13` East longitude. Pangandaran sub-district has an area of 82.65 km^2. It consists of eight villages stretching from south to north, consisting of three coastal villages and five non-coastal villages. Sidomulyo Village is the largest of the villages at 17.00 km^2 (22.38% of the total area of Pangandaran Sub-district), while Pananjung Village has the smallest area at 4.71 km^2 (5.70% of the total area of Pangandaran Sub-district). The Pangandaran Sub-district area is mostly in the form of sloping land, hills, and mountains. Pangandaran sub-district is between 0-137.00 meters above sea level. Air temperature varies and is also determined by altitude, averaging 23°C. Fishing activities in Pangandaran Regency are centered at PPI Cikidang. PPI Cikidang is located in Pangandaran District, precisely in Babakan village. PPI Cikidang is currently managed directly by the Government Center UPT and is applying to become an Archipelago Fishing Port (PPN). The area of PPI Cikidang reaches 5.4 ha. There are port facilities in the form of docks, harbor ponds that are sheltered from waves, fish auction sites, ice warehouses, and PPI Cikidang administrative offices. The main port facilities, such as the breakwater, dock, TPI, and administrative office at PPI Cikidang, play a role in supporting fishing business activities.

3.2. A profile of fishermen in Pangandaran Subdistrict

The respondents are fishermen who are members of KUB who go to sea in the Cikidang PPI area. The age of fishermen is dominated by productive age, 55%, which is 55-59 years old. The education level of fishers is dominated by basic education, which is 75%. Although most fishermen's work experience is more than 30 years, it can be stated that fishermen still need to be prosperous. Fishermen are not prosperous because their income is insufficient to meet their living needs, and working time fulfills their needs and work time [10]. About 55% of fishermen's income is IDR. 1,000,000 to IDR. 2,000,000, and only 5% of fishermen have an income of IDR. 4,000,000 to IDR. 5,000,000. This is less than the minimum wage in Pangandaran Regency. This year's West Java UMP increase is listed in Surat Keputusan Nomor 561/kep/752/Kesra on the West Java Provincial Minimum Wage for 2023, where starting January 1, 2023 the minimum wage in Pangandaran Regency is IDR 2,018,389.00. The average time spent at sea is 6-8 hours. Research conducted by Jeni and Nugroho, (2020) stated that the influence of fishermen's fishing time is correlated with fishermen's income. Moreover, from the identification of side jobs, around 85% of fishermen do not have side jobs, so their time is devoted to fishing. The characteristic profiles of age, education, length of time in business, income, fishing time, and side jobs of fishermen are presented in Figure 2 below.

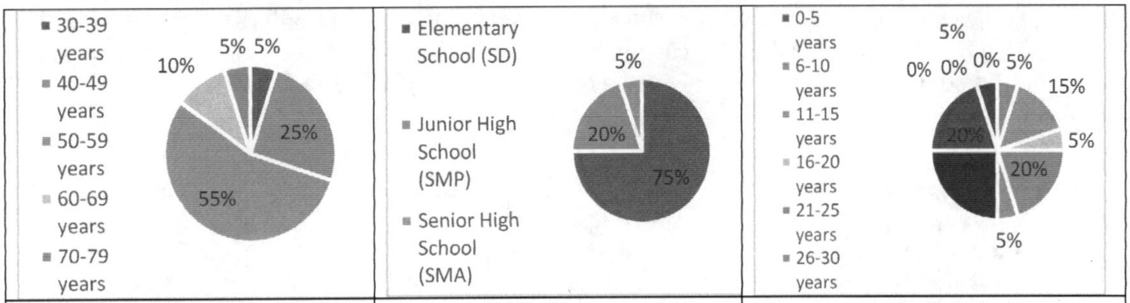

A B C

IOP Conf. Series: Earth and Environmental Science 1289 (2023) 012014 doi:10.1088/1755-1315/1289/1/012014

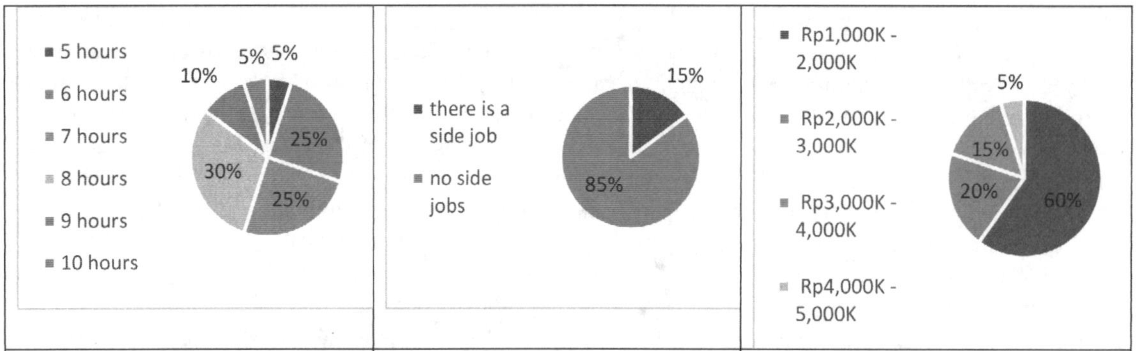

D	E	F

Figure 2. Profile of fishermen in PPI Cikidang (1A. Age of Fisherman; 1B. Educational level of fisherman; 1.C work experience of fisherman; 1D Daily working time per trip of fisherman; 1E Side jobs of Fisherman; 1F income per month of fisherman)

The profile of fishermen's catch based on the type of catch is divided into main catch and by-catch, as presented in Figure 2. During the fishing trip, the crew members (70%) mostly consist of 2 people per trip. The main catch is 10-30 kg per trip. The catch will vary per trip depending on several factors. Factors determining fishing productivity include the number of crew members, fishing trip, GT of the vessel, PK of the vessel's engine, and net length [12]. The volume of fish caught also depends on the season [13]. Most of the main catch is sold at fish auctions (TPI) (90%). The remaining fish were sold to middlemen. By-catch fish obtained an average weight of 5-10 kg per trip (75%).

In contrast to the main catch, by-catch fish are mostly sold to middlemen (60%), some to TPI, and the fishermen sell the rest. Fishermen still sell some of their fish catch to middlemen, not TPI. PPI Cikidang, as a type D port, has a marketing business facility, TPI. Fishermen do not fully utilize TPI. The TPI functions to sell and auction fish so that the price given by the seller becomes feasible.

Nevertheless, the fish auction (TPI) is not running due to the management of the fish auction needs to prioritize excellent service. The lack of funding for fishing makes fishermen have socio-economic closeness to non-formal institutions such as middlemen. In that case, there is no agreement between fishermen and buyers to carry out the auction, and there is no strict policy enforcement so that the auction is ongoing [14]. The middlemen even give low prices. Implement a comprehensive fish auction system so small-scale fishermen have a bargaining position in the fisheries trade system and get cash proceeds from their sales [15].

Interestingly, the fishermen at PPI Cikidang differentiate the sales location of the main catch and bycatch. By-catch fish have a low price and need to be handled thoughtfully, further reducing the selling value of the fish. Most of the by-caught fish from middlemen are made into salted fish. Images of various salted fish preparations from by-catch fish are presented in Figure 3 below.

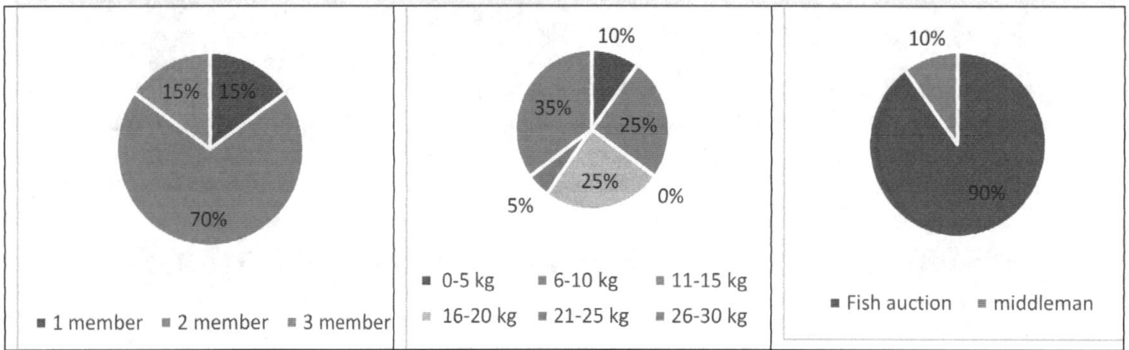

A	B	C

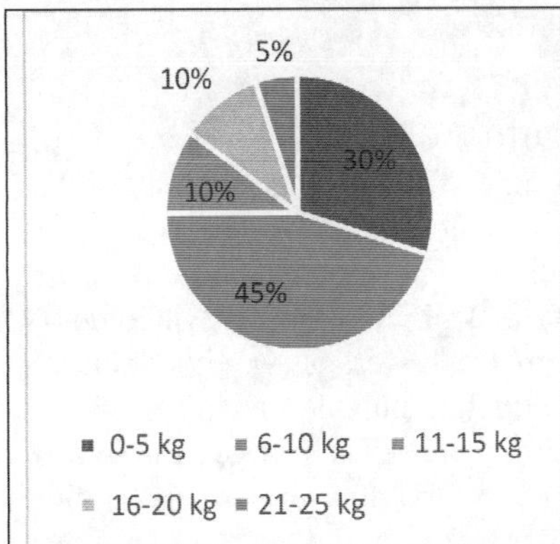

D

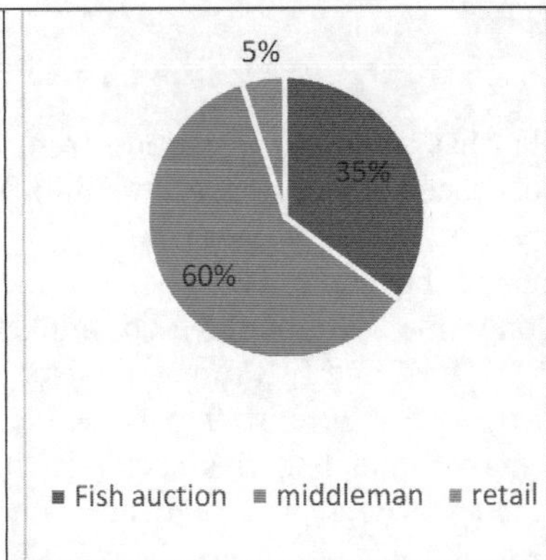

E

Figure 3. Profile of fish caught by Cikidang PPI fishermen: A. The number of crew members; B. Average main catch volume sold per trip; C. Points of sale of the main catch; D. Average volume of by-catch sold per trip; E. Points of sale of the main catch

Figure 4. Various processed salted fish from the by-catch of fish fishermen PPI Cikidang

3.3. The profile of fishing at PPI Cikidang

The fishing trend from 2017 to 2020 has decreased both in terms of weight (tons) and value (IDR.) (Figure 5). However, in 2021 there was an increase in fishing volume/weight. This decrease in fishing behavior is common in some parts of Indonesia's fishing grounds. The diversity of catches includes main catch and by-catch fish. The productivity of the dominant fish catches, such as calakang, layang, and tuna, using purse seine at Tamperan Beach Fishing Port, Pacitan, has decreased in tons/GT during the period 2016 to 2020 [12]. However, in other areas, the landed Lemuru (Sardinella lemuru) catch at the Kedonganan PPI, Bali Indah has an increasing trend from 2010-2019 [16]. The pandemic of Covid 19 in 2019-2021 greatly influenced the downward trend in fishing volume. The outbreak of the coronavirus and the implementation of regulations on PSBB (Large-Scale Social Restrictions) have reduced community activities outside the home, be it work, education, or even transportation. The decline in action, if it lasts for a long time, will undoubtedly impact the decline in the daily income of the community, especially local fishermen and fish farmers. Also, some fish storage warehouses have accumulated raw fish materials or overstock because they cannot be supplied outside the area as usual. The impact of Covid-19 on the marine and fisheries sector is the decline in fish prices in several regions, reaching a 50% decline due to decreased demand such as several fish export companies closing, many hotels and restaurants closed, which are the main buyers of fish and seafood [17]. On the other hand, the government's policy through Minister of Marine Affairs and Fisheries Regulation No. 2 of 2015, which prohibits the use of helicopter trawls and drag trawls, has had an impact on the volume of fishing in several areas on the island of Java [18]. The government's policies aimed at sustainable fisheries and conservation must be balanced with solutions to overcome the decline in fishermen's income.

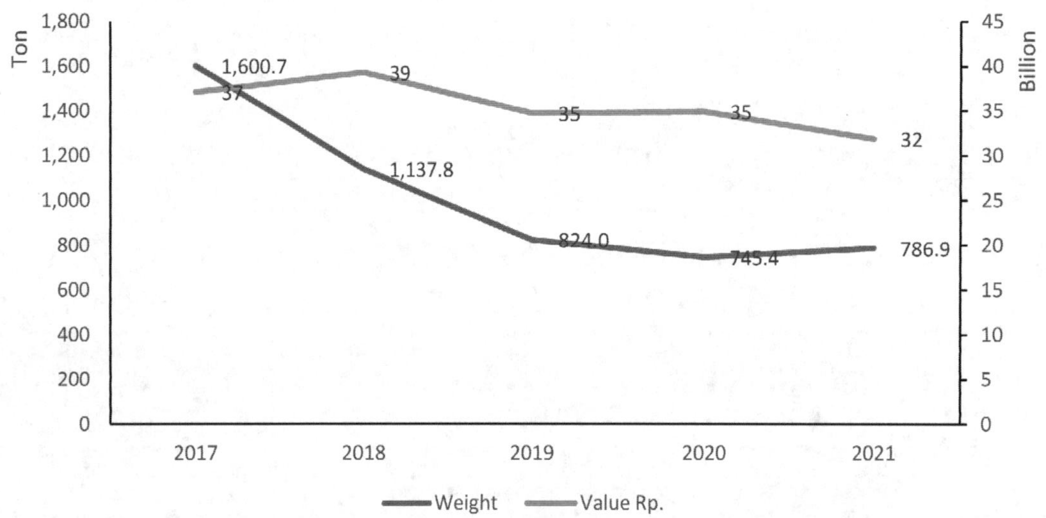

Figure 5. The Profile Fish catch at PPI Cikidang at 2017-2021

Sources: Data processing results from fisheries extension at Pangandaran sub-district (2022)

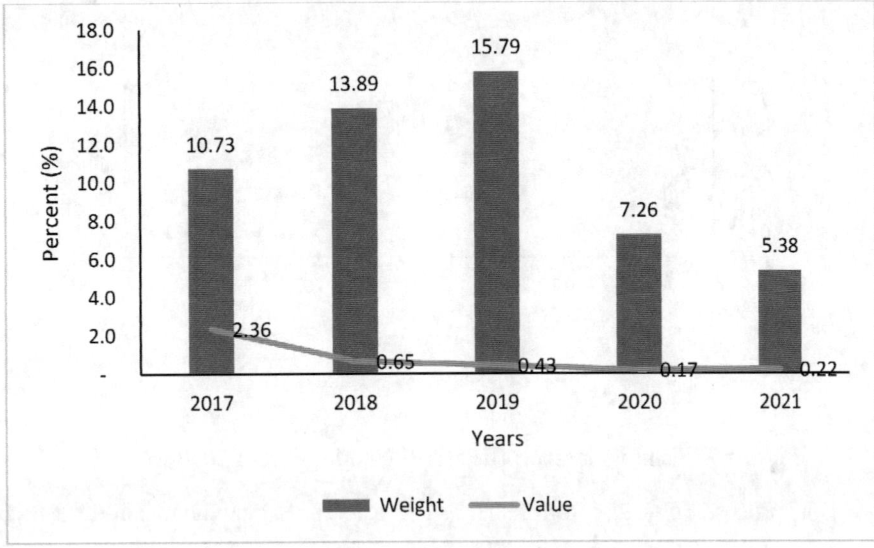

Figure 6. Percent of by-catch fish from the total fish catch at PPI Cikidang from 2017 to 2021

Sources: Data processing results from fisheries extension at Pangandaran sub-district (2022)

The percentage of by-catch fish to total fish caught varied from 5.4% to 15.4%. The lowest rate of by-catch fish was in 2021 at 5.4% of total fish caught, and the highest was in 2019. The downward trend in the volume of by-catch is linear with the overall decline in the volume of fish caught. Similarly, the selling value of by-catch fish is very low, only around 0.22% to 2.36%, not proportional to the volume of by-catch. The percentage of by-catch fish by-catch varies depending on the type of fishing gear. In tuna longline gear, the Benoa, Bali, fishery produces 69% main catch fish and 31% by-catch fish [19]. The trend of by-catch fish volume decreases along with the decrease in total fish catch from 2017 to 2020. However, in 2021 there was an increase in the volume of by-catch fish. The volume of by-caught fish has the lowest volume of around 24 tons in 2021 and the lowest value of IDR 47,000,000. Fishermen usually sell this by-catch fish through middlemen, not through the Cikidang PPI fish auction site. Some of the by-caught fish are used as raw material for salted fish. The price of fish is affected by the quality of the fish and the type of fish. The quality of fish will change and decrease along with the process of distributing fish from fishermen to consumers. Factors that affect fish quality during distribution are sanitation, hygiene, transportation, and handling by officers [20]. The following is Figure 7, which shows the downward trend in by-catch fish from 2017 to 2021 at PPI Cikidang.

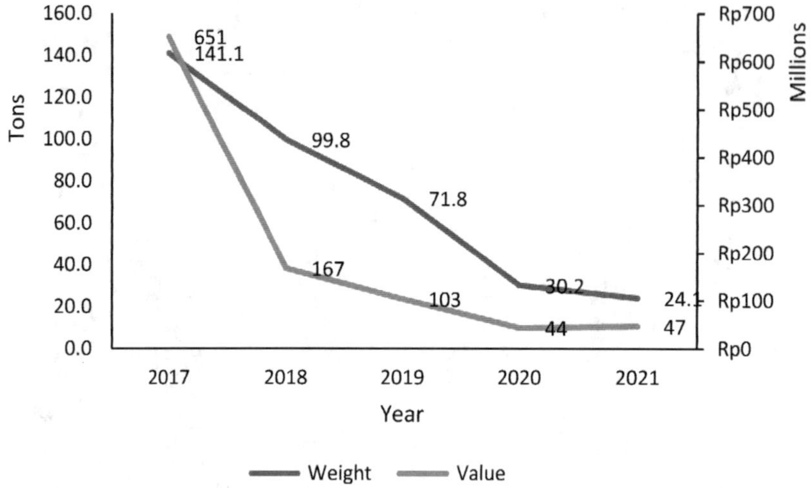

Figure 7. Fish by-catch profile at PPI Cikidang from 2017 to 2021

Sources: Data processing results from fisheries extension at Pangandaran sub-district (2022)

The dominant types of fishing gear and the most fishing volume from 2019-2021 at PPI Cikidang include Vessel Seine, Drift Net, Liong Bun, Hand Line, and seser, using outboard motor type vessels and motorboats less than 5GT. The main dominant fish catches that were rebon shrimp, krosok shrimp, jerbung shrimp, tiger shrimp, dogol shrimp, white pomfret, manyung fish, layur fish, tuna fish, tigawaja fish, tongue fish, bondolan peperek fish, mahi-mahi fish, gulamah fish, banyar fish and bycatch fish. Report from other research explained that the main capture fisheries commodities in 2019 in Pangandaran Regency include layur, black pomfret, white pomfret, kuwe, grouper, lobster, dogol shrimp, krosok shrimp, white shrimp, tiger shrimp, octopus, three waja, mackerel, japuh, julung-julung, layaran, yellow tail, golok-golok, cotton-cotton, and beloso. These commodities can support the economy of Pangandaran Regency [21]. By-catch fish are in the top 4 most abundant fish caught using vessel seine and drift net gear. As known, these bycatch fish have low value but have high volume. By-catch fish is a cheap source of fish raw materials, but it can still be utilized. In vessel seine gear, the species of bycatch fish usually include cumi-cumi, sotong, kepiting, rajungan, ikan buntal, pepetek, baji-baji [22]. Table 2-4 shows the four dominant types of fishing gear and fish catches from 2019-2021.

Tabel 2. The 4 top types of fishing gear and the majority of fish species caught at PPI Cikidang in 2019

Vessel Seine | Outboard Motors< 5 GT	
Fish Caught	Weight (kg)
Rebon Shrimp (*Acetes indicus*)	96,765.8
(By-catch)	53,993.7
Krosok shrimp (*Parapenaeopsis sculptilis*)	49,760.9
Tongue sole fish	20,802.9
Tigawaja fish (*Johnius dussmieri*)	6,875.1
Drift Net | Outboard Motors < 5 GT	
Fish Caught	Weight (kg)

Yellow tail layur (*Trichiurus lepturus*)	29,174.1
Pomfret fish	27,812.6
Tongue sole fish	20,376.8
(By-catch)	13,380.4
Tigawaja fish (*Johnius dussmieri*)	12,530.8
Trammel Net | Motor Ship < 5 GT	
Fish Caught	Weight (kg)
Dogol shrimp (*Metapenaeus monoceros*)	65,165.2
Jerbung shrimp (*Fenneropenaeus merguiensis* de Man)	30,437.2
Giant tiger prawn (*Penaeus monodon*)	9,743.2
Tongue sole fish	9,017.0
Krosok shrimp (*Parapenaeopsis sculptilis*)	5,053.9
Liong Bun | Motor Ship 5 - 10 GT	
Fish Caught	Weight (kg)
Mackarel (*Scomberomorus brasiliensis*)	12,448.8
Black fin layur (*Gempylus serpens*)	11,433.5
Gulamah (*Johnius trachycephalus*)	8,931.6
Banyar (*Rastrelliger kanagurta*)	6,568.6
Green Manyung (*Netuma thalassina*)	5,304.7

Tabel 3. The 4 top types of fishing gear and the majority of fish species caught at PPI Cikidang in 2020

Vessel Seine | Outboard Motors < 5 GT	
Fish Caught	Weight (kg)
Rebon Shrimp (*Acetes indicus*)	120,410.7
Krosok shrimp (*Parapenaeopsis sculptilis*)	60,299.9
(By-catch)	26,892.6
Tongue sole fish	18,134.8
Tigawaja fish (*Johnius dussmieri*)	2,829.2
Trammel Net | Motor ship < 5 GT	
Fish Caught	Weight (kg)
Dogol shrimp (*Metapenaeus monoceros*)	97,397.3
Jerbung shrimp (*Fenneropenaeus merguiensis* de Man)	43,147.4
Giant tiger prawn (*Penaeus monodon*)	15,391.0
Krosok shrimp (*Parapenaeopsis sculptilis*)	9,137.3
Tigawaja fish (*Johnius dussmieri*)	4,535.7
Drift Net | Outboard Motors < 5 GT	
Fish Caught	Weight (kg)
Pomfret fish	37,055.7
Yellow tail layur (*Trichiurus lepturus*)	18,861.2
Tigawaja fish (*Johnius dussmieri*)	7,916.7

IOP Conf. Series: Earth and Environmental Science 1289 (2023) 012014 doi:10.1088/1755-1315/1289/1/012014

Sand lobster (*Panulirus homarus*)	6,341.7
Peperek Bondolan (*Gazza minuta*)	4,850.2
Hand Line | Motor Ship 5 - 10 GT	
Fish Caught	Weight (kg)
Cakalang (*Katsuwonus pelamis*)	38,683.2
Black fin layur (*Gempylus serpens*)	25,059.2
Yellowfin tuna (*Thunnus albacares*)	22,486.9
Tongkol (*Euthynnus affinis*)	2,580.6
Lemadang-mahi-mahi (*Coryphaena hippurus*)	2,138.2

Tabel 4. The 4 top types of fishing gear and the majority of fish species caught at PPI Cikidang in 2021

Vessel Seine | Outboard Motors < 5 GT	
Fish Caught	Weight (kg)
Rebon Shrimp (*Acetes indicus*)	187,085.8
Krosok shrimp (*Parapenaeopsis sculptilis*)	64,875.5
(By-catch)	23,537.8
Tongue sole fish	17,105.6
Chicken feather fish (*Thryssa mystax*)	4,829.9
Trammel Net | Motor ship < 5 GT	
Fish Caught	Weight (kg)
Dogol shrimp (*Metapenaeus monoceros*)	85,084.1
Giant tiger prawn (*Penaeus monodon*)	23,091.6
Jerbung shrimp (*Fenneropenaeus merguiensis* de Man)	20,204.0
Krosok shrimp (*Parapenaeopsis sculptilis*)	5,890.8
Tigawaja fish (*Johnius dussmieri*)	1,926.2
Drift Net | Outboard Motors< 5 GT	
Fish Caught	Weight (kg)
Yellow tail layur (*Trichiurus lepturus*)	21,501.9
Pomfret fish	12,891.7
Tigawaja fish (*Johnius dussmieri*)	7,251.8
Tongue sole fish	4,086.8
Peperek Bondolan (*Gazza minuta*)	2,485.5
Seser | Outboard Motors < 5 GT	
Fish Caught	Weight (kg)
Sardinella (*Sardinella* sp.)	34,035.0
Decapterus spp.	24,732.0

4. Recommendations for strategies to utilize by-catch fish at PPI Cikidang

The Government of Indonesia, through the Ministry of Marine Affairs and Fisheries, supports the implementation of the Blue Economy Concept, which teaches how to create zero-waste products. This addresses the challenge of food insecurity through increased fish catch production, and contributes to increasing the country's foreign exchange by increasing the volume and value of exports of aquaculture and fish capture commodities. In addition, it has an essential role in improving the community's welfare and creating jobs. The principle of Blue Economy is not only environmentally friendly, but multiple cash flow means that there are various benefits economically because waste can be of economic value to produce other products, besides that Blue Economy must be able to empower the community through employment (social inclusive)" and utilization of by-catch fish.

Fishing profile at PPI Cikidang shows decreasing activity from 2017 to 2020, but starts to increase in 2021. In this fishing activity, the main target fish is caught according to the type of fishing gear, and bycatch fish are also caught. By-catch fish from fishing activities in Pangandaran Sub-district, centered at PPI Cikidang, is caught annually during fishermen's fishing trips. These fish do not have high economic value. So far, the sale of by-catch fish is mostly done by fishermen to middlemen. Middlemen buy cheap by-catch fish for IDR. 2000,- to IDR. 4000,- per kg. The price of fish will be linear with the quality and type of fish. On the other hand, fishermen have an income from the fishery field [23].

The blue economy concept when applied to fishing in PPI cikidang, that is through measured fishing and utilizing by-catch fish into value using the valorization principle where all parts of the fish body will be added to its economic value. Part of the fish can be utilized at every stage of extraction without producing waste.

PPI Cikidang, as a source of fish raw materials, has the potential to play a role in utilizing by-catch fish. The size of PPI Cikidang, along with the existing facilities for a level D fish landing port, makes it possible to build a fish processing plant for by-catch. The local government has a role in making policies and budgets that support such installations.

Some alternatives for utilizing by-catch fish that can provide added value include using fish for making surimi [24], calcium flour manufacturing [25], fish oil manufacturing [26], and manufacturing of fish protein hydrolysate [27].

5. Conclusion

The profiles of fishermen from PPI Cikidang, Pangandaran Sub-district, are typical of the profile of fishermen in Indonesia. Human resources that require improvement, resulting in low income. Fishery resources in the form of fish by-catch that have low economic value, have the potential to be developed into high-value fishery commodities (valorization). As a superior raw material, fish undeniably can be utilized for the economy and natural preservation by applying the blue economy..

Acknowledgement

Thank you for the research funding support from BRIN through the decision letter of the deputy for research and innovation No 65/II.7/HK/2022 concerning the recipients of the research and innovation program for Indonesia advanced (RIIM) batch 1 in 2022.

Reference

[1] Syauqi I Muninggar R and Oktariza W, 2020 The Role of Cikidang Fish Landing Base (PPI) in Drift Gillnet Fisheries Bussiness Oleh: *ALBACORE J. Penelit. Perikan. Laut* **3**, 3 p. 321– 330.

[2] Sari, Suci Arum; Sutjipto, Darmawan Ockto; Lelono T D, 2022, Analysis of Catch Data Collection at the Archipelago Fishing Port (PPN) Prigi Trenggalek Regency: Case Study of 2 Vessel Ring Trawl., Brawijaya.

[3] Gilman, Eric; Milani Chaloupka M and Booth, Hollie; Hall, Martin; Murua, Hilario; Wilson J, 2023 Bycatch-neutral fisheries through a sequential mitigation hierarchy *Mar. Policy* **150**, 105522.

[4] Yuniarti T *et al.*, 2021 *Fisheries raw material knowledge-Pengetahuan bahan baku perikanan* 1st ed. Medan: Yayasan Kita Menulis.

[5] Yuniarti T Prayudi A Supenti L Suhrawardan H and Martosuyono P, 2021 The Hydrolysis Protein Profile of the by-product of the Fresh Shrimp Processing Industry *J. Perikan. Univ.*

Gadjah Mada **23**, 1 p. 63.

[6] Sardenne, Fany; Bodin, Nathalie; Médieu, Anaïs Antha, Marisa; Arrisol, Rona; Le Grand, Fabiena; Bideau, Antoine; Munaron, Jean-Marie; Loc'h, François Le; Chassot E, 2020 Benefit-risk associated with the consumption of fish bycatch from tropical tuna fisheries *Environ. Pollut.* **267**, 115614.

[7] Manalu I N Bustari B and Zain J, 2022 Analysis of Main Catch and Bycatch on Catching Devices "Gombang" in Sungai Gayung Kiri Village Rangsang Subdistrict Kepulauan Meranti Regency *J. Coast. Ocean Sci.* **3**, 3 p. 187–191.

[8] An Najah R *et al.*, 2015 The Study of Catch Production Market Value in PPS Nizam Zachman and PPI Muara Angke *Mar. Fish.* **6**, 2 p. 155–167.

[9] Luthfiani L Ghofar A and Purwanti F, 2018 Bycatch Type Composition of Pushnet in The Tambak Lorok, Semarang *Manag. Aquat. Resour. J.* **7**, 3 p. 288–297.

[10] Anna Z Rizal A and Anitaningrum M, 2019 Analysis Fishermen Term of Trade in Pangandaran Subdistrict of Pangandaran Regency *World Sci. News* **117**, January p. 13.

[11] Jeni J and Nugroho F, 2020 Pengaruh Curahan Jam Kerja Pada Rumah Tangga Kabupaten Batu Bara Provinsi Sumatera Utara *J. Sos. Ekon. Pesisir* **1**, 3 p. 63–67.

[12] Muzayanah L Imron M and Baskoro M S, 2022 Productivity and Dominant Fishing Season of Purse Seine at the Tamperan Fishing Port, in Pacitan District *Mar. Fish. J. Mar. Fish. Technol. Manag.* **13**, 1 p. 31–43.

[13] Rahmawati M Wijayanto Fitri A D P and Dian, 2013 Analysis of catch per unit effort and the Pattern of anchovies (Stolephorus spp.) fishing season in Pemalang waters *J. Fish. Resour. Util. Manag. Technol.* **2**, 3 p. 213–222.

[14] Muninggar R Dinarwan D and Anggara C, 2013 Influential Factors of Fishermen Sell Their Catch Outside Lampasing Fish Auction: A Policy Review *Bul. PSP* **21**, 2 p. 274457.

[15] Lubis E Pane A B Muninggar R and Hamzah A, 2012 Besaran Kerugian Nelayan dalam Pemasaran Hasil Tangkapan : Kasus Pelabuhan Perikanan Nusantara Palabuhanratu *Maspari J.* **4**, 2 p. 159–167.

[16] Nurtira I Restu I W and Pratiwi M A, 2021 Produksi dan Pertumbuhan Ikan Lemuru (Sardinella lemuru) yang Didaratkan di PPI Kedonganan, Bali *Curr. Trends Aquat. Sci. IV* **4**, 2 p. 141–151.

[17] Mery Nova Sari Fransiska Yuliasara and Mahmiah, 2020 Dampak Virus Corona (Covid-19) Terhadap Sektor Kelautan dan Perikanan : A Literature Review *J. Ris. Kelaut. Trop. (Journal Trop. Mar. Res.* **2**, 2 p. 41.

[18] Nababan B O Solihin A and Christian Y, 2018 Dampak Sosial Ekonomi Kebijakan Larangan Pukat Hela Dan Pukat Tarik Di Pantai Utara Jawa *Indones. Mar. Fellows Progr. – Mfp* p. 80.

[19] Ghofar A *et al.*, 2016 Type and Size Distribution of By Catch Tuna LongLine landed at Benoa Port, Bali *DIPONEGORO J. MAQUARES* **5**, 4 p. 453–460.

[20] Afiyah N N Solihin I and Lubis E, 2019 The Effect of Distribution Chain on The Quality of Tuna Mackerel (Euthynus sp.) from "PPP Blanakan" to Consumer Area *J. Sos. Ekon. Kelaut. dan Perikan.* **14**, 2 p. 225.

[21] Kartika L Nurhayati A Dewanti L P and Rizal A, 2020 The Contribution Of Capture Fisheries In Supporting The Economy In Pangandaran District *Syntax Idea* **2**, 8 p. 50–57.

[22] Ronny I. W M. Fedi A. S Sugeng H. W and John H, 2008 The composition of target catch and the bycatch from small scale demersal trawl fisheries in the West Java Water's *Bul. PSP* **17**, 3 p. 306–314.

[23] Caruso G Floris R Serangeli C and Di Paola L, 2020 Fishery Wastes as a Yet Undiscovered Treasure from the Sea: Biomolecules Sources, Extraction Methods and Valorization *Mar. Drugs* **18**, 12.

[24] Sihono Purnomo A H Wibowo S and Dewi F R, 2021 Current (2021) status of surimi industry in Indonesia and possible solutions: A review *IOP Conf. Ser. Earth Environ. Sci.* **919**, 1.

[25] Kusumawati P Triwitono P Anggrahini S and Pranoto Y, 2022 Nano-calcium Powder Properties from Six Commercial Fish Bone Waste in Indonesia *Squalen Bull. Mar. Fish. Postharvest*

Biotechnol. **17**, 1 p. 1–12.
[26] Viruly L and Muzahar, 2014 Extraction of fish oil from fish caught (by catch) on Ungar Island, Kab. Karimun *Din. Marit. J. Pesisir, Kelaut. dan Perikan.* **4**, 2 p. 96–102.
[27] Prayudi A Yuniarti T Taryoto A and Supenti L, 2020 Chemical and amino acid composition of snapper scrap meat hydrolysate *AACL.Bioflux* **13**, 4 p. 2228–2241.

IOP Conf. Series: Earth and Environmental Science 1289 (2023) 012015 doi:10.1088/1755-1315/1289/1/012015

Prelimenary Study Technology Readiness Level and Techno-economic Analysis of Seaweed-Based Soft Capsule Prototypes

T. Hidayat*[12], K. Syamsu[1], T.C. Sunarti[1], M. Nurilmala[3], L.P. Manalu[2]
[1]Department of Agricultural Industrial Engineering, IPB University
[2]Agro-Industry Research Center, National Innovation Research Agency
[3]Department of Aquatic Product Technology, IPB University

*Corresponding author mail: tauf023@brin.go.id

Abstract

Product downstream to be marketable can be seen from the technological and techno-economic readiness level. This research aims to determine the level of technological and techno-economic readiness of the seaweed soft capsule prototype. The method used is to determine TRL (Technology Readiness Level) based on technometers and techno economics by looking at the sensitivity and feasibility of the business. The results show that the readiness level of the seaweed soft capsule prototype technology is at level 7, meaning that it has been scaled up but several parameters in industrial scale scale-up must be improved. For the feasibility techno-economic analysis of the seaweed soft capsule prototype, the Net B/C result was obtained with a value of 1.64. The NPV results obtained were 4,621,393,281, with an IRR of 33%. The payback period is estimated to be 3.01 years. The level of technological readiness and feasibility results show that the development of soft capsule prototypes for downstream commercialization has good prospects, as long as the availability of raw materials is guaranteed.

1. Introduction

Soft capsule is a medicinal preparation whose shell consists of a film sheet hermetically sealed. Soft capsule shells are used for drugs or internal active substances liquid or suspension form at a certain dose. Soft capsules can also reduce odor or unpleasant odor or taste of the active substances contained in the capsule. According to [1], in the pharmaceutical industry, almost all drug candidates developed have low water solubility, and the use of soft capsules can encapsulate insoluble oils of water. Currently, soft capsules are one of the most widely used oral dosage forms used because they are easy to swallow, flexible in changing doses, and easy to combine with drugs tailored to the patient's needs [1]. Soft capsule production data from year to year shows an increasing trend and is even predicted to reach an average annual growth (CAGR) of 5.4% to USD 316.6 billion in 20. 25 ([2]. Regionally, one of the rapidly growing market shares is the Asia-Pacific market share, including Indonesia. Traditional soft capsules are made of gelatin so they are often referred to as softgels. Gelatin is a type of protein produced from partial extraction of animal collagen. Of the total available gelatin production, the highest source of gelatin comes from pig skin (46%), cow skin (29.4%), and pork and beef bones (23.1%). Only about 1.5% of total fish gelatin is produced [3]. The use of gelatin as a soft capsule has some drawbacks such as having a distinctive odor and not being suitable for patients who have concerns about halal and patients with a vegetarian lifestyle [4]. In addition, gelatin has low stability, especially at high temperatures [5]. Gelatin that is not properly modified will be susceptible to cross-linking and will be a problem when releasing drugs in vitro. Migration of drugs or formula components (for example hydrophilic co-solvent) into the capsule shell, as well as the movement of water between the capsule shell and the drug formula, is also critical. After the encapsulation process, the gelatin shell has a fairly high water content, namely around 35% w/w. The water content can move from the capsule shell towards the contents and reduce the solubility of the drug resulting in precipitation in the filling, but this depends on the storage period. To overcome some of these shortcomings, there are non-gelatin-based soft-gel alternatives are important to develop.

The development of non-gelatinized soft capsules has also been carried out with various sources of raw materials. The research that has been done is based on modified starch with different types of plants [6], using iota carrageenan [7,8], as well as various combinations using polyvinyl alcohol [9]. The results of this study indicate that non-gelatin soft capsule prototypes can be produced so downstream and commercialization must be carried out. Assessment of seaweed-based soft capsule prototypes from technological and techno-economic readiness has never been done. Therefore, this research aims to determine the level of technological and techno-economic readiness of seaweed-based soft capsules.

2. Materials and Methods

2.1 Materials and Tools

The material used in this research was modified iota carrageenan starch, plasticizer, and fillers. The tools used are a moisture analyzer, disintegrator, and technometer excel.

2.2 Methods

This research was carried out by determining the performance of the seaweed soft capsule prototype by measuring the water content and disintegration time. The next step is to assess technological readiness using a technometer sheet. After doing the calculations, a business feasibility assessment process for seaweed-based soft capsules is carried out.

2.2.1 Research Procedure

2.2.1.1 Water content [10]

The water content is analyzed to determine the amount of water content in the sample. Samples of 1-2 grams were dried using an oven at 105°C for 3 hours, then cooled in a desiccator., then weighed until a constant weight was obtained. Calculation of water content is determined using the following formula:

$$\text{Water content} = \frac{B-C}{B-A} \text{ x } 100\% \text{ .... } 1)$$

Description:
A = Empty cup weight (g)
B = Cup weight + initial sample (g)
C = Cup weight + dry sample (g)

2.2.1.2 Disintegration

The disintegration test was carried out in accordance with the Korean pharmacopoeia. Briefly, each of the six sample capsules was placed in a disintegration tester glass tube (DIT 400, LabFine Co., Ltd., Seoul, Republic of Korea) containing distilled water at 37 ± 2◦C. In the tester, the capsule was stirred up and down (amplitude 53 – 57 mm) at a speed of 29 – 32 times per minute. Capsules are declared disintegrated if no part of the capsules remains on the screen, except for the fragments originating from the coating agent. Record the time it takes for each capsule to disintegrate.

2.2.1.3 Technometer

This technometer measurement is a quick measurement to assess the level of readiness technology on seaweed-based soft capsules. Soft capsules are analyzed by looking at indicators of TRL 1 to TRL 9. The analysis results are shown with the fulfilled. If the value is reached, then you can see the resulting TRL.

2.2.1.4 Techno-economic Analysis

Eligibility parameters that will be used as indicators in determining eligibility investment are NPV, IRR, Net B/C Ratio, and Pay Back Period.

Table 1. Indicators of Investment Feasibility Analysis

Criteria	Indicator
NPV > 0	The project/business is feasible to be implemented
Net B/C > 1	The project/business is feasible to be implemented
IRR > i	The project/business is feasible to be implemented
PBP	Relative

2.2.1.5 Break Even Point (BEP)

Break Even Point (BEP) is used to see sales volume, where the company has been able to cover all its costs without experiencing profit or loss. According to [11], the definition of BEP (*Break Even Point*) or break even point is the volume of sales in which the total revenue and total expenses do not differ, there is no net profit or loss. The calculation of BEP (break even point analysis) is as follows:

$$\text{BEP} = \frac{Fixed\ cost}{1-(Variable\frac{cost}{Total}revenue)} \ \ldots\ldots\ldots\ldots\ldots 2)$$

2.2.1.6 Net Present Value (NPV)

Net Present Value (NPV) is used to calculate the difference between the present value of cash inflows and the present value of cash outflows. If the NPV is positive, the business or project makes a profit, and vice versa. NPV can be calculated with the equation [11].

2.2.1.7 Internal Rate of Return (IRR)

Internal Rate of Return (IRR) is a rate of return on capital used in a business or project and expressed in percent per year. A project that is feasible to implement has an IRR value greater than the discount rate value. According to [11], IRR can be calculated by equality :

$$\text{IRR} = i' + \frac{NPV'}{(NPV'-NPV'')}(i'' - i') \ldots\ldots\ldots\ldots\ldots\ldots\ldots 3)$$

Description:

IRR = *Internal Rate of Return* (%/year)
i' = interest rate that is expected to be close to the sought IRR (%/year)
NPV' = NPV based on i' (Rp)
I" = interest rate that is expected to be close to the higher IRR (%/year)
NPV" = NPV based on I" (Rp)
The NPV' and NPV" calculation iterations are stopped when the NPV" reaches a negative value.

2.2.1.8 Benefit Cost Ratio (B/C R)

B/C is a comparison between costs and benefits incurred during the project. There are two ways of calculating B/C, namely Net B/C and Gross B/C, but Net B/C is more commonly used. Net B/C is calculated as follows (Gray, et.al., 2002):

$$Net \frac{B}{C} = \frac{+NPV B-C posit}{-NPV B-C negatij} \quad \ldots\ldots\ldots\ldots\ldots 4)$$

Description:

NPV(B-C) positive $= \sum_{i=1}^{n} \frac{Bt-C_i}{Bt-C_i}$ For all NPV(B-C) positive , and

NPV(B-C) negative $= \sum_{i=1}^{n}$

To see the effect of changes in various aspects of the project. Sensitivity analysis influence on decision-making, because it is closely related to parameter changes NPV, IRR, B/C dan PBP.

2.2.1.9 Assumptions Used

Assumptions and parameters used to analyze the financial aspects of the business soft capsule shell industry from seaweed extract are as follows:

a. Project duration : 10 years
b. Working hours : 8 hours/day
c. Working days : 20 day/month
d. Working days in a year : 12 month/year
e. Average production scale:
   - Processing yield : 52 %
   - Amount of solution/day : 100 liters/day
   - Production of soft capsules/hour : 12.500 capsules/hour

f. Raw material price:
   - Modified starch : Rp. 65.000 /kg
   - Iota : Rp. 1.114.000 /kg
   - Sorbitol : Rp. 50.000/liter
   - Glycerin : Ro. 70.000/liter
   - Water : Rp. 1.000 /liter
   - Sun oil : Rp. 300.000/liter

g. Selling price of soft capsules : Rp. 27.500/ bottle contains 60 soft capsules
h. Working capital interest rate : 13 %
i. Fixed investment rate : 10 %
j. Income tax according to UU No. 36 of 2008 : 30 %
k. The remaining value of the building is 50%, machinery and equipment 10 %.

2.3 *Data Analysis*

Data analysis used a descriptive assessment through a technometer worksheet and was processed using MS Excel.

3. RESULTS AND DISCUSSION

3.1 Soft Capsule Performance

The results of the soft capsules produced are seen from the water content and disintegration. The resulting water content is 13% with a disintegration time of 35 minutes. H This result is indeed above the industry's demand for water content, but this result is in accordance with BPOM standards with a minimum water content limit of 16%. However, for pharmacopoeial standards, the expected moisture content range is 9-12%. Disintegration time is the time it takes to release fillers in the body. The results of testing the disintegration time of seaweed-based soft capsules are still within the BPOM limit, which is no more than 60 minutes, but the disintegration time according to pharmacopoeial standards is still low, because the disintegration time limit is 20 minutes. The resulting soft capsule weight is 4 g with a filler volume of 200 mL. Photographs of seaweed soft capsules can be seen in Figure 1.

Figure 1. Photograph of seaweed soft capsules

3.2 Technology Readiness Level (TRL)

The results of the seaweed-based soft capsule technology readiness level are at level 7. This level 7 means that the soft capsule is already on an industrial scale. However, it is not yet ready to be marketed because several indicators at levels 8 and 9 have not been fulfilled. The obstacles faced downstream are capsules that are still not uniform and there are still leaks. Besides, the transition from lab to industry needs to be evaluated again by optimizing the formula on an industrial scale. TRL results using the Technometer can be referred in Table 2.

Table 2 Estimates of TRL using the technometer method

	S or % fullfilled ►						Indicator TRL 7
	[Put a cross (X) in the appropriate column]						
No	0	1	2	3	4	5	(0= not fulfilled; 1=20%; 2=40%; 3=60%; 4=80%; 5=100% or fulfilled)
1						x	Equipment, processes, methods and engineering designs have been identified
2						x	Equipment fabrication processes and procedures began to be tested
3						x	Process equipment and test/inspection equipment are piloted in the production environment
4						x	Draft design drawings are complete
5						x	Equipment, processes, methods and engineering designs have been developed and started to be tested.

6					x		The estimated cost calculation has been validated (*design to cost*)
7						x	The general fabrication process is well understood
8						x	Almost all functions can run in the operating environment/conditions
9						x	The complete prototype has been demonstrated in a simulated operational environment
10						x	The system prototype has been tested in field trials
11						x	Ready for initial production (*Low Rate Initial Production- LRIP*)
S	0	0	0	0	1	10	
S	98.2%						
Indicator TRL 7 =							FULFILLED

As shown in Table 2, the aspect of the transition to industry is an important point for commercialization. This is because the resulting soft capsule is still leaking. The results of the TRL indicator for the soft capsule percentage were 98.2% with the TRL indicator achieved being TRL 7.

3.3 Techno-economy

3.3.1 Investment Cost

The composition of fixed investment costs can be seen in Table 2. Percentage of the cost of machinery, equipment, and supporting facilities from fixed investment costs is 45.86% which is Rp. 2,541,000,000,-.

Table 3. Composition of Investment Costs (Capital) Fixed (Rp.000).

No	Fixed Capital	Value (Rp.)	
1.	Land		1.400.000,-
2.	Building		1.500.000,-
3.	Preparation		100.000,-
4.	Encapsulation machines and production tools		1.980.000,-
5.	Other equipment/supporting facilities		561.000,-
	Amount		5.541.000,-

Table 4. Composition of Working Capital per year (Rp.000).

No	Working Capital	Value (Rp.)	
1.	Variable costs :		
	- Soft capsule raw materzials	1.173.840,-	
	- Capsule filling material (Habbat oil)	2.880.000,-	
	- Direct labor	630.000,-	
	- Packaging and auxiliary materials		36.000,-
	- Electricity and water for production		36.000,-

	Subtotal	4,755,840,-	
2.	Fixed costs :		
	- Indirect labor	1,200,000,-	
	- Depreciation	301,700,-	
	- Maintenance	554,100,-	
	- Administration		30.000,-
	- Promotion and marketing		24.000,-
	- Indirect electricity and water		24.000,-
	Subtotal		2,133,800,-
	Total		6,889,640,-

Table 5. Details and Sources of Investment Funds (Rp. 000)

No	Component	Value (Rp.)
1.	Sources of fixed investment funds:	
	Credit bank	3.324.600,-
	Owner's equity	2.216.400,-
	Fixed amount of investment funds	5.541.000,-
2.	Source of working capital funds:	
	Credit bank	953.346,-
	Owner's equity	635.564,-
	Total working capital fund	1.588.910,-
3.	Source of total investment fund :	
	Credit bank	4.358.046,-
	Owner's equity	2.905.364,-
	Total investment funds	7.263.410,-

3.3.2 Revenue Projection
Revenue from this business is obtained from the sale of production during one year that is as many as 24,000,000 seaweed soft capsules, and each soft capsule contains 500 mg of Black Seed oil. Black Seed soft capsules are packaged in bottles and each bottle contains 60 soft capsules. The selling price of the product is set at Rp. 27.500,-/bottle. Therefore, the proceeds from the sale of 400,000 bottles of seaweed soft capsules containing 500 mg of sunflower oil are Rp. 11,000,000,000,-/year. In projected acceptance, the price of soft capsules is high when compared to gelatin, but it can still enter the market if it is in the vegan market segmentation [11].

3.3.3 Profit and Loss Projection
The results of calculating the Profit/Loss Projection show that in year 1 this business has earned a net profit of Rp. 2,931,575,514,- . This profit will increase in the following years due to a decrease in the loan interest cost component. The average net profit during the project period is Rp. 3,075,339,933,- per year with an average profit margin of 27.96%.

Taking into account fixed costs, variable costs, and sales of soft capsules, the average BEP per year during the project period for the seaweed soft capsule business containing Black Seed oil is: Rp. 3,462,857,177 per year (sales value BEP), or 125,922 bottles of soft capsules @ 60 items (production BEP).

3.3.4 Projection of Cash Flows and Feasibility Analysis

According to [12], cash flow can be grouped into three parts, namely the initial cash flow (initial cash flow), cash flow operational cash flow, and terminal cash flow.

Cash flow is calculated by subtracting cash inflows and cash outflows each year. Cash inflows consist of net profit, depreciation, residual capital, equity, and loans. While cash outflows are investment costs, working capital costs, facility replacement, and loan repayment costs.

3.3.5 Projection of Cash Flow and Feasibility Analysis

Net cash flow is obtained by subtracting cash in and cash out. Cumulatively, net cash flow until the 10th year reached Rp. 30,240,951,000. Based on the cash flow estimates that have been made, the seaweed soft capsule industry, the filling material of which is Black Seed oil, has generated a positive cash flow in the first year of Rp. 2,805,481,000,-.

Based on the cash flow analysis, the B/C ratio or Net is calculated as B/C, Net Present Value (NPV), Internal Rate of Return (IRR) dan Pay Back Period (PBP). A business based on the investment criteria above is said feasible if the B/C ratio or Net B/C > 1, NPV > 0, and IRR > discount factor. Table 7 shows the results of calculating the value of investment criteria from the industry seaweed soft capsule.

Table 6. Feasibility Criteria for the Soft Capsule Industry

No.	Eligibility assessment criteria	Value
1.	Net B/C	1,64
2.	NPV	Rp. 4.621.393.281
3.	IRR	32 %
4.	PBP *(pay back periode)*	3,01 year

The calculation results show that the soft capsule industry business is feasible to implement, even profitable, because at an interest rate (discount factor) of 15% per year, the Net B/C is 1.64 (> 1), and the NPV is Rp. 4,621,393,281,- (> 0). With an IRR value of 32% (> 15% DF), it means this business is feasible even though the interest rate (discount factor) reaches 40% per year.

From the table above it can also be seen that the period required to return all business investment costs (business PBP) is 3.01 years. Thus, this industry is feasible because the payback period of the entire investment is shorter than the project period planned [10,13].

3.3.5 Sensitivity Analysis

Sensitivity is analyzed by repeating the calculations that have been done with changes that have occurred or may occur. Calculations were made to see the effect of rising raw material prices and decreasing selling prices on investment criteria [14,15]. Sensitivity analyzes have been summarized in Table 7.

Table 7. Summary of Sensitivity Analysis of Raw Material Prices, Prices Selling, and Fixed Investment Costs

Criteria	NPV (Rp)	IRR	Net B/C	PBP
		(%)		(Year)
Prices of raw materials and fillers increase by 30%	(548,856,156)	15	0.93	5,55
Prices of raw materials and fillers increase by 25%	312,852,083	18	1,04	4,88
Selling price down 13 %	(41,868,951)	17	0,99	5,13
Selling price down 10 %	1,034,268,487	21	1,14	4,41
Investment costs keep going up 60 %	(71,213,553)	17	0,99	5,10
Investment costs keep going up 30 %	2,275,089,864	23	1,25	3,97

4. CONCLUSION

The readiness level of seaweed-based soft capsule technology is at level 7, where it is necessary to evaluate the transition from the lab to the industry by evaluating the formula. The soft capsule prototype has the potential to be downstreamed and commercialized by considering the price of raw materials and their availability.

Acknowladgment

This research funding by National Research and Innovation Agency, Organization Agriculutral and Food

References

[1] Poeloengasih CD, Pranoto Y, Anggraheni FD, & Marseno DW. Potential of sago starch/carrageenan mixture as gelatin alternative for hard capsule material", AIP Conference Proceedings 1823, **020035** (2017) https://doi.org/10.1063/ 1.4978108

[2] Schofield L. 2014. Deep Dive Report: Softgels' clear advantages. *Natural Product Insider*. New York: Informa Exhibition LLC

[3] Badii, F., Howell, N.K., 2006. Fish gelatin: Structure, gelling properties an interaction with egg albumen proteins. *Food Hydrocoll.* **20**, 630–640. doi:10.1016/j.foodhyd.2005.06.006

[4] Nazzal, S., Wang, Y., 2001. Characterization of soft gelatin capsules by thermal analysis. *Int. J. Pharm.*2**30**, 35–45. doi:10.1016/S0378-5173(01)00856-0

[5] Lipinski, C., 2002. Poor aqueous solubility - An industry wide problem in drug discovery. Am.Pharm.Rev. **5**, 82–85.

[6] Ock YO, Lim WS, Park GD, Lee MH, Park JH.2020. Physical and mechanical properties of plant-derived soft-shell capsules formulated with hydroxypropyl starches from different botanical sources. *Polymer Testing* **91**(2020)106871

[7] Hidayat T, Sari DDP, Mustawafi W, Putri RPG, Jaffar MD, Syamsu K, Manalu LP. 2023. Effect starch and plastilizer for sheet film based carrageenan. IOP. *In progress*

[8] Oishi S, Kimura SI, Noguchi S, Kondo M, Kondo Y, Shimokawa Y, Iwao Y, Itai. 2018. New scale-down methodology from commercial to lab scale to optimize plant-derived soft gel capsule formulations on a commercial scale. *International Journal of Pharmaceutics*

[9] Misic D, Muffler K, Sydow G, Kuentz M, 2012. Novel starch-based pva thermoplastic capsules for hydrophilic lipid-based formulations. *Journal of Pharmaceutical Sciences.*1-13.

[10][AOAC]. Association of Official Analytical Chemist. 2006. Edisi Revisi. Edisi 18 2005. Official Methods of Analysis. Washington DC.

[11]Gray, C., P. Simanjuntak, L. K. Sabur, P. F. L. Maspaitella, R. C. G. Varley. (2002). Pengantar Evaluasi Proyek. Jakarta, ID: PT Gramedia Pustaka Utama

[12]Husnan, S. dan Suwarsono, M. (2000). Studi Kelayakan Proyek. Yogyakarta, (ID) : UPP AMP YKPN

[13] Hidayat, T. 2021. Melaksanakan Penyelidikan Kesesuaian Parameter Kelayakan. Technical Note. TN-004/BBO/2021. Pusat Teknologi Agroindustri. Badan Pengkajian dan Penerapan Teknologi. BPPT; Jakarta

[14] Putri, R.P.G. 2021. Menetapkan Parameter Kelayakan. Technical Note. TN-004/BBO/2021. Pusat Teknologi Agroindustri. Badan Pengkajian dan Penerapan Teknologi. BPPT; Jakarta

[15] Simamora, H. (2002). Akuntansi Manajemen. Yogyakarta, (ID) : UPP AMP YKPN

The resilience of fishery production supply chain in West Java Province, Indonesia

A.Nurhayati [1], A.A.Handaka *, T. Herawati*, A. Yustiati *, Iskandar *

[1] Faculty of Fisheries and Marine Science, Universitas Padjadjaran.
Jl. Bandung- Sumedang KM.21, Hegarmanah, District. Jatinangor, Sumedang Regency, West Java Province 45363

Corresponding author e-mail: atikah.nurhayati@unpad.ac.id

Abstract. Industrialization of fishery production in Micro, Small, and Medium Enterprises (MSMEs) in its infancy, post-covid-19, and human activities are the causes of climate change. This research aims to analyze the resilience of fisheries production in West Java Province, Indonesia. The method used descriptive quantitative, using primary and secondary data. This research was conducted from November 2022 - March 2023 and used a purposive sampling technique of 46 respondents. The SWOT analysis tool is used for supply chain innovation development strategies and measures fisheries' supply chain innovation using risk consequence analysis. Based on the results of the supply chain innovation development strategy research from the aspect of strength is to create a transformation of the potential of fishery resources towards the utilization of fishery resources based on bio-ecoregion. Opportunities faced by management transformation with open market access and high competitiveness in fishery products. The challenges faced must be able to create a digital marketing-based fishing industry system. Threats faced by the transformation of the workforce towards digitalization. Based on the SWOT analysis, being in quadrant 2 means a diversification strategy. Types of supply chain risk for fishery products that cannot be replaced, namely micro, small, and medium fishery product businesses, and product handling when transporting fishery products from distributors. Supply chain innovation in the digital marketing era must be able to convince consumers of the products they buy.

1. Introduction

In Indonesia, climate change can be seen from the more extreme weather that disrupts economic activity. Fishery business actors, including fish cultivators and fishermen, can no longer determine the period of fish stocking, harvesting period, and time of going to sea that can produce optimal catches. Floods and droughts have occurred everywhere and tidal floods have been experienced by coastal communities. The influence of human activities can be seen in the increasing emission of greenhouse gases (carbon dioxide, methane, nitrogen oxides, and several industrial gases) from human activities. Several human activities and fisheries resources are expected to reduce the risk of climate change in the future.

Indonesia is a country that has the potential for the production and marketing of fisheries and marine products. Fishery production in Indonesia in 2020 was 1.6 million tons. West Java Province contributes to fishery production. [1]. One of the provinces that produce fisheries and marine products is West Java Province. The Ministry of Marine Affairs and Fisheries in 2020-2024 targets to increase the national fish consumption rate from 56.39 kg/capita/year in 2020 to 62.50 kg/capita/year in 2024.[2].

Fisheries are all activities related to the management and utilization of fish resources and the environment starting from preproduction, production, and processing to marketing carried out in a fisheries business system. [3]. Fish resources are the potential of all types of fish. The environment of fish resources depends on the bio-ecoregion concession of each region. Bio-ecoregion is a landscape that is within a stretch of ecological unity determined by natural boundaries, such as watersheds, bays, and currents [4].

The fisheries business is carried out in the fisheries business system, including preproduction of fisheries, process productions, fisheries processing, and fisheries marketing. Fisheries business in implementing the fishing business should pay attention to the quality standards of the fishery, particularly the processing of fisheries products marketing chain [5]. Industrialization of fishery product processing must be the main object of activity in the fisheries sector in its handling and development. Fishery products have the characteristics of perishable, and voluminous and are seasonal, so it is necessary to process fishery products and supply chain management systems that process starting from processing activities, and marketing distribution, to the desired products to reach the consumers. The long product flow process in the supply chain of fishery products from the production center to the end consumer will increase marketing costs, so it is necessary to study the promotion of the supply chain of fishery products in the industrial era 4.0 [6].

Therefore, a strategy is needed in the design of the most effective and efficient supply chain process in order to prevent the occurrence of concentrated distribution at one point [7]. Mapping the distribution network promotion and matters relating to technical distribution promotion along with its operational costs within a certain period also needs to be examined because there are changes in the price patterns in each supply chain to reach the hands of consumers [8].

Limitations experienced by fisheries business actors at the time of Covid 19 include limited market access, so innovation is needed in the supply chain fisheries. The Industrial Revolution 4.0, is the role of technology in the supply chain between individuals is growing and has a variety of ways and methods, especially promotion in fisheries marketing

The relationship between industry era 4.0 with the promotion of marketing fishery products with the principle of being integrated with one another. 4.0 Industrial Revolution including Industrial Revolution 1, the first time occurred after the discovery of steam engines, resulting in the development and great growth of industrial machinery in England in 1764. This revolution affected the agricultural and manufacturing sectors/industries, especially the textile industry, and its decline. The 2nd Industrial Revolution was at the beginning of the 19th century when improvements were made to the process side, where large-scale processes were also known thanks to innovations from Ford [9]. The inventions at this time were more towards electricity and transportation, this combination of production methods and developments influenced technology, especially for the iron, machinery, oil, chemical, vehicle, and other industries. The 3rd Industrial Revolution, known as the automation period, after the World War, was marked by the use of high-tech automatics, electronics, and information technology. This is what continues to this day, where machines and increased production capacity and software are the main supporters of the manufacturing process. The 4th Industrial Revolution which is currently confronted explains that machines and production capacity alone are not enough to make growth for a country. Along with that the development of current information technology, software, and hardware as well as the rapid development of internet usage.

The Industrial Revolution 4.0 has principles that enable each company to identify and implement a variety of Industry 4.0 scenarios, Among them are (1) Interoperability to connect and communicate with each other through the Internet of Things (IoT); (2) Information transparency with sensor data; (3) Technical support to help people collect data and make visualizations in order to make wise decisions; (4) Independent decree is the ability of the cyber-physical system to make decisions and perform tasks as independently as possible [10] .

This research in conjunction with the fisheries and marine commodities is also faced with various challenges, including in the era of global competition, namely the era of the 4th industrial revolution. The main issues raised in this era are the competitiveness and productivity of fisheries [10]. Challenges of increasingly fierce global trade competition, fisheries, and marine products must be highly competitive, with the availability of efficient technology inputs, guaranteed product quality, efficient production system chains from upstream to downstream, and the promotion of fisheries and marine products [5].

The Ministry of Maritime Affairs and Fisheries continues to encourage digital e-commerce-based fisheries and marine products that aim to: (1) shorten inefficient distribution chains because fish producers are closer to the retail market; (2) provide price certainty to fish farmers and consumers; (3) increase connectivity and eliminate distance, space and time limits to provide input and market facilities in the development of aquaculture industry; and (4) presenting an efficient aquaculture business model in the midst of the community.

The transformation of fisheries and marine industrialization era 4.0 towards automation and digitalization includes: (1) the transformation of exploitation of natural resources leads to the efficiency of natural resources, services and the increase in added value and productivity, (2) the transformation of the use of unskilled labor towards the creation of employment opportunities that really intended for trained human resources while employment for unskilled labor can be reduced, and (3) transformation from conditions of limited market access and low product competitiveness towards broad open market access, high competitiveness, and efficient management.

Information technology through promotion can streamline the marketing chain of fisheries and marine products. Supply chain analysis needs to be done in order to support the competitive advantage of the fishing industry to produce products in accordance with consumer desires in terms of quality, quantity, price, time, and place right research needs to be done based on the bio-ecoregion [11]. The supply chain has a dynamic nature but involves three constant streams which are product flow, financial flow, and information flow. However, the main purpose of the supply chain is still to meet consumer needs [12]. Horizontally, there are five main components or actors in the supply chain, namely suppliers, manufacturers, distributors, retailers, costumer. Vertically there are five main components of the supply chain, namely buyers, transporters, warehouses, sellers, and so on.

The characteristics of perishable fish commodities have a significant impact on losses, which is around 35%. From that number, the distribution process contributed quite a lot, which is around 10%. Supply chain analysis needs to be done in order to support the competitive advantage of the fishing industry and the fisheries business to be able to produce products in accordance with consumer desires in terms of quality, quantity, price, time, and place [13].

Supply chain management is a concept of the right approach to overcome the problem of meeting consumer demand. When delivering products to end consumers is demanded as efficiently as possible while maintaining product quality. In the supply chain, there are regulatory systems related to product flow, information flow, and financial flow [14]. Supply chain management is a set of activities and related decisions to integrate suppliers, manufacturers, warehouses, transportation services, retailers, and consumers efficiently. Thus, goods and services can be distributed in the right amount, time, and location to minimize costs to meet consumer needs [15].

In the decision-making process, consumers will go through the stages first before finally deciding to buy the product. There are many factors that influence consumer decision-making, for example, the level of consumer confidence in the product, price, and quality of services provided by the company. The purchasing decision-making process is influenced by the company's stimulation which includes the product, price, and place. In addition to these stimuli, there are also factors driving purchasing decisions, namely service quality and trust [16].

Every supply chain activity certainly does not always run well, there are conditions where the supply chain experiences one or several obstacles that make a risk that must be borne by the actors of the supply chain. Business actors who run their businesses or businesses can be sure to know what are the risks in the business, as well as new risks that will be borne by the supply chain network.

Supply chain risk can be defined as a loss assessed in terms of the likelihood of it occurring. If one of the actors experiences problems in the supply chain, it will affect both directly and indirectly the partners in the supply chain network [17]. Supply chain development can be analyzed using: network structure, chain objectives, chain management, chain resources, and chain business processes. Supply chain risk does not only lie in activities within the business community but can cause disruption of product flow to the next consumer or supplier of raw materials to be processed into a product that has added value. This situation is certainly very detrimental to all parties involved with the flow of goods and the flow of money in the supply chain so there will be times when consumers suffer from the scarcity of desired products [18].

One of the supply chain risks of fishery products is caused by the length of marketing channels from production to consumers. One solution is to shorten marketing channels, through product promotion using online media. Online promotion is also known as digital marketing. Types of digital marketing can be categorized as marketing via email, social media marketing such as Istanbul, Facebook, and WhatsApp, and many types of display advertisements including web banner ads, and mobile advertising [19]. Digital marketing is considered the most appropriate to be applied to micro small and medium businesses, especially the promotion of processed fishery products. One of the supporting factors that must be owned by small and medium micro businesses is mastery of information technology.

2. Research Methods

2.1. Research methods

The method used in this research is descriptive qualitative and quantitative. The data used are primary and secondary data. The respondent used was chosen by the snowball sampling method, which is taking respondents based on recommendations from the first respondent. The number of respondents was 46 people consisting of fishermen, fish auction place traders, collector traders, small traders, and consumers. When the research was conducted in November 2022 – March 2023.

2.2. Validity Test Data

A validity test is used to determine the level of validity of the questionnaire instrument used in data collection. Using valid and reliable instruments in data collection it is expected that research results will be valid and reliable. Reliability testing in this study was conducted by calculating the value of Cronbach's Alpha instruments from each of the research variables tested [20].

2.3. SWOT Analysis

SWOT analysis compares factors of external opportunities and threats with factors of internal strengths and weaknesses. Internal factors are entered into the matrix called Internal Strategic Factor Analysis Summary (IFAS). External factors are entered into the matrix which is called External Strategic Factor Analysis Summary (EFAS) [21]. S-O Strategy (Strength-Opportunities). This strategy is based on thinking company, namely by utilizing all power to seize and take advantage of the biggest opportunities. S-T Strategy (Strengths-Threats) Is a strategy to use the power that companies must overcome threats. W-O Strategy (Weaknesses-Opportunities). This strategy is applied based on utilization opportunities that exist by way of minimizing existing weaknesses. W-T Strategy (Weaknesses-Threats). This strategy is based on activities that are defensive and trying to minimize existing weaknesses as well as avoid threats. The threat is a strategy to use the power that companies have to overcome threats. W-O Strategy (Weaknesses- Opportunities). This strategy is applied based on utilization opportunities that exist by the way to minimize existing weaknesses. W-T Strategy (Weaknesses-Threats). This strategy is based on activities that are defensive and trying to minimize existing weaknesses as well as avoid threats.

2.4. Consequence Risk Supply Chain Innovations Micro, Small and Medium Size Enterprises (MSMEs) marine and fisheries

Analysis of supply chain management conditions. The product flow approach includes marketing objectives, implementation of supply chain management, and demand and supply. The financial flow approach includes capital and payments [22]. In abnormal supply chain conditions, there are risks that can threaten the sustainability of the supply chain, both those that are easy to replace and those that are not easy to replace quickly within their management timeframe. [23].

$$\alpha = \frac{\delta \text{ replace}}{\delta \text{ collapse}} \quad (1)$$

Information:

α	:	The consequences of the risk of a product in the supply chain
δ replace	:	The time it takes for a supply chain to replace a sub-product or time needed to handle interruption of a product flow and return to normal scheduling conditions with the same quality level.
δ collapse	:	The time of the sub-product failed to be completed before the supply chain experienced a loss at a critical point in service of the market

The following are the assessment indicators for supply chain innovations for fisheries in Table 1.

Table 1. Value of Risk Consequences

Consequence	Information	α
Important	Irreplaceable	1.0
Necessary	Not easy to replace	0.6
Needs	Easily replaced	0.3
Wanted	Easily replaced	0.1

3. Result

3.1. Research Location

This research was conducted in the coastal area of West Java Province. The Province of West Java is geographically located between 5°50'- 7°50' South Latitude and 104 °48'- 108° 48' East Longitude, with territorial boundaries, north with the Java Sea and DKI Jakarta, east with Central Java province, south with Indonesian ocean, West with Banten province. West Java Province has natural conditions with complex geological structures with mountainous regions in the middle and south and lowlands in the north [24]. The southern coastal region of West Java produces capture fisheries production. Fisheries are all activities related to the management and utilization of fish resources and the environment starting from preproduction, production, and processing to marketing, which is carried out in a fisheries business system

Coastal waters are seas bordering land, covering waters as far as 12 nautical miles measured from the coastline, waters connecting the coast and islands, estuaries, bays, shallow waters, brackish marshes, and lagoons. Coastal areas are transitional areas between terrestrial and marine ecosystems that are affected by changes in land and sea. The condition of the bio-ecoregion from physiographic aspects, the southern coast of West Java is part of the Southern Java Mountain Range extending from Ujung Kulon in the west to Segara Anak in the east. The condition of the bio ecoregion from the geographical aspect of the southern coast bordering the waters in the Indian Ocean has a coastal condition located opposite the shallow waters of the Java Sea, with coastal conditions varying between sandy, rocky, or reefed

The southern coastal region of West Java produces capture fisheries production. Fisheries are all activities related to the management and utilization of fish resources and the environment starting from preproduction, production, and processing to marketing, which is carried out in a fisheries business system

3.2. Characteristic Respondent

Coastal waters are seas bordering land, covering waters as far as 12 nautical miles measured from the coastline, waters connecting the coast and islands, estuaries, bays, shallow waters, brackish marshes, and lagoons. Coastal areas are transitional areas between terrestrial and marine ecosystems that are affected by changes in land and sea. The condition of the bio-ecoregion from physiographic aspects, the southern coast of West Java is part of the Southern Java Mountain Range extending from Ujung Kulon in the west to Segara Anak in the east. The condition of the bio ecoregion from the geographical aspect of the southern coast bordering the waters in the Indian Ocean has a coastal condition located opposite the shallow waters of the Java Sea, with coastal conditions varying between sandy, rocky, or reefed.

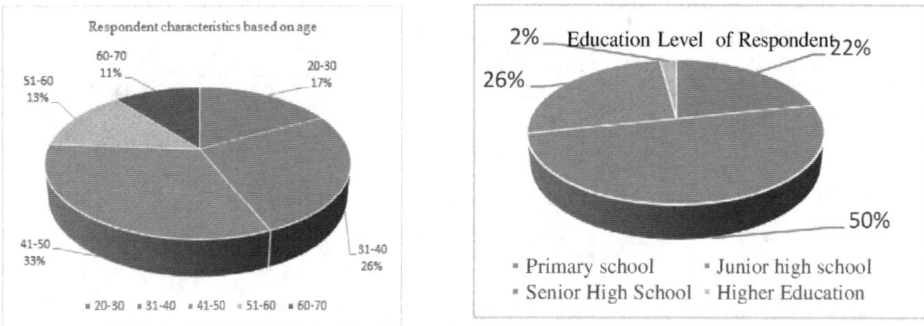

Figure 1 Characteristic Respondent

Based on interviews with 46 respondents, the average productive age is 89 % at age 20-50 years. The population of productive age is that age produces goods and services. Based on the results of research the level of education of respondents has an influence on decision making in running their business.

Table 2 Total Respondent

No	Respondent	Fisheries business actors	Total Respondents
1	Raw Material Provider	Fisherman	6
		Fish Auction Manager	3
		Collector trader	1
2	Fish Processing		7
3	Trader	Local Trader	15
		Inter-city traders	4
		Inter-district traders	4
		Inter-provincial traders	6
Total respondents			46

4. Discussion

4.1. Supply Chain Innovations for Micro, Small, and Medium Size Enterprises (MSMEs) Marine and fisheries

Fishery Products are any form of food products in the form of whole fish or products containing parts of fish, including products that have been processed in any way with the main raw material of fish. The National Fish Logistics System (SLIN) is a management system of fish and fishery product supply chains, materials, and production equipment, as well as information ranging from procurement, and storage, to distribution, as a unity of policies to increase capacity and stabilize the starting from upstream to downstream fisheries production system, control price disparity, and to meet domestic consumption needs. [25].

The national fish logistics system aims to: (1) increase the capacity and stability of the production and marketing of national fisheries; (2) strengthen and expand connectivity between upstream production centers, downstream production, and marketing efficiently: (3) improve the efficiency of management of fish supply chains, materials, and production equipment, as well as information from upstream to downstream. The application of digital technology in the fisheries sector has shown a positive trend for Indonesia. Through collaboration with international institutions, Indonesia has succeeded in increasing the productivity of capture fisheries from 6.54 million tons in 2016 to 9.45 million tons in 2018 (data as of October 2018)

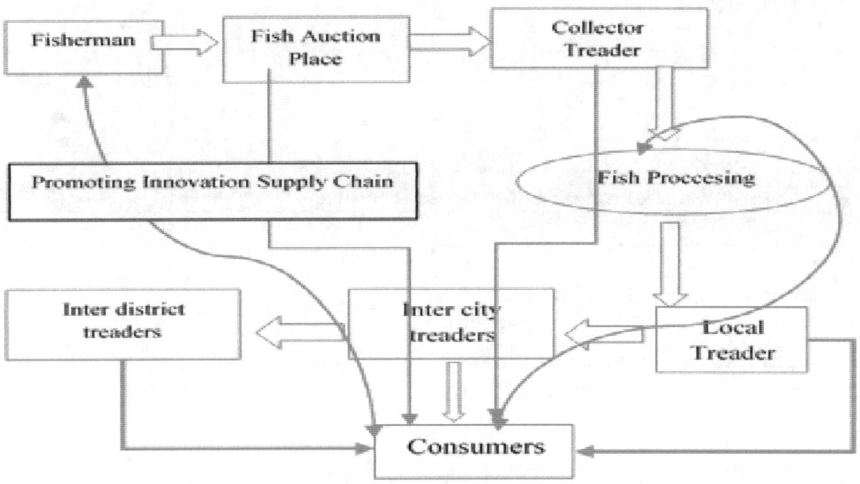

Figure 2 Supply Chain Innovations for a Fisheries

Based on Figure 2 we can find out the supply chain of fishery products. The supply Chain of captured fishery products enters the fish auction place, then to collecting traders and processed so that it has product value added. After the product has been processed, it will be marketed through local traders, inter-city traders, and inter-provincial traders, then to consumers. After promoting innovation supply chain marketing channels can be done directly between fishermen and consumers, fisheries product processors and consumers, through the use of internet marketing applications. Internet use applications for marketing fishery products have several strengths and weaknesses, which are as follows:

Table 3 Matrix Internal Strategic Factor Analysis Summary (IFAS) For Micro, Small and Medium Size Enterprises (MSMEs) marine and fisheries

No	Internal Strategy Factors	Weight	Ranting	Score
Strength				
1.	Increasing trends in fisheries productivity	0.18	4	0.36
2.	The comparative and competitive advantage of bio-ecoregion-based fishery products	0.25	3	0.75
3.	National fisheries company performance improvement through the use of Android-based digital technology	0.12	4	0.48

		0.55		1.56
Weakness				
1.	Transshipment prohibition as outlined in KKP Minister Regulation No. 57/2014.	0.20	4	0.60
2.	Education on the use of fisheries technology to fishermen and fishery product processors is still low	0.15	3	0.45
3.	Vertical integration has not been focused on the integration of information technology systems at various levels of production, processing fishery products to consumers	0.10	3	0.30
		0.45		1.35
		1		

Table 4 Matrix External Strategic Factor Analysis Summary (EFAS) Promoting 4.0 Supply Chain Innovations for a Fisheries

No	External Strategy Factors	Weight	Ranting	Score
Opportunity				
1.	Increasing and high demand for fish	0.16	3	0.48
2.	Increasing the added value of the national fishing industry	0.19	4	0.76
3.	Time efficiency and marketing costs for fishery products	0.14	4	0.56
		0.49		1.80
Threat				
1.	Illegal, Unreported dan Unregulated (IUU) Fishing	0.21	4	0.84
2.	Decreasing the number of workers in the fisheries sector, due to being replaced by machine automation	0.18	4	0.72
3.	Competitiveness of domestic and foreign fishery products	0.12	3	0.36
		0.51		1.92
		1		

Internal Analysis Coordinate, (Total strength score - total weakness score): 2 is 0.21. External Analysis Coordinates (Total opportunity score - total threat score): 2 is -0.06. So, the coordinates are located at (0.21; - 0.06).

Based on the SWOT analysis of Supply Chain Innovations for Micro, Small, and Medium Size Enterprises (MSMEs). Marine and fisheries are quadrant 2 in the diversified position, this shows that despite facing various threats in industrial technology 4.0, the fisheries sector still has internal strength. This strategy utilizes the strength of the fisheries sector which can reduce threats to the Industrial Revolution 4.0

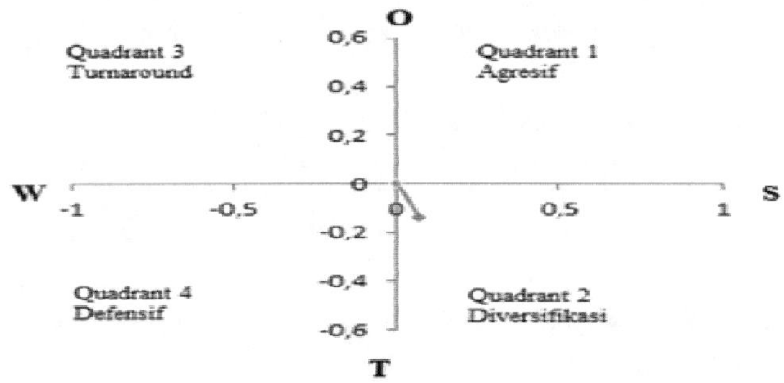

Figure 3 SWOT Analysis Diagram

IOP Conf. Series: Earth and Environmental Science 1289 (2023) 012016 doi:10.1088/1755-1315/1289/1/012016

Internal factors have a total score for strength of 1.56, and a total score for weakness of 1.35 so strengths are more dominant than weaknesses in determining development strategies promoting 4.0 supply chain innovations for a fishery. External factors have a total score for opportunity of 1.80, the total score for external opportunities is more dominating than the factor of the threats facing. The Cartesian diagram above shows the position in quadrant 2. This position has the power that can be utilized to seize the opportunities that exist. Alternative strategies that used all the power must take advantage of existing opportunities.

Table 5 SWOT Matrix Result

O / S	O1. Increasing and high demand for fish O2. Increasing the added value of the national fishing industry O3. Time efficiency and marketing costs for fishery products
S1. Increasing trends in fisheries productivity	Increasing trends in fisheries productivity through diversification and value-added processing of fishery products, so that it has high economic value (S1-O1,O2, O3)
S2. The comparative and competitive advantage of bio-ecoregion-based fishery products	The comparative and competitive advantage of bio-ecoregion-based fishery products through the development of fisheries production that is locally specific based on bio-ecoregion conditions in a particular region (S2-O1, O2, O3)
S3 National fisheries company performance improvement using Android-based digital technology	The Internet of Things is used to market fishery products (S3-O1, O2, O3).

Strategies supply chain innovations for Micro, Small, and Medium Size Enterprises (MSMEs) marine and fisheries through product flow, finance, and supply chain risk. The product flows in raw Materials for fishery products before processing the product. Supply chains include fishermen, fish auction places, traders, fishery product processors, local marketers, inter-city marketers, and inter-provincial marketers, with prices received from each different supply chain channel, based on fish species, length of the type of subscription fish, and travel time.

The financial flow in the supply chain of fishery products is the same as the financial flow in the supply chain of commodities or other goods where the financial flow moves from the bottom up where consumers make payments to processors and processors make payments to suppliers of raw materials. The flow of finance will not be stopped but there will be obstacles when the method of payment is not done in cash, but all methods of payment are based on agreements made previously.

Supply chain management of fishery products has not been going well as a whole starting from the procurement of raw materials to the hands of consumers. There are conditions that are not normal at some times, for example, influenced by natural conditions and other unexpected conditions that cause disruption in the supply chain. This situation can cause losses ranging from irreplaceable to irreplaceable.

4.2. Consequence Risk Supply Chain Innovations Micro, Small, and Medium Size Enterprises (MSMEs) marine and fisheries

Consequence risk supply chain innovations for Micro, Small, and Medium Size Enterprises (MSMEs) marine and fisheries used comparative between the time it takes for a supply chain to replace a sub-product or time needed to handle interruption of a product flow, and returns to normal scheduling conditions with the same quality level with the time of the sub-product failed to be completed before the supply chain experience a loss at a critical point in service the market.

Table 6. Consequence Risk Test Results Supply Chain Innovations For Micro, Small, and Medium Size Enterprises (MSMEs) Marine and Fisheries

No	Types of supply chain risks	Score	Results	Consequence	Information
1.	Micro, Small, and Medium Size Enterprises (MSMEs). marine and fisheries	$\frac{9}{9}$	1	Important	Irreplaceable
2.	Availability of fishery products	$\frac{3}{7}$	$0.4 \approx 0.6$	Necessary	Not easy to replace
3.	Availability of fisheries production inputs	$\frac{4}{7}$	$0.5 \approx 0.6$	Necessary	Not easy to replace
4.	Handling of fishery products after catching and processing	$\frac{1}{1}$	1	Important	Irreplaceable
5.	Handling of products when transporting fishery products from each distributor	$\frac{7}{14}$	$0.5 \approx 0.6$	Necessary	Not easy to replace
6.	Handling of unsold Fresh fish products for fish feed ingredients	$\frac{5}{14}$	0.3	Needs	Easily replaced
7.	Handling of unsold processing fish products for fish feed ingredients	$\frac{365}{1095}$	0.3	Needs	Easily replaced
8	Avoiding the risk of fraud in the marketing supply chain	$\frac{9}{9}$	1.0	Important	Irreplaceable

According to Table 6 can be seen consequence risk promoting 4.0 supply chain innovations for a fishery including Bio ecoregion conditions suitable for the fisheries sector, Handling of fishery products after catching and processing, Sales system with a consignment pattern, to reduce the risk of payment systems between distributors, Avoiding the risk of fraud in the marketing supply chain with the consequence is important. Types of supply chain risks are Availability of fishery products, Handling of products when transporting fishery products from each distributor with the consequence is necessary while handling of unsold Fresh fish products for fish feed ingredients with the consequence is needed. Consequence risk promoting 4.0 supply chain innovations for a fishery based on travel time in units of days in the supply chain of fishery products. Promoting 4.0 supply chain innovations for a fishery including procurement of fishery products, handling of fishery products, fisheries product transportation systems and distribution channels of fishery products are a unified whole to increase the added value of fishery products.

5. Conclusion

Based on the research results the fisheries supply chain with a digital technology approach needs to be done. Supply chain distribution channel innovations from fishermen or fish cultivators through digital platforms can directly connect with end consumers. Internal factors have a total score for strength of 1.56, and a total score for weakness of 1.35 so strengths are more dominant than weaknesses in determining development strategies promoting 4.0 supply chain innovations for a fishery. External factors have a total score for the opportunity of 1.80, and a total score for external opportunities is more dominating than the factor of the threats facing. The Cartesian diagram above shows the position in quadrant 2. Consequence risk supply chain innovations micro, small, and medium-sized enterprises (MSMEs) marine and fisheries must pay attention to aspects availability of fishery products, availability of fisheries production inputs, handling of fishery products after catching and processing, handling of products when transporting fishery products from each distributor, handling of unsold fresh fish products for fish feed ingredients, handling of unsold processing fish products for fish feed ingredients, avoiding the risk of fraud in the marketing supply chain. Supply chain innovation for fishery products needs to be adapted to changes in the market to meet consumer needs quickly and precisely

Contribution of individual authors to the creation of a scientific article (ghost-writing policy)

Atikah Nurhayati, Asep Agus Handaka Suryana main contributions were designing the research, doing field surveys, analyzing the result, and preparing the paper. Titin Herawati and Ayi Yustiati, and Iskandar main contributions were doing field surveys and conducting statistical analysis

Acknowledgments. We are grateful to the Faculty of Fisheries and Marine Science, Universitas Padjadjaran for providing facilities and support. This research was funded through Unpad internal resources with contract number: 1549/ UN6.3.1/ PT.00/2023. We also thank the anonymous referees who have given valuable comments and critiques for the earlier version of the manuscript.

References

[1]. Ministry of Marine Affairs and Fisheries.2020. Covid-19 Pandemic, Estimated Harvest of Aquaculture Reaches 450 Thousand Tons Throughout April to June 2020. Accessed 20 March 2021. https://kkp.go.id/artikel/18532-pandemi-covid-19

[2]. Ministry of Marine Affairs and Fisheries.2021.Targets Fish Consumption 56.39 KG. Accessed 12 March 2021. https://kkp.go.id/artikel/16451-2020-kkp

[3]. Republic of Indonesia Law No. 45 of 2009 concerning amendments to Law No. 31 of 2004 concerning fisheries

[4]. Republic of Indonesia Law No. 1 of 2014 concerning amendments to Law No. 27 of 2007 concerning the management of coastal areas and small islands.

[5]. Nurhayati.A., et.al, 2018. Promoting innovative business in the fishery sector in West Java, Indonesia. IOP Conf. Series: Earth and Environmental Science 137.

[6]. Suryaningrat IB. 2016. Raw Material Procurement on Agro-industrial Supply Chain Management: A Case Survey of Fruit Processing Industries in Indonesia. *Agricultural Science Procedia* Vol. 9, 2016: 253-257

[7]. Benassi C, Chirco A, Colombo C. 2017. Mixed Spatial Duopoly, Consumer's Distribution, and Efficiency. *Economics Letters* Vol. 156, 2017:74-77

[8]. Lee D, Yang SG, Kim K, Kim BJ. 2017. Product Flow and Price Change in an Agricultural Distribution Network. *Physica A* S0378-4371(17)30723-9

[9]. Hasmaidi.com. (2018). No Title. Retrieved September 10, 2018, from https://hasmaidi.com/pemahaman-revolusi-industri-4.

[10]. James Manyika.,et.al. 2017. Jobs Lost,Jobs Gained: Workforce Transitions In A Time of Automation. McKinsey & Company

[11]. Nurhayati.,A and Purnomo AH. 2018. Bioecoregion and Socio-Economic Connectivity Oriented Coastal Zone Management. *Applied Ecology and Environmental Sciences, 2018, Vol. 6, No. 1, 31-34*

[12]. Chopra S, Meindl P. 2007. *Supply Chain Management: Strategy, Planning & Operations,* 3rd Edition. Pearson Prentice Hall.

[13]. Nurhayati.A.,et.al. 2019. An Integrated Supply Chain Management Based Nila Nirwarna (Oreochromis niloticus) Seed Market Institution. Jurnal Perikanan, Universitas Gajah Mada. 2019, Vol 21 No 2
[14]. Tompodung E, Worang FG, Roring F. 2016. Tilapia Fish Supply Chain Analysis in Eris District, Minahasa Regency. *Jurnal EMBA* Vol.4, No.4, 2016: 279-290
[15]. Ling L. 2007. *Supply Chain Management: Concepts, Techniques, and Practices Enhancing Value Through Collaboration.* Singapore: World Scientific Publishing Co. Pte. Ltd
[16]. Kotler, P.dan G. Amstrong. 2016. Principles of Marketing Issue 13 Volume 1. Erlangga. Jakarta
[17]. Marimin & Magfiroh 2010. Technical Application of Decision Making in Supply Chain Management. IPB Press 281 page
[18]. Antarbangsa QB, Nurhayati A, Harahap SA. 2014. Risk Evaluation of Carp Farmers (Cyprinus carpio) Supply Chain in the Public Waters of the Malata Block Cirata Reservoir.Skripsi. Faculty of Fisheries and Marine Science. UNPAD. Jatinangor.
[19]. Ramadhanti.GCK., et.al. 2019. Analysis of Decisions to Purchase Abon Fish Tuna Products Through Online Media Distribution (Case Study of Bandung City). GSJ: Volume 7, Issue 10, October 2019, Online: ISSN 2310-9186.
[20]. Sugiyono. 2010. Educational Research Methods Quantitative, Qualitative, and R&D Approaches. Alfabeta
[21]. Freddy Rangkuti, 2004. SWOT Analysis of the Technique for Dissecting Business Cases., PT. Gramedia, Jakarta
[22]. Pawar, A., V. 2014. Study of the effectiveness of Online Marketing on Integrated Marketing Communication. Dissertation. DY Patil University.
[23]. Neureuther BD, Kenyon G. 2009. Mitigating Supply Chain Vulnerability. *Journal of Marketing Channels* 16: 245-263.
[24]. West Java Regional Development Planning Agency. 2019. Strategic Plan for 2018-2023. .Accessed 6 October 2019. http://bappeda.jabarprov.go.id/wp-content/uploads/2019/05/2.-Renstra-Bappeda-2018-2023.pdf
[25]. Regulation of The Minister of Marine and Fisheries the Republic of Indonesia Number 5/Permen-KP/ 2014 Concerning National Fish Logistic System.

IOP Conf. Series: Earth and Environmental Science 1289 (2023) 012017 doi:10.1088/1755-1315/1289/1/012017

The growth rate and antibacterial activity of lactic acid bacteria GMH2 and GMH3 in various salt concentration

A N Rosanto, R I Lestari and M M P Putra*

Fish Product Technology, Department of Fisheries, Faculty of Agriculture, Universitas Gadjah Mada
Jl. Flora, A4 Building, Bulaksumur, Yogyakarta, Indonesia

*Corresponding author e-mail: primaputra@ugm.ac.id

Abstract. Lactic acid bacteria (LAB) can be naturally found in fermented fish products that contain various salt concentrations. LAB is reported to produce proteinogenic antibacterial compounds named bacteriocin that can inhibit the growth of pathogenic and spoilage bacteria. In order to evaluate the performance of previously isolated LABs, namely GMH2 and GMH3, to inhibit the growth of these bacteria in fermented food systems, it is necessary to examine the effect of salt concentration on the growth rate and antibacterial activity. LAB GMH2 and GMH3 were inoculated on *de man rogosa and sharp broth* (MRSB) at various salt concentrations, including 0%, 7.5%, and 15%, followed by incubation at 37°C for 30 hours with observation every 6 hours. Bacterial growth was tested using the Total Plate Count (TPC) method, which was then further processed using the DMFit program to obtain growth rates, while antibacterial activity was tested using the macrodilution method against histamine-producing bacteria (HPB) *Klebsiella pneumoniae* (CK2). Both strains showed the highest growth rate at 0% NaCl, with 0.171 and 1.681 log CFU/hour for GMH2 and GMH3, respectively and remained steady until 7.5% salt concentration at 0,153 log CFU/hour for GMH2 and 0.044 log CFU/hour for GMH3. Antibacterial activity for both isolates showed the highest inhibitory value of 39.87±2.25% for GMH2 and 79.55±1.08 for GMH3. The addition of 7.5% and 15% interestingly did not show a significant reduction in antibacterial activity. The results of this study concluded that LAB GMH2 and GMH3 have the potency to be used as starters in fish fermented products, which possess up to 7.5% salt content.

1. Introduction

Fermentation is a technique used to improve product quality and extend shelf life. Fermented fishery products are susceptible to have a high content of biogenic amine, especially in spontaneously fermented products because naturally growing microbes play a role in producing amino acids which are the main factor in the formation of biogenic amines like histamine [1]. Histamine can cause poisoning known as histamine fish poisoning (HFP). Symptoms include nausea, vomiting, itching, headaches, burning in the throat, and swelling of the tongue. These symptoms will appear as early as 10 minutes to 2 hours after consuming fish with high histamine levels [2]. According to the Food and Drugs Administration (FDA), the allowed histamine level is no more than 50 mg/kg [3]. Inhibition of histamine accumulation in fermented fishery products can be done by adding lactic acid bacteria starter that are capable of producing antibacterial compounds, such as bacteriocins, hydrogen peroxide, and organic acids (lactic acid, acetic acid, and formic acid) [4].

Lactic acid bacteria (LAB) can naturally be found in fermented fish products that contain various salt concentrations. The concentration of salt used in the fermented fishery products determines the quality of the fish. This is because the addition of salt will change the type of microorganisms involved in the fermentation process. Fermented fishery products are generally added with 5%-20% salt [5]. Salt can cause bacterial metabolism to be disrupted due to lack of fluids, so it can reduce bacterial activity and even kill bacteria that produce the enzyme histidine decarboxylase [6]. This study used lactic acid bacteria GMH2 and GMH3 isolated from fermented fish product Pakasam [7]. These bacteria will later be added as a starter in fermented fishery products. Therefore, it is necessary to know the optimal salt concentration conditions for growth so that these bacteria can produce high antibacterial compounds to be added as starter in fermented fishery products.

2. Materials and methods

2.1. Preparation of lactic acid bacteria (LAB)
Lactic acid bacteria isolate GMH2 and GMH3 were prepared following the procedure in reference [8], but with some adjustments. LAB isolates GMH2 and GMH3 were taken from bacterial stock stored in glycerol solution at frozen conditions. A total of 1ose of bacterial stocks stored was inoculated and cultured on MRS agar medium with additional 1% $CaCO_3$ (w/v). After incubation for 48 hours at 37°C, LAB GMH2 and GMH3 were pre-cultured in MRS broth medium for 24 hours at 37°C until full growth. Subsequently, 1% (v/v) of fully grown LAB pre-culture was re-cultured in new MRS broth medium with adjusted salt content. The salt concentrations used in this study were 0%, 7.5%, and 15%, based on the categorization of halophilic bacteria based on [9].

2.2. Growth rate test of lactic acid bacteria (LAB)
The calculation of the amount of LAB GMH2 and GMH3 was carried out using the TPC (Total Plate Count) method which refers to SNI 01-2332.3-2015 which has been modified [10]. The TPC method procedure follows previous research by [11], in which pre-cultures of LAB GMH2 and GMH3 that have reached full growth were grown in Erlenmeyer containing MRS broth medium with salt treatment that has been homogenized. Dilutions were made up to 10^{-7}, but in this study only used odd dilutions of 10^{-1}, 10^{-3}, 10^{-5}, and 10^{-7}. Each dilution was inoculated on a petri dish containing MRS agar and CaCO3 medium, then incubated at 37°C for 24 hours. The number of colonies that developed during 24 hours incubation period at 37°C was counted according to the dilution used. Petri dishes that can be counted only contain 25 - 250 colonies and are spreader free. Calculation of bacteria by TPC can be calculated using the following formula:

$$N = \frac{\Sigma C}{[(1xn1)+(0,1xn2)]x(d)}$$

Description:

N : the quantity of product colonies, expressed in colonies per millilitre (mL) or colonies per gram (g)
ΣC : the quantity of colonies on all counted plates
n1 : quantity of plates at the first dilution counted
n2 : quantity of plates at the second dilution counted
d : the first dilution counted

Bacterial growth curves at various salt concentrations can be made by entering the TPC results data into the Microsoft Excel program. Meanwhile, to determine the growth rate of bacteria can be known based on DMFit analysis. DMFit is an Excel add-in that functions to create fit curves from the linear phase to the stationary phase in various microorganisms. The calculated parameters are bacterial growth rate (μ), initial bacterial count ($N_{initial}$), final bacterial count (N_{final}), and the highest bacterial count

(Nmax). The results of the growth rate were then validated to determine the accuracy of the DMFit correction using the Root Mean Square Error (RMSE) formula according to the equation expressed by [12].

2.3. Preparation of cell-free supernatant (CFS)
Cell-free supernatant (CFS) was prepared based on [13] with some modifications. LAB GMH2 and GMH3 isolates from glycerol stock were streaked on MRS agar and incubated at 37°C for 24 hours. Single colonies that grew were then pre-cultured into MRS broth at 37°C for 24 hours or until full growth. Re-inoculated 1% (v/v) into new MRS broth that had been given 0%, 7.5%, and 15% salt treatment, then incubated for 30 hours at 37°C. Every 6 hours, the bacterial culture was transferred into a microtube and centrifuged at 4000 x g at 4°C for 20 minutes. The centrifuged culture was then transferred to a new sterile microtube to separate the supernatant from the pellet. The CFS formed was adjusted to pH 7 using 1N NaOH and then heated in boiling water for 3 minutes.

2.4. Preparation of the tested bacteria
The tested bacteria used for the CFS antibacterial activity test were histamine-producing bacteria (HPB) namely *Klebsiella pneumoniae* CK2. *Klebsiella pneumoniae* CK2 was taken from glycerol stock and then cultured on tryptic soy agar (TSA) medium using the four-quadrant scratch plate method, then incubated at 37°C for 24 hours. One colony was continued to tryptic soy broth (TSB) medium and incubated for 24 hours at 37°C, followed by absorbance adjustment to OD_{600} 0.1 with UV-Vis spectrophotometer.

2.5. Antibacterial activity test
Antibacterial activity test of cell-free supernatant (CFS) was conducted by using the Macrodilution method based on [14]. CFS that has been adjusted to pH 7 and preheated at 90°C for 20 min is then divided into two into new microtubes for protease and non-protease treatments. Protease-treated CFS was added papain at a concentration of 4% (mg/ml), then incubated for 30 minutes at 37°C. CFS was then put into sterile TSB media with a volume proportion of 25% CFS of the total volume of 2.5 ml (v/v). Histamine-forming bacterial culture (HPB) was added to the TSB media at a volume of 1% (v/v) of the total volume. This test was carried out for both CFS, both protease and non-protease treated in 3 replicates. Kanamycin was used as positive control while MRS broth was used as negative. The inhibitory activity of CFS was measured by its absorbance value using a UV-Vis spectrophotometer at a wavelength of 600 nm. $\% \text{ inhibition} = \frac{OD\ negative\ control - OD\ CFS}{OD\ negative\ control} \times 100\%$

2.6. Statistical analysis
Statistical analysis was performed using a parametric two-way analysis of variance (ANOVA). The software used in the analysis was IBM SPSS Statistics 20.0.

3. Result and discussion

3.1. Growth Rate of lactic acid bacteria (LAB) GMH2 and GMH3 at various salt concentrations
The growth of GMH2 and GMH3 indicated to be affected by salt concentration and depicted in Figure 1. In general, as the salt concentration increases, there is a decrease in bacterial growth. This phenomenon is caused by the high osmotic pressure of salt, which can absorb water and disrupt microbial growth [15].

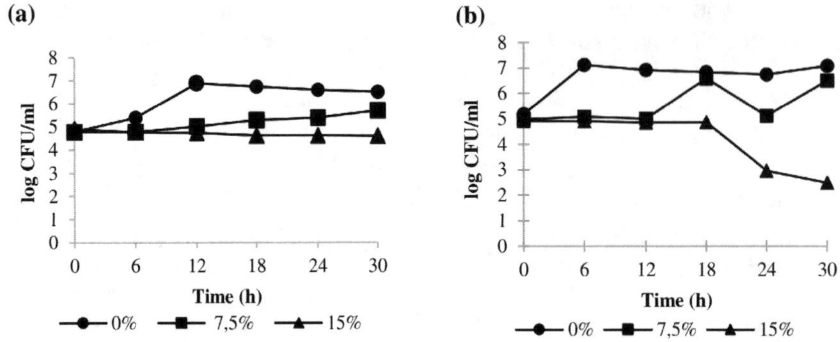

Figure 1. (a) Growth of LAB GMH2 at various salt concentrations (0%, 7,5%, 15%); (b) Growth of LAB GMH3 at various salt concentrations (0%, 7,5%, 15%).

Data from the calculation of total bacteria that grew were then analyzed using the DMFit program to determine the growth rates of GMH2 and GMH3 LAB isolates. The results of the growth rate of GMH2 and GMH3 can be seen in Table 1.

Table 1. Growth rate of LAB GMH2 and GMH3 at various salt concentrations

Isolate	NaCl (%)	μ (log CFU/h)[a]	Ninitial (log CFU/ml)[b]	Nfinal (log CFU/ml)[b]	Nmax (log CFU/ml)[b]	RMSE
GMH2	0%	0.171	4.79	6.52	6.89	0.21
	7.5%	0.153	4.78	5.70	5.70	0.10
	15%	-0.010	4.80	4.60	4.80	0.04
GMH3	0%	1.561	5.13	6.91	7.09	0.14
	7.5%	0.044	4.88	6.21	6.61	0.55
	15%	-0.118	5.28	2.69	5.01	0.45

a: obtained from DMFit calculation
b: obtained from TPC result data

The highest growth rate was detected at 0% NaCl, with 0.171 and 1.681 log CFU/hour for GMH2 and GMH3, respectively. The addition of 7.5% and 15% salt reduced the growth rate to 0,153 log CFU/hour and -0,010 log CFU/hour for GMH 2, while GMH3 was found to be 0.044 log CFU/hour and -0.118 log CFU/hour, respectively. The DMFit test showed positive and negative results on the growth rate (μ) indicating an increase and decrease in the number of bacterial cells. These results shows that high salt content affects the growth rate of LAB, which is also in accordance with the results of [16], where the salt content available as nutrients can affect the growth of certain types of bacteria. These results are also in accordance with the opinion of [17], where halophilic bacteria have slow growth.

3.2. Antibacterial activity of GMH2 and GMH3 cell-free supernatants

Cell-free supernatant (CFS) is a supernatant obtained after culturing specific bacteria in a liquid media that is centrifuged at high speed, so that the material with a higher density will settle to the bottom as a residue. The centrifuged culture was then transferred to a new sterile microtube to separate the supernatant from the pellet. CFS obtained from LAB culture contains organic acids, primary metabolites such as carbon dioxide, ethanol, diacetyl, and secondary metabolites such as hydrogen peroxide and bacteriocins [18]. The results of the antibacterial activity test of CFS isolates GMH2 and GMH3 against HPB CK2 tested in various salt concentration treatments can be seen in Table 2.

Table 2. Antibacterial activity against HPB CK2 using the macrodilution method

Isolate	NaCl (%)	Protease	HPB CK2 Inhibition Percentage (%)					
			0 hour	6 hours	12 hours	18 hours	24 hours	30 hours
GMH2	0%	Protease	15.77±5.14	19.62±1.53	27.30±4.91	26.57±4.95	32.35±2.54	31.79±4.50
		N-Protease	15.30±3.54	24.76±3.81	33.14±3.80	31.65±3.37	39.87±2.25	33.53±4.58
	7.5%	Protease	20.55±3.74	27.44±0.34	25.80±3.61	23.99±2.06	23.04±4.22	19.66±0.12
		N-Protease	17.78±3.34	26.17±1.87	22.13±1.16	22.90±2.45	19.08±3.43	16.58±0.75
	15%	Protease	30.09±0.82	32.71±1.08	33.20±2.84	35.43±1.30	33.16±4.90	34.89±2.45
		N-Protease	24.33±5.42	30.47±2.94	31.88±2.93	28.36±2.24	29.34±5.50	28.72±5.19
GMH3	0%	Protease	30.57±1.08	33.08±1.08	45.15±1.08	66.70±1.08	33.39±0.9	70.34±0.9
		N-Protease	28.59±0.8	30.96±1.08	39.37±1.08	78.08±0.8	40.28±0.8	73.07±0.9
	7.5%	Protease	29.99±1.08	34.07±0.8	35.06±0.8	65.13±0.8	37.19±1.08	71.24±0.8
		N-Protease	27.81±0.9	28.08±1.08	30.52±0.8	73.04±0.8	33.77±1.08	76.25±0.8
	15%	Protease	36.42±0.9	39.11±1.08	41.90±1.08	77.99±0.8	55.60±0.8	79.15±1.08
		N-Protease	32.07±1.08	36.44±0.8	40.77±1.08	79.55±1.08	49.38±0.9	74.40±0.9
Positive control (+)			100	100	100	100	100	100
Negative control (-)			0	0	0	0	0	0

The results of the antibacterial activity test of cell-free supernatant (CFS) LAB GMH2 and GMH3 showed inhibition against HPB CK2 in each salt concentration treatment with incubation times ranging from 0 hour to 30 hour. Antibacterial activity for both isolates showed the highest inhibitory value of 39.87±2.25% for GMH2 and 79.55±1.08 for GMH3. The addition of 7.5% and 15% interestingly did not show a significant reduction in antibacterial activity. The protease treatment in testing the antibacterial activity of CFS against HPB CK2 did not significantly affect the decrease in inhibitory activity. It can be seen from the average percentage value of HPB inhibition after adding papain which has higher results. This indicates that the inhibition results are not caused by the presence of bacteriocins, but can be caused by the presence of other LAB metabolites such as ethanol, fatty acids, and acetoin, or it could also be a type of Bacteriocin-Like Substance (BLS) because it is resistant to proteolytic enzymes [19].

4. Conclusion

Lactic acid bacteria (LAB) strain GMH2 and GMH3 isolated in this research growth well at 0% and remain good at 7.5% salt concentrations with fair antibacterial activities. These findings conclude that LAB GMH2 and GMH3 have the potential to be used as starters in fish fermentation products that have a salt content of up to 7.5%.

Acknowledgement

This research was part of ANR and RIL thesis and funded by Final Project Recognition Grant Universitas Gadjah Mada Number 5075/UN1.P.II/Dit-Lit/PT.01.01/2023 to MMPP. We thanks to Indun Dewi Puspita, Ph.D. for kindly providing *Klebsiella pneumoniae* CK2 isolates.

References

[1] Indriati N, Rispayeni R and Heruwati E S 2007 *Jurnal Pascapanen dan Bioteknologi Kelautan dan Perikanan* **1(2)** 117-123
[2] Mangunwardoyo W, Sophia R A and Heruwati E S 2010 *Makara J. Sci.* **11(2)** 104-109
[3] Food and Drug Administration 2011 *Fish and Fishery Products Hazards and Controls Guidance. 4th Edition US Department of Health and Human Services* (USA: Food and Drug Administration) p 113
[4] Indriati N, Setiawan I P D and Yulneriwani 2006 *J. Fish Sci.* **8(2)** 153-159
[5] Akbardiansyah A, Desniar D and Uju U 2018 *Jurnal Pengolahan Hasil Perikanan Indonesia* **21(2)** 345-355

[6] Romawati M D, Ma'ruf W F and Romadhon R 2014 *Jurnal Pengolahan dan Bioteknologi Hasil Perikanan* **3(1)** 80-88
[7] Mardinsyah A H 2022 *Isolasi Bakteri Asam Laktat Halofilik dari Pakasam dan Uji Aktivitas Antibakteri Supernatan Bebas Selnya terhadap Bakteri Pembentuk Histamin* [*Isolation of Halophilic Lactid Acid Bacteria from Pakasam and Antibacterial Activity Test of Cell Free Supernatant Against Histamin Forming Bacteria*] Undergraduate thesis (Universitas Gadjah Mada–Yogyakarta)
[8] Lim E S 2018 *Korean J. Microbiol.* **54(4)** 398-409
[9] Corral P, Amoozegar M A and Ventosa A 2019 *Mar. drugs 8*(**1**) 33
[10] Badan Standarisasi Nasional 2015 Standar Nasional Indonesia (SNI-2332.3-2015) *Cara uji mikrobiologi - Bagian 3: Penentuan Angka Lempeng Total (ALT) Pada Produk Perikanan* [*Microbiological test method - Part 3: Determination of Total Plate Number (ALT) in Fishery Products*] (Jakarta: Badan Standarisasi Nasional)
[11] Whitmire J M and Merrell D S 2012 *Helicobacter Species: Methods in Molecular Biology* **921** 17-27
[12] Dityanawarman A, Puspita I D, Ratnawati S E, Ekantari N and Tamplin M 2020 *Squalen Bull. Mar. Fish. Postharvest Biotechnol.* **15(1)** 1-9
[13] Rana S, Bhawal S, Kumari A, Kapila S and Kapila R 2020 *Microbial pathogenesis* vol 142 (Amsterdam: Elsevier) pp 104-105
[14] Putra M M P, Al-Hammam M Y, Ahsan G, Chandra K K B and Puspita I D 2022 *Journal of Fisheries Science and Technology* **18(1)** 13-18
[15] Thariq A S, Swastawati F and Surti T 2014 *Jurnal pengolahan dan bioteknologi hasil perikanan* **3(3)** 104-111
[16] Yuliana N 2008 *Jurnal Teknologi Industri dan Hasil Pertanian* **13(2)** 108-116
[17] Kiayi R, Naiu A S and Harmain R M 2014 *The NIKe Journal* **2(3)** 126-129
[18] Alvarado C, Garcia A B, Martin S E and Regalado C 2006 *Rev. latinoam. Microbiol.* **48(3-4)** 260-268
[19] Mani-López E, Arrioja-Bretón D and López-Malo A 2022 *Compr. Rev. Food Sci. Food Saf.* **21(1)** 604-641

Growth pattern of yellow finned barb (*Mystacoleucus marginatus,* Valenciennes 1842) in the Lau Belawan River, Pancur Batu District, North Sumatera Province

Desrita[1,2*], J S Hasibuan[1], A Fadhilah[1], D S Mulia[3], Nurfadillah[4]

[1]Aquatic Resources Management Program Study, Faculty of Agriculture, Universitas Sumatera Utara, Medan, Indonesia
[2]Biological Science Doctoral Program, Faculty of Mathematics and Natural Sciences, Universitas Sumatera Utara, Medan, Indonesia
[3]Freshgraduate of Aquatic Resources Management Program Study, Faculty of Agriculture, Universitas Sumatera Utara, Medan, Indonesia
[4]Aquaculture Program Study, Faculty of Marine Affairs and Fisheries, Universitas Syiah Kuala, Banda Aceh, Indonesia

*Coresponding author : desrita@usu.ac.id

Abstract. Yellow finned barb is one of the potential fish from the Lau Belawan River, North Sumatra. This study aims to determine the growth patterns of Yellow finned barb (*Mystacoleucus marginatus*) in the Lau Belawan River, Pancur Batu District, North Sumatera Province. The research was carried out from November 2022 until January 2023. Fish were caught using throwing nets with a mesh size of 0.2 inches with the help of local fishermen. The research location is divided into three stations with different characteristics for each station. During the study, the number of catches obtained was 114 (65 male fish and 49 female fish), with a total length of fish ranging from 51 to 162 mm. The catch at each station was different; stations II and III had 41 individuals, while stations I had only 32 individuals. The length of fish caught was dominated by 93–106 mm, and the lowest number caught was around 149–162 mm. The fish caught consisted of males and females, with a sex ratio 1:1.33. The growth pattern of male Yellow finned barb is negative allometric, while the female is positive allometric. Meanwhile, the condition factor value for male fish is 0.87; the female is 1.07.

Keyword: Yellow finned barb (*Mystacoleucus marginatus*), and growth, sex ratio

1. Introduction

Lau Belawan River is one of the rivers included (DAS) Belawan Watershed located in Pancur Batu District, Deli Serdang Regency, North Sumatera, Indonesia. Lau Belawan River is geographically located between 30 28.8' 31.01" LU and 980 34.8' 8.58" East. One of the potential in Lau Belawan River is Yellow finned barb fish.

Yellow finned barb fish are benthopelagic freshwater fish or bottom swimmers and can move vertically to the surface in inland public waters, especially rivers, streams, and flooded swamp areas. When the water level rises, Yellow finned barbwill move to lower waters to breed [11]. The classification of Yellow finned barb is as follows: kingdom Animalia, phylum Chordata, class Actinopterygii, order Cypriniformes, family Cyprinidae, genus Mystac locus, species *Mystacoleucus marginatus* [13].

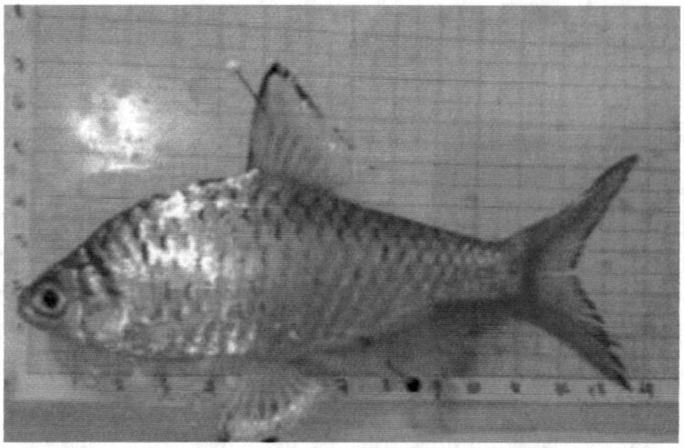

Figure 1. Yellow finned barb (*Mystacoleucus marginatus)*

Yellow finned barb fish (*Mystacoleucus marginatus*) belonging to the Cyprinidae family has the characteristics of a flat and long body shape with a raised back, small head, tapered snout, small mouth located at the tip of the nose, and very small snout. The color of the dorsal and caudal fins is gray and yellowish, the caudal fin is branched and has rounded lobes, the color of the pectoral fins is yellow, and the anal fin is bright. The scale structure has fingers that are straight and curved to the end, with little or no fingers on the sides [13]. Yellow finned barb fish can be found in the depths of riverbeds and river surfaces. They usually inhabit areas with sand or gravel from small to large rivers and can be found in rivers with clear, moving water. When the water level rises, Yellow finned barb will move to lower waters to breed. Yellow finned barb fish are benthopelagic fish that live in schools in the morning [17,4,5,6]. The distribution of Yellow finned barb fish is from large rivers, then tributaries to floodplains, especially during the rainy season. The natural distribution of Yellow finned barb fish has been recorded in the rivers of Sumatra, Java, Chao Phraya, Malay Peninsula, and Mekong [11].

Yellow finned barb fish is the most dominating fish in the Lau Belawan River. This fish is also very popular with the public as one of the favorite consumption fish. In line with this, with the increase in public demand, fishing intensity has also increased. Although Yellow finned barb fish is one resource that can recover (renewable resources), based on fishbase data, Yellow finned barb fish was included in The IUCN Red List of Threatened Species in 2011 as a species that gets little attention. Therefore, it is necessary to manage fisheries resources for Yellow finned barb fish. So far, data on the presence of Yellow finned barb fish in the Lau Belawan River has never been reported, so it is necessary to conduct research that can examine the Growth Pattern of Yellow finned barb fish in the Lau Belawan River, Pancur Batu District, North Sumatera.

2. Materials and Methods

2.1 Time and Location

This research was conducted in November 2022 - January 2023 with a sampling time interval of 1 time in a month in Lau Belawan River, Pancur Batu District, North Sumatera. Analysis of the Yellow finned barb fish obtained was carried out at the research site directly. The research location can be seen in Figure 2.

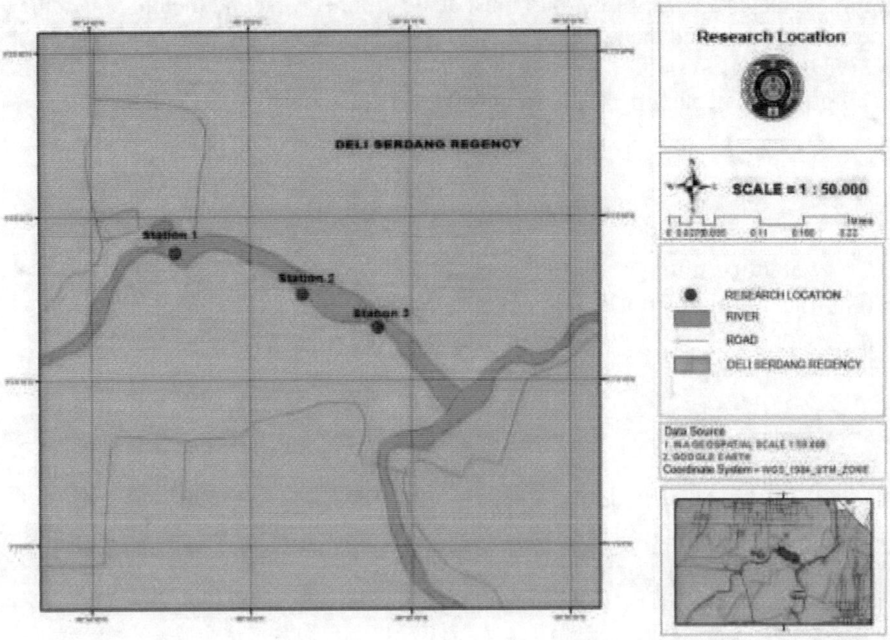

Figure 2. Map of research location

2.2 Tools and Materials

The tools used in this study are netted mesh size 0.2-inch analytical scales cool box and millimeter block. The materials used in this research are Yellow finned barb fish and ice cubes. Tolls used in this research is Microsoft Excel 2010.

2.3 Methods

The method of determining the sampling location used for the research station is Purposive Random Sampling, namely by selecting three research stations for activity tires around the river. Station I is part of the river waters with no activity with coordinates 30 29' 10.47" LU and 980 35' 10.98" BT. Station II is part of the river waters with sand dredging activity. This station is geographically located at 30 29' 12.62" LU and 980 35' 0.44" BT. Station III is part of the river waters where waste is generated from bathing or recreational activities and BWL (bathing, washing, latrines) with coordinates 30 29'17.67" LU and 980 34' 57.60" BT.

2.4 Data analysis

2.4.1 Length frequency ditribution

The procedure for creating a length frequency distribution is as follows [20] Determine the number of measurements group with the formula :

$$K = 1 + (3{,}32 \text{ Log } n)$$

Where :
K = number of size cllasses
N = number of observation

Determine the class width value of each measure group by using a formula :

$$\text{Class wide} = \frac{\text{Region}}{\text{Class}}$$

IOP Conf. Series: Earth and Environmental Science 1289 (2023) 012018 doi:10.1088/1755-1315/1289/1/012018

The lower limit value of the measure group must be determined first and then to obtain the upper limit From the next size group, add the width of the class minus one.

2.4.2 Weight length relationship

Weight length relationship of fish can find with used formula as follows [8]:

$$W = a\,L^b$$

Where :

W = fish weight (g/mm)
L = fish totaly length (mm)
a = intercept
b = koefesien regresi

2.4.3 Condition factor

Because of constant b ≠ 3 (get from weight length relationship), so factor condition is calculated using the formula that explained by Effendi [8] :

$$K = \frac{W}{a\,L^b}$$

Where :

K = condition factor
W = fish weight (gr)
L = fish totaly length
a = intercept
b = slope

2.4.4 Sex ratio

Sex ratio calculated with way compare the number fish male and fish female as follow :

$$RK = \frac{\sum J}{\sum B}$$

Where :

RK = sex ratio
$\sum J$ = number of male fish totaly (individual)
$\sum B$ = number of female fish totaly (individual)

to examine there is a difference between the two proportions, a chi-square test was performed used [20]:

$$X^2 = \sum_{i=1}^{n} \frac{(oi - ei)^2}{ei}$$

Where :

X^2 = chi-square (random variable values whose sample drawing distribution approximate the chi-square distribution
oi = frequency of male fish or female to-i that observation
ei = number of frequency hope from fish male and female fish that frequency male fish added frequency of female fish divided by two

3. Result and Discussion

3.1 Length frequency distribution

The results of the data analysis of the length and class interval of Yellow finned barb fish are known that Yellow finned barb fish with a total of 114 individual fish samples have a total length of 51 - 162 mm. The average length obtained is 106.5 mm, while the length that dominates in females with a class interval size of 93 - 106 is as many as 35 fish. Yellow finned barb fish (*Mystacoleucus marginatus*) has a maximum length of 200 mm [13]. While based on the total length, a long frequency distribution is obtained of Yellow finned barb fish (*Mystacoleucus marginatus*) obtained was 77 - 180 mm with a body weight of 26 - 77 gr [17], and stated that the range of yellow finned barb fish in the Brantas River was 110 - 190 mm [23]. The difference in catch size between the Seruai River, Lau Belawan River, and the Brantas River can occur due to differences in the aquatic environment [23].

The frequency distribution of length in (Figure 3) shows that the maximum length of Yellow finned barb fish is 51 - 162 mm. One of the factors affecting the length obtained is the mesh size used in throwing net fishing gear, which is 0.2 inches. The minimum size of Yellow finned barb fish obtained [12] showed that the mesh size of the net fishing gear used is 0.2 inches, so the total length obtained is 70 - 160 mm.

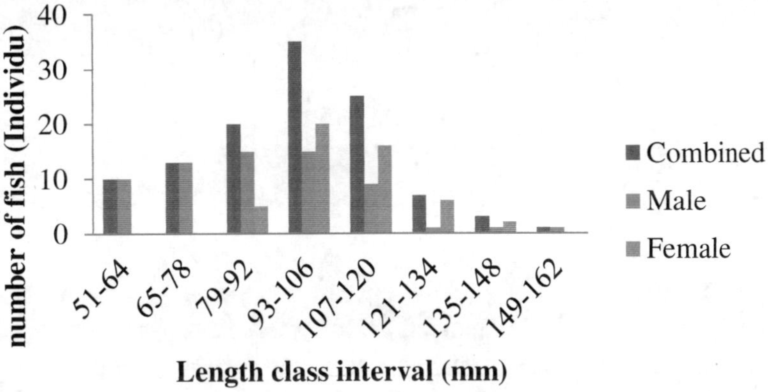

Figure 3. Distribution of yellow finned barb fish male, female, and combined based on length class interval

There were 114 Yellow finned barb fish were caught during the study in Lau Belawan River (65 male fish and 49 female fish). The number of Yellow finned barb fish caught at each station varies at stations II and III, which is the highest number of catches of 41 fish. At station II, there are 22 males and 19 females. While at station III, there are 24 males and 17 females. However, station I has the lowest number of catches, with 32 fish consisting of 19 males and 13 females. Differences in water quality and location characteristics. The research station caused a difference in the number of catches at each station. Station I is the confluence of two tributaries, namely the Lau Belawan River and the middle Lau Belawan River, so it is suspected that nutrients are abundant in this area; besides that, the high water level at this station causes Yellow finned barb fish to distribute to lower areas. According to Khadijah et al. (2019), the high and low water depths will affect fish distribution and degradation, and fish habitat will be lost. Meanwhile, stations II and III are areas favored by Yellow finned barb fish. Based on fishbase.org data, Yellow finned barb fish usually inhabit areas with sand or gravel from small to large rivers and can be found in rivers with clear and moving water. When the water level rises, Yellow finned barb fish will move to lower waters to breed.

3.2 Weight length relationship

The difference in length and weight of Yellow finned barb fish (Table 1) in male fish has a b value of

IOP Conf. Series: Earth and Environmental Science 1289 (2023) 012018 doi:10.1088/1755-1315/1289/1/012018

2.89 and females of 3.14. From the b value, The b value obtained indicates that the male fish is lower than female fish. This is influenced by differences in behavior between the two. Fish behavior is the main factor affecting the magnitude of the b value. Fish with active swimming ability have a lower b value than those with passive swimming ability. This refers to the amount of energy expended in movement and growth. Fish swimming in calm currents tend to have higher b values than fish in strong currents [22]. This is also supported [19] Different b values in growth patterns are also influenced by the movement of fish that move actively and migrate. In addition, differences in growth values can be caused by each fish's environmental and biological conditions. Current is another possible factor affecting fish growth patterns. Fish that live in waters with calm currents will have a greater b value [15]. Meanwhile, fish that live in waters with strong currents will have a smaller b value.

Tabel 1. Relationship of weight length yellow finned barb fish

Sex ratio	N	Formula	r	Thit	Ttab	Growth pattern
Combined	114	$W = 0{,}00001L^{2{,}96}$	0,96	-0,02	1,98	Negative allometric
Male	65	$W = 0{,}00002L^{2{,}89}$	0,95	-0,05	2,00	Negative allometric
Female	49	$W = 0{,}00006L^{3{,}14}$	0,96	0,09	2,01	Positive allometric

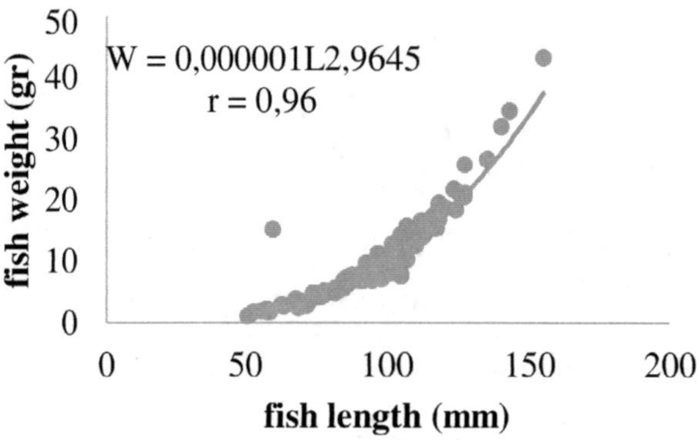

(a)

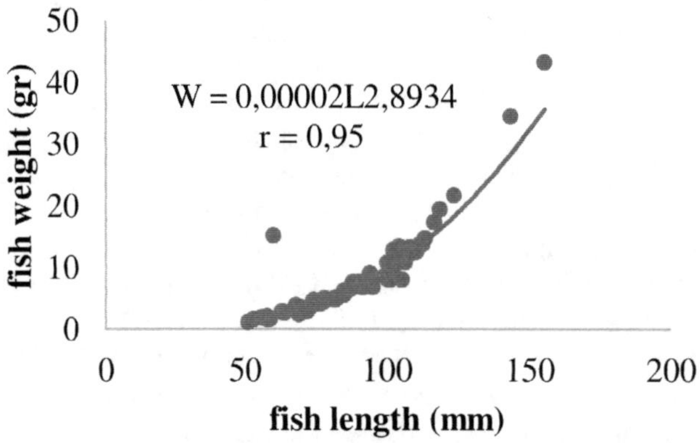

(b)

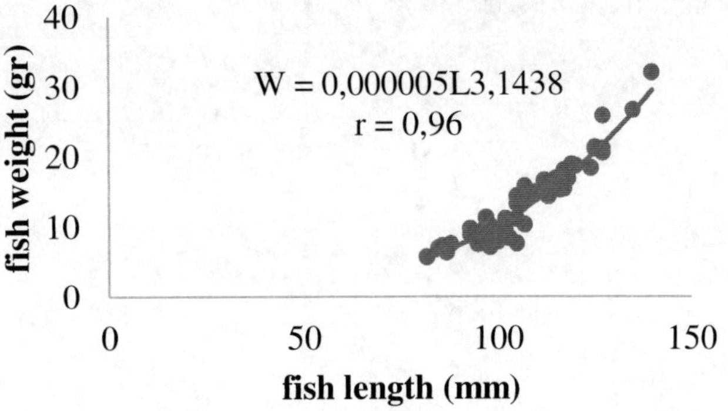

(c)

Figure 4. Weight length relationship graph yellow finned barb fish. (a) combined (b) male (c) female

Based on the length-weight relationship analysis (Figure 4), the correlation values (r) for male and female fish were 0.96, 0.95, and 0.96, respectively. This indicates that body length and body weight for both male and female fish have a very close relationship.The relationship between the length and weight of fish in males and females strongly correlates because the resulting r value is close to 1. The relationship shows that with the increase in the size of the fish length, the weight also increases [21]. This is also supported if the range of correlation coefficient values between > 0.75 - 0.99 indicates a very strong relationship between the two variables [16].

3.3 Condition factor

In line with the b value in the long-weight relationship data analysis, the b value in females is larger than male fish, so the condition factor of Yellow finned barb fish (Table 2) ranges from 0.87 - 1.26. Environmental factors and sufficient food sources cause differences in the range of condition factors in Yellow finned barb fish. The condition factor is also known as the index of plumpness, one of the derivatives of the length-weight relationship. The condition factor for Yellow finned barb fish is relatively stable, as seen from the range of numbers produced to reach more than 1 for both male and female fish. The results differ from research conducted on lame-duck fish from the Cyprinidae family, which has a condition factor value in males of 1.05 and 0.95 in females showing different body shapes [1].

Table 2. Value of condition factor male and female yellow finned barb fish

Sex ratio	Average	Standar deviation
Combined	1.26	0.67
Male	0.87	0.59
Female	1.07	0.14

The average value of the combined yellow finned barb fish condition factor obtained each research station ranges from 1.18 - 1.40 can be seen in Table 3.

Table 3. Value of condition factor yellow finned barb fish based on observation station

Observation station	Average	Standar deviation
Station I	1.40	1.24
Station II	1.18	0.16
Station III	1.22	0.14

Result average value of fish condition factor that got based on observation station that higher in station I and lower in station II can be seen in Figure 5.

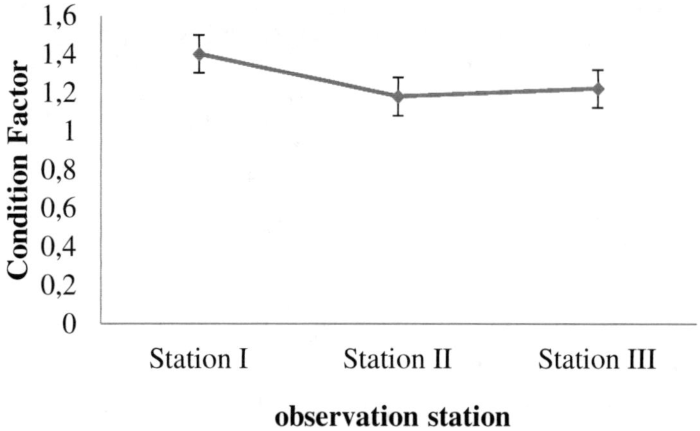

Figure 5. Value of condition factor yellow finned barb fish based on observation station

The analysis results of Yellow finned barb fish condition factors are based on observation stations (Table 3), namely at the station I obtained 1.40, station II obtained 1.18, and station III obtained 1.23. This shows that the highest condition factor value is at station I, while the lowest is at station II. The difference in the value of the condition factor is influenced by the current speed at the station. The speed of the current in the water body can affect the growth pattern of fish because the faster the current, the more fish movement that, makes the fish body size will be flatter [17]. Furthermore, the condition factor will affect the maturity of fish gonads. If the condition factor is higher, the fish is in gonadal development [9]. Different condition factor values can be influenced by behavior, environmental conditions, gonad maturity, age differences, and food availability [8].

3.4 Sex Ratio

The results of data analysis of sex ratio by month can be seen on November 1: 1.31, December 1: 1.41, and January 1: 1.26. Overall the sex ratio in each month is balanced. A sex ratio of 1:1 indicates equal numbers of male and female fish. If the number of male fish is more than female fish, then the sex ratio is greater than >1, while the sex ratio <1 indicates the number of female fish is more than male fish (dominant). The balanced sex ratio means that each male fish fertilizes one female fish. A balanced number of males and females ensures maximum fertilization during the breeding season. In general, differences in the number of female and male fish are closely related to the migratory behavior of these fish both in the spawning and foraging phases [10]. However, if there is an imbalance from the ideal condition (1:1), it is caused by differences in the distribution, growth, mortality rate, activity, and movement of these fish [1]. Furthermore, according to Omar (2015), it is suspected that high fishing pressure causes an imbalance in the number of male and female fish in a population.

Tabel 4. Sex ratio of yellow finned barb fish based on observation month

Month	Frequency		Sex ratio	X^2hit	Where
	Male	Female			
November	17	13	1 : 1,31	0.53	Balanced
December	24	17	1 : 1,41	1.20	Balanced
January	24	19	1 : 1,26	0.58	Balanced
Combined	65	49	1 : 1,33	2.25	Balanced

Where : x^2 table = 3,84; H0 = comparison between number male and female fish 1 : 1; H1 = comparison between number male and female fish not 1 : 1.

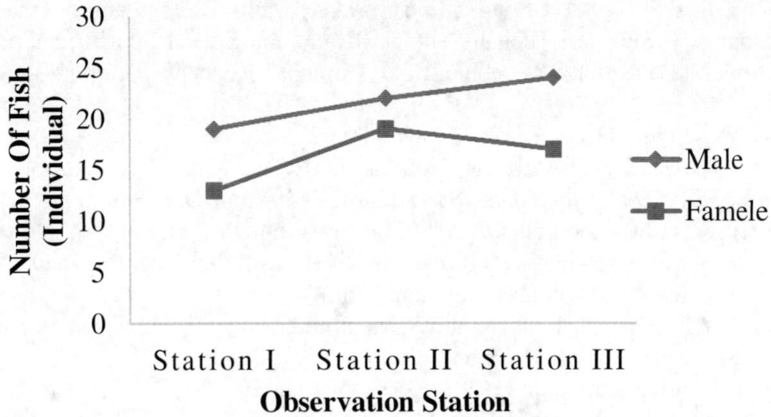

Figure 6. Sex ratio of yellow finned barb fish based on observation station

4. Conclusion

Growth pattern yelllow finned barb fish (*Mystacoleucus marginatus*) from Lau Belawan River for male fish classified as negative allometric and female as positive allometric. Condition factor yellow finned barb fish, each station for male and female fish, have plump figure difference in each station while sex ratio in each pickup is balanced.

References

[1] Aisyah S 2017 Pola Pertumbuhan dan Faktor Kondisi Ikan Lemeduk (*Barbodes schwanenfeldii*) di Sungai Belumai Deli Serdang Provinsi Sumatera Utara. [The Growth Pattern and Condition Factors of The Lemeduk Fish (*Barbodes schwanenfeldii*) *Thesis* (Medan: Universitas Sumatera Utara)

[2] Bone Q, and Moore R H 2008 Biology of fishes (New York: Taylor and Francis Group)

[3] Chadijah A, Sulistiono, Haryani G S, Affandi R, and Mashar A 2019 Distribusi Ukuran, Pola Pertumbuhan, dan Faktor Kondisi Ikan Endemik Opudi (*Telmatherina prognatha*) di Danau Matano, Sulawesi Selatan [The Size Distribution, Growth Pattern, and Condition Factors of Opudi Endemic Fish (*Telmatherina prognatha*) in Lake Matano, South Sulawesi] *JIPI* 24(**4**) pp 295 – 303

[4] Desrita, Muhtadi A, Tamba I S, Ariyanti J, Sibagariang R D 2018 Community Structure of Nekton in The Upstream of Wampu Watershed, North Sumatra, Indonesia *Biodiversitas* 19 (**4**) pp1366 – 1374

[5] Desrita, Muhtadi A, Leidonald R, Sibagariang R D, Nurfadillah 2020 Biodiversity of Nekton in Batangtoru River and Its Tributaries in North Sumatra, Indonesia Biodiversitas 21(**6**) pp 2344 – 2352

[6] Desrita, Rambey R, Muhtadi A, Onrizal O, Manurung V R, Hasibuan J S, and Tamba I S 2022 Biodiversity of Nekton in the Barumun Watershed, South Labuhanbatu District, North Sumatra, Indonesia *Biodiversitas* 23(**5**) pp 2426 – 2432

[7] Djumanto, Devi M I P, Yusuf I F, and Setyobudi E 2014 Kajian Dinamika Populasi Ikan Kepek *Mystacoleucus obtusirostris* (Valenciennes, in Cuvier and Valenciennes 1842) di Sungai Opak, Yogyakarta [The Population Dynamics Study of Kepek Fish *Mystacoleucus obtusirostris* (Valenciennes, in Cuvier and Valenciennes 1842) on The Opak River, Yogyakarta] *JII* 14(**2**) pp 145 – 156

[8] Effendi M I 2002 *Biologi Perikanan* [*Fisheries Biology*] (Yogyakarta: Yayasan Pustaka Nusantama) p 163

[9] Febriani L 2010 Studi Makanan dan Pertumbuhan Ikan Bilih (*Mystacoleucus padangensis*) di Danau Singkarak, Sumatera Barat [The Study of Food and Growth of Bilih Fish (*Mystacoleucus padangensis*) in Singkarak Lake, West Sumatra] *Thesis* (Bogor: IPB University)

[10] Hasanah N, Restiangsih Y H, and Nurdin M S 2019 Nisbah Kelamin dan Ukuran Pertama Kali

Matang Gonad Ikan Tongkol Lisong (*Auxis rochei*) yang Didaratkan di PPI Labuan Bajo, Kabupaten Donggala, Sulawesi Tengah [The Sex Ratio and Size at First Ripe Gonads of Lisong Tuna (*Auxis rochei*) Landed at PPI Labuan Bajo, Donggala Regency, Central Sulawesi] *JISA* 3(**1**) pp 1- 5

[11] http://www.fishbase.com/ (accessed on 30 November 2022)

[12]Kaban C L 2016 Efek Aktivitas Masyarakat Terhadap Keberadaan Ikan Cencen (*Mystacoleucus marginatus*) di Sungai Sibiru-Biru Kabupaten Deli Serdang Provinsi Sumatera Utara [The Effects of Community Activities on the Presence of Yellow finned barb (*Mystacoleucus marginatus*) in the Sibiru-Biru River, Deli Serdang Regency, North Sumatra Province] *Thesis* (Medan: Universitas Sumatera Utara)

[13] Kottelat M A J, Whitten S N, Kartika, and Wirjoatmodjo S 1993 *Freshwater Fishes of Western Indonesia and Sulawesi* (Jakarta: Periplus Edition)

[14] Mamangkey J J and Nasution S H 2014 Pertumbuhan dan Mortalitas Ikan Endemik Butini (*Glossogobius matanensis* Weber, 1913) di Danau Towuti, Sulawesi Selatan [The Growth and Mortality of Endemic Butini Fish (*Glossogobius matanensis* Weber, 1913) in Lake Towuti, South Sulawesi] *Berita Biologi* 13(**1**) pp 31–38

[15] Pongtiku J 2021 Hubungan Panjang Berat dan Faktor Kondisi Ikan Gulamah (*Johnius belangerii*) di Perairan Tarakan Kalimantan Utara [The Length Weight Relationship and Condition Factors for Gulamah Fish (*Johnius belangerii*) in Tarakan Waters, North Kalimantan] *Thesis* (Tarakan: Universitas Borneo Tarakan)

[16] Sarwono J 2006 *Metode Penelitian Kuantitatif dan Kualitatif* [*The Quantitative and Qualitative Research Methods*] (Yogyakarta: Graha Ilmu)

[17] Sihaloho P S A, Sitorus H, and Leidonald R 2017 Jenis dan Pola Pertumbuhan Ikan yang Tertangkap di Sungai Seruai Desa Namu Suro Kabupaten Deli Serdang Sumatera Utara [The Types and Growth Patterns of Fish Caught in the Seruai River, Namu Suro Village, Deli Serdang Regency, North Sumatera]

[18] Soekiswo A Y, Widyorini N, dan Solichin A 2014 Aspek Biologi Ikan Mendo (*Acentrogobius* Sp.) di Waduk Malahayu Kabupaten Brebes, Diponegoro [The Biological Aspects of Mendo Fish (*Acentrogobius* Sp.) in the Malahayu Reservoir, Brebes Regency, Diponegoro] *Maquares J.* 3(**4**) pp 154 – 160

[19] Utami M N F, Redjeki S, and Supriyantini E 2014 Komposisi Isi Lambung Ikan Kembung Lelaki (*Rastrelliger kanagurta*) di Rembang [The Composition Of Gastric Contents Of Male Mackerel (*Rastrelliger kanagurta*) in Rembang] *JMR* 2(**3**) pp 99-106

[20] Walpole R E 1992 *Probabality and Statistics for Engineers and Scientists* (San Antonio: University of Texas)

[21] Yudha I G, Rahardjo M F, Djokosetiyanto D, and Djamar T F 2015 Pola Pertumbuhan dan Faktor Kondisi Ikan Lumo *Labiobarbus ocellatus* (Heckel, 1843) di Sungai Tulang Bawang, Lampung [The Growth Patterns and Condition Factors of Lumo *Labiobarbus ocellatus* (Heckel, 1843) in the Tulang Bawang River, Lampung] 24(**1**) pp 29-39

[22] Zulfahmi I, Yuliandhani D, Sardi A, Kautsari N, and Akmal Y 2021 Variasi Morfometrik, Hubungan Panjang Bobot dan Faktor Kondisi Ikan Famili Holocentridae yang didaratkan di Pelabuhan Perikanan Samudra (PPS) Lampulo, Banda Aceh [Morphometric Variation, Long-Weight Relationship and Condition Factors of Fish of the Holocentridae Family Landed at the Lampulo Ocean Fishing Port (PPS), Banda Aceh] 24(**1**) pp 81-92

[23] Valen F S, Sambah A B, Wicaksono K P, Widodo M S, Soemarno and Hasan V 2020 Genetic diversity of Yellow Finned barb *Mystacoleucus marginatus* (Valenciennes, 1842) (Teleostei, Cyprinidea) in Brantas basin Upstream, Indonesia *Eco. Env & Cons* 27(**2**) pp 695 – 699

Acknowledgment

Thank you to Universitas Sumatera Utara for the TALENTA fund with number 007/UN5.2.3.1/SPP-TALENTA USU/2022 that has given finance for this research.

IOP Conf. Series: Earth and Environmental Science 1289 (2023) 012019 doi:10.1088/1755-1315/1289/1/012019

Molecular analysis of the blood cockle, *Tegillarca granosa* (Linnaeus 1758) from Indonesia

J I Castaneda[1], E Hardianto*[2], E Setyobudi[2], MD R Islam[3]

[1]National Institute of Advanced Industrial Science and Technology, Tokyo, Japan
[2]Department of Fisheries, Faculty of Agriculture, Universitas Gadjah Mada, Sleman, Yogyakarta, Indonesia, 55821
[3]Department of Fisheries and Marine Science, Noakhali Science and Technology University, Noakhali 3814, Bangladesh

Corresponding author e-mail: ekohardianto@ugm.ac.id

Abstract. The blood cockle, *Tegillarca granosa*, is one of the key member of shellfish communities that are ecologically and economically important specially in Indonesia. We collected samples from four sites in Indonesia: Banyuwangi, Bali Island, Maros and Buru Island. Nucleotide sequence analysis was performed on PCR-amplified mitochondrial DNA COI gene. In total, 556-558 base pair nucleotide sequences were obtained from 12 individuals. Molecular analyses revealed samples belonging to the *T. granosa* species. Phylogenetic analyses were carried out by comparing the gene sequences of *T. granosa* haplotypes in Indonesia with gene sequences of the same base pair length and with the of *Anadara tuberculosa* which was used as an outgroup. The phylogenetic trees were consistent and indicated in the two sub-clades. The presence of *T. granosa* and the phylogenetic status of the *T. granosa* Indonesian haplotype were reported. Clarifying the phylogenetic status of ecologically important species provides basic information studies for possible future conservation and control programs.

1. Introduction

The blood cockle, *Tegillarca granosa*, is a bivalve species which belongs to the family Arcidae. The species has very wide distribution in the Indo-Pacific Region including Several Sea in Korea, Japan, China and Southeast Asia [1]. *T. granosa* is ecologically warm water, inhabiting a sandy mud bottom comprising of a lot of mud and slime from intertidal zone to 10m in depth [1]. *T. granosa* has characteristics such as a rough surface of the shell, then a light gray color, black hairs growing on the shell, and the inside of the shell is white while the body has a blood-like color. In Indonesia, the species is significant commercial value, However, the recent catch statistics showed that production declined every year from 2000 [2].

Despite its ecological significance, this species frequently makes up a significant portion of fishermen's catches. Unregulated fishing practices may result in the mass extinction of this species, posing a major threat to marine biodiversity overall. Due to the strong demand, assessment of its populations in Indonesia found a high exploitation ratio. If preventive measures are not adopted, ongoing fishing pressure may ultimately expose these marine species to a significant danger of genetic and population bottlenecks [3][4][5]. In order to comprehend the status of the existing stock, baseline data must be established.

The previous research on DNA Barcode and phylogenetic tree has been carried out on marine organism like fish [6][7][8], shellfish [9][10][11], crustacean [12][13] and coral reef [14][15]. The ability to identify species at any stage of an organism's development (such as larval, juvenile, or adult stages), as well as at early stages that cannot be done using descriptions, is the most promising advantage of DNA barcoding and phylogenetic trees for species authentication, morphology and interspecies connections.

The blood cockle from Indonesia is the subject of this study in order to amplify nucleotide sequences of the mitochondrial COI gene, identify species, calculate genetic distances, and examine phylogenetic relationships. Therefore, utilizing the COI gene to identify the species is an attempt to gather genetic data that will serve as the foundation for future conservation efforts for the target species.

2. Materials and Methods

2.1 Sampling and DNA Extraction

Samples were collected in 2023 from Indonesia (Java Island; Banyuwangi, Bali Island; South Denpasar, Sulawesi Island; Maros, and Buru Island) Table 1. Individual *T. granosa* samples were fixed in 95% ethanol on site and then transported to the laboratory until used for genetic analysis. Muscle tissue (10 mg) from the samples was excised from near the shell using scissors and forceps. DNA was extracted using the Favorgen DNA Extraction Kit following the manufacturer's manual procedure.

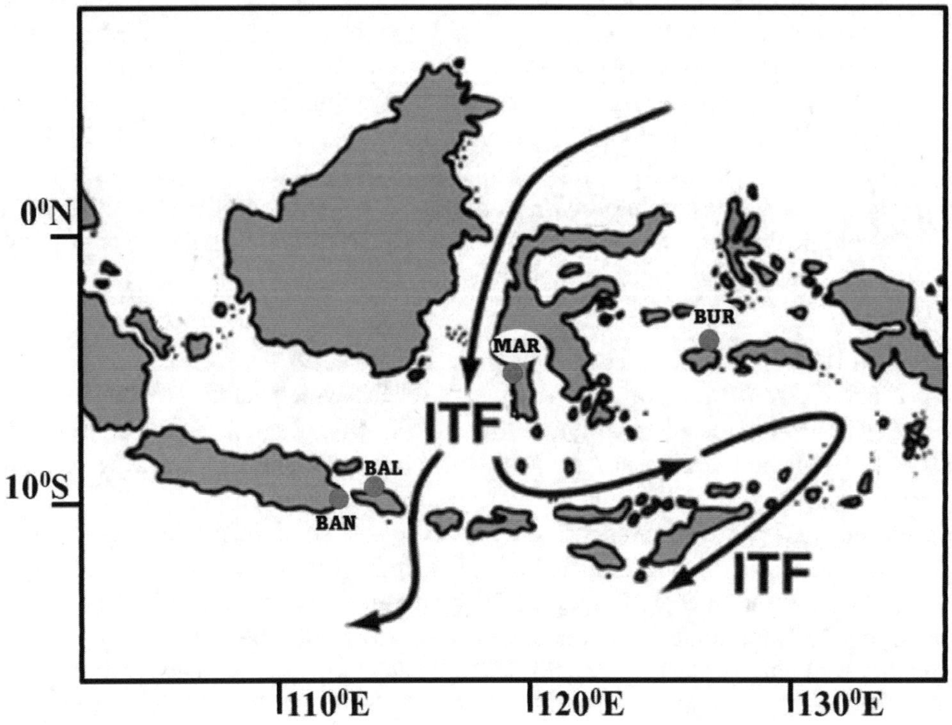

Figure 1. Map of sampling site of *Tegillarca granosa* in this study. ITF; Indonesian throughflow, BAN; Banyuwangi, BAL; Bali Island, MAR; Maros, BUR; Buru Island

2.2 Amplification and Sequencing

The partial mitochondrial cytochrome oxydase subunit I gene was amplified via polymerase chain reaction (PCR) using universal primers LCO1490 (5'-CGCCTGTTTATCAAAAACAT-3') and HCO2198 (5'-CCGGTTTGAACTCAGATCACGT-3') (Folmer et al. 1994). All reactions were carried out in a 25 µl total volume and the following reagents were added to each PCR microtube: 1 µl of template DNA, 25 pmol of each primer, 12.5 µl of KOD one master mix Taq DNA polymerase. Each sample was adjusted to 25 µl with distilled H_2O. PCR conditions consisted of a hot start (94°C, 180 s) followed by 30 cycles of denaturation at 94°C (10 s), annealing at 50-52°C for 5 s, extension at 72°C (10 s), and final extension at 72°C (420 s) using a GeneAmp 9700 thermal cycler (Applied Biosystems, Foster City, CA, USA). Verification of PCR was performed via electrophoresis in a 1% agarose gel. After electrophoresis, gels were stained with ethidium bromide and products checked on a transilluminator (Advanced Scientific Products Pty Ltd., Queensland, Australia). PCR products were purified using a PCR product pre-sequencing kit (USB Co., USA). Amplified DNA was sequenced on an ABI 3730xl DNA analyzer using the BigDye Terminator v3.1 cycle sequencing kit (Applied Biosystems, USA).

2.3 Data Analyses

COI Sequence data were analysis using muscle alignment software, implemented in MEGA X [16] using default alignment parameters, and the sequences were adjusted manually to avoid mismatches. The COI sequences were analyzed for species identity using BLAST (Basic Local Alignment Search Tool) in NCBI (National Center for Biotechnology Information). COI sequence data also calculated using Kimura's two-parameter (K2P) distance with 1,000 bootstrap replications, to analysis the pairwise genetic difference, nucleotide composition and to constructed phylogenetic tree by the Neighbor Joining (NJ) methods using MEGA X. Haplotype network analysis between the species also constructed by the analysis of minimum spanning network (MSN) using PopART [17].

3. Results

In the sequencing results, we obtained a 516–528 bp mtDNA COI gene sequence. The sequence was obtained after the alignment of eight specimens of *T. granosa* from four localities across Indonesia (Figure 1). Nucleotide sequences of all haplotypes were deposited in the DNA Databank of Japan.

Table 1 Details of four sampling sites, number of samples collected (n) and accession number for COI gene sequence.

Location site	n	Collecting date
Banyuwangi, East of Java	2	October 20, 2022
Kedonganan Beach, Bali Island	2	December 10, 2022
Maros, South Sulawesi	2	November 12, 2022
Buru Island, North Maluku	2	December 10, 2022

IOP Conf. Series: Earth and Environmental Science 1289 (2023) 012019 doi:10.1088/1755-1315/1289/1/012019

Table 2 BLAST Identity percentage of nucleotide of *Tegillarca granosa* based on COI gene sequence

No.	Species	Origin	Accession number	Identity (%)
Sample code: BAN-1, BAN-2				
1.	*Tegillarca granosa*	Malaysia	HQ896794.1	99.32
2.	*Tegillarca granosa*	Myanmar	MN119654.1	98.39
3.	*Tegillarca granosa*	India	KX608991.1	97.33
4.	*Tegillarca granosa*	China	MN608203.1	97.17
Sample code: BAL-1, BAL-2				
1.	*Tegillarca granosa*	Malaysia	HQ896794.1	99.19
2.	*Tegillarca granosa*	Myanmar	MN119654.1	98.71
3.	*Tegillarca granosa*	China	HQ699340.1	97.38
4.	*Tegillarca granosa*	Bangladesh	MF693398.1	97.20
Sample code: MAR-1, MAR-2				
1.	*Tegillarca granosa*	Malaysia	HQ896794.1	97.84
2.	*Tegillarca granosa*	Myanmar	MN119654.1	97.84
3.	*Tegillarca granosa*	China	HQ699340.1	97.67
4.	*Tegillarca granosa*	Bangladesh	MF693398.1	97.65
Sample code: BUR-1, BUR-2				
1.	*Tegillarca granosa*	Malaysia	HQ896794.1	97.82
2.	*Tegillarca granosa*	Myanmar	MN119654.1	97.74
3.	*Tegillarca granosa*	China	HQ699340.1	97.69
4.	*Tegillarca granosa*	Bangladesh	MF693398.1	96.79

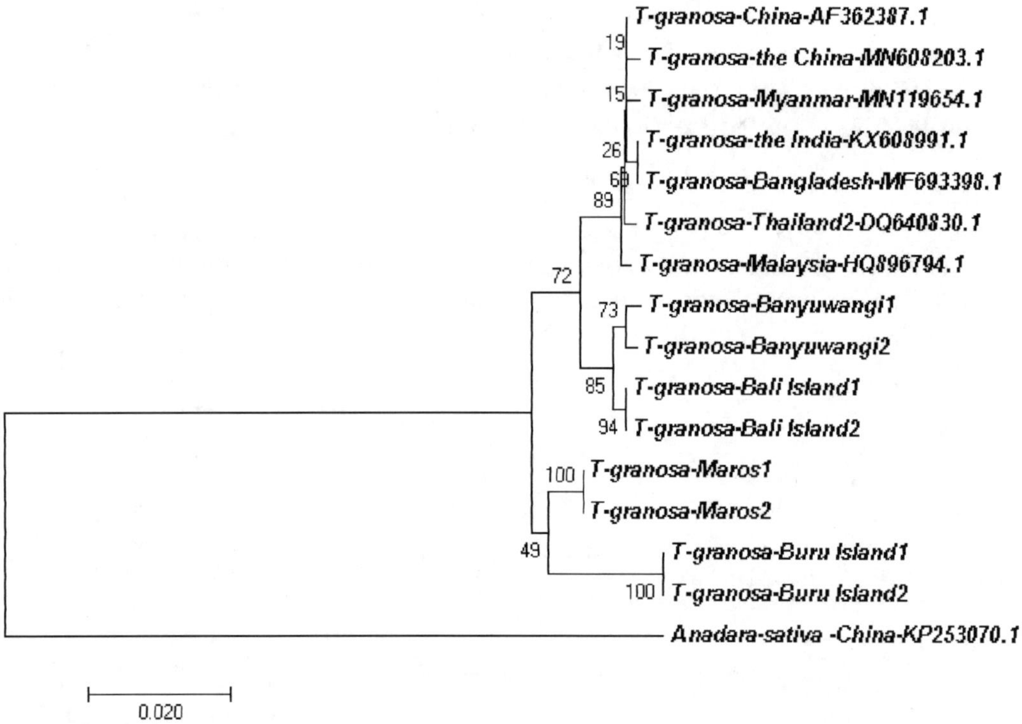

Figure 2. Phylogenetic relationships of *T. granosa* inferred by Neighbor Joining analysis (NJ) of COI gene regions.

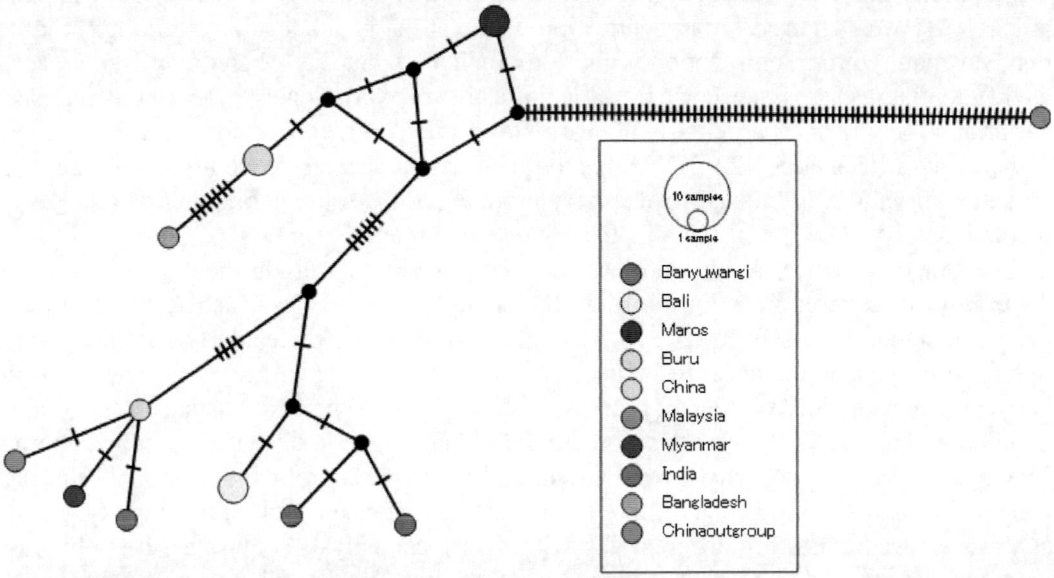

Figure 3. Haplotype network (implemented in PopART) of *Tegillarca granosa* based on COI gene sequences using minimum spanning network analysis.

BLAST analysis of the sequence results has a match range of 97% to 99% with *T. granosa* haplotypes from other DNA data banks (Table 2). The mean genetic distance in pairwise analysis was calculated to range from 0.01 to 0.04. Result of the haplotype network analysis showed that not many gaps between localities investigated (Figure 2). The most appropriate model was determined as K2+G+I as a result of base substitution model analysis. Bases ratios were calculated: Adenine ratio: 24.7%, Thymine ratio: 30%, Guanine ratio: 27.8%, and Cytosine ratio: 17.5%. Phylogenetic analyses indicated that our specimen was evolutionary closer among locality observed despite a two sub clade difference. Neighbor Joining trees showed the same topology and similar results, and they proved to be consistent (Figure 3).

4. Discussion

Previous research of *T. granosa* has been recorded in different countries in Asia. studies on its phylogenetic status are limited in the literature especially the phylogenetics status in Indonesia. In the previous study reported by [18], reported only Malaysian *T. granosa* were phylogenetically and DNA Barcoding studied. [19], compared the phylogenetic status of Bivalvia, but also only include the *T. granosa* from South Korea. Previous Study about the phylogenetic status of *T. granosa* in Malaysia has been clear [20]. Determining the phylogenetic status of an organism provides an important pool of information for conservation studies. *T. granosa* is a species that provides important information on environment quality and plays a role in the food chain of marine ecosystems. If the phylogeny of *T. granosa* is fully clarified, basic information will be supplied to the studies to be carried out in order to protect the balance of marine ecosystems.

In Table 2, the results of BLASTn studies showed that the nucleotide sequences from the four places in Indonesia were similar, with matches to *T. granosa* from Malaysia, Myanmar,

China, India, and Bangladesh ranging from 96% to 99%. At the species level, a resemblance of at least 97% was required for authentication, while at the genus level, a similarity of less than that was required for identification. This study indicated that DNA barcoding is quite accurate at differentiating across species. Because intraspecific variation is modest but interspecific variation is significant, particularly in nearby taxa, the COI gene provides an excellent species authentication technique [22]. DNA barcoding verifies species recognition for international trade, either validating the eating of fishery products or the trade in ornamental fish [23]. This method is very useful for sustainable fishery resource management.

The genetic distance in this study was also employed to ascertain the genetic link between *T. granosa* discovered in Indonesia and other samples from a DNA data bank. The range of genetic distance ranges from 0.01 to 0.04. [24] claimed that molecularly similar species are indicated by genetic distance differences of less than or equal to 3%. The genetic similarity between members of a group increases with decreasing genetic distance. In contrast, a population will be more varied the more an individual's genetic distance from the community. The species outgroup A. sativa from China and *T. granosa* from Indonesia had the biggest genetic distance (0.16 to 0.17). Similar to the genetic distance analysis, the haplotype network analysis showed not to big different gap within species analysis (Figure 2). It is obvious that there was less barcode variation within *T. granosa* than there was between species in closely related genera. Genetic separation between populations or species is indicated by their genetic distance [25]. Therefore, compared to the two species, a partial sequence of the COI gene with a lesser genetic distance provides an appearance that is more difficult to discern. Figure 3 showed the evolutionary tree of the blood cockle. This study identified a species' level of evolution and kinship, showing where two subclusters form when a species becomes split. A. sativa and *T. granosa* were divided using a bootstrap value/bv of 100%. *T. granosa* from Indonesia was divided into two sub-clusters within the species (bv of 68%).

The findings demonstrated that all of these samples were closely linked to the Asian *T. granosa*. The Wallace Line, which divides Indonesia into two physically distinct parts, has been demonstrated to have a significant influence on genetic variations in numerous marine species between the east and west region [3].[4][5]. Similar findings were found in a prior investigation that used sequencing data from four molecular markers (two mitochondrial and two nuclear) [26]. The data is considered to be reasonably stable when the bootstrap value is more than 70% [27]. The scale bar for the phylogenetic tree's creation was 0.02. A phylogenetic tree with a 0.01 scale bar, according to [11], displays a genetic distance with a nucleotide change once per 100 bp. The resulting phylogenetic structures show a genetic distance with nucleotide alterations occurring twice every 100 base pairs (bp). Many aquatic species have been studied using DNA barcoding technology, which has shown to identify greater variation between congeneric species than between conspecific individuals [21][22]. The Indo-Pacific region may so efficiently use it to differentiate a complex of physically dissimilar species [6][28].

5. Conclusion

In terms of species identification, molecular identification by phylogenetic analysis and DNA barcoding is more reliable and accurate than morphological identification. In the present work, the aim was to produce a definite species identification of the specimens obtained in Indonesia by phylogenetic analysis and DNA barcoding to clarify their phylogenetic position in order to give fundamental data for future biosecurity investigations. The analyses confirmed that the samples originate from the *T. granosa* species and are most closely related to the GenBank-registered haplotypes. In the light of these results, it was decided that the programs to be established in the places where phylogenetically linked haplotypes are identified can be included into the research in Indonesia.

Acknowledgments

We thank to Miss Dwi Lestari Widyaningsih, Mr. Jaenuri and Mr. Ali Syaifudin for helping us To Collected Samples.

Ethical Statement

Authors declare that the present study was conducted in an ethical, professional and responsible manner.

Conflict of Interest

The authors declare that there are no conflicts of interest that could be perceived as prejudicing the impartiality of the reported research.

Data Availability

All data analysed in this work are available in the article and will be shared by the corresponding author upon reasonable request.

REFERENCE

[1] Yoon, J-M. 2016. Genetic variation of three *Tegillarca granosa* populations investigated by PCR Technique. *Korean Journal of Malacology*, 32 (4): 255-261

[2] FAO (2006) Fishstat Plus: universal software for fishery statistical. Time series 1950–2004. Version 2.30. FAO Fisheries Department, Fishery Information, Data and Statistics Unit. www.fao.org/fi/statist/fisoft/FISHPLU-S.asp.

[3] Hardianto, E., Wijayanti, D. P., Shy, J-Y., Mather, P., Hughes, J., Imai, H. 2022a. Molecular ecology of the fiddler crab *Austruca perplexa* H. Milne Edwards, 1852: genetic divergence along a major biogeographical barrier, Wallace's Line. *Biological Journal of the Linnean Society*, *135*:310-321.

[4] Hardianto, E., Fukuchi, J., Hanamura, Y., Wijayanti, D. P., Sabdono, A., Imai H. 2022b. Molecular ecology of a shallow water shrimp Hansen 1919: evidence for strongly limited gene flow across the western Pacific. *Marine Ecology, 43*: e12695

[5] Hardianto, E., Hirayama, M., Wijayanti, D. P., Imai, H. 2023. Molecular ecology of the Javanese ricefish, *Oryzias javanicus* (Bleeker): Genetic divergence along the Indonesian Archipelago. *Marine and Freshwater Research*, 00:1-10

[6] Khansa, A. F., Farizky, H. S., Santanumurti, M. B., Jamal, M. T., Sani, L. M. I., Maduppa, H., Sari, P. D. W. 2023. First identification DNA barcoding of bronze featherback fish, *Notopterus notopterus* (Pallas, 1769) (Osteoglossiformes: Notopteridae), in Brantas River, East Java, Indonesia. *Genetics of Aquatics Organisms, 6* (1): GA549

[7] Nursalim, N., Trianto, A., Cahyani, N. K. D., Kholillah, N., Janarkho, G. F., Hardianto, E., Subagyo. 2022. Genetic and morphological variation of the redbelly yellow tail fusilier,

Caeseo cuning (Bloch, 1971) from the Nyamuk Waters, Karimunjawa Archipelago. *Indonesian Journal of Marine Science, 27* (4): 341-348.

[8] Syaifudin, M. Jubaedah, D., Taqwa, F. H., Octaviani, R. 2021. Phylogenetic of marble goby (*Oxyeleotris marmorata* BLKR.) in South Sumatrea based on cytochrome C oxidase subunit I (COI) gene. *Genetics of Aquatic Organisms, 6* (1): GA433

[9] Beden, E., Karahan, A. 2020. Phylogeographic resolution of the barnacle (*Chelonibia testudinaria*) from the North-Eastern Mediteranian Loggerhead Sea Turtkes epibiont community. *Genetics of Aquatics Organisms, 5* (1): 01-08

[10] Praipue, P., Klinbunga, S., Prasertlux, S., Janpoom, S., Rongmung, P., Ratdee, O., Ittarat, W., Boonphimpapha, P., Jarayabhand, P., Khamnamtong, B. 2022. Analysis of genetic diversity in wild and domesticated stocks of the tropical abalone *Haliotis asinine* by microsatellite polymorphism. *Genetics of Aquatics Organisms, 6* (2): GA501

[11] Hardianto, E., Satriyo, T. B. Molecular phylogenetic analysis of the commercially important Asian monsoon scallop, *Amusium pleuronecte* (Linneaus 1758) from Indonesia. *Journal Kelautan Tropis, 26 (2)*: 1-10

[12] Irwani, Wijayanti, D. P., Satria, A. W., Sabdono, A. 2020. Phylogeographic and molecular characterization of pronghorn spiny lobster (*Panilirus penicillatus* Oliver, 1791) in the Southern Coast of Java and Lombok, Indonesia. *Biodiversitas, 12* (12): 5690-5696

[13] Vella, A., Vella, N. 2022. The mitochondria genome of the freshwater crab *Potamon fluviatile*, the first sequence representative of subfamily Potaminae and its phylogenetic position within Potamidae. *Genetics of Aquatic Organisms, 6* (3. *Special issue*): GA460.

[14] Wijayanti, D.P., Indrayanti, E., Nuryadi, H., Rintiantoro, S.A., Sabdono, A. 2018. DNA barcode of *Acropora hyancinthus* of Karimunjawa Archipelago. *IOP Conference Series: Earth and Environmental Science, 139* (2018): 012017

[15] Wijayanti, D.P., Indrayanti, E., Nuryadi, H., Rintiantoro, S.A., Sabdono, A. 2017. Molecular identification and genetic diversity of *Acropora hyancinthus* from Boo and Deer Island, Raja Ampat, West Papua. *IOP Conference Series: Earth and Environmental Science, 116* (2018): 012065

[16] Kumar, S., Stecher, G., Li, M., Knyaz, C., Tamura, K. 2018. MEGA X: Molecular Evolutionary Genetics Analysis across Computing Platforms. *Molecular Biology and Evolution, 35*: 1547-1549.

[17] Leigh J W, Bryant D (2015) PopART: Full-feature software for haplotype network construction. *Methods in Ecology and Evolution,* 6: 1110-1116.

[18] Chee, S.Y., Devakie, M.N., Azizah, S. 2011. Phylogenetic study and barcoding of the blood cockle. *Tegillarca granosa,* found on west coast of peninsular Malaysia using COI gene. Genetic and Molecular Research, 10 (2): 1237-1244

[19] Yoon SH and Kim W (2005). Phylogenetic relationships among six vetigastropod subgroups (Mollusca, Gastropoda) based on 18S rDNA sequences. *Molecules and Cells* 19: 283-288.

[20] Tatsuya Y., 2020 Status of blood cockle, *Tegillarca granosa* (Linnaeus, 1758), aquaculture and sustainable production in Selangor, Malaysia. *Malaysian Fisheries Journal,* 19:1-20.

[21] Syaifudin, M. Jubaedah, D., Taqwa, F. H., Octaviani, R. 2021. Phylogenetic of marble goby (*Oxyeleotris marmorata* BLKR.) in South Sumatrea based on cytochrome C oxidase subunit I (COI) gene. *Genetics of Aquatic Organisms, 6* (1): GA433

[22] Ward, R. D., Zemlak, T. S., Innes, B. H., Last, P. R., & Hebert, P. D. N. (2005). DNA barcoding Australia's fish species. *Philosophical Transactions of the Royal Society of London. Series B, Biological Sciences*, *360*(1462), 1847–1857.

[23] Dahruddin, H., Hadiaty, R. K., & Hubert, N. (2016). DNA Barcoding: Foundations and Applications for Southeast Asian Freshwater Fishes. *Treubia*, *43*(December): 1–16.

[24] Hubert, N., Espiau, B., Meyer, C., & Planes, S. (2014). Identifying the ichthyoplankton of a coral reef using DNA barcodes. *Molecular Ecology Resources*, *15*(1): 57–67.

[25] Dogan, İ., Dogan, & Nurhan. (2016). Genetic Distance Measures: Review. *Turkiye Klinikleri Journal of Biostatistics*, *8*(1): 87–93.

[26] Mynhardt, G., Alejandro, A., Puslednik, L., Corralez, J., Serb, J. M. 2014. Shell shape convergence masks biological diversity in gliding scallops: description of *Ylistrum* n. gen. (Pectinidae) from the Indo-Pacific Ocean. *Journal of Molluscan studies, 80*: 400-411

[27] Lemey, P., Selemi, M., & Vandamme, A. M. (2009). *The Phylogenetic Handbook: A Practical Approach to Phylogenetic Analysis and Hypothesis Testing*. Cambridge-UK: Cambridge University Press.

[28] Last, P. R., Manjaji, B. M., & Yearsley, G. K. (2005). Pastinachus solocirostris sp. nov., a new species of Stingray (Elasmobranchii: Myliobatiformes) from the Indo-Malay Archipelago. *Zootaxa*, *1040*: 1–16.

IOP Conf. Series: Earth and Environmental Science 1289 (2023) 012020 doi:10.1088/1755-1315/1289/1/012020

The effect of continuous lighting on the rearing of yellowfin tuna (*Thunnus albacares*) larvae

Bejo Slamet*, Jhon H. Hutapea, Gunawan, and Ananto Setiadi

Central Research for Fishery, National Research and Innovation Agency
KKI Gondol, Ds Penyabangan, Gerokgak,81155, Singaraja, Bali, Indonesia

*Corresponding authors mail: bedjoselamet@yahoo.co.id

Abstract: Yellowfin tuna (*Thunnus albacares*) has important economic value in the world market. This study aims to determine the effect of continuous lighting on survival, growth, stomach contents, and histology of the eye of larvae. Larval rearing was conducted on 6 concrete tanks with 60 m^3 in volume that support the water circulation and aeration system with an initial stocking of 20 eggs/liter. The treatments tested were A: Larval rearing by a 40-watt fluorescent lamp (TL) which was positioned 1 m above the surface of the water surface for 24 hours and B: Without lights (control). Each treatment used 3 repetitions and data analysis with a t-test. The parameters observed were larval survival and growth stomach content and eyes histology. The results showed that the rearing of larvae in treatment B can with a survival of 0.26±0.0231% and a final total length of 3 6.2 ± 7.3 mm which is significantly higher (P<0.05) compared to treatment A (0.10±0.057% and 24.3 ± 5.5 mm). From the observation of stomach contents at night in treatment B still eats rotifers while in treatment A stomach is empty. The inner nuclear layer cell arrangement of larval eyes is more tenuous than control. From these results, it can be concluded that the provision of continuous lighting treatment can improve the survival, growth, stomach content, and vision ability of larvae.

1. Introduction

Tuna (*Thunnus albacares*) has an important economic value in the world market and contributes greatly to the export of fishery commodities in Indonesia [1]. The total annual catch of yellowfin tuna in the Indian Ocean from 1980 increased significantly between 2014 and 2018 with an average annual catch of around 404,655 tons [2].

Tuna larvae in nature experience a growth spurt under favorable environmental conditions, impacting the development of the notochord [3]. The availability of natural food, especially nauplii during the warm season, is sufficient for the relatively fast growth of tuna larvae. Warm temperatures are good for the hatching stage of eggs and absorption of the yolk of tuna larvae, but as the temperature increases it causes difficulties with larval metabolism [4]. Tuna is a large, highly migratory marine apex predator that emerges when temperatures rise to warm, Spawning and nature will provide plenty of natural food, and metabolism in the larvae must be perfect before the water temperature is further increased. [4]. Several

IOP Conf. Series: Earth and Environmental Science 1289 (2023) 012020 doi:10.1088/1755-1315/1289/1/012020

tuna hatchery studies began to be carried out, including the development of the digestive system tract [5], Juvenile tuna behavior, and anticipation of the movement of pounding the walls of the tub [6].

Efforts to develop tuna cultivation have been carried out in several countries, but the low survival of tuna larvae to reach juveniles for most farmed tuna species, especially deaths that occur in the first 10 days of larval rearing, is a limiting factor for mass cultivation. On average the larval phase of some marine fish including tuna is very critical, and increased mortality often occurs during the stadia [7]. High mortality in the first ten days is caused by egg quality, feed nutrition, environmental factors such as light and oxygen content, ammoniac and nitrite in water, and osmosis pressure of maintenance water media [8]. The size of the feed also determines the growth rate reported by [9] states that tuna larvae in the first 10 days grow slowly followed by a fast growth rate This is also associated with larger feed sizes and is often referred to as stadia "large prey-fast growth". The second factor occurs at the age of about 20 days due to the high rate of cannibalism among larvae [10]. Entering the juvenile stadia, there are often deaths due to hitting walls or jumping out of maintenance troughs. The nature of tuna larvae switches from planktivores to pisciforus in larval stages, where to maintain life and growth there must be fish larvae as feed rations [11].

At the Center for Marine Aquaculture Research and Fisheries Extension (Now the Fisheries Research Center, National RISE and Innovation Agency), tuna hatchery research has been started since 2003 and a series of studies ranging from egg harvesting to larval rearing and spawning began to occur but the eggs produced were mostly attacked by endoparasites, so some larvae, died at the beginning of stadia. In 2006 the results of research in the form of tuna fry as many as one hundred fry measuring 3-5 cm [12]. Starting in 2013, brood enlargement research activities for tuna spawning were carried out in floating net cages (KJA) with a diameter of 52 m and a depth of 9 m. Stocking tuna babies for 1 year can grow to a weight of 10-15 kg/head, with a survival rate of 90% [13]. In 2015, prospective tuna parents reached brood sizes with an estimated weight of 15-30 kg and began to lay eggs continuously with the number of eggs that can be harvested at 7,000 – 5,000,000 eggs/spawning. Many factors cause the number of eggs that can be harvested to vary greatly ranging from weather conditions, current direction, wind direction, and punctuality of egg harvesting [14].

Feeding in the form of newly hatched fish larvae can reduce the death of tuna larvae in D11-D15, but cannot reduce the death of larvae after D-15 until they become juvenile (Gunawan, 2016). The death of larvae after D15 is caused by cannibalism and also a death that occurs due to hitting the wall when tuna fry chase live feed in the form of milkfish seeds (nener) [13].

In addition, problems in the maintenance of tuna larvae are also related to lighting in larval rearing tubs. In the maintenance of tuna larvae, lighting with the application of lamps on the maintenance of larvae at night is aimed at improving the survival and growth of larvae. In tuna larvae, lighting at night can increase feed predation at night [8].

Some research results of adding lighting to the maintenance of fish larvae at night can increase feed consumption, survival, and larval growth Some research results of continuous (24-hour) larval lighting can increase feeding success when larvae are just starting to eat and increase the survival of early larval stadia. Retinal adaptation in tuna larvae is a specific trait that is an important parameter for larval cultivation and ecology, where feeding response is a behavior used to describe visual development [8]. Research on yellow-fin tuna larval rearing lighting needs to be carried out which is expected to improve larval survival and growth so that it can be used as a reference to improve yellow-fin tuna hatchery techniques. This study aims to determine the effect of lighting on the rearing tank of yellow-fin tuna larvae on the survival, growth, stomach content, and vision ability of larvae.

2. Material and methods

This research was conducted from October to November 2021 at the tuna hatchery and laboratory of the Central Research for Fishery, Gondol, Buleleng, Bali. Larval rearing research was carried out in 6 concrete tanks with a volume of 6 m^3 equipped with an aeration system to maintain the oxygen content in larval rearing water, and also a water change system. The cleaned larval tank is then filled with seawater that has passed through a sandy filter and filter bag. Eggs of yellow-fin tuna (*Thunnus albacares*) before stocking in the rearing tank are observed under a microscope to see their quality. Fertile eggs are then stocked in a larval-rearing bath with an egg density of 20 eggs/liter. Management of larval rearing and feeding in yellow-fin tuna larval rearing is shown in Table 1.

Table 1. Feeding, bottom cleaning, and water exchange management in the rearing of yellowfin tuna *(Thunnus albacares)* larvae

Feeding	Days after hatching																								
	1	2	3	4	5	6	7	8	9	10	11	12	13	14	15	16	17	18	19	20	21	22	23	24	25
Nannochloropsis sp.	-	-	-	-	-	-	-	-	-	-	-	-	-	-	-	-	-								
Rotifer		-	-	-	-	-	-	-	-	-	-	-	-	-	-	-	-								
Nauplii copepod				-	-	-	-	-	-	-	-	-	-	-	-	-	-	-	-	-	-	-	-	-	-
Grouper larvae (D0)										-	-	-	-	-	-										
Milkfish larvae (D0)												-	-	-	-	-	-	-							
Milkfish larvae (D20)																		-	-	-	-	-	-	-	-
Bottom cleaning													x		x		x	x	x	x	x	x	x	x	x
Water exchange					5		10		20		40		60		80		100		120		140			160	

Phytoplankton (*Nannochloropsis sp*) is given into the rearing bath when the larvae are 1 day old to the larvae are 17 days old (as long as the larvae are still consuming rotifers). *Nannochloropsis oculata* in addition to functioning as rotifer feed in the larval rearing basin, also functions as green water or shade in larval rearing water. Larvae are first fed in the form of rotifers (*Brachionus rotindiformis*) starting from two days old larvae (D-2) afternoon to 17 days old. Rotifers are given twice a day morning and evening depending on the amount of residue in the maintenance tank. The number of rotifers with an initial density of 10-20 ind./mL whose density increases with increasing age of the larvae.

Newly hatched grouper larvae (D-0) are given to tuna larvae from 10 to 15 days old, and newly hatched milkfish larvae (D-0) are given to larvae 12 to 18 days old. Newly hatched milkfish larvae have a total length of 3.2-3.5 mm which is larger than newly hatched grouper larvae which are 2.3-2.5 mm and are very suitable for the mouth opening of tuna larvae after D-12. Milkfish seeds (D-20) are given to D-18 larvae until the larvae become seeds and are ready to be harvested. As the age of tuna larvae increases, the size of the feed given is also getting bigger. The feeding of larvae and fish fry is adjusted to the density of tuna larvae in the rearing trough.

To maintain water quality in larval rearing basins, the water used in larval rearing has gone through a sandy filter process. This process aims to remove fine particles present in seawater and also to eliminate the seed source of disease present in seawater. To maintain larval rearing water quality in optimal conditions, water quality management should be maintained according to Table 1. Water management, cleaning of the tank bottom, and feeding fish oil in the rearing of yellowfin tuna larvae were used following studies conducted in the previous year. When the larvae hatch to four days old, fish oil is added to the surface of the maintenance water with a frequency of twice a day. This is so that the fat or mucus released when the larvae hatch can be pushed away from aeration and then after collecting taken with a beaker. The provision of fish oil is to reduce the death of larvae on the surface of the waters that occur due to the influence of surface tension. Water changes begin when the larvae are 5 days old (D-5) by 5% and continue to increase as the larvae age. Water turnover in the maintenance of tuna larvae is greater when compared to the maintenance of marine fish larvae in general (grouper, snapper, and milkfish), This is due to the nature of tuna larvae and fry that are always moving so that they require a greater supply of oxygen. Bottom cleaning of the trough (weaning) begins when the larvae are 17 days old. Tuna larvae on H-8 to D-15 tend to be at the bottom of the water, so at that age when cleaning the bottom of the tank (siphoning) the larvae will be sucked by siphoning.

The treatments tested are: A: Maintenance of larvae by irradiation of lamps, where each tub is installed with 1 40-watt fluorescent lamp (TL) which is positioned 1 m above the surface of the larval maintenance water, which lights 24 hours, and B: Without lights (control). Each treatment uses 3 repetitions. The parameters observed are survival and growth of larvae and predation of larvae (content of entrails). As supporting data, observations were made on the quality of maintenance water, including temperature, oxygen, pH, ammonia, nitrite, total bacteria, and light intensity.

A

B

Figure 1. Rearing tank of yellowfin tuna (*Thunnus albacares*) larvae A: With lamp and B Without lamp

Samples of yellowfin tuna eggs and larvae were observed under microscopes connected to cameras and computers. Capture of larvae using the ACT-1 program. The results of shooting with the ACT-1 program were then measured using the Win Roof V-5.0 program to determine the diameter, oil globule of yellowfin tuna eggs, and larval growth (total length and standard length). The growth data obtained were regression analysis between the growth rate (total length and standard length) with the age of the larvae. To determine the average seed produced and the best growth rate at the end of the study, data analysis with a t-test was analyzed. The research on the rearing of yellowfin tuna larvae does not require special permits because the larvae used come from hatching eggs from cultivated parents; and yellowfin tuna is not included in protected fish [2].

3. Results and discussion

Data of the number, diameter, diameter of oil globule and hatching ratio of stocked eggs, total length of larvae D1, total length of final juveniles, and survival of yellowfin tuna (*Thunus albacares*) larvae with continuous lighting and without continuous lighting can be seen in table 2. From data survival rate in the experiment during 21 days of rearing (Table 1) shows that there was a significant difference ($P<0.05$) among treatments. The better survival rate was on treatment A (with continuous lighting) with a survival rate of 0.26±0.0231% compared to treatment B (control) (0.10±0.057).

IOP Conf. Series: Earth and Environmental Science 1289 (2023) 012020 doi:10.1088/1755-1315/1289/1/012020

Table 2. The eggs number and diameter, diameter of oil globule and hatching rate of stocked eggs, total length of larvae D1, total length of final juveniles, and survival rate of larvae of yellowfin tuna (*Thunus albacares*) maintenance with continuous lighting and without continuous lighting.

	Treatment	
	A (with continuous lighting)	B (without continuous lighting)
Eggs number (pcs)	80.000	80.000
Eggs diameter (µm)	882 $\pm$ 23	882 $\pm$ 23
Oil globule diameter (µm)	202 $\pm$ 5	202 $\pm$ 5
Hatching rate (%)	45,5	45,5
Initial larval total length (mm)	3,16±0,18	3,16±0,18
Final larval total length (mm)	36,2 ± 7,3 [a]	24,3 ± 5,5 [b]
Survival rate (%)	0,26±0,0231[a]	0,10±0,057[b]

The results of observing the parameters of water quality of yellowfin tuna larvae rearing media (*Thunus albacares*) with continuous lighting and without continuous lighting, which include temperature, pH, DO, salinity, ammonia, nitrite, and light intensity can be seen in Table 3.

Table 3. Water quality parameters (temperature, pH, DO, salinity, ammonia, nitrite, and light intensity) rearing media for yellowfin tuna larvae (*Thunus albacares*) with continuous lighting and without continuous lighting.

Parameter	Treatment	
	A (with continuous lighting)	A (without continuous lighting)
Temperature (^{0}C)	28,5 – 30,0	28,0 – 29,5
pH	8,1-8,2	8,1-8,15
DO (mg/L)	6,6 – 7,5	6,2 – 7,2
Salinity (ppt)	33-34	33-34
Ammonia (mg/L)	0,225-1,045	0,255-1,142
Nitrite (mg/L)	0,01-0,19	0,01-0,025
Light intensity	1200 lux (night time) 2500-300lux (day time)	2500-3000 lux
Total Bakteria (CFU/mL x 1000)	8,4-22,26	9,6-24,15

Grafik development of the total length of yellowfin tuna (*Thunus albacares*) larvae in rearing with continuous lighting and without continuous lighting, can be seen in Figure 2.

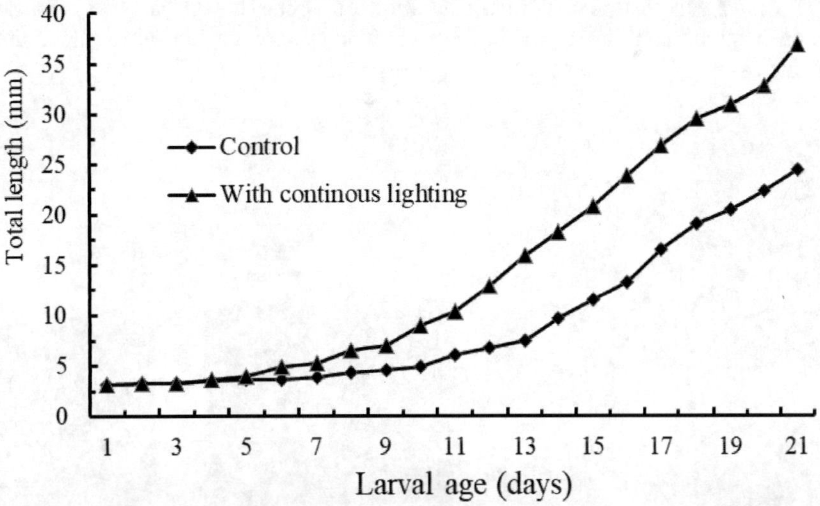

Figure 2. The development of the total length of yellowfin tuna (*Thunus albacares*) larvae in rearing with continuous lighting and without continuous lighting

The results of observations of the development of the number of rotifers in the stomach of yellowfin tuna (*Thunus albacares*) larvae in maintenance with continuous lighting and without continuous lighting, can be seen in Figure 3.

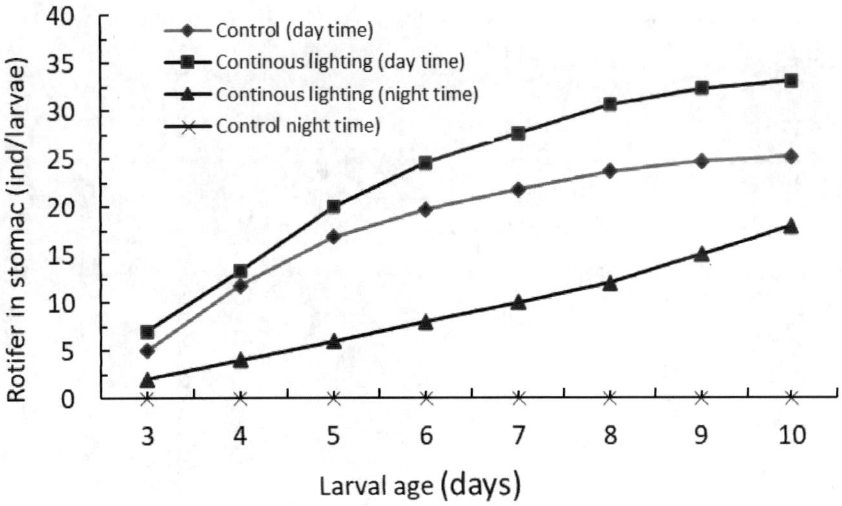

Figure 3. The development of the number of rotifers in the stomach of yellowfin tuna (*Thunus albacares*) with continuous lighting and without continuous lighting

The results of observations of the development of larval histology (Figures 4, 5, 6, and 7) show that the larval containment device has begun to develop starting from D-3 larvae. A comparison of eye histology starting from D-3 larvae in treatment with the addition of lights shows that in the inner nuclear layer tissue, the arrangement of cells is relatively tenuous. The difference in cell arrangement in the inner nuclear layer is increasingly evident as the larval age develops.

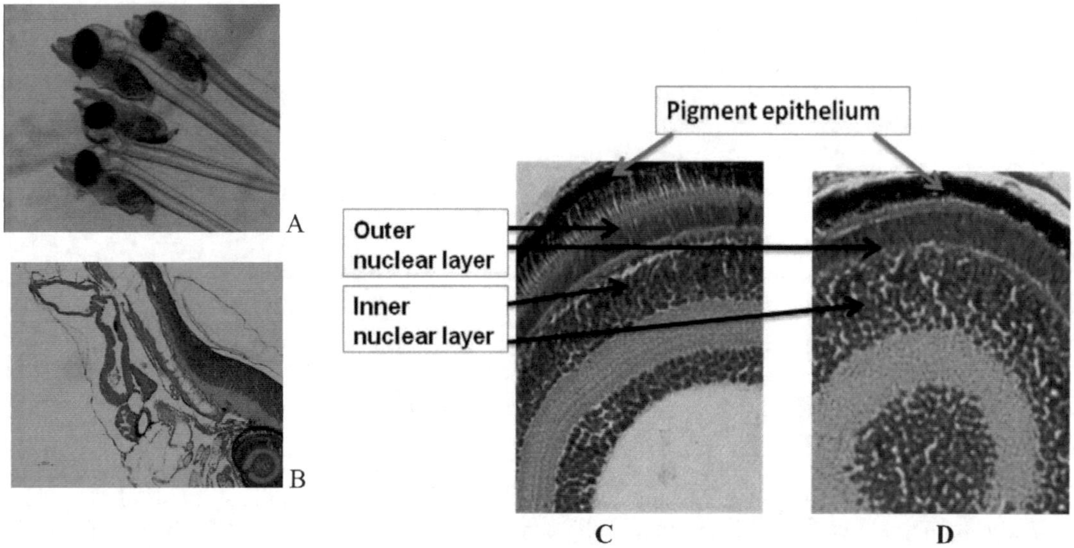

Figure 4. Morphological of larvae D-3 (A), longitudinal cut (B), and histology of eye slices of yellowfin tuna larvae on larval rearing without continuous lighting (C) and with continuous lighting (D)

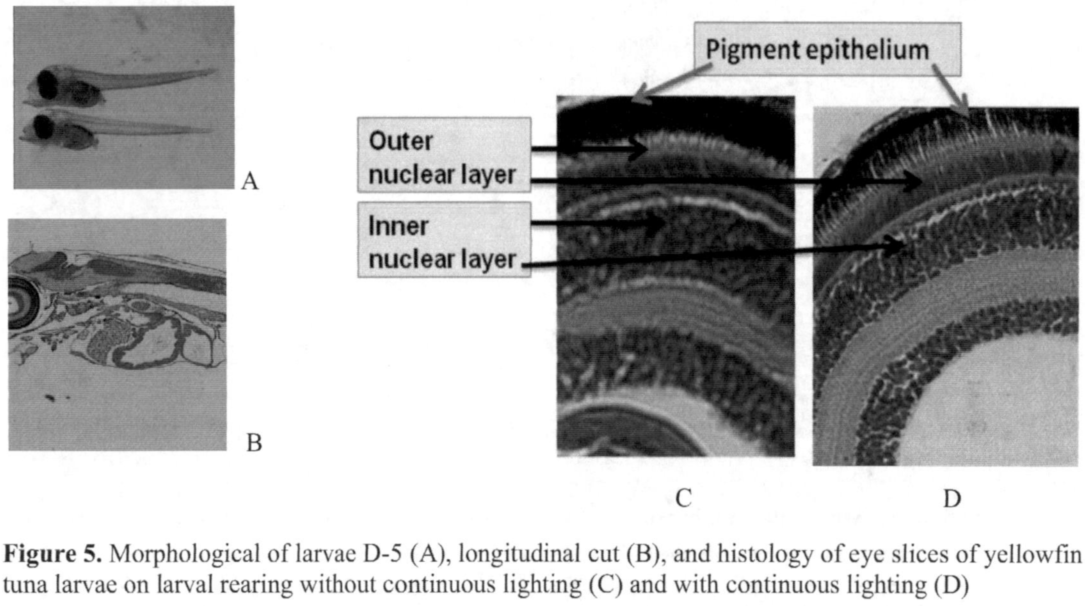

Figure 5. Morphological of larvae D-5 (A), longitudinal cut (B), and histology of eye slices of yellowfin tuna larvae on larval rearing without continuous lighting (C) and with continuous lighting (D)

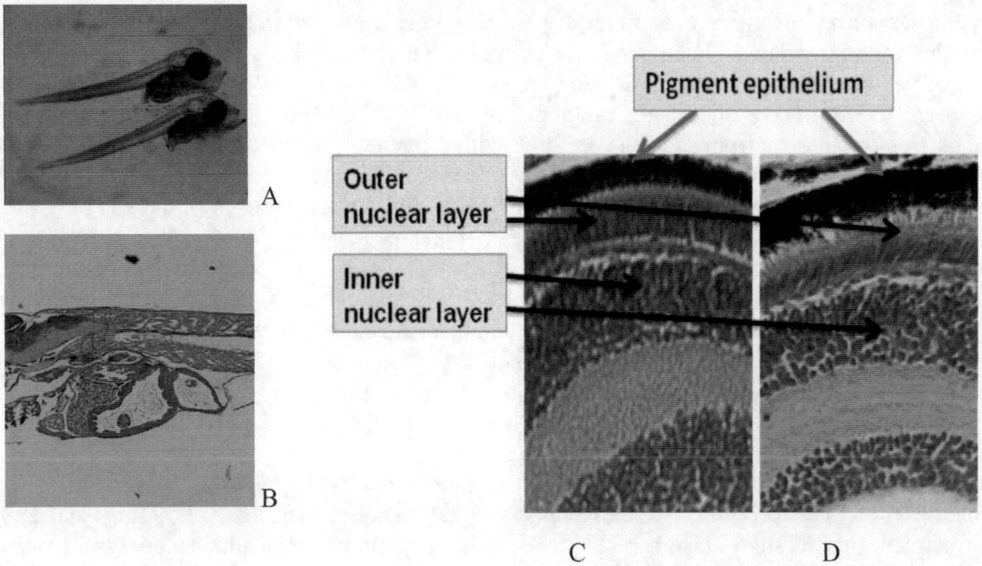

Figure 6. Morphological of larvae D-7 (A), longitudinal cut (B), and histology of eye slices of yellowfin tuna larvae on larval rearing without continuous lighting (C) and with continuous lighting (D)

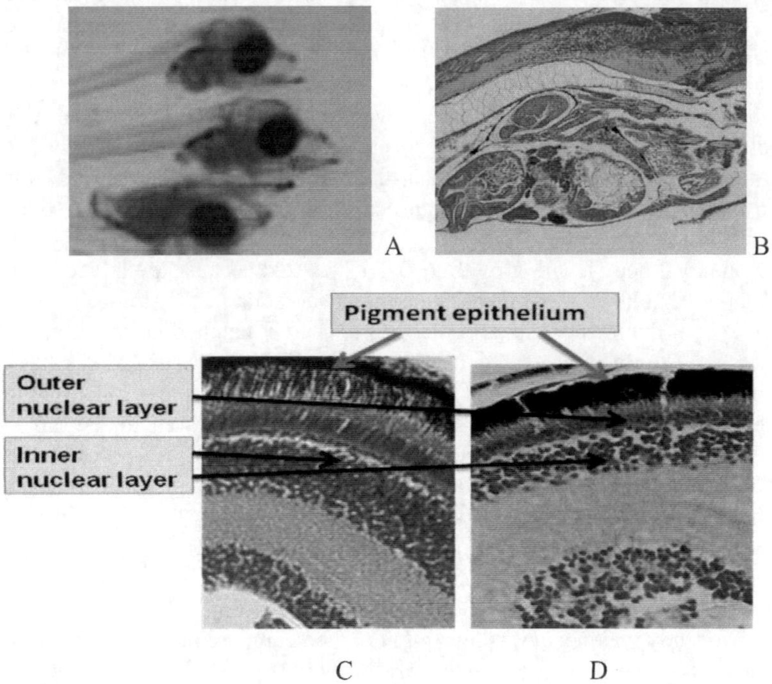

Figure 7. Morphological of larvae D-9 (A), longitudinal cut (B), and histology of eye slices of yellowfin tuna larvae on larval rearing without continuous lighting (C) and with continuous lighting (D)

From the observations, the average diameter of yellowfin tuna eggs used in this study was 882±23 µm and oil globules were 202±5 µm. The size of yellowfin tuna eggs in this study is almost the same as the average size of yellowfin tuna spawned in floating net cage 891±15 µm and oil globules 204 + 5 µm [13], eggs from parents raised in controlled baths ranging from 880 + 15 µm and oil globules 210 + 4 µm 12] and also eggs from mother spawning in KJA in 2016 with a diameter of 875 + 23 µm and oil globules measuring 206 + 8 µm [13]. The size of the egg diameter and oil globules of yellowfin tuna in this study was slightly smaller when compared to the diameter of yellowfin tuna eggs raised in the hatchery of the Inter-American Tropical Tuna Commission's Achotines Laboratory Panama with egg diameters of 0.85 – 1.13 mm and oil globule diameters of 0.22 mm [6].

Based on the report of Wexler et al., 2003 in [13] the size of yellowfin tuna eggs is very uniform with a diameter of 1 mm, as well as observations on Bluefin tuna (Thunnus thynnus) with an average diameter of 989 µm and 970 µm. Based on this data, it can be said that the diameter of eggs spawned by yellowfin tuna in Indonesia is slightly smaller than existing reports. This difference may have something to do with the age of the brood as well as environmental factors, in particular, the water temperature is higher than in Japan and Panama.

From the results of research that has been done, it is known that tuna eggs used in this study have a hatching rate of 45.5%. The hatchability of these eggs is low. The growth of yellowfin tuna larvae from both treatments in this study (Gambar 2) showed that the beginning of maintenance was very slow, but the growth of these larvae increased very quickly after the D-13 larvae. Rapid larval growth after D-13 is due to yellowfin tuna larvae already consuming newly hatched fish larvae (grouper larvae and milkfish larvae) which have a larger size compared to the previous feed (rotifers and copepod naupli).

The larvae of yellowfin tuna at the age of one day (D-1) in this study had a total body length of 3.16±0.18 mm. Tuna larvae in D-1 in this study are almost the same as studies conducted in the previous year which have a total body length of 3.24 + 0.11 mm total body length of 3.38±0.07 mm [15]. There were no significant differences in larval growth (total length and standard length) from the two treatments until the end of the study. This is because the feed given in both treatments is the same. The larvae of tuna fish move actively in pursuit of given natural feed prey (rotifers and nauplii copepods) in a living state. Likewise, when tuna larvae are given live feed in the form of grouper larvae and milkfish larvae, tuna larvae will move actively in pursuit of prey. Significant growth increases after D-16 larvae where larvae have begun to transform into juvenile stadia.

The difference in larval body length growth in D-16 is caused because the larvae have consumed newly hatched milkfish larvae that have been given. From the observation of the stomach contents of tuna larvae in D-12, it is also known that there is feed in the form of newly hatched larvae in most tuna larvae in all treatments.

The total length growth of yellowfin tuna larvae at the end of the study (D-21) was slightly higher in the lamp (A) treatment, with a total length of 26.2±0.93 mm, compared to the lamp (B) treatment of 24.3±8.5 mm. The results of measuring the total length of tuna fry at the end of this study showed that there was no difference in larval length growth in treatment A and treatment B. From statistical analysis of the two treatments the growth rate was not significantly different, ($P > 0.05$). From the results of this study, it is known that the growth of yellowfin tuna larvae in these two treatments is higher when compared to research conducted by Kobayashi et al., (2015) on the same species at the age of 20 days conducted in Panama which has a total body length of 10.40 ± 0.70 mm. The total length of yellowfin tuna larvae from these two study treatments was also higher when compared to the total body length of Atlantic bluefin tuna larvae at D-20 which was approximately 17 mm [16].

Survival at the end of the study on treatment A was 0.026±0.0231%, and treatment B was 0.10±0.057%. The survival of seeds produced in treatment B was slightly higher than in treatment A, but from statistical analysis, there was no real difference ($P>0.05$). Survival in treatment A is smaller than in treatment B, this is due to death caused by larvae having difficulty sleeping due to continuous light. Actually, until D-8 the number of larvae in the maintenance trough in both treatments is still quite large. But when the larvae begin to eat fish larvae (D-10) mortality begins to increase, and from the age of 18 days, mortality increases dramatically, due to approaching the change to juvenile until it has become juvenile there are many deaths caused by cannibals and also deaths from hitting the walls of the maintenance basin. Cannibalism in tuna larvae is due to the nature of tuna larvae that switch from planktivores to pisciforus in larval stadia, where to maintain life and growth there must be fish larvae as feed rations. These cannibals in larvae and juveniles occur not only due to variations in size but also occur in larvae and juveniles of similar size. Smaller larvae or juveniles will be eaten by larger larvae or juveniles, but larvae or juveniles of relatively the same size die from being bitten but not eaten. Death due to hitting the wall after the larvae become juvenile usually occurs when chasing feed in the form of live milkfish seeds given as feed. Milkfish fry in the tuna larval tank tend to swim on the edge of the tank.

In the rearing of bluefin tuna larvae, lighting with the provision of lights for the maintenance of larvae at night can increase the survival and growth of larvae [17]. Several other studies have shown opposite results, where prolonged or continuous illumination (24 hours) decreases larval survival and growth [18]. In some cases, increased exposure periods in larval rearing can cause morphological damage increase morphological abnormalities, or lead to increasingly aggressive and cannibalistic larval behavior [19]

The number of rotifers in the stomach can be known after the larval stomach is squeezed and the number of rotifers in the stomach is observed. The number of rotifers in the abdomen of yellowfin tuna larvae (Figure 3) continues to increase as the age of yellowfin tuna larvae increases. In all treatments the number of rotifers in the larvae's stomach began to decrease after D11, this is because the larvae have begun to consume feed in the form of fish larvae given to D-9. From the observation of stomach contents, it is known that larvae begin to consume newly hatched larvae on D-10. In the rearing of larvae of some types of marine fish, lighting at night can increase feed predation at night [20].

Some research results of adding lighting to the maintenance of fish larvae at night can increase feed consumption, survival, and growth of larvae [21] [18]. Some research results of continuous (24-hour) larval lighting can increase feeding success when larvae are just starting to eat and increase survival of early larval stages [22] [23]. The number of rotifers in the stomach can be known after the larval stomach is squeezed and the number of rotifers in the stomach is observed. The number of rotifers in the abdomen of yellowfin tuna larvae (Figure 3) continues to increase as the age of yellowfin tuna larvae increases. In all treatments the number of rotifers in the larvae's stomach began to decrease after D11, this is because the larvae have begun to consume feed in the form of fish larvae given to D-9. From the observation of stomach contents, it is known that larvae begin to consume newly hatched larvae on D-10.

The results of temperature measurements in the rearing water of yellowfin tuna larvae in this study ranged from 28.5 – 30.0 0 C in all treatments. The highest water temperature occurs during the day when the larvae are 2 - 4 days old and when water changes have not yet started. The water temperature stabilizes with the age of the larvae as the water change gets bigger. The temperature range in this larval-rearing medium is still within normal limits for the life of marine fish larvae in general, including yellowfin tuna larvae [24].

The pH range in the larval rearing media of the two treatments is not much different, namely 8.01-8.2, this is still included in the normal limits for the life of marine fish larvae in general [24]. The results of measuring dissolved oxygen (DO) levels in the rearing water of yellowfin tuna larvae from both treatments were relatively the same at values ranging from 6.2 – 7.5 mg / L. The DO value in this study is still within the limits of the DO level quality for tuna larval rearing [24]. The high value of DO In the second maintenance basin, this treatment is due to the addition of aeration in the tub and also the water change system (water circulation) during the larval rearing period [24]. This DO value is higher when the larvae enter the juvenile stadia due to the greater percentage of water turnover and also an increase in the speed of aeration from which the larvae enter the juvenile stadia. For salinity values in water, the maintenance of larvae of 33-34 ppt is still following the standard tuna larval rearing [24]. The range of ammonia in the larval rearing medium of the two treatments was not much different, namely 0.225-1.142. The ammonia content in larval rearing water exceeds the normal limit for tuna larval rearing [25]. The source of ammonia in larval rearing water comes from wasted feed residues, larval excrement, and also from the remaining fertilizer contained in nannochlorosis given in the larval tank. Nitrite levels from the two treatments in the maintenance bath were not much different, namely 0.01-0.25 mg / L. Nitrite levels from the two treatments in the maintenance bath were not much different, namely 0.04-0.19 mg / L. and still in good conditions for the maintenance of tuna larvae [24]. The total bacteria in the larval rearing media from the two treatments were not much different, namely (8.4-22.26) x103 CFU/mL in treatment A and (9.6-24.14) x103CFU/mL The total bacteria in larval rearing water is still within the normal limit range, where the total bacteria in the waters ranges from 103 - 106 CFU/mL. and still in good condition for the maintenance of tuna larvae [26]

4. Conclusion

The addition of continuous lighting for the larval rearing of yellowfin tuna can increase larval predation of natural food and reduce mass mortality due to a drastic decrease in light intensity during the change from day to night, thus increasing larval survival and growth compared to without continuous lighting.

References

[1] Setyadji B and H. Hartaty 2022 Relative abundance of yellowfin tuna *T. albacares* (Bonnaterre, 1788) in the tropical area of the north-eastern Indian Ocean *Indian J. Fish.* **69** 36-43

[2] Asrial E, Fathurriadi, Arapat Y, Hadi U K, Tilar L A T, Kalih W S, Rizal L S, Liliyanti M A, Rosadi A, Isfatu R, Khasanah, Setyohadi D, Junaidi M, and Rathnayake I N 2021 Population dynamics of bullet tuna from the Indian Ocean, west nusa tenggara, Indonesia *J. Ilm. Pelikan. dan Kelaut.*, **13** 24-35

[3] Russo S, Torri M, Patti B, Musco M, Masullo T, Natale M V D, Sarà G and Cuttitta A 2022 Environmental Conditions along Tuna Larval Dispersion: Insights on the Spawning Habitat and Impact on Their Development Stages *Water* **14** 10

[4] Fiksen O and Reglero P 2022 Atlantic bluefin tuna spawn early to avoid metabolic meltdown in larvae *Ecology* **103** 1

[5] Gunawan G, Hutapea J H, Setiadi A, and Mahardika K 2019 Perkembangan saluran dan sistem pencernaan pada larva ikan tuna sirip kuning *Thunus albacasre J. Ris. Akuakultur* **13** 309-316

[6] Hutapea J H, Gunawan G, Setiadi A, and Slamet B 2019 The Inner Wall of The Tank Painted in Black is Good for Larval Rearing of Yellowfin Tuna *T. albacares Aquac. Indones.* **20** 102.

[7] Abdussamad E M, Koya S K P, Rohit P, Joshi K K, Ghosh S, Elayathu M N K, Prakasan D, Sebastine M, Beni M and Rao G S 2012. Fishery of yellowfin tuna *Thunnus albacares* (Bonnaterre, 1788) in the Indian EEZ with special reference to their biology and population characteristics. Indian J. Fish., **59** 3 43-51

[8] Hilder P E, Battaglene S. C., Hart N S., Collin S. P., and Cobcroft J. M. 2019 Retinal adaptations of southern bluefin tuna larvae: Implications for culture *Aquaculture* **507** 222-232

[9] Reglero P, Blanco E, Ortega A, Fiksen O, De La GáNdara F, Seoka M, Javier Viguri F, and Folkvord A 2015 Prey selectivity in piscivorous bluefin tuna larvae reared in the laboratory *Journal of Plankton Research* **37** 1 2-5

[10] Miyashita S, Sawada Y, Okada T, Murata O, and Kumai H 2001 Morphological development and growth of laboratory-reared larval and juvenile T*hunnus thynnus* (Pisces: Scombridae) *Fish. Bull.* **99** 4 601-616

[11] Puncher G N, Alemany F, Arrizabalaga H, Cariani A, and Tinti F 2015 Misidentification of bluefin tuna larvae: a call for caution and taxonomic reform *Rev. Fish Biol. Fish*, **25** 3 485–502
[12] Hutapea J H, Permana G N, and Setiadi A 2016 Pemeliharaan induk ikan tuna sirip kuning *Thunnus albacares* dalam bak terkontrol *J. Ris. Akuakultur* **5** 1 79
[13] Gunawan M, and Hutapea J H Perbedaan waktu awal pemberian larva ikan sebagai pakan dalam pemeliharaan larva ikan tuna sirip kuning *Thunus albacares Journal Fish. Mar. Res.* **4** 1 135–143
[14] Tridjoko, Hutapea J H, Setiadi A, Gunawan, and Selamet B 2021 Maintenance and spawning on yellowfin tuna broodstock reared in floating net cage *IOP Conf. Ser. Earth Environ. Sci.* **890** 1
[15] Hutapea J H and Setiadi A 2016 Perkembangan morfologi dan pertumbuhan larva ikan tuna sirip kuning *Thunnus albacares* hasil budidaya *Prosiding Forum Inovasi Teknologi Akuakultur* **493** 499
[16] Yúfera M, Ortiz-Delgado J B, Hoffman T, Siguero I, Urup B, and Sarasquete C 2014 Organogenesis of digestive system, visual system and other structures in Atlantic bluefin tuna (*Thunnus thynnus*) larvae reared with copepods in mesocosm system *Aquaculture* **426** 126-137
[17] PartridgeG J., Benetti D D, Stieglitz J D, Hutapea J, McIntyre A, Chen B, Hutchinson W, Scholey V P 2010 The effect of a 24-hour photoperiod on the survival, growth and swim bladder inflation of pre-flexion yellowfin tuna (*Thunnus albacares*) larvae *Aquaculture* **318** 3-4
[18] Blanc E, Reglero E P, Ortega A, de la Gándara F, Fiksen O, and Folkvord A 2017 The effects of light, darkness and intermittent feeding on the growth and survival of reared Atlantic bonito and Atlantic bluefin tuna larvae *Aquaculture* **479** 233-239
[19] Jayakumar R, Sakthivel M, Abdul Nazar A K, Tamilmani G, Rameshkumar P, Samal A K, Anikuttan K K, Anbarasu M, Balamurugan V, Thiagu R, Sirajudeen S, & Gopakumar G 2018 Impact of increase in temperature and light intensity on development and metamorphosis of hatchery reared silver pompano *Trachinotus blochii* (Lacepede, 1801) larvae *Indian Journal of Fisheries* ***65*** 2 133-137
[20] Li F, Cai M, Lin M, Huang X, Wang J, Ke H, Zheng X, Chen D, Wang C, Wu S, and An Y 2019 Differences between motile and nonmotile cells of *Haematococcus pluvialis* in the production of astaxanthin at different light intensities. *Marine Drugs* **17** 1 1-14
[21] Di Z, Li K, Li T, Yan L, Jiang H, and Liu L 2023 Effects of light intensity and photoperiod on the growth performance of juvenile Murray cods (*Maccullochella peelii*) in recirculating aquaculture system (RAS). *Aquaculture and Fisheries* **8** 3 274-279
[22] Villamizar N, Blanco-Vives B, Migaud H, Davie A, Carboni S, and Sánchez-Vázquez F J 2011 Effects of light during early larval development of some aquacultured teleosts: A review *Aquaculture* **315** 2 86-94
[23] Alejos M S and Serrano A F 2018 Continuous illumination improves growth and survival in the early stage of snub nose pompano *Trachinotus blochii AACL Bioflux* **11** 5 1557-1563
[24] Nakagawa Y, Kurata M, Sawada Y, Sakamoto W, and Miyashita S 2011 Enhancement of survival rate of Pacific bluefin tuna (*Thunnus orientalis*) larvae by aeration control in rearing tank *Aquat. Living Resour.* **24** 4 403-410
[25] Nakase G, Honryo T, Guerra L, Perz D, Cano A, Margulies D, Scholey V P, and Sawada Y 2013 Addition of *Nannochloropsis* sp. to Pre-rearing Water Improves Survival of Yellowfin Tuna *Thunnus albacares* Larvae *Aquaculture Science* **61** 4 395-398
[26] Gatesoupe F J, Covès D, Ortega A, Papandroulakis N, Vadstein O, and de la Gándara F 2013 A spatiotemporal study of bacterial community profiles associated with Atlantic bluefin tuna larvae, *T. thunnus* l., in three Mediterranean hatcheries *Aquaculture Research* **44** 10 1511-1523

IOP Conf. Series: Earth and Environmental Science 1289 (2023) 012021 doi:10.1088/1755-1315/1289/1/012021

Survival Rate and Growth of Red Tilapia (*Oreochromis* sp.) Seed Hybridized from Four Different Strains

Farkhana Laily*, Afida Sabilil Hikma, and Rustadi

Department of Fisheries, Faculty of Agriculture, Universitas Gadjah Mada Jl. Flora, Bulaksumur Yogyakarta, Indonesia

*Corresponding author e-mail: farkhanalaily@mail.ugm.ac.id

Abstract. Breeding red tilapia is one of the solutions to produce the best tillers based on their growth. The strains used are Cangkringan, Janti, Sukabumi, and Sukamandi. Each strain was spawned to produce 16 crosses, both inbreeding and outbreeding. This study aims to determine the results of crosses that have the best growth based on survival rate, growth, and feed conversion ratio. This research lasted for 90 days from 1 September 2022 to 30 March 2023 at the Aquaculture Technology Development Center DIY, Cangkringan, Sleman. Spawning was done by hybridization. The study was conducted in a completely randomized design using a 1x1 meter pond. Feed is given 2 times a day with a 10% biomass ratio. This study was conducted by two nurseries. Observation of the weight and length of individuals was carried out by sampling, as well as water quality every two weeks. Survival rate, growth, and feed conversion ratio data were analyzed by analysis of variance and tested by Tukey's Multiple Comparison Test. Water qualities were tested descriptively. The results obtained: survival rate ranged from 73-98%, specific weight 1.35-6.451%, and feed conversion ratio 1.11-1.68. From the results of the research, it can be concluded that the Janti-Sukamandi cross breeding has the best quality for the first nursery and the Cangkringan-Sukabumi cross breeding has the best quality for the second nursery. The water quality meets the requirements for cultivating red tilapia.

1. Introduction

Red tilapia is a type of fish hybridized between O. *Mossambicus* and O. *Niloticus* or O. *Aureus*. This type of tilapia is popular all over the world, especially in Southeast Asia [1]. Red tilapia (*Oreochromis* sp.) is an important aquaculture commodity because it has an attractive shape and appearance. In addition, red tilapia also has a tolerance to various environmental conditions such as high salinity and density [2].

One of the efforts to increase the productivity of red tilapia cultivation in Indonesia is in the form of the official release of several superior strains resulting from breeding to the public for cultivation. Some of these superior strains include NIRWANA (Nila Ras Wanayasa), BEST (Bogor Enhancement Tilapia Strain), GESIT (Genetically Supermale Indonesian Tilapia), NILASA (Nila Satria) and Pandu. Apart from these strains, several tilapia strains were introduced and developed in cultivator communities. The introduced tilapia strains that are quite popular include GIFT (Genetically Improvement of Farmed Tilapia) tilapia and NIFI (National Inland Fishery Institute) red tilapia [3].

Currently, there are problems in the development of red tilapia seeds. In addition, there are indications of inbreeding which causes a decrease in genetic quality. Genetic inheritance is characterized by slow growth, faster gonadal maturation and high mortality due to weak disease and environmental resistance. Alternatives to overcome this problem can be done by breeding, one of which is hybridization [4]

There are four types of tilapia based on origin and/or strain used in this study, including nilasa (Cangkringan), red nifi (Sukabumi), pandu (Janti) and tilapia from Sukamandi. Nilasa (Nila strain satria) is a type of red tilapia as a result of individual selection from the Aquaculture Technology Development Center DIY, Cangkringan, which was introduced in 2012 based on the Decree of the Minister of Maritime Affairs and Fisheries of the Republic of Indonesia Number KEP.47/MEN/2012. Nilasa has fast growth and can survive in an environment with a salinity up to 18 ppt. Red NIFI (National Inland Fisheries Institute) is a fish imported from the Philippines in 1981. Red nifi can survive up to a salinity of 35 ppt [5]. Red nifi used in this study was imported from Sukabumi. Based on the Decree of the Minister of Maritime Affairs and Fisheries of the Republic of Indonesia Number KEP.48/MEN/2012 concerning the Release of Male Pandu Nile Tilapia and Female Kunti Tilapia, the Pandu Tilapia used in this study came from the Janti Freshwater Fish Hatchery and Cultivation Center, Central Java. Tilapia strain Pandu can survive up to 40 ppt salinity. Another advantage is that it has fast growth with a low FCR [6]. Red tilapia from sukamandi is red tilapia originating from the Fish Breeding Research Institute, Sukamandi, Subang, West Java. This red tilapia started research on breeding red tilapia in 2016. The productivity of red tilapia from Sukabumi continues to increase compared to its parents [7].

2. Materials and Methods

This study was conducted at the Aquaculture Technology Development Center DIY, Cangkringan, Sleman for 90 days. Tilapia were reared in 1x1 meter square net. Four strains of tilapia-Cangkringan (CK), Janti (JT), Sukabumi (SB), and Sukamandi (SM) were hybridized and produced sixteen offspring. Besides that, individual synthetic selection was also carried out which is referred to as TOP. The nursery was carried out 2 times. In the first nursery, seeds have an average size of 0.67 g and 3.22 cm. Stocking density of 200 individuals per meter. In the second nursery, seeds have an average size of 4 g and 6 cm. Stocking density of 50 individuals per meter.

Sampling was conducted every two weeks to measure the weight of each population. Every pond was taken 32 fish then measured using digital scales. Water quality was carried out 2 times a month. The water quality parameters that are measured are temperature, pH, dissolved oxygen (DO), CO_2, alkalinity, and ammonia (NH_3). Water quality checker was used to measure the physics parameters and titration is used to measure the chemical parameters.

The following equation was used for calculating the survival rate:

$$SR = Nt/No \times 100\%$$

Where, SR= Survival rate (%)
Nt= Number of live fish at the end of rearing
No= Number of live fish at the start of rearing

The following equation was used for calculating the growth rate:

$$SGR = (\ln Wt - \ln Wo)/T \times 100\%$$

Where, SGR= Specific growth rate (%)
Wt= average weight of biomass at the end of observation (g)
Wo= average weight of biomass at the beginning of observation (g)
T= cultivation time

The following equation was used for calculating the feed conversion ratio

$$FCR= F/(Wt-Wo)$$

Where, FCR= feed conversion ratio
F= Amount of feed consumed (g)
Wt= average weight of biomass at the end of observation (g)
Wo= average weight of biomass at the beginning of observation (g)

Each experimental breeding was reared in a completely randomized design. Statistical analyses in this study were conducted using SPSS Statistical. Data were subject to one-way ANOVA and the Tukey's Multiple Comparison Test applied to evaluate differences between means at $p<0.05$. Water quality parameters were tested descriptively by comparing literature. The fish caught do not require special permits because they are in the Least Concern status based on the list on the IUCN Redlist.

3. Results

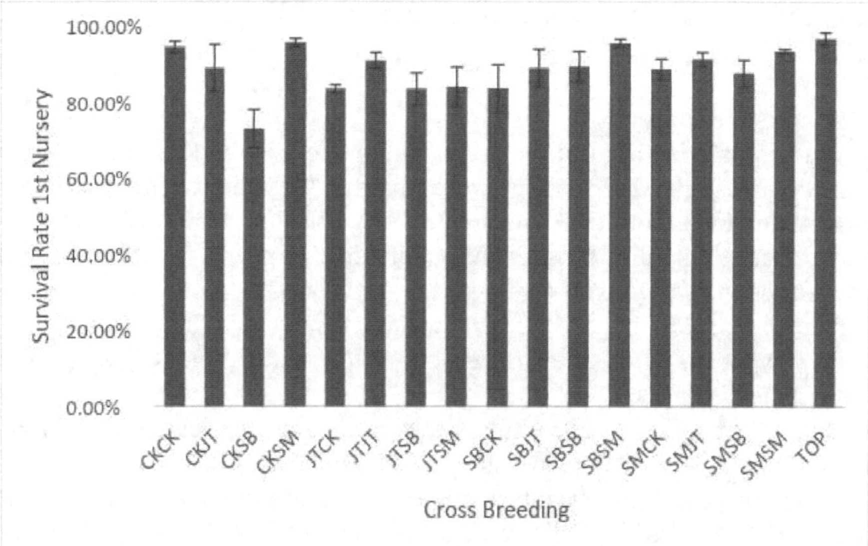

Figure 1. Survival rate 1st nursery of red tilapia

The survival rate of the 1st nursery of red tilapia is presented in Figure 1. Based on observations during rearing, the best survival rate was obtained in TOP with a value of 96.83%. The average survival rate was 88.82%±0.031. The lowest survival rate was obtained in CKSB with a value of 73%. The survival rate between crosses was different from the other treatments.

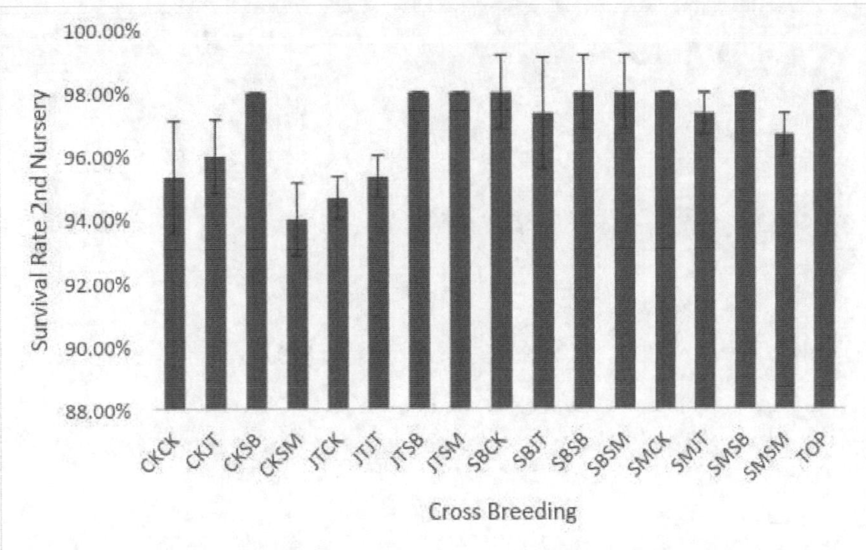

Figure 2 Survival rate 2nd nursery of red tilapia

The survival rate of the 2nd nursery of red tilapia is presented in Figure 2. Based on observations during rearing, the best survival rate was obtained in CKSB, JTSB, JTSM, SBSB, SBSM, SMCK, SMSB, and TOP with a value of 98%. The average survival rate was 96.98%±0.007. The lowest survival rate was obtained in CKSM with a value of 94%. The survival rate between crosses was different from the other treatments.

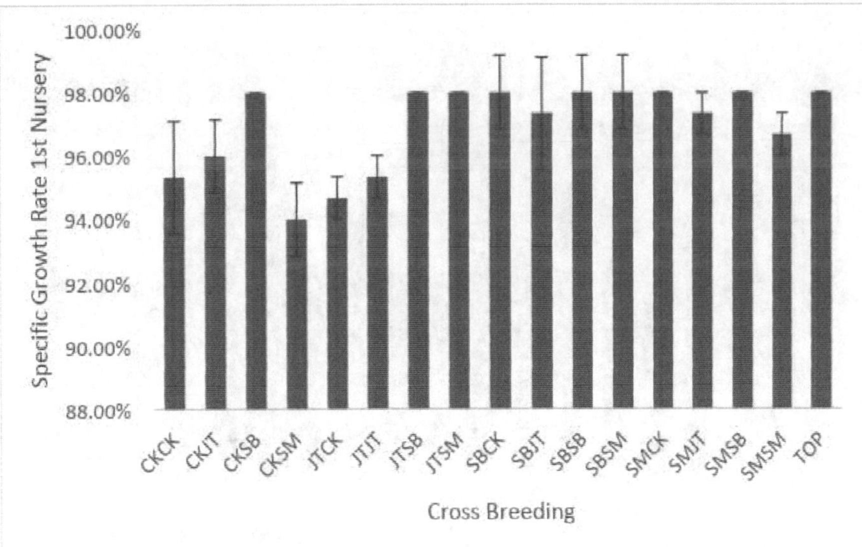

Figure 3 Specific growth rate 1st nursery of red tilapia

The specific growth rate of the 1st nursery of red tilapia is presented in Figure 1. Based on observations during rearing, the best specific weight growth was obtained in JTSM with a value of 6.41%. The average specific weight growth obtained was 4.51% ± 0.001. The lowest specific weight growth was obtained in SMSB with a value of 3.41%. The specific weight growth obtained between crossbreed fish was different from the other treatments.

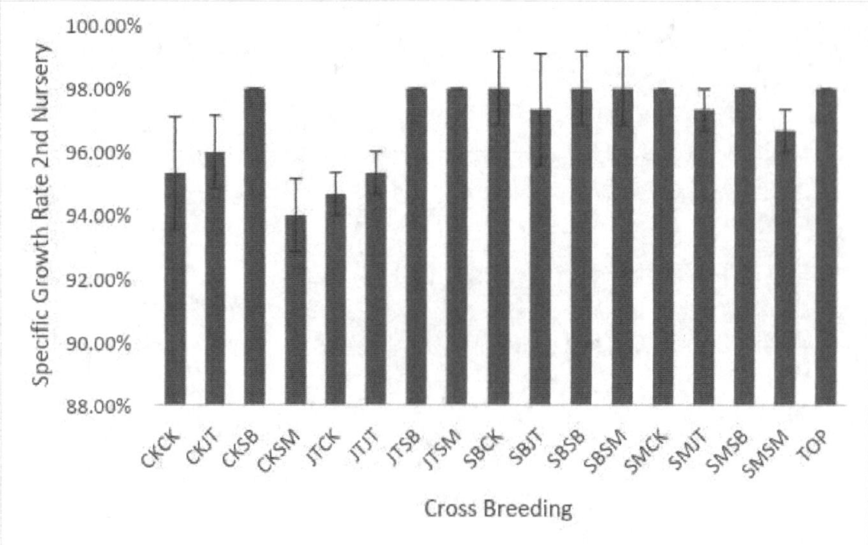

Figure 4 Specific growth rate 2nd nursery of red tilapia

The specific growth rate of the 2nd nursery of red tilapia is presented in Figure 2. Based on observations during rearing, the best specific weight growth was obtained in CKSB with a value of 2.83%. The average specific growth rate obtained was 2.32% ± 0.001. The lowest specific weight growth was obtained in TOP with a value of 1.35%. The specific growth rate obtained between crossbreed fish not have a significantly different effect from the other treatments.

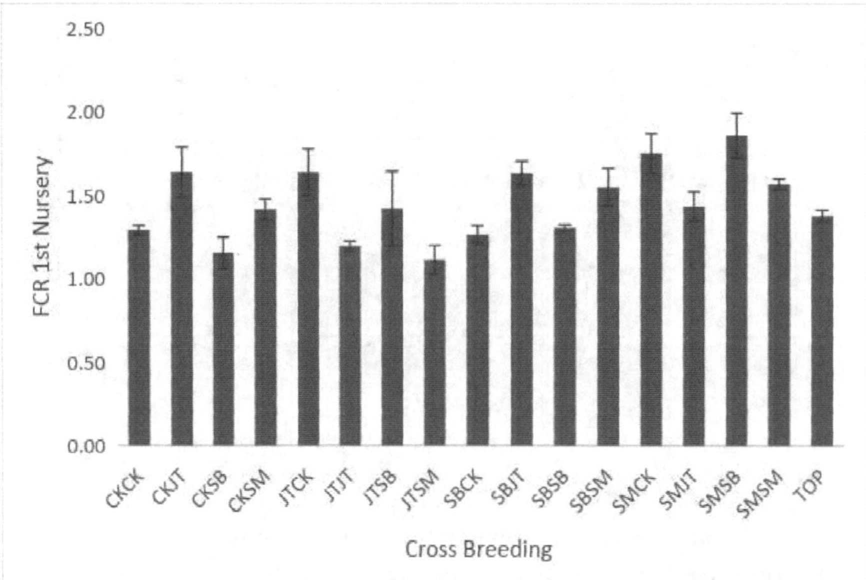

Figure 5 Feed conversion ratio 1st nursery of red tilapia

The feed conversion ratio of the 1st nursery of red tilapia is presented in Figure 5. Based on observations during rearing, the best feed conversion ratio was obtained in JTSM with a value of 1.11. The average feed conversion ratio was 1.45±0.087. The lowest specific feed conversion ratio was obtained in SMSB with a value of 1.86. The feed conversion ratio between crosses was different from the other treatments.

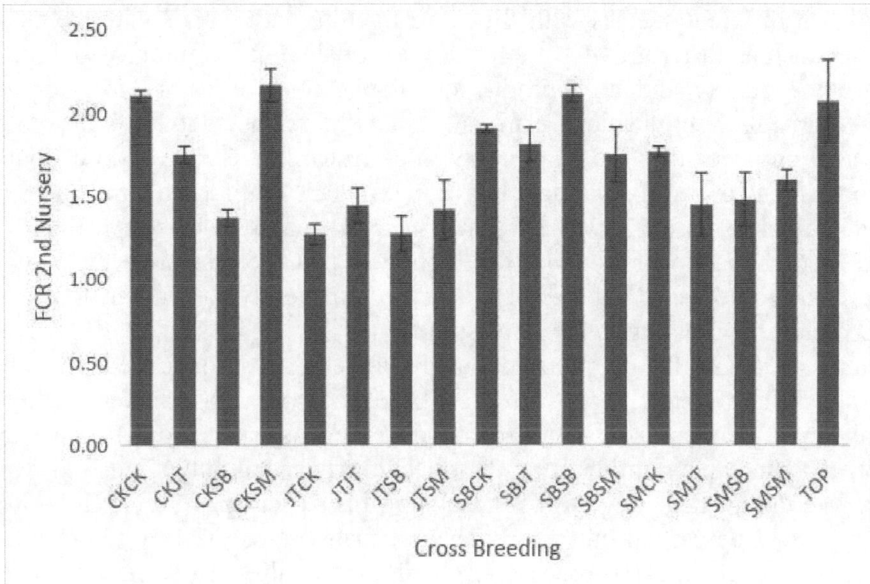

Figure 6 Feed conversion ratio 2nd nursery of red tilapia

The feed conversion ratio of the 2nd nursery of red tilapia is presented in Figure 6. Based on observations during rearing, the best feed conversion ratio was obtained in JTCK with a value of 1.26. The average feed conversion ratio was 1.68±0.101. The lowest specific feed conversion ratio was obtained in CKSM with a value of 2.16. The feed conversion ratio between crosses was different from the other treatments.

Parameter	First Nursery	Second Nursery	Appropriateness
Temperature (^{0}C)	26.5 – 30	28.26-31.78	25 – 32 *
pH	6.5 – 8.5	6.51-7.90	6.5 – 8,5 *
DO (ppm)	3.3 – 8.0	5.29-10.76	≥ 3 *
CO_2 (ppm)	0 -13.2	0-20.00	≥ 5 **
Alkalinity (mg/L)	72 – 124	80-108	30 – 300 ***
NH_3 (mg/L)	0.97 – 0.122	0.097-0.243	≤ 0.02 ****

Information :
* *SNI 7550 : 2009.*
** *SNI 6141 : 2009.*
*** Rustadi, (2018).
**** *PP. No 82 tahun 2001*

The temperature range obtained during the study was 226.5-31.78°C. The pH value obtained is in the range of 6.5-7.9. DO values ranged from 3.3 to 10.76 mg/l. CO2 values obtained during rearing ranged from 0-20 mg/l. The alkalinity values obtained during rearing ranged from 72-124 ppm. Ammonia values were obtained from measurements at the beginning, middle and end of the study. The ammonia range obtained during the study was 0.097-0.87 mg/l.

4. Discussion

Hybridization is an important part of fish genetics and breeding. This activity combines the characters of different species with the aim of getting individuals with the desired good characteristics such as faster growth, improved meat quality, and immunity to disease [8]. In order to provide quality and high production tilapia, it is necessary to improve growth performance through hybridization, especially in fish strains that have the potential for cultivation and are needed by the community. This research were carried out four different hybridizations strains of red tilapia including Cangkringan, Janti, Sukabumi, and Sukamandi. To find out the best crosses, it was necessary to carry out several tests such as survival, growth and feed conversion ratio. The result showed that there are two types of breeding, outbreeding and inbreeding. Outbreeding is the crossing of two different types of strains, while inbreeding is the crossing of the same strains.

Pongthana *et al.* (2010) also carried out a synthetic selection called a synthetic base population consisting of individuals with the best phenotypic appearance regardless of their genetic makeup [9]. In this study, the selection process is carried out by considering the resulting phenotypic value. Furthermore, the selection is carried out as much as 5% -10% of the value above the average weight. The results of this selection are referred to as TOP and further analysis will be carried out regarding the genetics possessed by the results of these crosses. The individual selection process to obtain high quality crosses has been carried out by BPTKP Yogyakarta. The synthetic population was formed by crossing Chitralada, Philippines, NIFI, and Singapore tilapia strains and then the best 10% of the population was taken. The results obtained statistically showed that as a result of the selection process there was a 25% decrease in diversity from F-0 to F-4 [4].

The sex of the mated female and male parents can affect the ploidy of the hybrid offspring [8]. The crosses that were carried out gave a significantly different effect on the survival rate in the first nursery. This shows that different strains have different survival rates. while the second nursery did not have a significant effect. This shows that all red tilapia crosses in this study were able to survive with the density and quality of water used. The highest passing life in the first nursery is TOP. The highest passing life in the first nursery is CKSB. JTSB, JTSM, SBSB, SBSM, SMCK, SMSB, and TOP. The selection process is carried out to get the best individual in a population. Parameters observed were the phenotype of the fish in the form of length and weight. The results of this selection are referred to as the TOP population. Furthermore, the results of this study are in accordance with the statement of Pongthana *et al.* (2010) that outbreeding has an impact on performance in its environment [9]. The results of this study were able to solve the problem of decreasing the quality of tilapia growth from inbreeding results.

Growth performance is an important indicator to determine the success of aquaculture [8]. The hybrid progeny with the best growth performance in the first nursery was found in a cross between the female Janti tilapia and the male Sukamandi tilapia or called JTSM. There is a significant correlation between outbreeding and growth performance. This shows that improvement of red tilapia can be done by crossing various strains of red tilapia [9]. In the second nursery, the best growth results occurred between the female Cangkringan tilapia and the male Sukabumi tilapia or called CKSB as outbreeding crosses.

The best FCR in the first nursery was found in a cross between the female janti tilapia and the male sukamandi tilapia or called JTSM. It is the same result with the growth rate that indicates the best growth rate has a good FCR too. best growth and FCRwere found in the outbreeding crosses. The FCR value can be used as a measure of success in cultivation. Janti and Sukamandi crosses produced the best FCR 1.11. This comparison shows that every 1.11 g of feed given can be converted into 1 g of meat.

Heterosis reflects dominance effects. Crosses of individuals with genetically different characters such as strains or breeds can produce greater heterosis than crosses of individuals with the same character [9].

The results obtained at the first nursery and the second nursery had differences due to differences in seed size and stocking density. The first nursery still uses larval seeds with a density of up to 200 individuals/m3. While in the second nursery using a larger seed size. The stocking density used is also smaller. The second nursery uses 50 individuals/m3 due to adjusting the size of the pond so that it is expected to achieve optimal growth.

Water quality is an important factor affecting the performance of red tilapia. Several parameters were observed, such as physical parameters (temperature, pH, and DO) and chemical parameters (CO2, alkalinity, and total ammonia). The results obtained indicated that the treatment given did not have a significant effect on temperature, pH, DO, CO2, alkalinity, and total ammonia. This is because the treatment of the water media used, the intensity of sunlight, and the aerator used are the same. Based on SNI 7550: 2009, the appropriateness of temperature value is 25 – 32. The range temperature of the first and second nursery was 26.5-31.78. The appropriateness of pH value is 6.5-8.5. The range pH value of the first and second nursery was 6.5-8.5. The appropriateness of dissolved oxygen is ≥ 3. The use of an aerator also helps in maintaining the availability of dissolved oxygen in the water so that it can be utilized by red tilapia. The range pH value of the first and second nursery was 3.3-10.76. Based on SNI 6141: 2009, the appropriateness of CO2 value is ≥ 5. The range CO2 value of the first and second nursery was 0-20. According to Rustadi (2018) the appropriateness of alkalinity is 30-300. The range alkalinity value of the first and second nursery was 72-124. *PP No 82 Tahun 2001* states that the appropriateness of NH3 value is $\leq$ 0.02. The range NH3 value of the first and second nursery was 0.097-0.243. Increasing the amount of commercial feed used causes an increase in ammonia produced in the aquatic environment.

5. Conclusion

Seeds from crossing 4 strains had a significantly different effect ($P<0.05$) on the survival rate, growth, and feed conversion ratio of red tilapia seeds. The best performance of red tilapia seeds was obtained from the outbreeding treatment of Pandu females-Sukamandi Male (JTSM).

6. Acknowledgements

In future studies, genetic tests are needed to determine heterozygosity and a salinity challenge test for brackish water aquaculture.

7. References

[1] Hien T. T. T, Tu T. L. C., Haga Y., Phu T. M. 2021 Combined effects of elevated salinity and temperature on growth performance and feed utilization in hybrid red tilapia fingerlings (Oreochromis mossambicus x O. niloticus). AACL Bioflux 14(6):3310-3321.

[2] Banuelos-Vargas, I., Rodríguez-Montes de Ocaa, G.A., Martínez-Montano, E., Perez-Jimenez, A., Mendoza-Gamboa, O.A., Estrada-Godínez, J.A., Hernandez, C. 2021. Antioxidant and

immune response of juvenile red tilapia (*Oreochromis* sp) cultured at different densities in sea water with biofloc plus probiotics. Aquaculture 544, 1-8.

[3] Ariyanto, D., Listiyowati, N. 2015. Interaksi Genotipe dengan Lingkungan, Adaptabilitas, dan Stabilitas Penampilan Fenotipik Empat Varietas Unggul Ikan Nila (*Oreochromis niloticus*). Jurnal Riset Akuakultur 10(1):1-9. https://doi.org/10.15578/jra.10.1.2015.1-9

[4] Nugroho, E., Rustadi, Priyanto, D., Sulistyo, H., Susila, Sunaryo, Wasito, B. 2014. Penurunan Keragaman Genetik pada F-4 Ikan Nila Merah "Cangkringan" Hasil Pemuliaan Dideteksi dengan Marker Genetik. Jurnal Riset Akuakultur, 9(1), 25–30.

[5] Ariyanto, D., Listiyowati, N., Imron, I. 2011. Analisis Truss Morfometrik Beberapa Varietas Ikan Nila *Oreochromis niloticus*. Jurnal Riset Akuakultur, 6(2), 187. https://doi.org/10.15578/jra.6.2.2011.187-196.

[6] Nasution, A. S. I., Basuki, F., Sri, H. 2014. Analisis Kelulushidupan dan Pertumbuhan Benih Ikan Nila Saline Strain Pandu (*Oreochromis niloticus*) yang Dipelihara di Tambak Tugu, Semarang dengan Kepadatan Berbeda. Journal of Aquaculture Management and Technology, 3(2): 25–32. http://ejournal-s1.undip.ac.id/index.php/jfpik

[7] Lamanto, Gunadi, B., Fansuri, N. 2020. Peningkatan Produktivitas Larva Ikan NIla Merah Generasi Pertama (G1) terhadap Populasi Dasar (G0) di Balai Riset Pemuliaan Ikan. In Buletin Penelitian Pemuliaan Ikan (Mengenal Populasi, Vol. 9). Balai Penelitian Pemuliaan Ikan.

[8] Hu, F., Zhong, H., Yu, P., Fan, J., Wu., Wang, S., Gong, D., Sun, Y., Gao, X., Wen, M., Tang, C., Tao, M., Liu, S. 2023. Comparative analysis of growth performance, pharyngeal teeth and intestinal traits in F1 hybrids of female *Megalobrama amblycephala* x male *Culter mongolicus*. Aquaculture 562.

[9] Pongthana, N., Nguyen, N. H., Ponzoni, R. W. 2010. Comparative performance of four red tilapia strains and their crosses in fresh- and saline water environments. Aquaculture 308, 109-114.

IOP Conf. Series: Earth and Environmental Science 1289 (2023) 012022 doi:10.1088/1755-1315/1289/1/012022

Leaf extract effectiveness of pecut kuda (*Stachytarpheta jamaicensis* L. Vahl) on barramundi (*Lates calcarifer*) eggs hatchability

P Hartami[1*], Mahdaliana[1], Mainisa[1], S Adhar[1], M Hatta[1], E Ayuzar[1], DF Hanum[2], Rasidi[3]

[1] Department of Aquaculture, Faculty of Agriculture, Malikussaleh University, Kampus Utama Reuleut, Jl. Cot Tgk. Nie Muara Batu, Indonesia
[2] Graduate Student of Aquaculture Department, Faculty of Agriculture, Malikussaleh University, Kampus Utama Reuleut, Jl. Cot Tgk. Nie Muara Batu, Indonesia
[3] Center for Fisheries Research, National Research and Innovation Agency, Gedung Biologi, Jl. Raya Bogor KM. 47 Nanggewer Mekar Bogor West Java, Indonesia

Corresponding author mail: prama.hartami@unimal.ac.id

Abstract. Barramundi is one of the preferred fishes in great demand because of its delicious taste and high nutritional content. One factor that influences the success of culture is hatchability and high larval survival. In the business of barramundi aquaculture, problems are often encountered, namely the presence of fungi that attack fish eggs. Precautions and treatment use of continuous synthetic compounds with improper concentration will cause new problems. The traditional pecut kuda anti-fungal, this plant is also easily obtainable and easy to use in the prevention and treatment of fish diseases. The purpose of this study was to analyze the success rate of using pecut kuda (*Stachytarpheta jamaicensis* L. Vahl) in increasing the hatchability of barramundi eggs and observing the growth of barramundi larvae. This research was conducted on July 25 – August 16, 2022. Located at Brackish Water Cultivation Fishery Center, Ujoeng Batee, Aceh. This research method used a completely randomized design with 5 treatments and 3 replications, namely P1 (0 ml/control), P2 (2 ml/L), P3 (4 ml/L), P4 (6 ml/L), and P5 (8 ml/L). Observation of the data collected during the research was presented in the form of tables and graphs. The results showed that pecut kuda leaf extract at different doses affected the hatchability of eggs, percentage of fungus attack, and mortality of barramundi larvae, while it did not affect the abnormality and growth length of barramundi larvae. The best treatment at P3 with a dose of 4 ml/L was 90.33% of the hatching rate with a low mortality value of P3 was 11.80%.

1. Introduction

Barramundi (*Lates calcarifer*) is one of the premium marine and brackish water aquaculture products in Indonesia because of its relatively fast growth rate and ability to adapt to fluctuations in environmental change [1], this makes barramundi suitable for small and large-scale aquaculture companies. It has been proven that barramundi can be cultured in ponds or floating net cages in brackish water or seawater [2], [3]. Barramundi also has market demand that continues to increase both domestically and internationally [4]. In addition, barramundi is also one of the types of fish loved by many people because of its delicious taste and high nutritional value. The successful development of barramundi aquaculture is largely determined by the supply of good quality and quantity broodstock, eggs and larvae [6]. One of the factors affecting the success of the farming crop is the high egg-hatching rate and larval survival rate, so barramundi seeds are always available [1]. Obstacles to the availability of barramundi seed often include the existence of fungi that attack fish eggs, both unfertilized and fertilized, leading to reduced hatchability and survival rates. of larvae. [7]. The fungus that commonly attacks barramundi (*Lates calcarifer*) eggs is *Saproglenia* sp, a fungal infection of eggs commonly known as fish mold [8]. Fungi can attack eggs and reproduce there because of wounds caused by bacterial attack or mishandling [9]. According to [10], preventive and therapeutic measures often use synthetic compounds that have been shown to have antifungal effects to improve egg quality. But on the other hand, continuing to use chemicals and antibiotics at inappropriate concentrations will lead to new problems, including increased resistance to fungi and negative impacts on the environment and humans.

The use of synthetic chemicals and antibiotics in large-scale farming is ineffective due to the relatively high cost of antibiotics. Therefore, it is necessary to use alternative natural materials to treat and prevent diseases, which are more effective and less expensive [12]. This alternative material is not mold-resistant and is environmentally friendly. One of the medicinal plants with high properties that can be used as a natural fungicide is pecut kuda leaves [13], [14]. These ingredients contain active phytochemicals in the form of alkaloids, phenolic compounds, tannins, terpenoids, saponins and flavonoids that act as antioxidants, bactericides, fungicides and anti-inflammatory agents [15]. There is currently no information on the use of extracts of these natural ingredients to prevent and inhibit fungal infections as well as increase the hatchability of barramundi (*Lates calcarifer*) eggs. Therefore, research must be done to determine the ideal concentration of pecut kuda leaf in the incubation of barramundi eggs and possibly provide a solution to combat the low production of barramundi seeds. This study aims to test, analyze and evaluate the effectiveness of phytopharmaca raw material extract (pecut kuda leaves) applied to the incubation environment with different doses on egg hatching and survival rate of barramundi fry (*Lates calcarifer*).

2. Research method

2.1 Time and place

This experiment was done from August to September 2023 at the Brackish Water Cultivation Fisheries Center in Ujong Batee Aceh Besar.

2.2 Research methodology

This research utilized a fully randomized design with five treatments and three replications. The dosage used refers to research [16] for treatment using pecut kuda leaf without changing water up to 10 days of rearing after hatching. The treatment used in detail was presented as follows.

P1 : Control (0 ml/L)
P2 : 2 ml/L
P3 : 4 ml/L
P4 : 6 ml/L
P5 : 8 ml/L

2.3 Research procedure

2.3.1 Extraction processing

The preparation of pecut kuda leaf extract was carried out at the Hatchery and Cultivation Technology Laboratory, Faculty of Agriculture, University of Malikussaleh, using the maceration method. The obtained leaves are first dried and then ground into powder and the dry weight is weighed as 500 grams of powder. In addition, each ingredient is soaked with 96% ethanol at a ratio of 1: 5 at room temperature 20-25^{0}C, leave in a glass jar for 3 days and stirred well every day with a spoon. Additionally, evaporation was performed at a temperature of 40–50°C to obtain a concentrated extract using a rotary evaporator [17]. Once the extract was obtained, it was stored in a 500 ml glass container and then stored in the freezer before testing at predetermined doses.

2.3.2 Container preparation

Before starting the study, the first step was to sterilize the equipment to be used by first washing it with detergent and then drying it at room temperature. After drying, each plastic tub-shaped container with a capacity of 2 litres for each treatment was filled with water using a filter, placed in the incubation room and provided with an aeration system.

2.3.3 Preparation of barramundi eggs
The eggs used in this study were obtained from the spawning of barramundi at the Brackish Aquaculture Fishery Center, Ujong Batee Aceh Besar, which looked clear and floated as a sign that 100 eggs had been fertilized in the container. According to [18], eggs develop healthily, are transparent, and have a bright colour so they are easy to distinguish from unfertilized eggs.

2.3.4 Soaking test
The extract of each ingredient was taken according to the dose of each treatment, put into each container, diluted and added with water until each container reached a volume of 1 litre. Furthermore, each selected egg was put into a test container for observation.

2.3.5 Larvae rearing
Larvae rearing during the study was carried out in 2 litres of incubation media with 1 litre of water volume. Barramundi larvae were observed for 10 days, during maintenance no water changes and siphoning were carried out because the larvae were still very vulnerable. Barramundi larvae 1-2 days old were not given feed because there was still a yolk sack available in the larvae's body. Three-day-old larvae were given rotifers until they were 10 days old at a density of 2-3 ind/ml every morning starting at 08.00 WIB. According to [19], Rotifers were chosen as feed because they have good digestive enzymes for larvae and complete nutritional content.

2.4 Observation parameters
2.4.1 Hatching rate
Hatching Rate (HR) was the hatchability of eggs or the number of eggs that hatch, the unit value of the Hatching Rate was expressed in percent (%). For the calculation results of egg hatchability using the formula [20] specifically.

$$HR\ (\%) = \frac{\text{Number of hatched eggs (individual)}}{\text{Total number of eggs}} \text{ x } 100 \quad (1)$$

2.4.2 Fungal infection percentage
To know the level of fungal infection in percent, prevalence calculations were carried out based on the formula [21].

$$Prevalence\ (\%) = \frac{\text{Number of infected eggs}}{\text{The number of eggs observed}} \text{ X } 100 \quad (2)$$

2.4.3 Larvae abnormality
Observation of larval abnormalities based on abnormalities or defects in head shape, body shape, and tail shape. These physical abnormalities were documented in the form of pictures recorded and then discussed descriptively. Calculations were made to determine the magnitude of the abnormality as stated by [22], with the following formula.

$$Abnormalities\ (\%) = \frac{\text{Abnormal number of larvae}}{\text{The number of hatched larvae}} \text{ x}100 \quad (3)$$

2.4.4 Larval growth
Growth measurements were carried out at the beginning of rearing and the end of rearing larvae using a micrometre on a microscope. Growth in length was calculated by the formula [23].

$$\text{SGR (\% per day)} = \frac{\text{In } Lt - \ Lo}{\text{t}} \text{ X } 100 \quad (4)$$

Information:
SGR : Specific Growth Rate (%
InLt : Average body length at the end of rearing (individual)
InLo : Average body length at the start of rearing (individual)
t : Rearing period (day)

2.4.5 Survival rate

Survival Rate (SR) was the comparison of fish survival at the beginning to the end of the study. SR calculation was done by using the formula [24].

$$SR\ (\%) = \frac{Nt}{N0} \times 100 \tag{5}$$

Information:
SR : Survival Rate (%)
Nt : Number of larvae living in the final stage (individual)
N0 : Number of larvae living in the early stage (individual)

2.4.6 Water quality parameters

Physical and chemical water parameters observed at the time of hatching and rearing include salinity, temperature, pH, dissolved oxygen, and ammonia [25]. Temperature measurements were taken every morning using a thermometer, while measurements of dissolved oxygen, salinity, pH, and ammonia were taken at the starting and the end of this study.

2.5 Data analysis

Observation data was presented in tables and graphs, and then several stages of testing were carried out, namely the normality test and homogeneity test. If the data analyzed was normally distributed and homogeneous, a parametric statistical test was carried out using one-way ANOVA. If the significance value indicates a difference, then Tukey's further test was carried out to find out the difference in each treatment.

3. Result

3.1 Hatching rate

The results showed that pecut kuda leaf extract with different doses was able to increase the hatchability of barramundi fish eggs as shown in Figure 1 below.

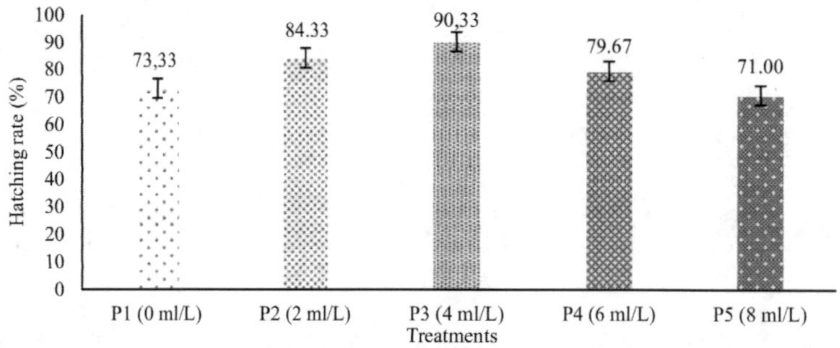

Figure 1. Graph of egg hatchability in each treatment.

It can be seen from Figure 1 that each treatment had different egg-hatching rates. The best hatching rate was observed in the treatment P3 with a dose of 4 ml/L, which was 90.33%, while P5 with a dose of 8 ml/L was the treatment with the lowest hatching value, namely 71%. Based on the ANOVA test, it showed that administration of pecut kuda leaf extract had a significant effect on the hatchability of barramundi fish eggs with a F_{count} of 19.06 > $F_{table\ (0.05)}$ of 3.48. Tukey's further test results showed that treatment P3 had a significant effect on other treatments, but was not significantly different from P2.

3.2 Fungus attack percentage

The results of the study of giving pecut kuda leaf extract to barramundi fish eggs showed the percentage of fungal attacks as shown in Table 1 below.

Table 1. Percentage of fungal attack in each treatment.

Treatments	Number of egg (grain)		Percentage (%)
	Infected	Checked	
P1 (0 ml/L)	25	300	8.33
P2 (2 ml/L)	21	300	7.00
P3 (4 ml/L)	17	300	5.66
P4 (6 ml/L)	16	300	5.33
P5 (8 ml/L)	10	300	3.33

Table 1 describes the highest percentage of fungal attacks in treatment P1 (0 ml/L) with a value of 8.33%, and P5 (8 ml/L) is the treatment that has the lowest percentage of fungal attacks, namely 3.33%. The following is a graphical picture of the level of resistance to fungal attack in each treatment.

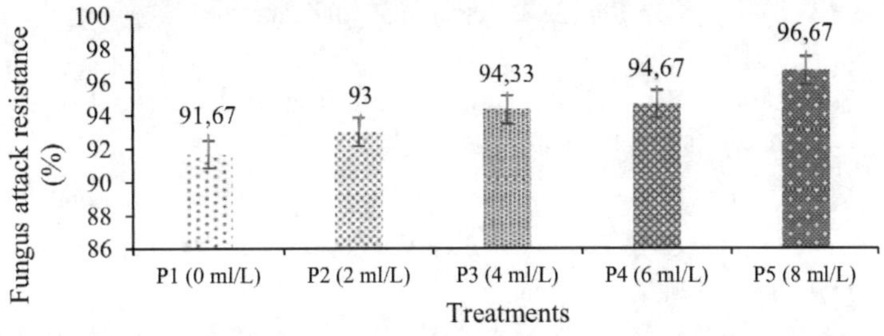

Figure 2. Fungus attack resistance level graph.

Figure 2 shows the highest level of fungal attack resistance in treatment P5 (8 ml/L) with a value of 96.67%, then P1 (0 ml/L) is the treatment that has the lowest fungal attack resistance level of 91.67%. The results of the Anova test showed that administration of pecut kuda leaf extract had a significant effect on the level of resistance to the fungal attack on barramundi eggs with a Fcount of 9.906 > Ftable (0.05) of 3.48. Tukey's further test results showed that treatments P5, P4, and P3 had a significant effect on other treatments. The type of fungus identified at the time of observation was *Saprolegnia* sp as shown in Figure 3 below.

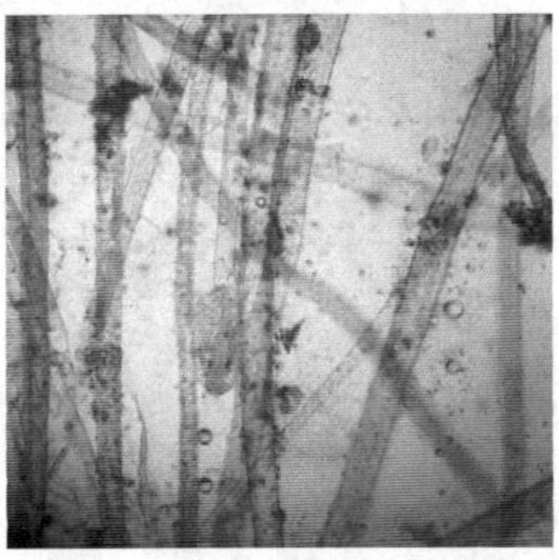

Figure 3. *Saprolegnia* sp which attacks barramundi fish eggs

The fungus that was found to be seen with the naked eye is white like cotton. After being observed using a microscope, it can be seen that the fungus *Saprolegnia* sp. is long like threads and is grey-white in colour.

3.3. Larvae abnormalities

The results of the study of giving pecut kuda leaf extract to barramundi eggs showed that there were no abnormal barramundi larvae which can be seen in Figure 4 below.

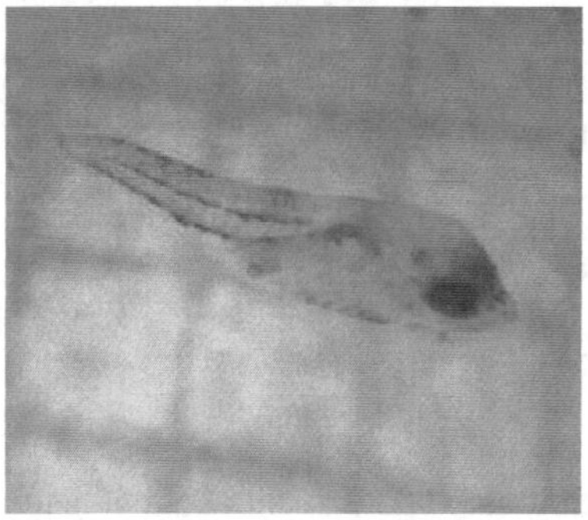

Figure 4. Barramundi fish larvae

It can be seen in Figure 4, that the D10 age barramundi larvae looked normal for all treatments. In the picture, it can be seen that the tail of the larvae is straight, the head size is not enlarged, the eyes are normal size, and the colour of the white snapper larvae is also bright.

3.4. Growth length

The results showed that pecut kuda leaf extract at different doses did not affect the growth of barramundi larvae as shown in Figure 5 below.

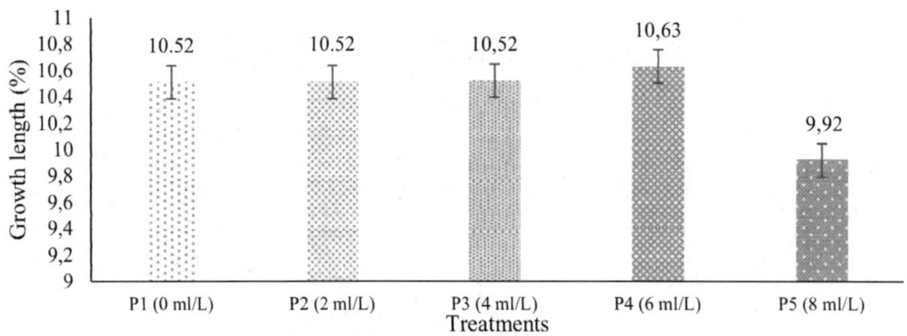

Figure 5. Graph of specific growth rate for each treatment

It can be seen in Figure 5, that the highest percentage of specific growth rate during observation was in treatment P4 (dose 6 ml/L) which was 10.63% per day, then P5 with a dose of 8 ml/L was a treatment that had a specific growth rate percentage the lowest is 9.92% per day. Based on ANOVA, it showed that administration of pecut kuda leaf extract did not affect the specific growth rate of barramundi with F_{count} 1.391 < $F_{table\ (0.05)}$ 3.48.

3.5. Larvae mortalities

The results showed that larval mortality had differences for each treatment as shown in Table 2, and the survival rate is presented in Figure 6 below.

Table 2. Larvae mortality in each treatment.

Treatments	Number of larvae (Individual)		Mortalities (%)
	Hatching	Dead	
P1 (0 ml/L)	220	68	30.90
P2 (2 ml/L)	253	38	15.01
P3 (4 ml/L)	271	32	11.80
P4 (6 ml/L)	239	72	30.12
P5 (8 ml/L)	213	117	54.92

Table 2 shows that the highest mortality was in treatment P5 with a dose of 8 ml/L which was 54.92% and P3 with a dose of 4 ml/L was the treatment that had the lowest mortality value, namely 11.80%. The Survival Rate graph for each treatment can be seen in Figure 6.

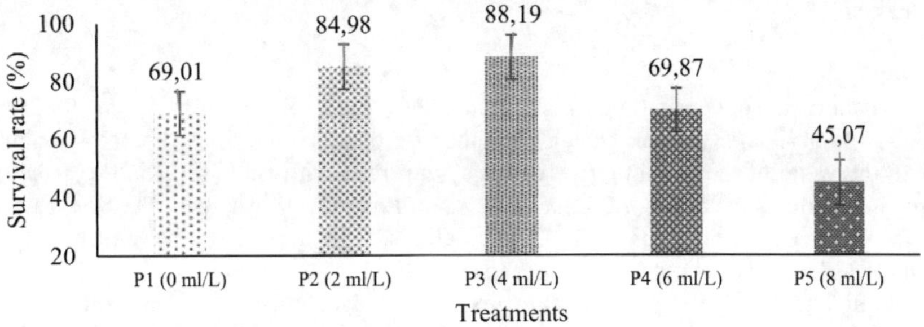

Figure 6. Larvae survival rate.

As shown in Figure 6, the best survival rate was observed in treatment P3 with a dose of 4 ml/L which was 88.19% and P5 with a dose of 8 ml/L was the treatment that had the lowest survival value, namely 45.07%. Based on statistical analysis using the Anova test, it was shown that administration of pecut kuda leaf extract had a significant effect on the survival rate of barramundi larvae with F_{count} 43.004 > $F_{table\ (0.05)}$ 3.48. Tukey's further test results showed that the P2 and P3 treatments had a significant effect on the other treatments.

3.6. Water quality parameters

The results of measurements of salinity, temperature, pH, and DO in research for egg-hatching media and larval-rearing media are presented in the following table.

Table 3. Data on the measurement of water quality for hatching eggs

Treatments	Average water quality data			
	Temperatures (°C)	pH	Dissolved oxygen (ppm)	Salinity (ppt)
P1 (0 ml/L)	27.00-27.80	7.80-8.00	5.00-5.50	28.00
P2 (2 ml/L)	27.80-27.90	7.60-8.00	4.67-5.30	28.00
P3 (4 ml/L)	27.50-28.00	7.80-8.10	4.28-5.20	28.00
P4 (6 ml/L)	27.40-27.90	7.90-8.10	4.27-5.21	28.00
P5 (8 ml/L)	27.80-28.10	7.90-8.20	4.29-5.20	28.00

Table 4. Data on the measurement of water quality for rearing eggs

Treatments	Average water quality data			
	Temperatures (°C)	pH	Dissolved oxygen (ppm)	Salinity (ppt)
P1 (0 ml/L)	27.00-29.60	8.00-8.20	4.30-5.00	28.00-30.00
P2 (2 ml/L)	26.90-29.70	7.90-8.30	3.60-4.00	28.00-30.00
P3 (4 ml/L)	27.00-29.80	8.10-8.30	3.80-4.00	28.00-31.00
P4 (6 ml/L)	27.00-29.80	7.90-8.20	3.50-3.90	28.00-31.00
P5 (8 ml/L)	27.00-29.60	8.00-8.40	3.08-3.84	28.00-31.00

Tables 3 and 4 show that the water quality in the egg and larvae observation media is included in the optimal category, namely temperature, pH, and salinity. Meanwhile, dissolved oxygen tends to decrease, so it does not meet the optimal category for larval rearing even though it has been equipped with aeration in each container.

4. Discussion

The highest hatching rate of barramundi eggs within 14 hours of observation in treatment P3 with a dose of 4 ml/L was 90.33%. The value of the egg hatching rate obtained in this study almost coincides with 92.33% with the value of the hatching rate of eggs from research results [26] on the effectiveness of using chemical disinfectants to Prevent fungal attacks on *Labeo bata* fish eggs. The dosage of 4 ml/L is the dosage containing active ingredients such as alkaloids, flavonoids and other antifungal substances that are completely optimal and suitable for barramundi, so at this dosage the eggs hatch well. The content in pecut kuda leaves is not only an active antifungal agent but also contains some toxic substances in large doses. Thus, at this dosage, the alkaloid active ingredient is an antifungal compound that has good effects, while the tannic acid in the form of an acid that inhibits egg hatching has no harmful effect because it is still at a low dosage.

Egg hatching is influenced by several factors, while egg failure can be caused by many factors such as light intensity, temperature, pH, turbidity, dissolved oxygen content and ammonia. [27]. The 8 ml/L dose contains more active ingredients so this concentration may affect the eggs. This is the opinion [28] that hatching is not successful because the egg is highly permeable, so when fertilization begins, the egg membrane absorbs water and the egg hatches quickly. If the incubation medium has a higher ion concentration than the egg, the egg will be damaged because the liquid it contains will be absorbed by the more concentrated medium. According to [29], in fertilized eggs, there is fluid that fills the space around the spore, helping to provide space and protection for the developing embryo. The egg's hatchability will decrease if the egg swells due to promoted osmosis. When the osmotic concentration in the Perivitelline space is greater than the concentration of extracellular water surrounding the egg, the extracellular water will move into the perivitelline space by osmosis, causing the egg to swell. Then, the liquid inside the egg will flow out, the egg will shrink and die. Therefore, the P5 dose concentration is not optimal for barramundi egg incubation.

The high hatching rate was also mainly determined by the low fungal attack rate in treatment P3, which was 5.66%. The fungal attack percentage value in the extract treatment in this study was better than the study by [30] who used clove leaf extract in African catfish hatchery at a dose of 11 ppm, or 50.77%. Through observation, the fungus found in this study is *Saprolegnia* sp with cotton fibre-like characteristics, white to grey in colour. It is suggested [31] that *Saprolegnia* sp. fibrous, cotton-like, white to grey and brown. The discovery of *Saprolegnia* sp during research supports the view [32] that one of the fungi that attack barramundi eggs is *Saprolegnia* sp, commonly known as fish mold because it attacks fish and eggs. fish. This fungus creates tangled excess mycelium, causing the living eggs around the dead eggs to die. The fungus will interfere with the egg's respiration process and will eventually die before hatching. There is an opinion [8] that when white barramundi eggs are attacked by fungi, they will show signs of milky white eggs or smooth cotton-like fibres around the eggs.

Saprolegnia sp is characterized by the ability to move freely (have two flagella), the body's cells produce filaments (hyphae do not have septa or branches), are filamentous and have no segments.

The ability of fungi to attack eggs decreased when increasing the concentration of pecut kuda leaf extract used in the incubation media. This shows that the concentration of the extract affects fungal attacks. According to [33], active compounds in the form of flavonoids, saponins, tannins, quinones, steroid compounds, triterpenoids and alkaloids have good antifungal effects, so fish eggs are protected from fungal attack because the fungal cells become sensitive to changes in the environment which causes die. For P5 treatment, an extraction dose of 8 ml/L is the best dose to overcome fungal attack. This is due to the content in pecut kuda leaves in the form of active alkaloids that protect the body's organs from toxic substances, both in the form of parasites and fungi, as well as fungicidal antibacterial substances has an inhibitory effect of some enzymes, leading to disturbances in the metabolism of fungal cells, thereby inhibiting the elongation process of fungal hyphae. The fungal hyphae cannot be fragmented because the mycelial tissue is damaged, so the fungal cells easily die [34]. However, the dose of P5 (8 ml/L) has a good inhibitory effect on fungi but does not inhibit egg hatching. The higher extract concentration, not only has a more effective antifungal effect, but also kills barramundi eggs.

In addition to its ability to increase egg hatching rate and inhibit fungal attack, pecut kuda leaf extract is also safe for barramundi larvae and therefore does not cause abnormalities during the hatching process of the larvae. Indeed, pecut kuda leaf extract contains active compounds, including flavonoids. It has been suggested [35] that flavonoid compounds can increase bone density, which is attributed to the content of ethyl p-methoxycinnamate which belongs to the flavonoid group and is part of the phytoestrogen group of compounds. The mechanism of action of phytoestrogen compounds in regulating bone density may be similar to that of estrogen. Estrogen influences bone regeneration by controlling the activity of osteoblasts and osteoclasts, including maintaining a balance in the activity of the two cells by regulating the production of osteoclasts. paracrine factors [36]. Estrogen also affects bone breakdown by inhibiting the maturation of osteoclasts, which in turn can inhibit bone resorption (35).

Another benefit of pecut kuda leaf extract is that it does not harm the development of barramundi larvae. Larval development in all treatments remained relatively normal, although there was a tendency to decrease with increasing doses. The growth rate is influenced by competition for food and varying levels of fish appetite. This is evident in statement [30] which states that growth can be influenced by biotic and abiotic factors. Biological factors include age and the fish's ability to meet dietary requirements, and abiotic factors include food availability and fish habitat quality. The highest SGR found in this treatment was 10.63%, this value is higher than the results obtained in [30] on the effect of soaked dose of clove leaf extract (*Syzygium aromaticum*) on catfish (*Clarias gariepinus*) egg hatching, the best specific growth rate was achieved at the average of 0.252%.

Research results showed that pecut kuda leaf extract also had an impact on the low mortality rate of barramundi larvae, especially in treatment P3, which was 11.80%. The highest mortality rate in this study was 54.92%, this value is better than the results obtained by [37] regarding the effectiveness of meniran plant extract (*Phyllanthus niruri* L.) against fungal attack (*Saprolegnia* sp.) on gouramy (*Osphronemus gouramy*) egg hatchability, namely at a dose of 3 ml/l of 74.67%. The mortality rate of larvae is estimated due to the administration of pecut kuda leaf extract which contains high tannin compounds along with increasing doses. This follows the opinion [38] which states that there is an effect of the tannin compounds contained in the leaf extract in the egg-hatching process, the tannin compounds are acidic so that they can trigger the softening of the chorion layer which can cause the larvae to hatch with a state of weakened immunity so they die easily. Thus, reducing fungal infection in eggs at high doses that affect embryo viability will result in reduced egg hatchability.

Other factors affect the mortality rate of larvae, specifically the amount of dissolved oxygen during rearing is relatively low, the optimal amount of dissolved oxygen for reared larvae is at least 4, while the amount of dissolved oxygen obtained during observation is <4. This is consistent with the dissolved oxygen value in seawater aquaculture, which must be greater than 4 mg/L [39]. According to [25], if the amount of dissolved oxygen is unbalanced, the fish will be stressed because the brain is not provided with enough oxygen and die due to lack of oxygen (anorexia) because the body does not receive enough oxygen. [40] states that water quality is one of the factors that influence the production of aquaculture activities, where the temperature parameter is a parameter that has a complex relationship with other water qualities so that it can affect growth, reproduction, and survival.

The decrease in dissolved oxygen value was due to not changing the water during the research process, so the ammonia concentration in the farming environment became high. It has been suggested [41] that in aquaculture systems, for all water quality parameters, ammonia is the second limiting factor after oxygen. At high concentrations, ammonia is toxic, significantly reducing oxygen supply and causing unwanted changes in aquatic ecosystems. The water exchange or storage was not carried out because the larvae were still too young, namely D1-D10. This is by the opinion [42], that early-stage larvae from one to ten days old are sensitive to environmental factors as a result of water changes. Environmental factors can be in the form of water quality parameters such as dissolved oxygen, temperature, salinity, pH, and ammonia.

The temperature of the water media during egg rearing for all containers was stable and not too high, namely around 27.00-28.10^{o}C, and the temperature during larval rearing was around 26.90-29.60°C, this corresponds to the temperature range for barramundi which is 27.00 -30.00°C [43]. [44], added that egg hatching will be faster at high temperatures because at high temperatures metabolic processes will occur more quickly so that embryo development will also be faster and the movement of embryos in the shell will be more intensive so that hatching will be faster. The pH of the media during the study for all containers was in the range of 7.6-8.4 which is considered optimal, which is under the pH for barramundi according to [39] that is, for the stages of rearing larvae in tubs and raising eggs are the same, ie, 7.00-8.50. [29] said that a low pH can cause the work of the chorionase enzyme to be disrupted because this enzyme cannot work optimally at a more acidic or lower pH.

5. Conclusion

The conclusion that can be drawn from this study was that administration of pecut kuda leaf extract (*Stachytarpheta jamaicensis* L. Vahl) with different doses affects egg hatchability, percentage of fungal attack, and larval mortality in barramundi (*Lates calcarifer*). The use of pecut kuda leaf extract can reduce the egg mortality rate, thus giving a good value to the hatching rate. The use of pecut kuda leaf extract with different tests did not affect the specific length growth rate of barramundi. The use of pecut kuda leaf did not influence the advancement of barramundi larvae, so no abnormal larvae were found. Providing pecut kuda leaf extract had no impact on changing the water quality of the barramundi egg or larval culture environment.

References

[1] A. R. S. H. Saputra and S. H. Samara, "Performance analysis of white snapper (*Lates calcarifer*) nursery at BBPBAP Jepara," in *IOP Conference Series: Earth and Environmental Science*, Institute of Physics, 2022. doi: 10.1088/1755-1315/1036/1/012118.

[2] D. S. Fielder, W. J. Bardsley, G. L. Allan, D. S. Fielder, and P. M. Pankhurst, "Effect of photoperiod on growth and survival of snapper *Pagrus auratus* larvae," *Aquaculture*, vol. 211, no. 1–4, pp. 135–150, Aug. 2002, doi: 10.1016/S0044-8486(02)00006-6.

[3] N. M. Bacheler, K. W. Shertzer, B. J. Runde, P. J. Rudershausen, and J. A. Buckel, "Environmental conditions, diel period, and fish size influence the horizontal and vertical movements of red snapper," *Sci Rep*, vol. 11, no. 1, Dec. 2021, doi: 10.1038/s41598-021-88806-3.

[4] D. M. Cawthorn, C. Baillie, and S. Mariani, "Generic names and mislabeling conceal high species diversity in global fisheries markets," *Conservation Letters*, vol. 11, no. 5. Wiley-Blackwell, Sep. 01, 2018. doi: 10.1111/conl.12573.

[5] Hardoko and S. Utami, "Chemical-physical properties characterization of white snapper fish skin rambak crackers based on boiling and drying duration," *Jurnal Ilmiah Perikanan dan Kelautan*, vol. 12, no. 1, pp. 122–130, 2020, doi: 10.20473/jipk.v12i1.14842.

[6] Z. R. Paniza. "Broodstock management and seed production of marine fishes." *SEAFDEC Asian Aquaculture*, Vol. XXIII May-August 2001, no. 3-4.

[7] M. E. Bunker, G. Elliott, H. Heyer-Gray, M. O. Martin, A. E. Arnold, and S. L. Weiss, "Vertically transmitted microbiome protects eggs from fungal infection and egg failure," *Anim Microbiome*, vol. 3, no. 1, Dec. 2021, doi: 10.1186/s42523-021-00104-5.

[8] F. Fadaeifard, M. Raissy, H. Bahrami, E. Rahimi, and A. Najafipoor, "Freshwater fungi isolated from eggs and broodstocks with an emphasis on Saprolegnia in rainbow trout farms in west Iran," *Afr J Microbiol Res*, vol. 5, no. 22, Oct. 2011, doi: 10.5897/ajmr11.385.

[9] P. Kalatehjari, M. Yousefian, and M. A. Khalilzadeh, "Assessment of antifungal effects of copper nanoparticles on the growth of the fungus *Saprolegnia* sp. on white fish (*Rutilus frisii* kutum) eggs," *Egypt J Aquat Res*, vol. 41, no. 4, pp. 303–306, Dec. 2015, doi: 10.1016/j.ejar.2015.07.004.

[10] R. Sun, J. Chen, C. Pan, Y. Sun, B. Mai, and Q. X. Li, "Antibiotics and food safety in aquaculture," *J Agric Food Chem*, vol. 68, no. 43, pp. 11908–11919, Oct. 2020, doi: 10.1021/acs.jafc.0c03996.

[11] X. Liu, J. C. Steele, and X. Z. Meng, "Usage, residue, and human health risk of antibiotics in Chinese aquaculture: A review," *Environmental Pollution*, vol. 223, pp. 161–169, 2017, doi: 10.1016/j.envpol.2017.01.003.

[12] Y. Fujaya, A. A. Hidayani, D. K. Sari, S. Aslamyah, N. Rukminasari, A. Muthalib, S. Cristianto, E. Defista, H. Fazhan, and K. Waiho, "The optimal dosage of fermented herbal extract on growth and feed efficiency of nile tilapia (*Oreochromis niloticus*)." *Tropical Life Sciences Research*, no. 1-22, November 2022.

[13] P. M. Liew, and Y. K. Yong, "*Stachytarpheta jamaicensis* (L.) Vahl: from traditional usage to pharmacological evidence," *evidence-based complementary and alternative medicine*, vol. 2016. Hindawi Limited, 2016. doi: 10.1155/2016/7842340.

[14] P. D. Yadav, K. P. Modi, and M. B. Shah, "Phytochemistry, pharmacology, and botanical aspects of Stachytarpheta species-A review," *International Journal of Green Pharmacy*, vol. 15, no. 2, p. 114, doi: 10.22377/ijgp.v15i2.3078.

[15] O. Zs, O. Oo, K. Se, and A. Oo, "*Stachytarpheta jamaicensis* leaf extract: Chemical composition, antioxidant, anti-arthritic, anti-inflammatory and bactericidal potentials," *Journal of Scientific and Innovative Research*, vol. 6, no. 4, pp. 119–125, 2017, [Online]. Available: www.jsirjournal.com

[16] L. A. Al_husnan and M. D. F. Alkahtani, "Impact of Moringa aqueous extract on pathogenic bacteria and fungi in vitro," *Annals of Agricultural Sciences*, vol. 61, no. 2, pp. 247–250, Dec. 2016, doi: 10.1016/j.aoas.2016.06.003.

[17] Q. W. Zhang, L. G. Lin, and W. C. Ye, "Techniques for extraction and isolation of natural products: A comprehensive review," *Chinese Medicine (United Kingdom)*, vol. 13, no. 1. BioMed Central Ltd., Apr. 17, 2018. doi: 10.1186/s13020-018-0177-x.

[18] M. A. Haque, M. I. Hossain, S. Aftabuddin, A. Habib, and M. A. M. Siddique, "First onboard fertilization of Asian seabass, *Lates calcarifer*: Effects of egg stocking density on the fertilization, hatching and survival rate," *Sci Afr*, vol. 12, Jul. 2021, doi: 10.1016/j.sciaf.2021.e00841.

[19] M. Kailasam, A. R. Thirunavukkarasu, A. G. Ponniah, S. Selvaraj, and P. Stalin, "Recent advances in rotifer culture and its application for larviculture of finfishes," in *Advances in Marine and Brackishwater Aquaculture*, Springer India, 2015, pp. 17–23. doi: 10.1007/978-81-322-2271-2_2.

[20] M. A. Rahman, M. H. Rahman, S. M. Yeasmin, A. Al-Asif, and D. Mridha, "Identification of causative agent for fungal infection and effect of disinfectants on hatching and survival rate of bata (*Labeo. Bata*) larvae," *Advances in Plants & Agriculture Research*, vol. 7, no. 4, Aug. 2017, doi: 10.15406/apar.2017.07.00264.

[21] S. Elbashir, M. Jahncke, A. DePaola, J. Bowers, J. Schwarz, A. J. Punchihewage-Don, B. Min, T. Rippen, and S. Parveen, "Prevalence and abundance of bacterial pathogens of concern in shrimp, catfish and tilapia obtained at retail stores in Maryland, USA," *Pathogens*, vol. 12, no. 2, Feb. 2023, doi: 10.3390/pathogens12020187.

[22] E. Shuba, D. Adriaens, and P. Bossier, "Variations in body shape of European seabass larvae (*Dicentrarchus labrax*) reared under xenic and axenic conditions," *Ethiopian Journal of Science and Technology*, vol. 7, no. 2, p. 85, Jul. 2015, doi: 10.4314/ejst.v7i2.2.

[23] A. E. Sallam, M. M. M. El-feky, M. S. Ahmed, and A. T. Mansour, "Potential use of whey protein as a partial substitute of fishmeal on growth performance, non-specific immunity and gut histological status of juvenile European seabass, *Dicentrarchus labrax*," *Aquac Res*, vol. 53, no. 4, pp. 1527–1541, Mar. 2022, doi: 10.1111/are.15688.

[24] A. T. Mukti, O. Carman, A. Alimuddin, M. Zairin, and M. A. Suprayudi, "Growth performance, survival rate, flesh, and proximate composition of sex-grouped triploid and diploid Nile tilapia (*Oreochromis niloticus*)," *Turk J Vet Anim Sci*, vol. 44, pp. 290–298, 2020, doi: 10.3906/vet-1905-79.

[25] A. Bhatnagar, and P. Devi, "Water quality guidelines for the management of pond fish culture," *Int J Environ Sci*, vol. 3, no. 6, pp. 1980–2009, 2013, doi: 10.6088/ijes.2013030600019.

[26] M. A. Rahman, M. H. Rahman, S. M. Yeasmin, A. Al-Asif, and D. Mridha, "Identification of causative agent for fungal infection and effect of disinfectants on hatching and survival rate of bata (*Labeo bata*) Larvae," *Advances in Plants & Agriculture Research*, vol. 7, no. 4, Aug. 2017, doi: 10.15406/apar.2017.07.00264.

[27] M. Gomes, A. Correia, L. Pinto, C. Sá, V. Brotas, and M. Mateus, "Coastal water quality in an atlantic sea bass farm site (Sines, Portugal): a first assessment," *Front Mar Sci*, vol. 7, Apr. 2020, doi: 10.3389/fmars.2020.00175.

[28] T. Myosho, H. Takahashi, K. Yoshida, T. Sato, S. Hamaguchi, T. Sakamoto, and M. Sakaizumi, "Hyperosmotic tolerance of adult fish and early embryos are determined by discrete, single loci in the genus Oryzias," *Sci Rep*, vol. 8, no. 1, Dec. 2018, doi: 10.1038/s41598-018-24621-7.

[29] M. Kasiri, M. Sudagar, and S. A. Hosseini, "Effect of water hardness on egg hatchability and larval viability of angelfish (*Pterophyllum scalare* (Schultze, 1823)," 2011. [Online]. Available: http://www.urpjournals.com

[30] B. N. D. Yulihastiana, N. Cokrowati, and A. R. Scabra, "Pengaruh dosis perendaman ekstrak daun cengkeh (*Syzygium aromaticum*) terhadap daya tetas telur ikan lele dumbo (*Clarias gariepinus*)," *Jurnal Perikanan Unram*, vol. 11, no. 1, pp. 89–97, May 2021, doi: 10.29303/jp.v11i1.251.

[31] R. S. M. Chong, "Saprolegniasis," *Aquaculture Pathophysiology: Finfish Diseases: Volume I*, vol. 1, pp. 645–650, Jan. 2022, doi: 10.1016/B978-0-12-812211-2.00053-6.

[32] K. Chukanhom and K. Hatai, "Freshwater fungi isolated from eggs of the common carp (*Cyprinus carpio*) in Thailand," *Mycoscience*, vol. 45, no. 1, pp. 42–48, Feb. 2004, doi: 10.1007/s10267-003-0153-9.

[33] A. A. El-Shahir, D. A. El-Wakil, A. A. H. A. Latef, and N. H. Youssef, "Bioactive compounds and antifungal activity of leaves and fruits methanolic extracts of *Ziziphus spina-christi* L.," *Plants*, vol. 11, no. 6, Mar. 2022, doi: 10.3390/plants11060746.

[34] B. Basir, and K. Kaharuddin, "Effectiveness of ketapang leaf extract (*Terminal Capta* L) in water media on crude egg hatching," *International Journal on Advanced Science, Education, and Religion*, vol. 3, no. 1, pp. 33–39, Mar. 2020, doi: 10.33648/ijoaser.v3i1.44.

[35] A. Budiarti, D. N. Hidayati, and N. Aini, "Identifikasi awal potensi ekstrak etanol rimpang jahe merah (*Zingiber officinale* Rosc. var. rubrum) sebagai obat osteoporosis."

[36] D.G. Monroe, F.J. Secreto, and T.C. Spelsberg, "Overview of estrogen action in osteoblasts: Role of the ligand, the receptor, and the co-regulators," *J Musculoskel Neuron Interact,* vol. 3, no. 4, pp. 357-362, 2003. PMID: 15758323.

[37] H. Andika, E. Dewantoro, and E. I. Raharjo, "Perendaman telur ikan gurame (*Osphronemus gouramy*) dengan ekstrak meniran (*Phyllanthus niruri* L) sebagai anti jamur," *Jurnal Ruaya: Jurnal Penelitian dan Kajian Ilmu Perikanan dan Kelautan*, vol. 1, no. 1, Jan. 2013, doi: 10.29406/rya.v1i1.231.

[38] A. Baharudin, M. Bahrus Syakirin, and T. Yusufi Mardiana, "Pengaruh perendaman larutan teh terhadap daya tetas telur ikan lele sangkuriang (*Clarias gariepinus*)," 2016.

[39] BSN, "Penetapan revisi 7 (tujuh) Standar Nasional Indonesia" 1. 14243_SK SNI no 148-KEP-BSN-9-2014.

[40] B. R. Swistock, W. E. Sharpe, and T. McCarty, "Water quality concerns for ponds," p. 6, 1998.

[41] I. Joseph, S. Joseph, B. Ignatius, G. S. Rao, K. S. Sobhana, D. Prema, and M. Varghese, "A pilot study on culture of Asian seabass *Lates calcarifer* (Bloch) in open sea cage at Munambam, Cochin coast, India," *Article in Indian Journal of Fisheries*, 2010. [Online]. Available: https://www.researchgate.net/publication/289202792

[42] L. J. Magnoni, O. Gonçalves, P. G. Cardoso, F. Silva-Brito, R. O. A. Ozório, M. Palma, and I. Viegas, "Growth performance and quality indicators of European seabass (*Dicentrarchus labrax*) fed diets including refined glycerol," *Aquaculture*, vol. 569, May 2023, doi: 10.1016/j.aquaculture.2023.739377.

[43] R. B. Hasibuan, H. Irawan, and T. Yulianto, "The effect of temperature on hatched of seabass egg," *Intek Akuakultur*, vol. 2, no. 2, pp. 49–57, Oct. 2018, doi: 10.31629/intek.v2i2.539.

[44] M. Papadaki, M. Papadopoulou, I. Siggelaki, and C. C. Mylonas, "Egg and sperm production and quality of sharpsnout sea bream (*Diplodus puntazzo*) in captivity," *Aquaculture*, vol. 276, no. 1–4, pp. 187–197, Apr. 2008, doi: 10.1016/j.aquaculture.2008.01.033.

IOP Conf. Series: Earth and Environmental Science 1289 (2023) 012023 doi:10.1088/1755-1315/1289/1/012023

Performance Of Maintaining Brood Stock of *Clarias gariepinus* With Different Densities

G Prabowo[1], M G E Kristiany[2*], E Marlina[2], A Suhermanto[1], C A Adi[1]

1 Karawang Politechnic of Marine and Fisheries, Tanjungpura road, Karangpawitan, Karawang Regency West Java, Indonesia
2 Jakarta Technical Unversity of Fisheries, AUP Road, Pasar Minggu, South Jakarta, Jakarta, Indonesia

* correspondent author email : eny.kristiany@gmail.com

Abstract. African catfish is a type of fish that is favored by Indonesian people. In the cultivation, availability of broodstock determines the success rate of production. A well-managed of broodstock will produce high productivity. Productivity is measured from the time interval of broodstock required to produce eggs after spawning, the fecundity and hatching rate. The density of broodstock will affect on productivity. This study aims to determine the right broodstock density in order produce eggs continously. The main container used is a cement tank with dimensions of 1 x 1 m^2 and 1,2 m depth. The broodstock were reared in a recirculating system with stocking densities of 5, 10 and 15 fish/m^2. Each treatment was repeated in 3 times. During maintenance, the number of broodstock was fed with floating pellets at 3% per day with the provision of 2 times every day. The results of water quality measurements as supporting data include temperature, pH, ammonia, nitrite and nitrate. Productivity was measured during 3 months of rearing to determine the egg production cycle. The productivity of broodstock with a stocking density of 15 fish/m^2 resulted in a faster gonadal maturity level. However, the resulting fecundity, Fertilization Rate and Hatching Rate were better at the number of stocking 10 and 5 fish/m^2. Keywords: *catfish broodstock, stocking density, performance*

1. Introduction

African catfish is one type of freshwater fish that is very much in demand by the public, because it has a soft meat texture and delicious meat taste and has a rather large body shape than other types of catfish. This type of fish is very popular in Indonesia and very widely cultivated.

[9] state that good broodstock management is one of the efforts to improve the quality of catfish broodstock, to anticipate inappropriate spawning cycles and low fecundity and future developments to anticipate low brood productivity. Maximizing fry productivity in hatcheries is the ultimate goal of broodstock management. The gonad maturity of catfish is influenced by internal and external factors. The environment and fish population are also external factors that affect gonad maturity in catfish [7]. [4] state that genetic, physiological, environmental and broodstock nutrition can affect gamete production in terms of quality and quantity.

Increased production of catfish must be supported by increasing the number and quality of fry, one way to increase productivity is by using intensive rearing technology. Intensification is characterized by an increase in stocking density and the addition of artificial feed which will affect the survival and growth of fish [11].The low stocking density affects production yields and broodstock of African catfish quality. Increased stocking density can increase production if accompanied by good feed and environmental management [12]. This study aims to determine the good density of African catfish female broodstock management in order to produce eggs continuously.

2. Material and Method

2.1 Material

The fish used in this study were female African catfish originating from the Fish Cultivation Laboratory at Jakarta Technical Unversity of Fisheries (Mutiara strain) with a weight of 1.3 + 0.12 kg and a length of 46.5 + 6.5 cm. The feed was given broodstock pellets (Broodstock Feed Vitality/10 Cargill) which were first fermented using EM4 probiotics. Water quality measurements include dissolved oxygen, temperature, pH, ammonia and nitrite.

2.2 Methode

This study was conducted on January 2023 at the Jakarta Techincal University of Fisheries Aquaculture Laboratory and it has obtained approval from the research ethics committee. This study used a completely randomized design, the container used was a cement tank with dimensions of 1 x 1 m^2 and 1.2 m depth. The water level in each tank is 1 m. Each tank was lined with softly nett to make it easier to lift the broodstock during observation. The broodstock were reared in a recirculation system with 5, 10 and 15 fish/m^2 treatments. Each treatment was replicated 3 times. During rearing the broodstock was given broodstock pellets of 3% of the biomass per day with a frequency of 2 times/day. Every week observations were made to determine the gonad development of the female broodstock. Fish who has matured gonads then performed semi-artificial spawning, by stimulating male and female broodstock using ovaprim. The dose of ovarim used was 0.3 ml/kg. Broodstock spawned semi-artificial with fine waring as a shelter for eggs. Broodstock productivity was measured for 3 months of rearing to determine the egg production cycle. Water quality measurements carried out include dissolved oxygen, temperature, pH, ammonia, nitrite and nitrate.

2.3 Performance data of broodstock measurement

The broodstock performance measured included gonadal maturity level, fecundity, egg fertilization rate and hatching rate [10]. The level of gonad maturity in female catfish was observed visually by looking at the shape of the abdomen and feeling the lower of the abdomen. The number of mature gonads was calculated based on the broodstock that had shown the characteristics of being ready for spawning [15]. Fecundity is the number of eggs produced by the broodstock, and the calculation by gravimetric method, the Fertilization rate (FR) is the percentage of fertilized eggs from a number of eggs successfully released by the female broodstock. Fertilization rate (FR) taken is by counting the number of fertilized eggs which are clear or transparent in color. Meanwhile, unfertilized eggs are white or cloudy. To determine the level of fertilization is the number of fertilized eggs / number of sample eggs multiplied by 100% [17].

Fertilized eggs appeared clear in color, while unfertilized eggs would be cloudy white. By counting the unfertilized eggs and dividing by the number of eggs produced, the FR value will be known. The resulting D1 larvae were then immediately transferred to the larval rearing tank by siphoning the eggs and collecting them. The resulting D1 larvae were then calculated as the hatching rate (HR) value. Broodstock that can produce high values of fecundity, FR, HR and even larvae are then used as a measure of their performance. Fish fecundity was determined using the gravimetric method wich reprecented by the formula [5]:

$$F = G/Q \times N$$

Note :
F = Fecundity (egg)
G = fish body weight (g)
Q = sample gonad weight(g)
N = Number of eggs in the sample gonads (egg)

The hatching rate is determined by the formula [5] :

$$HR = a/F \times 100\%$$

Note :
HR = Hatching Rate
a = number of eggs hatched (egg)
F = number of eggs released

Fertilization Rate [3]

$$FR = \frac{\text{Number of fertilized eggs}}{\text{Total number of eggs}} \times 100\%$$

3. Results and Discussion

3.1 Number of mature gonads

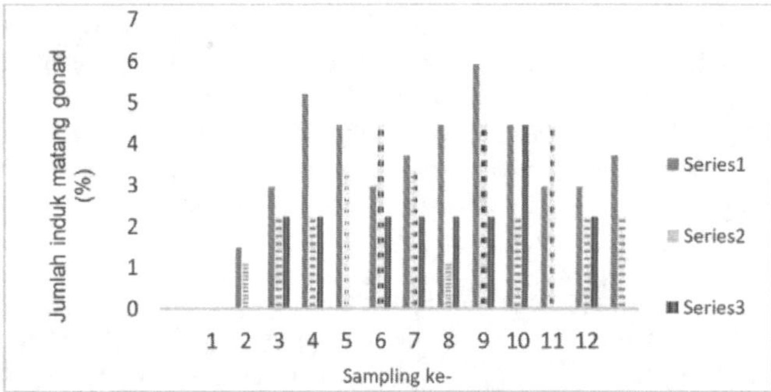

Figure 1. Graph of the number of mature gonads produced each week.

Figure 1 shows the percentage of gonadal mature broodstock sampling results during the study. It is known that broodstock in tanks with stocking numbers of 15 fish/m^2 and 10 fish/m^2 always produce mature gonad. However, at the number of stocking 5 fish/m^2 at week 1, 4, 10 and 12 there were no mature brood gonads. The number of broodstock matured to pectoral gonads at 15 fish/m^2, the number of broodstock matured to gonads was also higher than at stocking numbers of 10 fish/m^2.

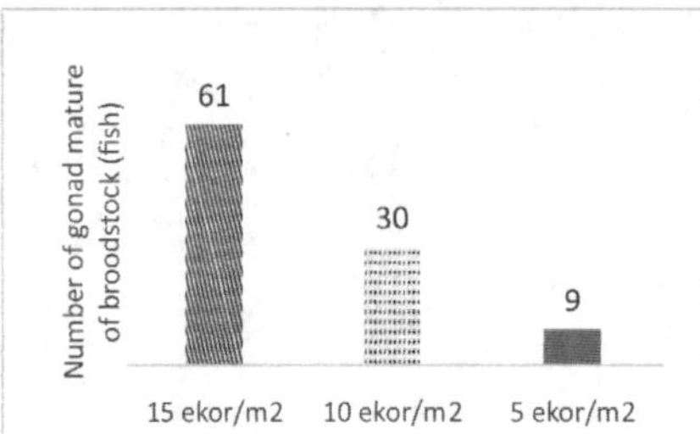

Figure 2. Graph of the number of mature gonads of broodstock during the study.

In Figure 2, the stockings number 15 fish/m^2, produce the number of gonadal mature broodstock reached 61 fish (135.5%). So it was assumed that the broodstock in all replications in this treatment had matured gonads and 35.5% of them experienced the next period of gonadal maturity. At the stocking number 10 fish/m^2 the number of broodstock that matured to gonads reached

100% or it was assumed that all broodstock reached the gonadal mature in 3 months. At the number of stocking 5 tails/m2, the number of broodstock is only 53% so that there are 47% broodstock that are not gonadally mature. African catfish are fish that like to live in groups and this makes the fish's appetite high[13].

On visual observation, at a stocking number 15 fish/m^2, the broodstock showed a calmer movement, brighter color and no injuries were found on the body. Meanwhile, at the stocking number 5 fish/m^2, the broodstock tended to stay at the bottom of the tank, were paler in collor and found several injured broodstock which was indicated as a result of fighting between broodstock. Broodstock catfish react based on instinct and conditioning. Instinct centers on some basic concepts for breeding, feeding, defending, and attacking. The behavior of the broodstock of catfish is centered on that instinct and environmental conditions that are imprinted in the cerebellum [14]. Associated with the number of stocking different broodstock, this shows the influence of the broodstock on the response of the surrounding environment. The nature of catfish in groups also stimulates the parent to respond well and stimulates the development of the gonads. Based on the results of the ANOVA test at a confidence level of 0.05, the number of mature gonads in the three treatments was significantly different. So it is assumed that the different stocking number of broodstock affects the number of fish mature gonads produced. Meanwhile, based on the Tukey HSD test, the stocking number 15 fish/m^2 and 10 fish/m^2 produced gonad mature of broodstock which were not significantly different. While the stocking number of 5 fish/m^2 produced significantly different from the broodstock with the stocking number of 15 fish/m^2 and 10 fish/m^2. The rematuration time of the male broodstock catfish is about two weeks, while the female broodstock is 1.5 months.

3.2 Fecundity

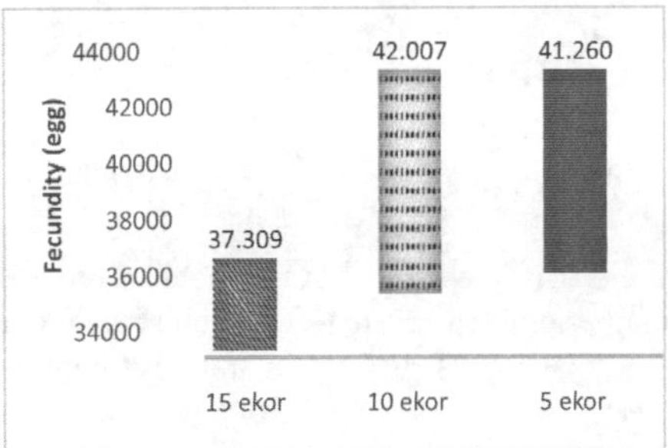

Figure 3. Graph of the mean fecundity of broodstock (eggs produced per brood during the study).

In Figure 3 it can be seen that the stocking number 10 fish/m^2 resulted in the highest mean fecundity value per brood compared to the number of 5 and 15 fish/m^2. It is known that the number of mature gonads is not always accompanied by the number of fecundity. The eggs formed are also influenced by the activity of prostaglandins. More and more vitellogenin is carried to the gonads, which will increase the number of eggs formed in the gonads [16]. It was further explained that the increase in catfish fecundity was also influenced by the quality of the female broodstock and the amount of feed nutrients and the efficiency of their use. Broodstock female of catfish are about three years old or weigh at least 1.36kg can be expected to spawn 40.000-60.000 eggs [8].

Based on the ANOVA test, the number of broodstock stocked had an effect on the average value of fecundity (at $P<0.05$). Furthermore, based on the Tukey test, the stockings number of 15 fish/m^2 was significantly different from the stockings number 10 and 5 fish/m^2. While the fecundity of feed is also one of the causes of gonadal maturation. This is because the basic ingredients in the formation cells of egg and sperm come from the metabolism of feed given especially to female fish, this maturation process is known as the vitellogenesis process. The basic ingredients in the gonad maturation process consist of carbohydrates, fats and proteins.

3.3 Fertilization Rate

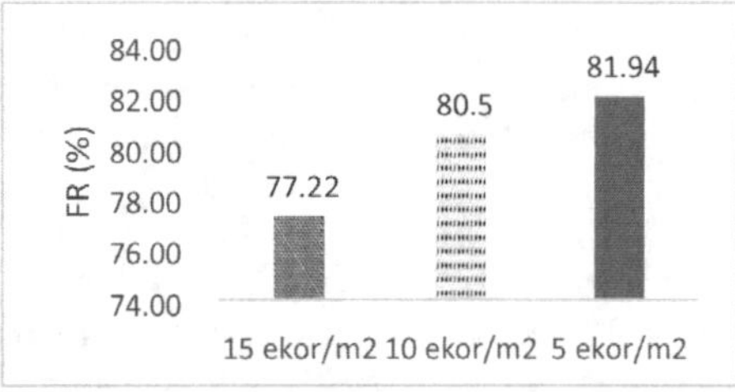

Figure 4.Graph of the percentage of fertilization rate in the three treatments.

Figure 4, showed that the average percentage of fertilized eggs per brood (FR) was highest at stocking number 5 fish/m2. While at 15 fish/m2, the lowest percentage of fertilized eggs was (77.22%). Based on the ANOVA test, the number of broodstock tested affects the percentage of fertilization rate.

3.4 Hatching Rate

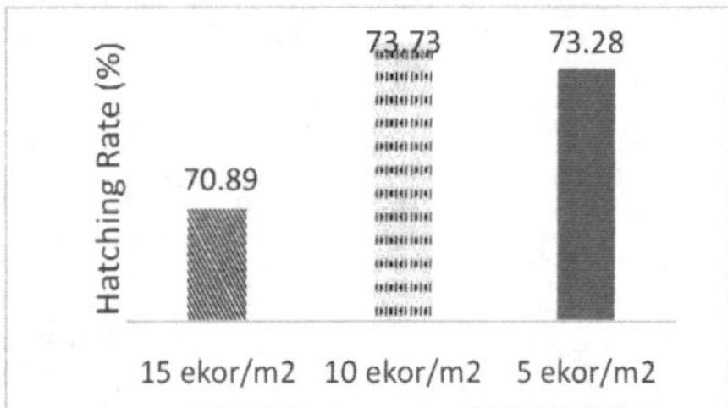

Figure 5.Graph of the percentage of the average hatching rate in the three treatments.

However, in Figure 5 it is known that the number of broodstock stocked in the rearing tank does not affect the number of hatcing rate (HR). In the pond or lake, catfish will not breed if fish do not have an adequate spawning structure. an attractive habitat for catfish to produce eggs and grow. Ovulation response and latency time the ovulation is a response or indicate where the fish has ovulated and is ready to spawn the eggs [14].

3.5 Performance of catfish broodstock

Table 1. Performance of catfish broodstock reared at different densities in recirculating tanks.

	Treatment		
	15 fish/m^2	10 fish/m^2	5 fish/m^2
Maturation gonadal of broodstock (fish)	61	30	8
Ratio maturation gonadal of female broodstock (%)	135,5	100	53
Number of eggs	2.239.220	1.292.812	331.480
Fecundity (egg/fish)	36.708	43.094	41.385
Diameter of egg (mm)	1,4 – 1,5	1,4 – 1,5	1,4 -1,5
FR average (%)	77.22	80.5	81.94

HR average(%)	70.89	73.73	73.28

In Table 1 it is known that the 15 fish/m^2 treatment showed the best broodstock performance results based on the number of mature gonads and the number of eggs produced, although the FR and HR values were lower than the 10 fish/m^2 and 5 fish/m^2 treatments. [1] state that a dense number of eggs can be caused due to the stickiness of the eggs where the eggs agglomerate together when released into the water, resulting in low fertilization and hatching rates. The gonadal maturation ratio of female broodstock in the 15 fish/m^2 treatment was also higher than the 10 fish/m^2 and 5 fish/m^2 treatments. During the observation period, all of female broodstock experienced gonadal maturity and 35% of the female broodstock in the 15 fish/m^2 treatment experienced a second gonadal maturity.

3.6 Water quality measurement

Table 2. The result of water quality measurement.

Parameters	Treatment		
	15 fish/m^2	10 fish/m^2	5 fish/m^2
pH	7 – 7.2	7 – 7.2	7 - 7.2
DO (mg l^{-1})	3.4 – 5.4	3.4 – 5.8	3.2 – 5.8
Themperature (°C)	27 - 29	27 – 29	27 – 29
Ammonia	0 - 0.25	0 – 0.25	0 - 0.25
nitrit	0 – 0.25	0 – 0.25	0 - 0.25

Based on the results of observations of water quality during the study, such as temperature, DO, and pH. The temperature ranges from 27 – 29°C. DO or dissolved oxygen ranged from 3,2 - 58 mg/l. The pH value ranges from 7 – 7.2. Observations of water quality are still within the feasible range for rearing broodstock, stating that the appropriate range of water quality for the maintenance of catfish is a temperature of 22–34 °C, for DO or dissolved oxygen >1 mg/l and for pH value between 6–9 [6]. The optimum range of water quality and tends to be the same is due to the broodstock being reared in a recirculation tanks.

4. Conclusion

The density broodstock of catfish in a concrete tank with a recirculation system has an effect on gonad maturity and the next gonadal maturation cycle, fecundity, number of fertilized eggs. However, on the number of eggs that hatched, the broodstock density tested had no effect. Broodstock that produces eggs more quickly does not necessarily produce high fecundity, fertilization rate and hatching rate as well. To optimize broodstock productivity, it is recommended that broodstock rearing with a stocking density of 15 fish/m^2 be developed. It is also necessary to study broodstock management related to feed to increase broodstock performance.

References

[1] Asraf, Amirul, Zainal A. Muchlisin, and Siti Azizah 2013 Removal of Eggs Adhesiveness of African catfish (*Clarias gariepinus*) at Different Concentrations of Urea Solution Aceh International Journal of Science and Technology,2 (3): 94-97 December 2013.

[2] Ataguba, Gabriel Arome, Victor Tosin Okomoda and Martins Chukwuemeka Onwuka 2013 Relationship Between Broodstock Weight Combination And Spawning Success In African Catfish (*Clarias gariepinus*) Corelation. Journal of fisheries, 2013, 71, 176-181

[3] Azizati, VL, Gung Sudaryono and Tristianya Yuniarti 2015 Pengaruh Penambahan Kombinasi Omegasqua Dan Klorofil Terhadap Fekunditas, Daya Tetas Dan Kelulushidupan Larva Ikan Lele Sangkuriang (*Clarias* sp.). Journal of Aquaculture Management and Technology Volume 4, Nomor 4, Halaman 136 - 140
[4] Brooks, S., Tyler, C.R. & Sumpter, J.P. 1997 Egg quality in fish: what makes a good egg Rev. Fish Biol. Fish., 7, 387–416
[5] Effendie, I 2002 Pengantar Akuakultur Penebar Swadaya, Jakarta.187 hlm
[6] Fahrian, AI 2017 Manajemen Kualitas Air Pada Kolam Pembesaran Calon Induk Ikan Lele Dengan Metode Akuaponik Di Balai Pengembangan Teknologi Perikanan Budidaya Cangkringan, Sleman, Di Yogyakarta. Fakultas Perikanan dan Kelautan. Surabaya.
[7] Goos, HJT dan Carel JJ Richter 1996 Internal And External Factors Controlling Reproduction In African Catfish, *Clarias gariepinus*. Aqual Living Resour. Vol 9. Hors serie, 45-58.
[8] Iswanto Bambang, Rommy Suprapto, Huria Marnis, dan Imron 2016 Performa Reproduksi Ikan Lele Mutiara (*Clarias gariepinus*) Media Akuakultur, 11 (1), 2016
[9] Mastego LO.,La Onu La Ola, dan Akhmad Mansyur 2020 Kajian Produktivitas Usaha Budidaya Ikan Lele Dumbo Jani Jaya Di Desa Lambusa Kecamatan Konda Kabupaten Konawe Selatan J. Sosial Ekonomi Perikanan FPIK UHO, ISSN 2502-664X: 5(2) Mei 2020.
[10] Muchlisin, Z.A. Damhoerim, A. Fauziah, R. Muhammadar. dan Musman, M. 2003. Pengaruh Beberapa Jenis Pakan Alami terhadap Pertumbuhan dan Kelulushidupan Larva Ikan Lele Dumbo (*Clarias gariepinus*). Universitas Syiah Kuala. Banda Aceh. ISSN 08537240. Biologi, 3 (2) : 105-113.
[11] Naggar, Gamal O. El; George John, Mahmoud A. Rezk, Waheed Elwan, and Mohammed Yehia. 2006. Effect of varying density and water level on the spawning response of African catfish *Clarias gariepinus*: Implications for seed production. Aquaculture Volume 261, Issue 3, 1 December 2006, Pages 904-907
[12] Nopras, YF., Ricky Hjauhari dan Irawadi Gunawan. 2020. Kajian Padat Tebar Pada Teknologi Bioflok Skala Laboratorium. Prosiding Seminar Nasional Perikanan dan Kelautan ISSN 2655-8947 Fakultas Perikanan Dan Kelautan Universitas Lambung Mangkurat.
[13] Ridho, Ari., Usman M,Tang, Iskandar Putra. 2014. The Growth of Catfish (*Clarias gariepinus*) with aquaponic system. Jurnal Online Mahasiswa (JOM) Bidang Perikanan dan Ilmu Kelautan.
[14] Salim, R., Farid Mudlofar, Suparmin, Rizal Hutagalung, Samila dan M. Taufik. 2020. Effectiveness Of Alternative Feed Accelerated Maturity of Male Gonads Of Sangkuriang Catfish (*Clarias* sp). Journal of Aquaculture Development and Environment Volume 3 No 1 Maret 2020. 117
[15] Tasyah, Novira Nanda, Mugi Mulyono, Moch Farchan, Amyda Suryati Panjaitan and Effi Aftiani Thaib. 2020. Performa Budidaya Ikan Lele Sangkuriang (*Clarias Gariepinus*) Sistem Bioflok Dengan Intervensi Grading. Jurnal Agroqua Volume 18 No. 2
[16] Tondang,H. Rita Rosita, Lintang Permata Sari Yulandi dan Ujang Subhan. 2019. Pematangan Gonad Ikan Lele Dumbo (*Clarias gariepinus*) Menggunakan Tepung Biji Kecipir.
[17] Wibowo Trio, Netti Aryani and Nuraini. Konsentrasi Larutan NaCl pada Semen Terhadap Fertilisasi dan Penetasan Telur Ikan Lele Sangkuriang (*Clarias gariepinus* var Sangkuriang). Jurnal Ilmu Perairan (Aquatic Science) Volume 10 No.1 (2022): 42-47

Towards a Customizable and Cost-Effective Monitoring System for Vannamei Shrimp Cultivation

Aynuddin[1*], B Triyatmo[2], and A Rakhmanda[3]

[1] Industrial Waste Treatment Department, AKA Bogor Polytechnic, Bogor, Indonesia
[2] Aquaculture Department, Gadjah Mada University, Yogyakarta, Indonesia
[3] PT. Delos Teknologi Maritim Jaya, Jakarta, Indonesia

*Corresponding author mail: aynuddin@apps.aka.ac.id

Abstract. The advancement of technology has revolutionized the operational practices in vannamei shrimp (L. vannamei) cultivation, shifting towards a science-based and data-driven approach. The success of shrimp cultivation heavily relies on making informed technical decisions throughout the process. To facilitate optimal decision-making, an integrated monitoring system encompassing feed-growth data, water quality data, and material inputs is imperative. This research aims to design a customizable system that optimizes available resources and offers easily modifiable and adaptable program scripts for users. The development process involved constructing a prototype device, script development, and system implementation testing. The prototype incorporated three sensors (temperature, pH, and dissolved oxygen) and utilized floats for flexible customization based on specific needs and available sensors or materials. Data analysis and visualization were achieved through various free options. Remarkably, the system only required 2-3 concise scripts to operate effectively. The experimental results demonstrated the system's potential for broader application and future development. The proposed monitoring system provides a cost-effective solution, offering flexibility, reliability, and valuable data-driven insights to enhance decision-making in vannamei shrimp cultivation.

1. Introduction

Vannamei shrimp (*L. vannamei*) is a significant economic commodity in Indonesia which produced 892,000 tonnes in 2020, the fourth most producer in the world [1]. However, along with the economic gains, there also comes a high level of risk, particularly in the technical aspects of shrimp cultivation. Unstable water quality in shrimp ponds increases the risk of diseases and syndromes occurring, posing a threat to the overall success of the shrimp farming industry [2]. To mitigate these risks and enhance decision-making, one crucial approach is to monitor water quality levels effectively.

In Indonesia, intensive or large shrimp farms often have dedicated laboratory units to support the decision-making process of farm technicians. However, the manual labor involved in monitoring each pond's water quality multiple times a day for daily parameters (such as temperature, dissolved oxygen (DO), pH, and salinity), and every 3-7 days for biological (plankton, total vibrio count, and total bacterial count) and chemical parameters (ammonia, nitrite, nitrate, phosphate, and carbon) can be labor-intensive and time-consuming.

Real-time water quality monitoring system can reveal gaps in data [3] throughout the aquaculture process. The weakness of manually conducted monitoring lies in its grab technique, which may not accurately represent the actual conditions over a 24-hour period. On the other hand, a monitoring system utilizing sensors and logger technology fills these gaps by providing dynamic data reflecting current field conditions. The adoption this technology also enables denser data series collection with minimal labour consumption.

Hence, the adoption of technology to make the laboratory unit's work more efficient has become imperative. Informed technical decision-making based on data-driven insights plays a critical role in achieving successful shrimp cultivation [4]. Looking ahead, integrated databases that encompass fry data, feed-growth information, water quality parameters, material inputs, and records of modules, batches, and decisions will become even more important for sustainable shrimp farming.

The primary aim of this research is to design a customizable monitoring system that optimizes available resources and offers modifiable and adaptable program scripts for users. By developing an efficient and user-friendly system, we intend to enhance the precision and reliability of water quality monitoring, ultimately leading to improved decision-making in vannamei shrimp cultivation.

2. Methods

The research was conducted in four distinct phases: prototyping, script development, data visualization and analysis, and exploring a practical workflow for the pond monitoring system. The experiment was conducted in 2018, spanning one cultivation cycle (3 months), at ponds located in the southern coast of Garut Regency, West Java.

Three ponds with relatively similar size and treatments were selected as the installation sites for the monitoring equipment. In each pond, three sensors (temperature, DO, and pH) were installed, equipped with Arduino Mini and LoRa module for data transmission to the receiver (see Figure 1). The receiver subsequently sent the data to Google Cloud services for storage and analysis. Temperature, DO, and pH were monitored due to the importance of the parameter as pond water quality indicator.

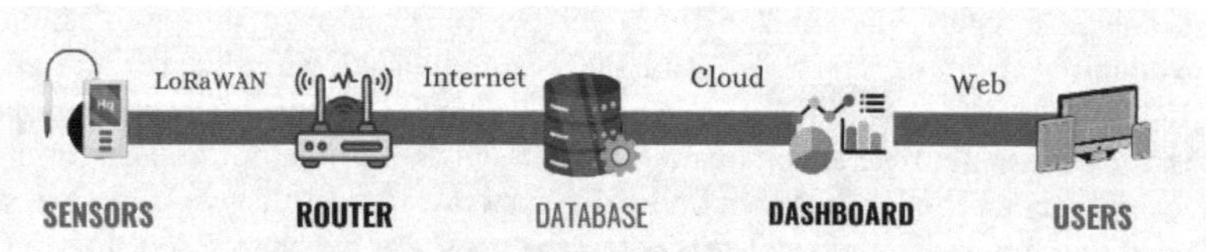

Figure 1. Monitoring system device.

2.1. Tools and Materials

The research utilized specific tools and materials to implement the monitoring system effectively [5]. For the microcontrollers, we employed Arduino technology, using three Arduino Nano boards for in-pond applications and one ESP32 board for data aggregation. Each Arduino Nano was equipped with three sets of sensors, including a temperature sensor, a Dissolved Oxygen (DO) sensor (DFROBOT SEN0237-A), and a pH sensor (SEN0161-V2). Additionally, a LoRa module was integrated into each Arduino Nano to enable efficient data transmission.

The ESP32 board, responsible for data aggregation, also featured a LoRa module for receiving the data transmitted by the in-pond Arduino Nanos. It was connected to a stable WiFi network, facilitating the seamless uploading of data to Google Sheets. Google Sheets was chosen as the platform for data storage and visualization due to its widespread familiarity and user-friendly interface. This choice allowed us to easily expand the monitoring record with additional columns or time-series data, making the dataset more comprehensive.

To ensure accurate data collection, the devices in the ponds were securely attached to floating devices [6] placed or anchored in the vicinity of the death zone. This strategic placement allowed us to capture crucial minimum values of each parameter, which is of utmost importance from the technical decision-making perspective. By leveraging these tools and materials, we created an efficient and customizable monitoring system that optimizes available resources and offers valuable insights for informed technical decisions in vannamei shrimp cultivation.

2.2. Research Phases

In the prototyping phase, we created a set of microcontrollers equipped with sensors and data transmission modules for each pond, referred to as "slaves." Additionally, we established another microcontroller with a data receiving module and a WiFi module to aggregate data and transmit it to the internet, referred to as the "master."

Three types of scripts were written to support different functionalities. The first script was developed to program the slave microcontrollers, while the second script was designed for the master microcontroller. The third script was developed for data visualization purposes. For data visualization, we opted to utilize Google Datastudio and RStudio for data analysis, allowing us to create a comprehensive and user-friendly dashboard.

During the research, we also explored other potential options that would offer ease of customization and high practicality, especially for shrimp cultivation businesses that typically lack IT support [7]. By implementing these methods, we aimed to develop a customizable and efficient pond monitoring system, providing valuable insights to support informed technical decisions in shrimp cultivation while being accessible to a broad range of users in the industry.

3. Result and Discussion

3.1. Device Prototype

The implementation of our monitoring system followed a straightforward workflow. Sensor data was transmitted via LoRa and organized into an array, which was then pushed to Google Sheets through Google AppScript (Figure 2a). The data was made accessible to users through a dashboard link created using Google Datastudio. An advantage of using Google Sheets as the data storage platform was its collaborative feature. Different users, including those from the aquaculture division (providing feed-growth data), warehouse division (supplying supporting materials data), and laboratory division (providing weekly biological and chemical parameter reports), could modify the Google Sheet rows and columns without disrupting the automated data collection process.

(a) (b)

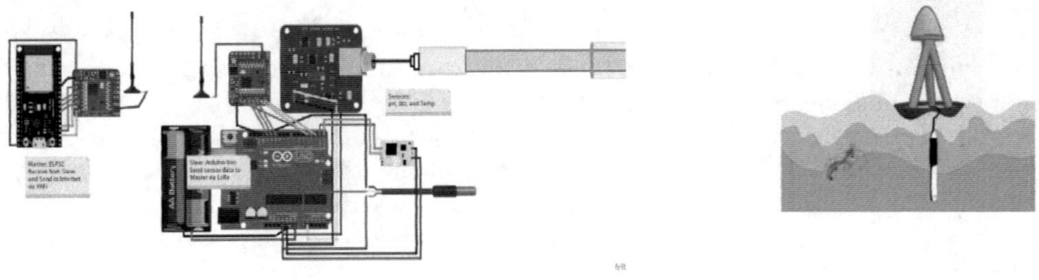

Figure 2. Device prototype: (a) MCU design, and (b) MCU with bouy in each pond.

To optimize the microcontroller unit (MCU) performance, we included additional components in the setup, such as a power source (lithium battery), antenna, buoy, and anchor (Figure 2b). These elements ensured the optimal functioning of the data logger. To maintain accurate sensor operation, regular maintenance of the slave devices was essential. Cleaning was recommended to performed biweekly or monthly, meanwhile calibration-verification were scheduled every 1-2 cultivation cycle.

The application of this technology significantly reduced the workload of the laboratory team by more than 60% due to the automation of daily tasks. This allowed laboratory personnel to focus on more critical aspects, such as chemical and biological parameter analysis. From an economic standpoint, the cost of three slave devices and one master device was approximately IDR 18-19 million, which is comparable to the cost of commercial Hanna or Hach DO-meters excluding the operator cost. It is worth noting that one master device could serve multiple slaves by efficiently scheduling data transmissions. The cost of the devices was primarily attributed to the sensors, accounting for more than 80% of the total expense. To strike a balance between sensor durability, accuracy level, and economic feasibility, careful consideration of the sensor requirements and budget constraints was crucial in the decision-making process.

3.2. Code and Script

The implementation of our monitoring system required two main parts of code. The first part involved programming the Arduino microcontrollers, both for the slave and master devices (Figure 3a and 3b). The second part of the code was a script responsible for pushing the collected data array to the Google Sheet (Figure 3c). In essence, the code for the Arduino was designed to read analog signals from the connected sensors, transmit this data to the master device using LoRa communication, and then collect and organize the data into an array. The master device, equipped with an appropriate script, would then send the data array to the Google Sheet using Google AppScript.

Figure 3. Program coding: (a) slave code, (b) master code, and (c) Google AppScript.

It is important to note that the code may require adjustments when using different hardware or sensors. The flexibility of our system allowed for easy modifications to accommodate various sensor types or hardware configurations. For instance, if we were using a WiFi-enabled MCU like the ESP32 as a slave, the code could be updated over the air (OTA) to incorporate new functionalities. This adaptability and programmable nature of our monitoring system offered versatility and ease of customization to suit the specific requirements and preferences of shrimp cultivation operations. It allowed users to employ different sensor types or even upgrade the system with new features without the need for extensive reprogramming.

3.3. Data Processing

Throughout the 90-day cultivation period, a comprehensive dataset was recorded in the Google Sheet, consisting of 22,159 rows of data, with a file size of approximately 910 kB (Figure 4). To enhance efficiency and loading speed, it is advisable to transform the database from its current long format into a wide format, with each parameter occupying a separate column.

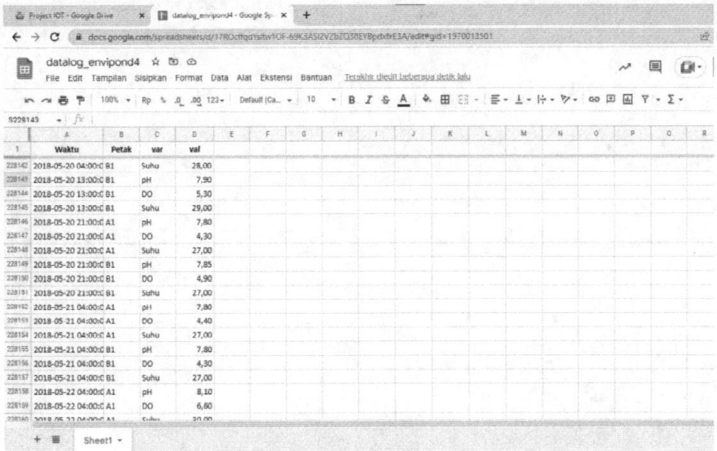

Figure 4. Long database format in Google Sheet.

Table 1. Data summary of pond's pH, DO, and temperature.

Pond	Parameter	Min	Max	Mean	SD
A1	PH (unit)	7.3	8.50	7.88	0.18
B1		7.4	8.60	7.90	0.17
C1		7.2	9.20	7.87	0.23
A1	DO (ppm)	4.0	7.1	5.55	0.55
B1		4.3	8.0	5.60	0.53
C1		3.9	7.0	5.28	0.53
A1	Temp. (°C)	26	34	29.35	1.82

Analyzing the sensor data (Table 1), we observed that the pH values ranged from 7.2 to 9.2, with an average value of 7.89 and a standard deviation of 0.19. The dissolved oxygen (DO) levels varied between 3.9 and 8.0, with an average of 5.34 and a standard deviation of 0.54. The temperature sensor readings showed fluctuations between 26 to 35 degrees Celsius, with an average value of 29.62 and a standard deviation of 1.74.

Through data visualization using RStudio, distinct patterns emerged from the water quality fluctuations during shrimp cultivation (Figure 5). Initially, in the early stage of cultivation, the water quality exhibited an unstable and irregular pattern. As the cultivation progressed, this irregularity transformed into a more stable fluctuation. However, with further aging of the culture, we observed a return to irregular water quality patterns. The insights gained from data processing and visualization shed light on the dynamic nature of water quality during shrimp cultivation. Understanding these fluctuations is crucial for making informed decisions and optimizing the overall cultivation process for higher productivity and efficiency.

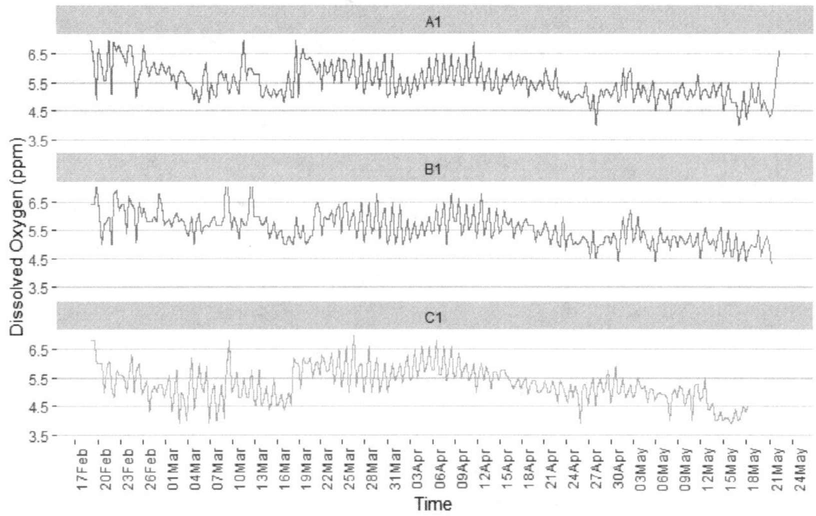

Figure 5. Timeseries data of Dissolved Oxygen of pond A1, B1, and C1.

3.4. More Options and Potential Workflow

While our basic workflow serves as a solid foundation, there are numerous options and potential enhancements available based on specific needs and available resources. The versatility of the system allows for customization to suit different requirements and budgets of a shrimp farm (Table 2). For instance, instead of using Arduino, one could opt for Raspberry Pi as the microcontroller, or choose different data transmission technologies like GSM or Zigbee, depending on the communication range and data transfer requirements. Additionally, the Google Sheet could be replaced with a dedicated server, offering more control over data management and security. Alternatively, utilizing Free IoT project slots can optimize the dashboard presentation and visualization. Despite the hardware and program changes, the underlying logic of the system remains consistent. The main focus is on monitoring and aggregating critical data to aid in decision-making and enhance shrimp cultivation practices.

Table 2. More possible options for every components.

MCU	Sensor	Data Transmission	Database	Dashboard	Analytics
• Arduino ESP32 • Raspberry Pi	• Standard grade • Industrial grade • Laboratory grade	• Radio • Zigbee • WiFi • LoRa • GSM	• Google sheet • OpenDB • Own server • Commercial	• Google datastudio • Open IoT Project • Own dashboard	• Python • R

Moving towards the future, we envision an evolved workflow that integrates an AI & Suggestion System (Figure 6). This advanced system would take advantage of GitHub for seamless code version updates and establish a community database to accelerate big data aggregation. Such an approach allows for collective insights and continuous improvement. Our proposal for future use involves partnering with the feed industry to integrate this community server and system as part of their technical support for shrimp farms using their feed products. By adopting this system, the feed industry can significantly improve their support to the farms, streamline operations, and develop an AI-powered suggestion system to assist technicians in making well-informed decisions.

Figure 6. Potential workflow for future development.

In conclusion, the potential workflow and further advancements hold great promise for the aquaculture industry, facilitating data-driven, efficient, and intelligent decision-making for enhanced shrimp cultivation practices.

4. Conclusion

Our integrated monitoring system, comprising microcontrollers, sensors, data transmission, database, dashboard, and data analytics, has proven to be highly promising and easily customizable. This system not only automates processes but also ensures accurate and comprehensive data recording, leading to valuable insights and informed technical decisions in shrimp cultivation.

The system's high flexibility offers a wide range of options, allowing for tailored adaptations based on specific needs and available resources. The seamless integration of various components provides a holistic view of water quality parameters and feed-growth data, empowering users to optimize decision-making throughout the cultivation process. As we move forward, we envision further advancements in sensor technology. The development of more affordable yet reliable sensors would significantly benefit the aquaculture industry, making it accessible to a broader range of farmers and further enhancing the effectiveness of our monitoring system.

In conclusion, our research has demonstrated the potential of data-driven and technology-driven approaches in improving shrimp cultivation practices. By harnessing the power of modern technology, we can usher in a new era of informed decision-making, increased efficiency, and ultimately, sustainable and profitable shrimp cultivation.

References

[1] The State of World Fisheries and Aquaculture 2022 [Internet]. FAO 2022 [cited 2023 Aug 6]. Available from: http://www.fao.org/documents/card/en/c/cc0461en

[2] Emerenciano MGC, Rombenso AN, Vieira FDN, Martins MA, Coman GJ, Truong HH 2022 Intensification of Penaeid Shrimp Culture: An Applied Review of Advances in Production Systems, Nutrition and Breeding *Animals* **12**(3):236.

[3] Das B, Jain PC 2017 Real-time water quality monitoring system using Internet of Things. In: 2017 International Conference on Computer, Communications and Electronics (Comptelix) *IEEE* p78–82. Available from: http://ieeexplore.ieee.org/document/8003942/

[4] Rahman A, Xi M, Dabrowski JJ, McCulloch J, Arnold S, Rana M, et al 2021 An integrated framework of sensing, machine learning, and augmented reality for aquaculture prawn farm management *Aquacultural Engineering* **95**:102192.

[5] Lin JY, Tsai HL, Lyu WH 2021 An Integrated Wireless Multi-Sensor System for Monitoring the Water Quality of Aquaculture *Sensors* **21**(24):8179.

[6] Lu HY, Cheng CY, Cheng SC, Cheng YH, Lo WC, Jiang WL, et al 2022 A Low-Cost AI Buoy System for Monitoring Water Quality at Offshore Aquaculture Cages *Sensors* **22**(11):4078.

[7] Ulhaq I, Pham NTA, Le V, Pham HC, Le TC 2022 Factors influencing intention to adopt ICT among intensive shrimp farmers. *Aquaculture* **547**: 737407

Gelling characteristics of fish surimi gel added with calcium hydroxyapatite (HAp)

M R Hasan[1,2], **C A C Abdullah**[3], **N A Mustapha**[1], **M S M Ghazali**[4], and **M Noranizan**[1*]

[1]Department of Food Technology, Faculty of Food Science and Technology, Universiti Putra Malaysia, 43400 Serdang, Selangor, Malaysia
[2]Faculty of Fisheries and Food Science, Universiti Malaysia Terengganu, 21030 Kuala Nerus, Terengganu, Malaysia
[3]Nanomaterial Synthesis and Characterization Laboratory (NSCL), Institute of Nanoscience and Nanotechnology, Universiti Putra Malaysia, 43400 Serdang, Selangor, Malaysia
[4]Faculty of Science and Marine Environment, Universiti Malaysia Terengganu, 21030 Kuala Nerus, Terengganu, Malaysia

*Corresponding author email: noraadzahan@upm.edu.my

Abstract
Surimi is widely used as the main ingredient for various innovative seafood products (ISP), e.g., imitation crab meat, fish balls, fish sausage, kamaboko, fish sticks, and paupiette. Sodium chloride (NaCl), a primary ingredient of surimi, plays an important role in gelling and flavor. A reduced-NaCl product is preferred by consumers due to health-related reasons. This study focuses on evaluating the gelling characteristics of fish surimi gel formulated with different levels (0.5, 1, and 1.5%) of calcium hydroxyapatite (HAp). Surimi gel added with HAp showed a good gel-forming ability ($p<0.05$). The texture profile analysis indicated an acceptable value. As for the water-holding capacity (76.88%), the highest value ($p<0.05$) was recorded for 1% of HAp addition. On the contrary, the cooking loss (21.09%), expressible moisture content (23.48%), and whiteness index (70.20) for samples with 1% HAp were the lowest compared to other samples. This preliminary finding suggests that the addition of calcium HAp at 0.5 to 1% could be an alternative ingredient to NaCl for maintaining gelling characteristics of surimi. It also adds nutritional value to the product. This finding could be a reference to the industry in improving the processing of surimi-based products.

1. Introduction
Surimi, a concentrated fish myofibrillar protein, is widely used as the main ingredient for various seafood products including imitation crab meat, fish balls, fish sausage, kamaboko, fish sticks, and paupiette [1]. The recent increasing demand for surimi-based seafood products is attributable to its innovative characteristics *i.e.*, low fat, low cholesterol, and high nutritional and textural characteristics [2]. Global production and market value of surimi surged to 820, 000 million tons and $4.06 billion in 2018, respectively [1].

Salt generally contributes to flavor and textural alterations in surimi products. Reducing salt content for nutritional purposes may affect the characteristics of surimi products. The water-holding capacity, gelation ability and whiteness of surimi indicate the gel value [3]. The gelation process is responsible for the textural quality, which is primarily regulated by sodium chloride salt (NaCl) content. Gel attributes of surimi goods are governed by NaCl (>3%), which is liable for myosin solubilization of myofibrillar protein [4]. However, low-sodium products are preferred by many consumers for lessening the chance of cardiovascular disease [4]. Sodium intake in a high proportion leads to various risks of cardiovascular diseases (CVD) *i.e.,* high blood pressure and hypertension [5]. Sodium intake from 3 to 5 g, increases the blood pressure from 0.74 to 2.58 mm Hg, respectively [6]. The sodium intake <2.3 g per day is therefore recommended for maintaining a stable blood pressure [7]. The development of low-salt surimi products (< 2.5% NaCl) with similar quality has recently attracted many researchers via processing approaches and alternative-salt materials [8].

The incorporation of various additive ingredients was recently reported on surimi gel, *i.e.*, protein hydrolysate [9, 10], inulin [11], transglutaminase [11, 12], sodium bicarbonate [13], Micro-algae [14, 15], hydrocolloids [16], fish oils [17], coconut husk extract [12, 18], cellulose nanofibrils [1], fibers [3, 19-21], fatty acid [22], camel milk [23], vegetable oils [24] and amino acids [5]. Some other studies have reported on the addition of calcium from fish bone for textural improvement of silver carp fish sausage [25] and sardine (*sardinella fimbriata*) surimi gel [26] under normal-salt condition. However, additive-protein interaction change as different calcium and fish species is formulated, particularly under reduced-salt condition.

Calcium hydroxyapatite (HAp) was applied extensively for food, pharmaceutical, cosmetic, as well as dentistry [27]. It is a naturally occurring mineral in the form of calcium phosphate with a calcium-to-phosphorus stoichiometric ratio of 1.67 [28]. Apart from providing nutritional value, it has been successfully employed in improving many products due to its bioactivity and biocompatibility [27]. This calcium-based material has the potential to increase the quality characteristics of the surimi-based product during the gel formation process through filling and inducing cross-linking principles [30]. Calcium addition was reported to increase the overall acceptability of surimi gel, with no effect on flavor [26]. The recommended dietary allowance (RDA) of calcium-based food fortification ranges from 0.4 to 2.5 g [30].

Gel formation (setting) of surimi gel is normally done by incubating at 25 to 40°C, depending on the species of fish [31]. The surimi gelation process is thermo-irreversible with several chemical bonds that involve connecting the myofibrillar protein molecules [17]. Calcium addition could enhance endogenous transglutaminase (TGase) activity, responsible for catalyzing cross-links, *i.e.*, epsilon-(gamma-glutamyl) lysine bridge, between myofibrillar protein molecules via non-disulfide covalent [31], hence altering the textural characteristics of surimi gel. The rate of the TGase enzyme depends on the available amount of Ca^{2+} ions. Hence increasing the level of the aforesaid ions may provide a desirable space for its activity, leading to gel quality improvement. Snappers (*Lutjanus sp.*) fish are among the vital categories of tropical marine fishes for aquaculture, owing to its well-growth and increasing market demand [31]. The composition of the fish approximately contains 15-22% of protein, 73-75% moisture, 1-5% of fats, and 1-1.8% ash [32]. It is commonly harvested in Southeast Asia as a by-catch in large quantities. Due to its relatively thick skin, the fish is rarely consumed as a fresh product, and thus preferable for surimi production. There was no study found reporting on the surimi gel improvement, through calcium addition. The gel-forming ability of red snapper fish surimi gel under reduced-salt condition was investigated. The gels were analyzed for whiteness, water-holding capacity (WHC), cooking loss rate (CLR), expressible moisture content (EMC), and texture profile analysis.

2. Materials and methods

2.1 Materials

Snapper fish (*Lutjanus sp.*) surimi (grade AA) was obtained from Manja Food Sdn. Bhd, Selangor, Malaysia. Sodium chloride (NaCl) was acquired from a local supplier (Hunka Trading Sdn. Bhd., Taiping, Perak). Pure hydroxyapatite ($Ca_{10}(PO_4)_6(OH)_2$) with 99% purity was purchased from a supplier (Thermo Fisher Scientific Inc., Massachusetts, USA).

2.2 *Surimi gel preparation*

The gel preparation was done according to the technique of Yu et al. [33] with slight modifications. The frozen block of surimi was then placed at 4±1 °C for 24 h before cutting into fragments. The surimi was chopped for 1 min using a blender mixer (Panasonic MX-800S, Panasonic Malaysia Sdn. Bhd., Selangor, Malaysia). The NaCl was added at 2.4% (w/w) of the prepared gel formulation. An adjustment of the moisture to 80% (v/w) was made with a cooled water addition. The mixture was maintained below 10°C, whereby a thermometer (Taylor BI-THERM®, Taylor Scientific, USA) was used for temperature measurement. Further chopping was done for 3 mins before being filled into a cattle bone collagen casing (2.5 cm in diameter). A 30-minute incubation was conducted at 40±3 °C, with a subsequent 20-minute heating using water bath Iso temp 220 (Thermo Fisher Scientific Inc., Massachusetts, USA). An iced water was used for cooling (30 mins), and the gels were stored in chiller condition (4±1 °C) for 24 h.

2.3 *Addition of calcium hydroxyapatite in surimi gel*

The NaCl and hydroxyapatite (HAp) were added according to the previously described range [25], which was divided into control (0% HAp, 2.4% NaCl), and treatment groups, *i.e.*, SHAp-0.5 (0.5% HAp, 2.4% NaCl), SHAp-1 (1% HAp, 2.4% NaCl), and SHAp-1.5 (1.5% HAp, 2.4% NaCl). These concentrations are also within the RDA value [29].

2.4 *Water holding capacity*

The water-holding capacity (WHC) was evaluated following to technique by Piao et al. [1]. The gel was cut into smaller sizes (2.5 cm x 1.5 cm) using an adjustable cutter and weighed (W1). Two filter papers were used Whatman® No.1 (Whatman International, Ltd., Maidstone, England) and the gel was placed in between, before being mounted in centrifuge tubes (50 mL). A 20-mins centrifugation of the gels was conducted at 4±1°C, 8000 rpm, using micro refrigerated centrifuge Kubota 3740, (Kubota Corp., Tokyo, Japan), before reweighed (W2). WHC was estimated using the equation:

$$WHC\ (\%) = \frac{W2}{W1} \times 100\% \tag{1}$$

2.5 *Cooking Loss Rate (CLR)*

The cooking loss rate (CLR) was evaluated following the technique of Song et al. [24]. The gels with 2.5 cm of diameter surimi gel were weighed (*G*1). The gels were put in a polyethylene zip-lock bag and sealed. A 20-mins cooking was done (90±2°C) to the gels by placing in a water bath Isotemp 220 (Thermo Fisher Scientific Inc., Massachusetts, USA). The gel was dried with Whatman® No.1 (Whatman International, Ltd., Maidstone, England), and the final weight of the gels were taken (*G*2). The CLR was estimated using the equation:

$$CLR\ (\%) = \frac{G1-G2}{G1} \times 100\% \tag{2}$$

2.6 *Expressible moisture content*

A previously described technique was used to evaluate the expressible moisture content (EMC) of the gels [12]. The gels with 2.5 cm of diameter were then weighed (*X*) and put between filter papers Whatman® No.1 (Whatman International, Ltd., Maidstone, England), *i.e.*, 3:4 top-bottom ratio. A standard 5 kg of weight was used with a 2-minute measurement. The final weight was taken (*Y*). EMC was estimated using the equation:

$$EMC\ (\%) = \frac{X-Y}{X} \times 100\% \tag{3}$$

2.7 *Whiteness Index*

The whiteness index (WI) of the gels was analyzed following the technique by Buda et al. [3]. The color parameters, *i.e.*, L* (black/white), a* (red/ green) and b* (yellow/blue), of the gel were determined using a colorimeter (Konica Minolta CR-400, Konica Minolta Inc., Japan). The values were expressed in WI using the equation:

$$WI=100-\sqrt{(100-L*)^2+a*^2+b*^2} \quad (4)$$

2.8 *Texture profile*
The texture profile analysis was evaluated based on the technique described by an earlier study [12] with slight modification, using a TA. HD plus texture analyzer (Stable Micro System Ltd., Surrey, UK). A 2.5 cm height of gels and a 75-mm probe (cylinder) was used for the measurement.

2.9 *Statistical analysis*
The significant difference between treatments was evaluated using One-way Analysis of Variance (ANOVA). The significant level was set at $P < 0.05$.

3. Results and discussion
3.1 *Water-holding capacity of surimi gel*
The influence of hydroxyapatite (HAp) on water holding capacity (WHC) was investigated. The WHC indicates the ability of the myofibrillar protein to imbed moisture through gel network structure, a common tool for evaluating the quality of gel [3]. It is an imperative attribute in altering the gel quality, by affecting various organoleptic and consumer acceptance rates [1]. The WHC value of the gel is shown in **Figure 1**.

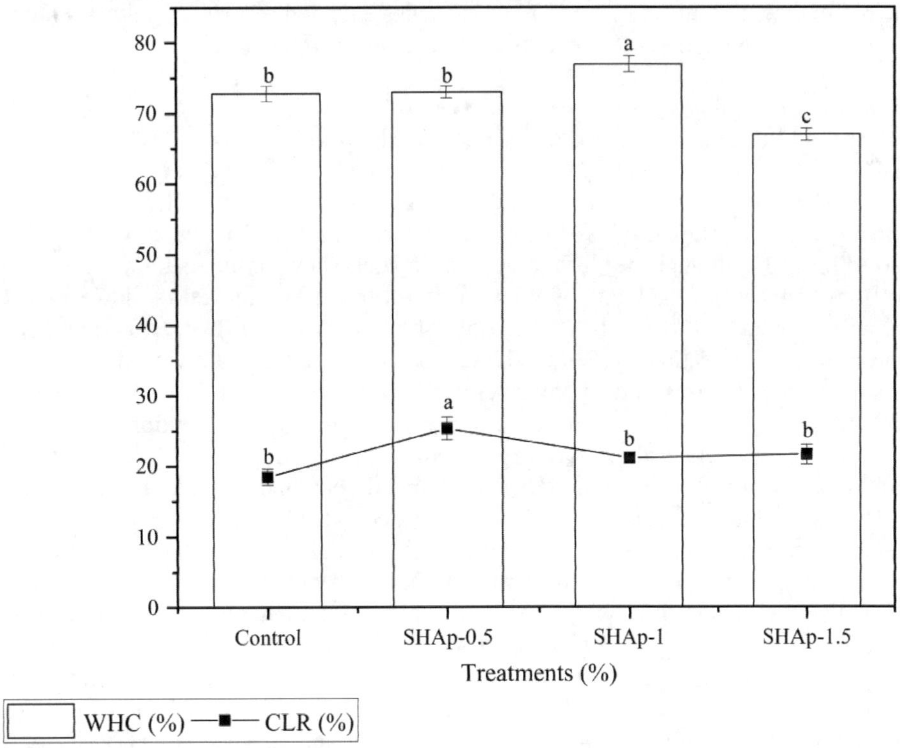

Figure 1. Water-holding capacity and cooking loss rate values of gel added with hydroxyapatite at different concentrations. Each point indicates the mean ± standard deviation. Different lowercase letters denote significant differences ($P < 0.05$) between treatments.

WHC of the gels increased with HAp concentration, then decreased at 1.5% of HAp. Control treatment showed the second lowest WHC (72.78%), with surimi gel containing the highest level of HAp (1.5%) showing the lowest WHC (66.96%). The water in both treatments is unstable.[1]. Adding 1% HAp to surimi gels increased the WHC by 76.88%, which wasn't observed in other treatments ($P < 0.05$). This can be attributed to the ability of HAp to effectively trap moisture within the matrix, thereby enhancing the overall stability of the system. [3]. The higher value of WHC for 1% of HAp addition could also be attributed to a high specific surface area and absorption capacity of the HAp particles. Wei et al. [20] previously observed that the curdlan incorporation escalated the WHC of Alaskan Pollock surimi, owing to the ability of the polysaccharide to absorb free water when heated, and entraps into the gel through interactions; hydrogen bonds, hydration, and hydrophobic. When the highest concentration of HAp (1.5%) was added to the surimi gel, the WHC value considerably decreased to 66.96% ($P < 0.05$). This can be attributed to the excessive level of HAp, separating the myofibrillar protein network structure, hence degrading the aggregation of the myofibrillar protein [3]. The value of WHC found in this study was slightly lower than other earlier studies [3, 1] for treatment groups, however, a comparable value was obtained for the control (70-77%). Singh et al. [13] observed that the sodium bicarbonate (SB) increased the WHC of Asian bass minced fish gel, due to protein denaturation by-product, hence multi-layer of water was entrapped. They added that the addition of SB at an appropriate level (0.1%) sufficiently develops a gel structure to absorb moisture effectively, facilitated by the high pH of the SB, hence increasing ionic interactions between charged protein chains and water molecules. The high pH value of HAp in this study could also contribute to the high WHC. Higher pH exposes the functional groups for the endogenous transglutaminase in the myofibrillar protein, enhances interactions and aggregation, and thus increases gelling properties [13]. This result is matched with the cooking loss rate (CLR) of surimi gel (**Figure 1**), showing that the treated gel with 1% HAp addition showed the lowest CLR value compared to other treatments. Therefore, it is suggested that the WHC of gel was enhanced by HAp, with 1% of HAp showing the highest value of WHC, hence stronger gel.

3.2 Cooking loss rate of surimi gel

The influence of HAp on the gels was investigated based on the cooking loss rate (CLR). The CLR indicates the quality of the surimi gel network structure based on water loss during the heating and cooking process. It is closely associated with the WHC in which, the high WHC value of surimi gel is prone to have a low value of CLR, hence less loss of water during heating. The values of the CLR of the gel are shown in **Figure 1**. Although the CLR value of surimi gels showed fluctuation with increasing HAp addition, it is consistent with the value of WHC. Control treatment showed the lowest CLR of 18.44%, with surimi gel containing HAp at 1% and 1.5% showing lower CLR of 21.09 and 21.60%, respectively compared to other treatments ($P < 0.05$), suggesting that a three-dimensional structure is formed through various cross-linking, induced by calcium HAp. The low value of CLR could also be attributed to the high absorptivity of calcium HAp, hence facilitating the prevention of water loss through a protein-calcium network system. Surimi gels with 0.5% addition showed the highest ($P < 0.05$) CLR (25.34%), showing that the calcium HAp addition aided in binding water in the surimi gel network system. The high value of CLR (0.5% HAp) might be due to the lower absorptivity of moisture in the protein-calcium structure. Gao et al. [5], stated that the higher value of CLR causes more components, *i.e.*, organic component and moisture, to be removed, hence the low capability of the gel in regulating immobilized water during cooking. The value of CLR found in this study (18 to 25%) was higher than in a recent study (9 to 13%) conducted by Song et al. [24]. However, it was lower than that was found by Gao et al. [5] who reported the CLR value of Amour sturgeon surimi gel under reduced-salt condition ranged from 6 to 42%. Adding 1% HAp to surimi gel improved its CLR characteristics ($P < 0.05$), and aligned with WHC (**Figure 1**). Therefore, the CLR of gel was maintained by HAp, with 1% of HAp showing the lowest value of CLR, better entrapment, and network stability, thus reducing loss during cooking.

3.3 Expressible moisture content of surimi gel

Further investigation on the gelling attributes was done on the expressible moisture content (EMC). The EMC measures the capacity of the protein network to imbibe moisture in which, a three-dimensional gel structure formed during the setting has a high capability of holding water [12, 13]. The values of the EMC of the gel are presented in **Figure 2**.

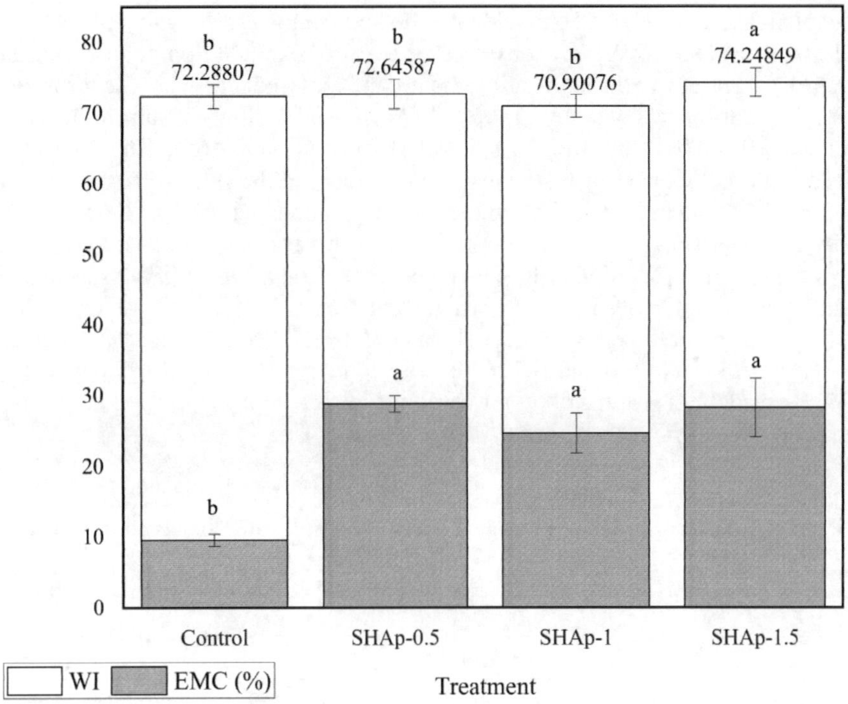

Figure 2. Whiteness index and expressible moisture content of the gels. Each point indicates the mean ± standard deviation. Different lowercase letters denote significant differences ($P < 0.05$) between treatments.

EMC value of surimi gels showed a fluctuation pattern but agreed with CLR and WHC analyses (**Figure 1**). Control treatment showed the lowest EMC of 9.55%, with surimi gel containing the HAp containing 0.5% and 1.5% showed a high value of EMC, indicating that the surimi gel has low capability in holding water. Surimi gels with 1% addition showed a lower value of EMC (23.48%) compared to other treatments ($P < 0.05$), indicating that HAp addition at 1% could retain more water within the structure. The low value of EMC in surimi gel with 1% HAp addition was attributed to the association of ionic interactions between water and charged protein chains at higher pH [13]. Dissociation of the actomyosin complex may occur via the negatively charged domain of the protein surface, thus providing space for water entrapment [13]. The value of EMC found in this study (9-29%) was comparable with the one observed by Sing et al. [12], where the EMC value of spotted golden goatfish surimi gel incorporated with ethanolic coconut husk extract and microbial transglutaminase ranged from 4.97 to 20.84%. Both additives (ethanolic coconut husk extract and transglutaminase) decreased the level of EMC, due to the highly ordered-structure gel network which absorbs more water, hence juicy texture with elasticity. This result is in line with CLR and WHC of surimi gel (Figure 1), showing that the treated surimi gel with 1% HAp addition showed the lowest EMC value compared to the treatment groups. Therefore, it is suggested that a sufficient level of HAp addition (1%) maintained the EMC of surimi gel, hence imbibing water with a stronger gel network.

3.4 *Whiteness index of surimi gel*

The WI indicates the gel value according to desired specifications [3]. It is usually influenced by the protein network structure and additive materials. The values of the WI of the surimi gels are shown in **Figure 2**. WI value of surimi gels showed a decreasing pattern when HAp was added before increasing at a level of 1.5 of HAp addition. 1% of HAp addition showed the lowest WI value of 70.90, followed by control (72.29). A low value of WI was observed on 1% of HAp addition and control due to the network change to a denser structure through various bonds and cross-linkage which absorbs light, hence expressing low WI [13]. Surimi gel containing HAp at 1.5% showed the highest value (74.25), markedly higher than the control ($P < 0.05$), followed by 0.5% of HAp addition (72.65). The high value of WI could be influenced by the color of HAp and the loosened structure of the gel. The finding here is in line with WHC, EMC, and CLR values which showed the surimi gel added with 1% of HAp exhibited good gel-forming ability characteristics. The value of WI found in this study (70 to 74) was slightly lower than a study by Mi et al. [22] observed that the WI of gel from *Nemipterus virgatus* surimi added with various starch-fatty acid complexes showed values ranged from 75 to 77. However, the WI value reported in this study was found to be higher than Buda et al. [3] and comparable with Sing et al. [12] who reported to range from (56 to 65) and (67 to 76), respectively. Therefore, it is suggested that the surimi gel with 1% HAp addition is at the optimum level due to an acceptable level of desired whiteness level with a good gel properties.

3.5 *Texture profile analysis and gel strength*

The textural attributes of the surimi gels were assessed and the results are displayed in **Table 1**.

Table 1. Texture characteristics and gel strength of surimi gel added with different concentrations of HAp.

	Control	SHAp-0.5	SHAp-1	SHAp-1.5
Hardness (g)	2632.81±757.87^{Aa}	3214.51±318.70^{Aa}	2390.52±212.32^{Aa}	2306.41±1015.36^{Aa}
Springiness	0.83±0.045^{Aa}	0.80±0.005^{Aa}	0.83±0.005^{Aa}	0.76±0.012^{Aa}
Cohesiveness	0.59±0.047^{Aa}	0.65±0.014^{Aa}	0.60±0.016^{Aa}	0.45±0.036^{Bb}
Gumminess (g)	1570.30±571.81^{Aa}	2077.28±159.69^{Aa}	1430.70±164.43^{Aa}	1048.62±536.27^{Aa}
Chewiness (g)	1283.08±400.87^{Aa}	1657.94±116.72^{Aa}	1180.60±127.98^{Aa}	800.81±420.28^{Aa}
Resilience	0.32±0.016^{Aa}	0.32±0.008^{Aa}	0.32±0.013^{Aa}	0.20±0.036^{Bb}

[a]Each point indicates the mean ± standard deviation. Different lowercase letters denote significant differences ($P < 0.05$) between treatments.
[b]Control: gels without hydroxyapatite; SHAp-0.5: gels added with 0.5% of hydroxyapatite; SHAp-1; gels added with 1% of hydroxyapatite; SHAp-1.5 represents surimi gels added with 1.5% of hydroxyapatite.

The hardness value usually indicates the protein structure of the gel. An increase in hardness was observed on the gel with 0.5% of HAp, however, a decreasing pattern was observed at 1 to 1.5% of HAp. The high value of hardness for 0.5% of HAp addition might be due to the appropriate level of HAp addition which acts as a filler that aids the gel in retaining more water, increases WHC with multi-layer water system in denser and network structure, hence better gel. This level (0.5% of HAp) could also trigger the formation of a denser network structure, due to cross-linking catalyzed by endogenous transglutaminase [30]. However, there was no change observed on other treatments ($P < 0.05$) from the control, suggesting that the addition of HAp (regardless of the level) promotes an acceptable gel network structure at reduced-salt conditions. The hardness value found in this study (2306 to 2632 g) was within the range of hardness reported by Buda et al. [3] which ranged from 1362 to 7438 g. Springiness measures the capability of the surimi gel to return to its initial form after the first compression [12].

There is no huge effect on springiness value except for 1.5% of HAp addition which exhibited a slightly lower value. This can be attributed to the loose network structure formed, disrupted by excessive filler (HAp) in the network system, hence diminishing the gel formation. This lower value of springiness for 1.5% of HAp addition is in line with the hardness. A similar trend to the result obtained by other earlier studies showed no significant effect of springiness on surimi gels added with different types of additives [13, 14, 1]. Cohesiveness indicates the energy needed for masticating a solid food [3]. A slightly increased cohesiveness value was observed with the addition of HAp at 0.5 to 1% level, then decreased at 1.5%. An analogous trend was obtained for gumminess and chewiness. Resilience measures the restoration capability of the gel to its initial form. No changes in resilience were observed for all treatments except for 1.5% of HAp addition which showed a slightly lower value (0.20). The low value of resilience is in line with other texture profile characteristics, suggesting that the excessive level of HAp addition (1.5%) could disrupt the protein-HAp network system, hence forming a loose network structure. Therefore, this result here indicates that the gel-forming ability and texture attributes of the gels were slightly improved with HAp at 0.5 to 1% level of incorporation. This value is under the minimum level of calcium-based food fortification (> 0.4 g) and maximum tolerance upper intake level (ULs) (2.5 to 3g), recommended by the United States Food and Nutrition Board of the Institute of Medicine [29].

4. Conclusion

The incorporation of hydroxyapatite (HAp) in the surimi formulation improved the gelling ability of the surimi under reduced-salt condition at a level of 0.5 to 1% of addition. This study showed that the addition of 1% HAp increased water-holding capacity compared to the control ($P < 0.05$). Texture profile analysis displayed that the addition of 0.5-1% of HAp addition improved the textural characteristics of the surimi gel, apart from providing additional nutritional value.

Acknowledgments

The authors would like to express an appreciation to the Faculty of Food Science and Technology (FSTM) for providing the necessary facilities to conduct the experiments. This work was financially supported by the Universiti Putra Malaysia under the grant with Project No. GPB/2018/9668400.

References

[1] Piao X, Li J, Zhao Y, Guo L, Zheng B, Zhou R, and Ostrikov K 2022 *Food Hydrocoll.,* **133** 1-12 https://doi.org/10.1016/j.foodhyd.2022.107893
[2] Zhao Y, Kong H, Zhang X, Hu X, and Wang M 2019 *Food Sci. Nutr* **7** 2083–2090 https://doi.org/10.1002/fsn3.1049
[3] Buda U, Priyadarshini M B, Majumdar R K, Mahanand S S, Patel A B, and Mehta N K 2021 Int. J. Biol. Macromol **175** 123–130 https://doi.org/10.1016/j.ijbiomac.2021.01.191
[4] Yuan L, Kong Y, Leng W, Wang Y, Jin W, and Gao R 2021 *Food Biosci.* **40** 1-8 https://doi.org/10.1016/j.fbio.2021.100886
[5] Gao R, Wang X, Shi T, Wijaya G Y A, Bai F, Wang J, Jin W, and Yuan L 2021 *J. Food Process. Preserv* **45** 1-11 https://doi.org/10.1111/jfpp.15887
[6] Mente A, O'Donnell M J, Rangarajan S, McQueen M J, Poirier P, Wielgosz A, Morrison H, Li W, Wang X, Di C, Mony P, Devanath A, Rosengren A, Oguz A, Zatonska K, Yusufali A H, Lopez-Jaramillo P, Avezum A, Ismail N, Lanas F, Puoane T, Diaz R, Kelishadi R, Iqbal R, Yusuf R, Chifamba J, Khatib R, Teo K, and Yusuf S 2014 *N. Engl. J. Med* **371** 601-611 https://doi.org/10.1038/s41598-021-83049-8
[7] World Health Organization 2016 https://www.who.int/en/news-room/fact-sheets/detail/salt-reduction.
[8] Zhu Y, Lu Y, Ye T, Jiang S, Lin L, and Lu J 2022 *Gels* **8** 1-12 https://doi.org/10.3390/gels8010010

[9] Hajfathalian M, Jorjani S, and Ghelichi S 2020 *J. Food Sci. Technol* **57** 1439–1448 https://doi.org/10.1007/s13197-019-04179-6
[10] Bharane P R, Bethi C M S, and Kudre T G 2020 *J. Food Meas. Charact* **14** 1391–1401 https://doi.org/10.1007/s11694-020-00389-1
[11] Huang J, Ye B, Wang W, Li J, Yi S, Li X, Xu Y, and Mi H 2020 *J. Food Meas. Charact* **15** 1–11 https://doi.org/10.1007/s11694-020-00604-z
[12] Singh A, Prabowo F F, Benjakul S, Pranoto Y, and Chantakun K 2020 *Food Hydrocoll.* **109** 1-10 https://doi.org/10.1007/s13197-022-05429-w
[13] Singh A, Buamard N, Zhou A, and Benjakul S 2022 *J. Food Sci. Technol* **59** 3109–3119 https://doi.org/10.1016/j.foodhyd.2020.106107
[14] Atitallah AB, Hentati,F, Dammak M, Hadrich B, Fendri I, Ayadi M-A, Michaud P, Abdelkafi S, and Barkallah M 2019 *Appl. Sci* **9** 1-12 https://doi.org/10.3390/app9091830
[15] Jannat-Alipour H, Rezaei M, Shabanpour B, Tabarsa M, and Rafipour F 2019 *J. Food Sci. Technol* **56** 3777–3789 https://doi.org/10.1007/s13197-019-03846-y
[16] Chen J, Deng T, Wang C, Mi H, Yi S, Li X, and Li J 2020 *J. Sci. Food Agric* **100** 2252–2260 https://doi.org/10.1002/jsfa.10254
[17] Yan W, Yin T, Xiong S, You J, Hu Y, and Huang Q 2021 *Int. J. Biol. Macromol* **191** 925–933 https://doi.org/10.1016/j.ijbiomac.2021.09.167
[18] Vate N K, and Benjakul S 2017 *Int. J. Food Eng* **13** 869 https://doi.org/10.1515/ijfe-2016-0163
[19] Gore S B, Xavier K M, Nayak B B, Tandale A T, and Balange A K 2022 *Bioact. Carbohydr. Diet. Fibre* **27** 100305 https://doi.org/10.1016/j.bcdf.2021.100305
[20] Wei Y, Zhang T, Yu F, Xue Y, Li Z, Wang Y, and Xue C 2018 *Int J Food Prop* **21** 1778–1788 https://doi.org/10.1080/10942912.2017.1306557
[21] Yan B, Jiao X, Zhu H, Wang Q, Huang J, Zhao J, Cao H, Zhou W, Zhang W, Ye W, Zhang H, and Fan D 2020 *Food Hydrocoll.* **105** 1-9 https://doi.org/10.1016/j.foodhyd.2020.105779
[22] Mi H, Su Q, Chen J, Yi S, Li X, and Li J 2021 *Food Chem* **362** 1-8 https://doi.org/10.1016/j.foodchem.2021.130253
[23] Eddin A S, Adegoke S C, Ibrahim S A, and Tahergorabi R 2021 *J. Aquat. Food Prod. Technol.* **30** 535–548 https://doi.org/10.1080/10498850.2021.1900970
[24] Song C, Lin Y, Hong P, Liu H, and Zhou C 2022 *Food Sci. Nutr* **10** 2935–2946 https://doi.org/10.1002/fsn3.2889
[25] Hemung B-O, Yongsawatdigul J, Chin K B, Limphirat W, and Siritapetawee J 2018 *J. Aquat. Food Prod. Technol.* **27** 305–315 https://doi.org/10.1080/10498850.2018.1432733
[26] Murthy L N, Phadke G G, Jeyakumari A, and Ravishankar C N 2021 *J. Food Sci. Technol* **58** 427–436 https://doi.org/10.1007/s13197-020-04551-x
[27] Hasan M R, Ghazali M S M, and Mohtar N F 2021 *J. Mech. Eng. Sci* **15** 7792–7806 https://doi.org/10.15282/jmes.15.1.2021.15.0615
[28] Jeong J, Kim J H, Shim J H, Hwang N S, and Heo C Y 2019 *Biomater. Res.* **23** 1-11 https://doi.org/10.1186/s40824-018-0149-3
[29] Food and Agriculture Organization of the United Nations 2006 *Guidelines on Food Fortification with Micronutrients* L Allen *et al* (France: WHO) pp 137-177
[30] Yin T, Reed Z H, and Park J W 2014 *LWT - Food Sci. Technol.* **58** 412–416 https://doi.org/10.1016/j.lwt.2014.03.037
[31] Velamala G R, Babu K R, and Naranji, M K 2017 *J. Appl. Ichthyol* **35** 815-817 https://doi.org/10.1111/jai.13854
[32] Ahmed I, Jan K, Fatma S, and Dawood M A O 2022 *J Anim Physiol Anim Nutr* **106** 690–719 https://doi.org/10.1111/jpn.13711

IOP Conf. Series: Earth and Environmental Science 1289 (2023) 012025 doi:10.1088/1755-1315/1289/1/012025

[33] Yu N, Xu Y, Jiang Q, and Xia W 2017 *Int J Food Prop* **4** 1–14 https://doi.org/10.1080/10942912.2016.1248846

IOP Conf. Series: Earth and Environmental Science 1289 (2023) 012026 doi:10.1088/1755-1315/1289/1/012026

Pathogenic bacteria contaminant and heavy metal content in peeled green mussels sold in the traditional market in Yogyakarta Indonesia: a case study in 2022

W. Suwito[1*], **Andriani**[2] and **G. R. Barokah**[1]

[1]Research Center Processing Food Technology, National Research Innovation Agency (BRIN), Gunungkidul 55861, Yogyakarta, Indonesia

[2]Veterinary Research Center, National Research Innovation Agency (BRIN), Bogor 16124, West Java, Indonesia

[*]Corresponding author email: widodo.suwito@yahoo.com

Abstract. Food is needed to prevent foodborne illness. This study aimed to detect pathogenic bacteria contaminants and heavy metal content in peeled green mussels sold in the traditional market of Yogyakarta, Indonesia. This research is a descriptive study with random sampling. A total of 33 samples examined *E. coli.*, and *Salmonella* sp., contamination based on the *most probable number* (MPN) method whereas heavy metal content by *atomic absorption spectrophotometry* (AAS). Contamination of *E. coli.*, and *Salmonella* sp., were 33.3% and 9.09% respectively, exceeded by SNI 7388:2009 limits of microbial in food contamination (MPN *Salmonella* sp., negative/25g and MPN *E. coli.*, <3/g). The levels of Pb and Hg in the food sample were found to be above the limit set by the Food and Drug Supervisory Agency of the Republic of Indonesia in 2018, which is greater than 0.5mg/kg. However, the levels of As and Cd were still within the limit allowed by the regulatory agency, which is less than 0.1mg/kg. It's important to ensure that food products meet the established safety standards to protect consumers' health and well-being. Peeled green mussels sold in traditional market in Yogyakarta, Indonesia contaminated with pathogenic bacteria and contains Pb and Hg metal.

1. Introduction

Currently, halal food and safe food are needed for human consumption. It has become a consumer demand in line with the increase of public awareness, and progress of the decade. Food with good taste, and high levels of vitamins and nutrients will become unsafe for consumption if contaminated with pathogenic bacteria or harmful chemicals. It shows that food safety is important in preparing food consumption. Food safety refers to the necessary measures taken to prevent food contamination by biological, chemical, or other harmful objects to ensure that it is safe for human consumption [1]. These efforts should not contradict any religious beliefs or cultural practices that may exist within a community [1]. It's important to be aware of the potential risks when it comes to food safety. There are three main types of hazards that can pose a threat: microbiological, chemical, and physical. Microbiological hazards are particularly concerning as they can include harmful bacteria, viruses, and parasites [2]. It's important to take steps to minimize these risks and ensure that food is safe to consume.

Peeled green mussels are fishery products with good taste, affordable prices, and a source of animal protein. I have come across some studies that indicate that green mussels may contain heavy metals and contaminated pathogenic bacteria [3, 4, 5]. This is definitely a cause for concern for our community. It is important to note that the heavy metals that are commonly found in fishery products include lead (Pb), mercury (Hg), cadmium (Cd), copper (Cu), and arsenic (As) [6]. It is crucial that we take the necessary precautions to ensure that we are not consuming any contaminated seafood. Reports show that the most common pathogenic bacterial contamination in fishery products are *E. coli* and *Salmonella* sp [7]. These bacteria are known to cause foodborne diseases [8].

Heavy metal is poisoning in humans and usually occurs chronically. Heavy metals can accumulate in the body and lead to degenerative diseases, including cancer. This has been established by scientific research [9]. Heavy metals enter the body through food and accumulate continuously over a long period of time, causing nervous system disorders, paralysis, premature death, and a decrease in children's intelligence levels [9]. It is quite harmful to human health [9].

Recently the symptoms of an *E. coli* infection can vary from person to person [10]. However, it seems that diarrhea is a common symptom that usually appears 3-4 days after exposure to *E. coli*. In addition, diarrhea an *E. coli* infection can also cause severe abdominal pain, cramps, nausea, vomiting, flatulence, loss of appetite, chills, fever, dizziness, and muscle aches. It's worth noting that *E. coli* strains, especially *E. coli* O157:H7, can lead to serious health problems like hemorrhagic colitis (HC), hemolytic uremic syndrome (HUS), and thrombocytopenia purpura (TPP) [11].

Salmonellosis is an infection caused by the bacteria *Salmonella* sp [12]. The symptoms of this condition include sudden fever, abdominal pain, diarrhea, nausea, and sometimes vomiting. Typically, these symptoms appear 6-72 hours after consuming *Salmonella* sp and can last from 2-7 days [12]. The symptoms of salmonellosis are usually mild, some patients, especially children and elderly patients, may experience severe dehydration, which can be life-threatening. Although large outbreaks of *Salmonella* sp., often get media attention, 60-80% of all cases are not recognized as part of a known outbreak and are classified as sporadic cases or are not diagnosed at all [12].

Peeled green mussels in Yogyakarta are sold at the traditional markets or itinerant vegetable vendors. The safety level of peeled green mussels is very low because the trader's place does not separate with chicken meat whereas chicken meat can be the source of *E. coli* and *Salmonella* sp. In addition, the water that uses chicken meat is also used to wash peeled green mussels. In addition, information about the source of peeled green mussels is not known with certainty because the sellers are only deposited by fish traders. Even though, several publications stated that green mussels are high in heavy metals contamination of peeled green mussels in the traditional market in Yogyakarta is limited information. Therefore, this study aimed to detect pathogenic bacteria contaminants and heavy metal content in peeled green mussels in the traditional market of Yogyakarta, Indonesia.

2. Materials and methods

2.1. Research Design

The research design is descriptive with an observational method conducted from January to February 2022 in the traditional market of Yogyakarta Indonesia. A total of 33 peeled green mussel samples were collected in the morning at 7:00-10:00 AM placed into an ice box and brought to the laboratory.

2.2. Three-tube Most Probable Number (MPN) *E. coli*

Three-tube MPN *E. coli* was performed by [13]. A total of 25 g samples were mixed using a stomacher machine for 10 minutes. One gram of sample was homogenized into 9 ml lactose broth (LB; CM:0137, Oxoid Ltd., Basingstoke, United Kingdom) medium to produce a 10^{-1}, 10^{-2}, 10^{-3} dilution and incubated at 37 °C aerobically overnight. All of the tubes were incubated for 48 hours at 37°C and checked for gas formation. Bacteria will ferment lactose and produce gas. If gas is not formed within 24 hours, incubation is continuing up to 48 hours. The test was said positive if gas in the tube which was indicated by the floating Durham tube. *E. coli* confirmation test was carried out on a positive tube by culture on eosin methylene blue agar medium (EMB; CM:0069, Oxoid Ltd., Basingstoke, United Kingdom), and then incubated at 37 °C aerobically overnight. Metallic sheen colonies on (EMB; CM:0069, Oxoid Ltd., Basingstoke, United Kingdom) medium suspected *E. coli* followed by biochemical tests [13].

2.3. Three-tube Most Probable Number (MPN) *Salmonella* sp

Three-tube MPN *Salmonella* sp., performed by [14, 15]. A total of 25 g sample was weighed and entered in 250 ml lactose broth (LB; CM:0137, Oxoid Ltd., Basingstoke, United Kingdom) and incubated at 37 °C aerobically overnight. After incubation, 1 ml, 0.1 ml, and 0.01 ml samples entered in 9 ml Rappaport Vassiliadis broth (RV; CM:0669, Oxoid Ltd., Basingstoke, United Kingdom) to produce a 10^{-1}, 10^{-2}, 10^{-3} dilution. All of the tubes were incubated at 37 °C aerobically overnight and checked for gas formation. If gas was not formed within 24 hours, incubation is continuing up to 48 hours. Gas in the tube is indicated by the floating Durham tube. *Salmonella* sp., confirmation test was carried out on a positive gas tube by culturing on xylose lysine deoxycholate agar (XLD; CM:0469b, Oxoid Ltd., Basingstoke, United Kingdom) and then incubated at 37 °C aerobically for overnight. Two typical *Salmonella* sp., colonies were streaked into triple sugar iron (TSI; CM:0277, Oxoid Ltd., Basingstoke, United Kingdom) slants and incubated at 37 °C aerobically overnight. Presumptive *Salmonella* sp., results from TSI (TSI; CM:0277B, Oxoid Ltd., Basingstoke, United Kingdom) slants were confirmed with a biochemical test. Total of *Salmonella* sp., estimated with MPN analysis (MPN/g) $=P/(NT)^{1/2}$, where P is number positive tubes, N is total quantity sample (in grams) in all negative tubes, and T is total quantity sample (in grams) in all tubes [15]. Because of values were small, they were multiplied by 100 to give the MPN per 100 g. The limit detection (1.2 MPN/100 g) was determined by calculating the value for an MPN of 0/3, 0/3, and 1/3 for the 25/g, 2.5/g, and 0.25/g samples, respectively [15].

2.4. Analysis of Heavy Metal

Firstly, the sample weighed about 500 grams. Heavy metal content in peeled green mussels was determined by using nitric acid (HNO_3) and perchloric acid ($HClO_4$) solution with the digestion method, while oxidized by acid and then was dissolved [16]. This process is called destruction. Standard and calibration solution prepared according to the HNO_3 and $HClO_4$ digestion methods. Analysis of heavy metal Pb, Hg, As, and Cd in samples was carried out in the laboratory using *atomic absorption spectrophotometry* (AAS) [16].

2.5. Data Analysis

All of data was analysed by descriptively. Data was calculated in absolute values and percentages using Microsoft Office Excel 2010 software and presented in the table.

3. Results and Discussion

Peeled green mussels in the traditional markets in Yogyakarta contaminated *E. coli* (33.33%) and *Salmonella* sp., (9.09%) were exceeded by [17] limits of microbial in SNI food contamination (Table 1). This result accordance with found [18] contaminant *E. coli* in the green mussels up to 100%, then lower in *Salmonella* sp. in Lorok fishpond, Semarang, green mussels contaminated *Salmonella* sp., as much as 30% [19]. Unlike the researcher [20] stated that green mussels from Tanjung Kait Tangerang, Banten were not contaminated *E. coli* and *Salmonella* sp. Difference results are caused by differences in sampling locations. Therefore, if the samples taken directly from the fishpond or waters have not been contaminated with domestic waste may be no detection of contamination pathogenic bacteria.

Table 1. *E. coli* and *Salmonella* sp., in peeled green mussels from the traditional markets in Yogyakarta.

		MPN			
		E. coli		*Salmonella* sp	
No	Source of sample	< 3 MPN/g	>3 MPN/g	negative/25 g	positive/25g
1	Sleman (n=9)	n=5	n=4*	n=7	n=2*
2	Bantul (n=6)	n=4	n=2*	n=6	-
3	Kulonprogo (n=5)	n=3	n=2*	n=5	-
4	Gunungkidul (n=6)	n=4	n=2*	n=6	-
5	Yogyakarta (n=7)	n=6	n=1*	n=6	n=1*
	Total (n=33)	n=22	n=11*	n=28	n=3*

*Exceed by SNI 7388:2009

Several studies have shown that mussels contaminated *E. coli* and *Salmonella* sp., although with different percentages [21, 22]. It is in line with [23] finding that green mussels are easily contaminated with pathogenic bacteria if they are not handled properly after being caught from fishponds. It was caused by domestic waste [21]. *E. coli* and *Salmonella* sp., are coming bacteria from animal or human feces. In addition, it caused by change in chemical-physical factors in water, so levels of *E. coli* and *Salmonella* sp., increase, especially during rains because of an increase in organic matter concentrations, temperature, and salinity [24].

E. coli is an indicator bacteria to see the level of hygiene of food products, while *Salmonella* sp., is a pathogenic bacteria whose presence in food must be zero. In this study, *E. coli* and *Salmonella* sp., contamination can occur because the traders do not separate peeled green mussels from chicken meat. Whereas, chicken meat can be the source of transmission *E. coli* and *Salmonella* sp., through feces. It is in line with [25] that *Salmonella* sp., and *E. coli* are two most critical food-borne pathogens of public health interest incriminated in poultry meat worldwide. In the traditional market in Surabaya, Indonesia shows that all of chicken meat samples were positive for *Salmonella* sp., (48.3%) and *E. coli* (40%) [26].

E. coli and *Salmonella* sp., contamination can also come from water that is used to wash peeled green mussels because their water is also used to wash chicken meat. Water which is used to wash chicken meat can be a source of *Salmonella* sp [25]. In chicken meat, *Salmonella* sp., contamination occurred during processing, facility's equipment (e.g., grinders, belts, and saws), food handler's contact (e.g., hand contact and knives), and exposure to other environmental sources (e.g., air and water) [25]. *E. coli* and *Salmonella* sp., contamination occurred due to poor sanitation, such as unclean selling points, placing green mussels that are not separated from chicken meat, and not using cooler box [25]. Place of washing chicken meat with ceramics causes water to retain and contaminate peeled green mussels and it called by cross-contamination. Therefore, sanitation places washing for traders selling peeled green mussels and chicken meat must be separated.

Peeled green mussels sold in the traditional markets in Yogyakarta contain Pb as much as 72.72% (>0.5 mg/kg) and Hg at 39.39% (>0.5 mg/kg), while As and Cd are still within the limits allowed by Food and Drug Supervisory Agency (BPOM) Republic Indonesia in 2018 [27] that is <0.1mg/kg (Table 2). The contents of Pb and Hg in green mussels from various waters and the traditional market in Indonesia have been reported [28, 29, 30, 31, 32].

Table 2. Level of Pb, Hg, As and Cd in peeled green mussels from the traditional markets in Yogyakarta.

		Pb (mg/kg)		Hg (mg/kg)		Cd (mg/kg)		As (mg/kg)	
No	Source of sample	<0.5	>0.5	<0.5	>0.5	<0.1	>0.1	<0.1	>0.1
1	Sleman (n=9)	2	7*	4	5*	9	-	9	-
2	Bantul (n=6)	1	5*	5	1*	6	-	6	-
3	Kulonprogo (n=5)	2	3*	4	1*	5	-	5	-
4	Gunungkidul (n=6)	1	5*	4	2*	6	-	6	-
5	Yogyakarta (n=7)	3	4*	3	4*	7	-	7	-
	Total	9	24*	20	13	33	-	33	-

* Limit of Food and Drug Supervisory Agency of the Republic of Indonesia in 2018 recommendation.

High levels of Pb and Hg in peeled green mussels caused by the fishpond or waters have been contaminated by waste from various industries. Pb and Hg come from industrial waste which is dumped into the sea even though a series of processes have been carried out, so the factory waste can be disposed in the sea [21]. Pb and Hg will be distributed in various components of the marine ecosystem for samples of green mussels, and then accumulate. Green mussels are a marine biota that functions to filter organic and inorganic materials [21]. Therefore, it is very possible that contains Pb and Hg. Cordova's research [32] showed that the content of Pb and Hg metals in green mussels from the Jakarta Gulf has a very high effect on pollution from Jakarta rivers. Hence, the source of green mussels is closely related to high levels of Pb and Hg. Source of peeled green mussels which are sold in the traditional markets in Yogyakarta is not certainly known, and high levels of Pb and Hg indicate its source from waters or fishpond has been contaminated by industrial waste.

4. Conclusions

Peeled green mussels sold in the traditional market in Yogyakarta are contaminated pathogenic bacteria with content levels of Pb and Hg over the limit recommended by the Food and Drug Supervisory Agency Republic Indonesia (BPOM) in 2018.

Acknowledgements

This research was not granted but was partially supported by a fishseller in Sleman Yogyakarta, Indonesia, Research Center Veterinary (PRV), National Research, and Innovation Agency (BRIN), Cibinong, Bogor, Indonesia. The authors are highly thankful to Mr Widada, Mr. Sri Widada, Mrs. Laksmi, Mrs. Niken Pawestri, and Mrs. Dyan are highly appreciated for their technical assistance in the laboratory.

References

[1]. Peraturan Pemerintah Republik Indonesia No. 86. Tahun 2019 tentang Keamanan Pangan.
[2]. Undang-Undang Republik Indonesia No. 18. Tahun 2012 tentang Pangan.
[3]. R. Yulistiani, E.A. Saputro, and D. Raharjo. 2021. Bacteria contamination and Cadmium heavy metal content of blood cockle (*Anadara granosa Linn*) satay on street vendors in Surabaya, Indonesia. International Conference on Science and Technology. *E3S Web of Conferences* **328**:1-13.
[4]. Y. Lin, J. Lu, and J. Wu. 2021. Heavy metals pollution and risk assessment in farmed scallops: low level of Cd in coastal water could lead to high-risk seafood. *Ecotox and Environ Safety* **208**:1-10.

[5]. R. Siahaan, J. Silalahi, and Z. Lubis. 2022. Analysis of heavy metals in shellfish in Tanjung Balai waters. *J Food Pharm Sci* **10**:610-619.

[6]. A. Yulyana, A.A.M.B. Hastuti, A. Rohman, B. Setiawan, F. Khasanah, and Irnawati. 2023. Heavy metal levels in fish products in Indonesia: a survey. *Food Res* **7**:74-84.

[7]. L. Sheng, and L. Wang. 2021. The microbial safety of fish and fish products: Recent advances in understanding its significance, contamination sources, and control strategies. *Compr Rev Food Sci Food Saf* **20**:738–786.

[8]. T. Bintsis. 2017. Review: foodborne pathogen. AIMS Microbiology. http: www.aimspress.com/journal/microbiology.

[9]. M. Jaishankar, T. Tseten, N. Anbalaga, Blessy B. Mathew, and K. N. Beeregowda. 2014. Toxicity, mechanism, and health effects of some heavy metals. *Interdiscip Toxicol* **7**: 60–72.

[10]. World Health Organization. 2018. *E. coli*. https://www.who.int/news-room/fact-sheets/detail/e-coli.

[11]. S.U. Chinyu, and L.J. Brandt. 1995. Review on *E. coli* O157:H7 infection in humans. *Ann Intern Med* **123**:698–707.

[12]. World Health Organization. 2018. *Salmonella (non-typhoidal)*. https://www.who.int/news-room/fact-sheets/detail/salmonella-(non-typhoidal).

[13]. Badan Standardisasi Nasional (BSN). 2015. SNI 2332.1:2015 Cara uji mikrobiologi-Bagian 1: Penentuan koliform dan *Escherichia coli* pada produk perikanan. Jakarta, Indonesia: BSN.

[14]. Badan Standardisasi Nasional (BSN). 2006. SNI 01-2332.2-2006 Cara uji mikrobiologi-Bagian 2: Penentuan *Salmonella sp* pada produk perikanan. Jakarta, Indonesia: BSN.

[15]. A. Bansal, T.M. Jones, J. Abd. Shirin, M.D. Danyluk, and L.J. Harris. 2010. Most-Probable-Number Determination of *Salmonella* Levels in Naturally Contaminated Raw Almonds Using Two Sample Preparation Methods. *J Food Prot* **73**:1986–1992.

[16]. Badan Standarisasi Nasional (BSN). 2021. SNI 2354-23:2021 Cara Uji Kimia-Bagian 21: Penentuan Kadar Logam Penentuan logam berat: merkuri (Hg), timbal (Pb), kadmium (Cd), arsen (As) dan timah putih (Sn) pada produk perikanan, Jakarta, Indonesia: BSN.

[17]. Badan Standarisasi Nasional (BSN). 2009. SNI 7388:2009 Batas maksimum cemaran mikroba dalam pangan, Jakarta, Indonesia: BSN.

[18]. C. Papadopoulou, E. Economou, G. Zakas, C. Salamoura, C. Dontorou, and J. Apostolou. 2006. Microbiological and pathogenic contaminants of seafood in Greece. *J Food Quality* **30** :28-42.

[19]. D. Putriyani. 2017. Identifikasi *Salmonella sp.* pada kerang hijau (*Perna viridis*) yang dijual di tambak Lorok, Semarang. *Diploma III thesis.*

[20*]. V. Mustahal, Shiva, F. R. Indaryanto, and S. D. Soraya. 2019. Analisis Mikrobiologi Lingkungan budidaya kerrang hijau (Perna Viridis) di perairan Tanjung Kait, Tangerang, Banten. *J Perikanan Kelautan* **9**:194-201.

[21]. H. Mannas, R. Mimouni, N. Chaouqy, F. Hamadi, and J. Martinez-Urtaza. 2014. Occurrence of *Vibrio* and *Salmonella* species in mussels (*Mytilus galloprovincialis*) collected along the Moroccan Atlantic coast. *Springerplus* **3**:1-11.

[22]. M. R. Katon, A. Solichin, and O. E. Jati. 2020. Analisis pendugaan bakteri *Escherichia coli* pada kerrang (*Perna Viridis*) di Morosari, Demak. *J Maquares* **9**:40-46

[23]. P. Masniyom. 2011. Detrioration and shelflife extention of fish and fishery products by modified atmosphere packaging. *Songklanakarin J Sci Tech* **33**:181-192.

[24]. L.I. Sutiknowati. 2016. Bioindikator Pencemaran Bakteri *Escherichia coli*. *Oseana* **41**: 63-67.

[25]. G. T. Adeyanju, and O. Ishola. 2014. *Salmonella* and *Escherichia coli* contamination of poultry meat from processing plant and retail markets in Ibadan, Oyo State, Nigeria *Springerplus* **3**:139-145.

[26]. D. K. Wardhana, D.A. Safitri, S. Annisa, N. Harijani, A. T. S. Estoepangestie, and L. Maghfiroh. 2021. Prevalence of extended-spectrum beta-lactamases in producing *Escherichia coli* in beef sold in traditional markets in Surabaya, Indonesia. *Biodiversitas* **22**:2789-2793.

[27]. Badan Pengawas Obat dan Makanan Republik Indonesia. 2009. Penetapan batas maksimum cemaran mikroba dan kimia dalam makanan. *Regulasi Pangan* BPOM no HK.00.06.1.52.4011:1-28.

[28]. B. Yulianto, P. S. Oetari, S. Februhardi, T. W. C. Putranto, and A. Soegianto. 2019. Heavy metals (Cd, Pb, Cu, Zn) concentrations in edible bivales harvested from Northern Coast of Central Java, Indonesia. *IOP Conf. Series: Earth and Environmental Science* **259**:1-11.

[29]. D. Nurhayati, and D. A. Putri. 2019. Bioakumulasi logam berat pada kerang hijau (*Perna viridis*) di Perairan Cirebon berdasarkan musim yang berbeda. *J Akua Indonesia* **4**:6-10.

[30]. A. Andayani, I. Koesharyani, U. Fayumi, Rasidi, and K. Sugama. 2020. Akumulasi logam berat pada kerang hijau di perairan pesisir Jawa. *Osean Limnol Indonesia* **5**: 135-144.

[31]. A.A.I.M. Dharma Dewi, and I.G.A.G. Wiadyana. 2019. Analisis kandungan logam berat timbal (Pb) dan cadmium (Cd) pada kerang hijau (Perna viridis) yang beredar di pasar Badung. *J Emasains* **8**:161-169.

[32]. M. R. Cordova, N. P. Zamani, and F. Yulianda. 2011. Akumulasi logam berat pada kerang hijau (*Perna viridis*) di perairan teluk Jakarta. *J Moluska Indonesia* **2**:1-8.

Quality of skipjack surimi (*Katsuwonus pelamis*) with addition of calcium carbonate ($CaCO_3$)

M F Hidayat and Ustadi*

Department of Fisheries, Faculty of Agriculture, Universitas Gadjah Mada, Jl. Flora, Bulaksumur, Yogyakarta, Indonesia

*corresponding author e-mail: ustadi@ugm.ac.id

Abstract. This study aims to determine the effect of the addition of CaCO3 as a whitening agent on the quality properties of the skipjack surimi. The functional properties were evaluated using physical and chemical tests, including white degree analysis, folding test, texture test, water holding capacity, and fat content analysis. Skipjack surimi was carried out in four treatments: 0%, 0.5%, 1%, and 1.5% CaCO3 in a completely randomized design (CRD). The results revealed that adding 1.5% CaCO3 achieved the highest white degree value of 54.22. The folding test values increased with higher CaCO3 concentrations, ranging from 1.57 (control) to 2.05 (1.5% CaCO3). Surimi fat content analysis showed no significant differences ($P>0.05$) between treatments, varying from 1.16% in the control to 1.51% in the 1.5% CaCO3 treatment. Furthermore, the WHC of skipjack surimi was significantly different ($P<0.05$) between treatments and increased with the addition of CaCO3: 52.44% (0% CaCO3), 66.42% (0.5% CaCO3), 74.57% (1 % CaCO3), and 79.96 % (1.5% CaCO3). However, sensory color tests showed a significant difference ($P<0.05$) between the treatments. In conclusion, the incorporation of 1.5% CaCO3 as a whitening agent resulted in an optimal treatment with a folding test value of 2.05 and a WHC value of 79.96%. These findings suggested the potential use of CaCO3 to enhance the functional properties of skipjack surimi during production.

1. Introduction

Surimi is a concentrated myofibril protein from fish, produced through several process stages, including separating the flesh from the skin and bones, grinding, washing, adding salt, adding cryoprotectant, and followed by freezing [1]. The raw material for surimi is generally selected from white-fleshed marine fish species, as they are considered capable of producing surimi with good gel quality and color [2]. Commonly used fish species for surimi production in Indonesia are low-value economic fish such as threadfin bream, bigeye scad, Indian mackerel, bullet tuna, and frigate tuna. These demersal fish species are abundant in coastal to deep-sea environments (abyssal zone) and are mostly found in the nearshore environments. Demersal fish have low oil content (one to four percent of their body mass) compared to pelagic fish, which can have up to 30 percent oil content. Therefore, demersal fish are classified as white-fleshed fish [3].

Demersal fish resources in shallow waters are often targeted for exploitation due to their relatively high market value and easy accessibility of fishing grounds [4]. In Indonesia, especially in the Malacca Strait, Makassar Strait, and the Netherlands Sea, demersal fish stocks have experienced a significant decline due to overfishing [5]. The shortage of white-fleshed fish as raw materials has presented a challenge for some surimi industries. To overcome this, surimi development using pelagic fish as an alternative raw material has been explored, even though most of these fish have red or dark-colored flesh. According to [6], dark-fleshed fish account for 40-50% of the total global fish catch, including fish like skipjack.

Skipjack is an important fishery commodity in Indonesia and is managed separately under the tuna, skipjack, and mackerel fisheries management. Indonesian waters, covering all existing provinces, have the potential for significant skipjack resources, with skipjack production reaching 432,851 tons in 2021. Skipjack production from 2011 to 2021 showed fluctuating trends, with the highest production occurring in 2019 at 512,846 tons. Based on the region, the highest skipjack production in 2021 was in North Sulawesi Sea [7].

Skipjack possesses several advantages that make it an optimal choice as an alternative surimi raw material. It has thick flesh, high yield, making it capable of producing a considerable amount of flesh for product utilization. Additionally, skipjack exhibits a gregarious living behavior in large groups, making it easier and more efficient to catch [8], reducing operational costs. The size of skipjack is like mackerel, making it convenient for industrial-scale processing. Consequently, skipjack is a suitable choice for surimi raw material due to its thick flesh, high yield, ease of capture, and uniform size.

Skipjack is a pelagic fish species with high myoglobin content, resulting in red to dark red flesh color [9]. Fish with dark red flesh has limited appeal in the local market, leading to reduced market demand. Fish flesh can be processed into various food products such as meatballs, nuggets, sausages, and other processed foods. Fish flesh can also be added to other products like donuts, biscuits, bread, etc. [10]. However, skipjack is still under-diversified in Indonesia, typically sold raw or processed into smoked skipjack. Another way to increase market appeal for this fish is by diversifying it into semi-finished products, such as surimi, which can then be further processed into various derivative products.

In this study, a bleaching agent is added because skipjack has dark flesh, one of the main components that cannot be applied to surimi products. To address this issue, Calcium Carbonate ($CaCO_3$) is used as the bleaching agent. Calcium carbonate is a whitening agent commonly used as an additive in surimi [11]. Besides influencing the color, it turned out that calcium carbonate as a bleaching agent could enhance protein-protein interactions through the formation of salt bridges between myofibrils and divalent calcium ions [12].

2. Materials and Methods

The research process was conducted in several stages, which included the preparation of surimi with the addition of calcium carbonate, chemical analysis, physical analysis, and sensory testing. The study was carried out from January to April 2023 in the Fish Technology Laboratory, Food Quality and Safety Laboratory, and Nutrition Laboratory of the UGM Fisheries Department.

2.1. Sample Preparation

The production of wet surimi, as per [13], began with the preparation of the fish sample. The fish was first subjected to a gutting process, involving the removal of the head and gut contents, followed by washing. After washing, the fish was filleted, separating the flesh from the skin and bones. Once the sample was prepared, the flesh was minced and then washed four times with cold water at a temperature of 5-10°C, using a fish-to-ice ratio of 1:3 and adding 0.1% salt during the final wash. After the washing process, the fish was pressed with a cloth press. Once the fish was pressed, the flesh was minced again and weighed. Then, it was mixed with cryoprotectants at 0.3% for each type: sodium tripolyphosphate (STPP), salt, and trehalose. Calcium carbonate was also added during the cryoprotectant mixing in several treatments. In the final step, the samples were packaged and placed into a freezer at -18°C. The surimi production process involved thawing the sample and subjecting it to two steamings. The first steaming was conducted at 40°C for 30 minutes, and the second steaming was done at 90°C for 20 minutes.

2.2. Proximate composition

The total protein content was determined through three replicates using the micro-Kjeldahl analysis method [14]. The water content was determined using a moisture analyzer instrument (Ohaus MB120). The ash content and fat content were determined in three replicates [15].

2.3. Whiteness

The measurement of whiteness was conducted using a colorimeter (Minolta CR 400 Chroma-meter). The values L* (lightness/darkness), a* (redness/greenness), and b* (yellowness/blueness) were recorded in five replicates, and the whiteness was calculated using the following equation [16].

$$\text{Whiteness} = 100- [100 - L^*)^2 + a^{*2} + b^{*2}]^{1/2}$$

2.4. Water holding capacity

The water holding capacity testing on skipjack surimi was conducted using the Centrifuge method. Firstly, a finely ground surimi sample weighing 1 gram was weighed and placed into a centrifuge tube. Subsequently, each centrifuge tube was supplemented with 1 ml of distilled water (aqueous solution). The next step involved centrifugation, where the centrifuge tube containing the surimi sample was run at a speed of 3000 rpm for 30 minutes. After the centrifugation process had been completed, the contents of the centrifuge tube were poured into a funnel equipped with filter paper. Next, the residue of the surimi sample remaining on the filter paper was weighed.

2.5. pH

The pH measurement was carried out using the Ohaus 3100 pH meter. Before conducting the measurements, the pH meter had to be calibrated by immersing the probe into pH 4 buffer solution and then into pH 7 buffer solution. The sample analysis was performed by weighing 5 grams of the sample and homogenizing it in 45 ml of cold distilled water. After homogenization, the pH of the sample was measured using the pH meter [17].

2.6. Gel Strength

The gel strength was evaluated using the LLOYD Texture Analyzer machine, specifically the LLOYD 1000S model (England). The analysis included measuring three parameters: hardness, cohesiveness, and chewiness of the gel sample.

2.7. Sensory test

The sensory test was based on and measured the level of suitability of a product using evaluation sheets. Generally, the sample evaluation specifications included texture, aroma, and color. The test ratings were based on the preference level of the panelists [18].

2.8. Folding test.

The folding test was conducted by slicing the surimi sample with a thickness of 5 mm and then folding it into a half-circle and a quarter-circle [19]. Subsequently, the presence or absence of cracks on the sample was observed, and the values were recorded on the scoresheet based on SNI 2964:2013.

2.9. Bite test

The bite test was performed by cutting (biting) the surimi sample between the upper and lower incisors [19]. Panelists evaluated the level of resilience of the surimi product according to the predetermined scoring format based on the scoresheet of SNI 2964:2013.

3. Result and Discussion

3.1. Skipjack yield

The amount of remaining raw material after undergoing processing is referred to as yield. The yield percentage determines the amount of fish flesh that can be utilized in surimi production. In this study, the surimi yield percentage of skipjack fish is calculated as the initial weight of the raw material minus the weight of the sample after the fish flesh grinding process, expressed as a percentage. The yield of skipjack fish flesh obtained in this research is 54.05%. According to [20], skipjack fish has a higher proportion of flesh yield compared to bones and offal. The skipjack fish yield comprises 58% flesh, 25% bones and head, and 17% viscera.

3.2. Proximate composition

Table 1. The proximate composition of skipjack surimi is influenced by calcium carbonate

Content	$CaCO_3$ 0% (%)	$CaCO_3$ 0.5% (%)	$CaCO_3$ 1% (%)	$CaCO_3$ 1.5% (%)
Moisture	67.24 ± 1.30^{a}	66.28 ± 1.69^{ab}	64.05 ± 2.52^{ab}	62.96 ± 2.18^{b}
Ash	0.737 ± 0.203^{a}	1.059 ± 0.799^{b}	1.603 ± 0.148^{c}	2.082 ± 0.151^{d}
Fat	1.16 ± 0.166^{a}	1.30 ± 0.202^{a}	1.46 ± 0.257^{a}	1.51 ± 0.208^{a}
Protein	30.22 ± 2.322^{a}	29.68 ± 1.755^{a}	28.53 ± 1.796^{a}	27.96 ± 1.622^{a}

Note: The data was the average of triplication ± standard deviation.
Different superscript on the same column indicates significantly at level of α 0,05

Based on the results of moisture content (Table 1), it can be observed that the average moisture content of skipjack surimi with different concentrations of calcium carbonate ranges from 62.92% to 67.24%. The ANOVA test results indicate that $P < 0.05$, H0 is rejected, indicating a significant difference between the treatments (P0, P1, P2, and P3) concerning the water content of skipjack surimi with various concentrations of calcium carbonate. Based on the Duncan's Multiple Range Test (DMRT), a significant difference was found between P0 and P3, while there was no significant difference between treatments P1 and P2. The highest moisture content value was observed in treatment P1 with 0% added calcium carbonate, amounting to 67.24%, while the lowest moisture content value was observed in treatment P4 with 1.5% added calcium carbonate, amounting to 62.96%. From these results, it can be concluded that the addition of calcium carbonate concentration affected the water content, where higher concentrations of calcium carbonate led to a decrease in the water content of skipjack surimi. According to the Indonesian National Standard [18], the maximum allowable water content in surimi is 80%. Based on the research by [21], the water content of skipjack during the woodfish processing was 67.40% before processing, 63% after boiling, and 23.80% after processing. The high concentration of calcium carbonate added to skipjack surimi influences the water content, causing the surimi to absorb more water during initial heating. However, during the second heating, the absorbed water was significantly evaporated. Ca2+ can interact with the proteins in skipjack meat, forming cross-links and making the skipjack surimi less able to absorb water [22].

Based on the ash content test results (Table 1), it was observed that the average ash content of skipjack surimi with different concentrations of calcium carbonate ranged from 0.737% to 2.082%. The ANOVA test results showed that $P < 0.05$, indicating that the null hypothesis (H0) was rejected, signifying a significant difference among treatments (P0, P1, P2, and P3) concerning the ash content of skipjack surimi with various concentrations of calcium carbonate. The Duncan's Multiple Range Test (DMRT) revealed the highest ash content in the treatment with the addition of 1.5% calcium carbonate, which amounted to 2.082%, while the lowest ash content was observed in the control treatment (0% calcium carbonate) with a value of 0.737%. The results indicated significant differences among treatments P0, P1, P2, and P3. DMRT analysis demonstrated that the addition of calcium carbonate significantly influenced the ash content, wherein higher concentrations of calcium carbonate resulted in increased ash content in skipjack surimi. The average ash content in skipjack surimi treated with calcium carbonate did not meet the recommended quality requirement of 1% [18]. The increase in ash content in surimi processed with calcium carbonate was due to the presence of calcium carbonate as a mineral component containing calcium. The increase in ash content may also have been caused by a decrease in water content, leading to an increase in the proportion of protein and ash content [23]. During the heating process, calcium carbonate underwent a chemical reaction, producing calcium oxide and carbon dioxide. Calcium oxide reacted with water to form calcium hydroxide, which accelerated the formation of ash during heating [24].

Based on the results of the fat content test (Table 1), it was observed that the average fat content of skipjack surimi with different concentrations of calcium carbonate ranged from 1.16% to 1.51%. The ANOVA test results showed that $P > 0.05$, indicating that the null hypothesis (H0) was accepted, and there was no significant difference among treatments (P0, P1, P2, and P3) regarding the fat content of skipjack surimi with various concentrations of calcium carbonate. From these results, it was concluded that the addition of calcium carbonate did not have a significant effect on the fat content but slightly increased the fat content in the surimi. In the study by [25], it was found that the fat content in sardine surimi before the addition of a whitening agent was 7.24%, and the fat content after the addition of the whitening agent, calcium carbonate, was 4.77%. Calcium carbonate in the surimi directly bound to fat. The increase in fat content in surimi could be attributed to the protein component in the surimi. Proteins in surimi could form a network or matrix that physically bound to fat, thereby increasing the fat content in surimi. Calcium in calcium carbonate could play a role in the formation and maintenance of protein structures in surimi. The proteins in the network acted as fat binders, causing the fat to be trapped within the protein network [26].

Based on the results of the protein content test (Table 1), it was observed that the average protein content of skipjack surimi with different concentrations of calcium carbonate ranged from 27.96% to 30.22%. The ANOVA test results showed that $P > 0.05$, indicating that the null hypothesis (H0) was accepted, and there was no significant difference among treatments (P0, P1, P2, and P3) regarding the protein content of skipjack surimi with various concentrations of calcium carbonate. From these results, it was concluded that the average protein content in skipjack surimi treated with calcium carbonate met the quality and safety requirements for the product. The minimum protein content limit for surimi based on the Indonesian National Standard was 12% [18]. The protein content in skipjack surimi was higher compared to the findings of [27], who reported a protein content of 19.14% in wet baronang fish surimi. Protein in surimi played a crucial role as it influenced other physical and chemical properties. The addition of calcium carbonate to skipjack surimi did not have a significant effect on the protein content, which may have been due to the insufficient concentration of calcium carbonate to increase the protein in surimi. According to the study by [25], calcium could increase the protein content because calcium ions could form cross-linking bridges between proteins, known as calcium bridges.

3.3. Whiteness

Based on the results of the whiteness test (Figure 1), it was observed that the average whiteness of skipjack surimi with different concentrations of calcium carbonate ranged from 51.80% to 54.23%. The ANOVA test results showed that $P > 0.05$, indicating that the null hypothesis (H0) was accepted, and there was no significant difference among treatments (P0, P1, P2, and P3) regarding the whiteness of skipjack surimi with various concentrations of calcium carbonate.

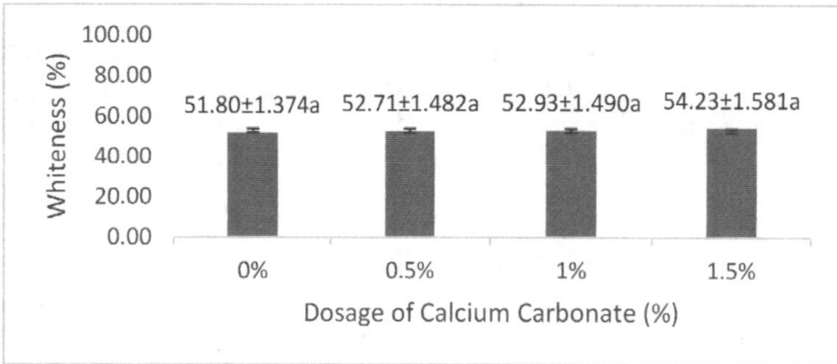

Figure 1. The whiteness content of skipjack surimi is influenced by calcium carbonate

From these results, it was concluded that the addition of calcium carbonate did not have a significant effect on the whiteness value but slightly increased the brightness of surimi. The effect of adding calcium carbonate to skipjack surimi aligned with the study by [11], where calcium carbonate contributed to making the surimi brighter, although the resulting color change was not significant. The findings were similar to the study by [28], which found that adding calcium carbonate in the concentration range of 0-2% could not whiten the gel of kurisi fish surimi to resemble pollock surimi's color. Insoluble calcium carbonate particles were believed to have a light-scattering effect, making the product appear whiter. The increase in whiteness value could be attributed to the rise in water content, making the surimi appear whiter and more yellow [28]. Surimi with dark flesh could be brightened using other whitening agents such as soybean oil and titanium dioxide, but these whitening agents might adversely affect the surimi's deformation value [11].

3.4. Water holding capacity

Based on the results of the water holding capacity test (Figure 2), it was observed that the average water holding capacity of skipjack surimi with different concentrations of calcium carbonate ranged from 52.44% to 79.96%. The ANOVA test results showed that $P < 0.05$, indicating that the null hypothesis (H0) was rejected, and there was a significant difference among treatments (P0, P1, P2, and P3) regarding the water holding capacity of skipjack surimi with various concentrations of calcium carbonate.

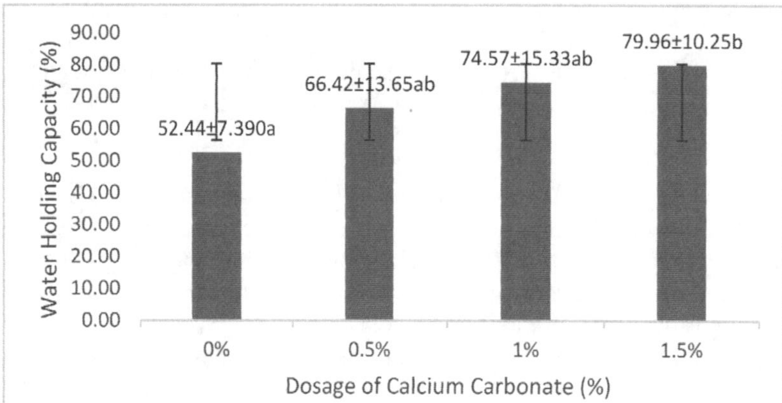

Figure 2. Effect of the addition of calcium carbonate on water holding capacity of cakalang surimi

Based on the results of Duncan's Multiple Range Test (DMRT), the highest water holding capacity (WHC) value was obtained in the treatment with the addition of 1.5% calcium carbonate, which was 79.96%, while the lowest ash content was found in the control treatment (0% calcium carbonate) with a value of 52.44%. The further Duncan's test showed that P0 significantly differed from P3, but P2 and P3 did not significantly differ from P0 and P3. The presence of calcium carbonate in surimi can react with chloride ions in fish flesh and form calcium chloride salt. Calcium chloride can help enhance the water-binding capacity of the meat, as it has good hydration properties. Additionally, the calcium ions from calcium carbonate can reduce the electrostatic repulsion between water molecules, making them more easily bound to the protein in the fish meat [29].

3.5. pH

Based on the results of the pH test (Figure 3), the average pH of surimi skipjack with different concentrations of calcium carbonate ranged from 6.5 to 6.7. The ANOVA test results indicated that $P > 0.05$, meaning that H0 was accepted, and there was no significant difference among the treatments (P0, P1, P2, and P3) in terms of the pH of the surimi skipjack with various concentrations of calcium carbonate.

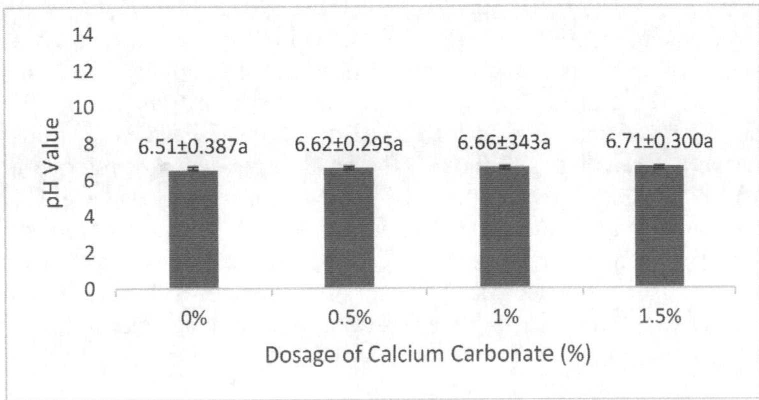

Figure 3. Effect of the addition of calcium carbonate on acidity (pH) of skipjack surimi

Based on the results, it can be concluded that the addition of calcium carbonate does not have a significant effect on the whiteness value but slightly increases the pH of the surimi. The pH level reflects the presence of hydrogen ions associated with acidity. When calcium carbonate interacts with carbon dioxide in water, bicarbonate (HCO3-) is formed. Bicarbonate can exhibit both acidic and basic properties as it can undergo hydrolysis, producing hydroxide ions (OH-) and acts as a buffer [30].

3.6. Gel strength

Based on the hardness test results (Table 2), the average texture value for the hardness parameter ranged from 22.6643 to 35.2767 (gr). The ANOVA test showed that $P < 0.05$, indicating a significant difference among the treatments (P0, P1, P2, and P3) in the texture of skipjack surimi with various concentrations of calcium carbonate. Further Duncan's Multiple Range Test (DMRT) on the Hardness parameter revealed that P0 and P3 were significantly different from P1, but P2 did not show a significant difference compared to P0, P1, and P3. Hardness is a mechanical measurement that describes the force required to induce deformation in a substance. This force was necessary to compress the material between the molars in a solid form or between the tongue and the palate in a semi-solid form [31]. The addition of calcium carbonate affects the level of hardness in skipjack surimi. The reduction in hardness in P2 and P3 may be due to the excessive concentration of calcium carbonate added to the surimi, leading to a very compact texture. However, the water content in P3 is significantly low, which may contribute to the decrease in hardness. Calcium can act as a cation that interacts with the free carboxyl groups of proteins, forming calcium bridges that result in a more compact texture [32]. Based on [33], the addition of surimi and potato flour to fish burger products can reduce the hardness value of the fish burger product.

Table 2. Effect of the addition of calcium carbonate on gel strength of skipjack surimi

Perlakuan	$CaCO_3$ 0%	$CaCO_3$ 0,5%	$CaCO_3$ 1%	$CaCO_3$ 1,5%
Hardness (gr)	22.79±1.770a	35.28±9.465b	30.46±4.467ab	22.66±4.467a
Cohesiveness (gr)	0.503±0.028a	0.460±0.043ab	0.473±0.079ab	0.348±0.104b
Chewiness (gr)	19.73±8.288a	19.29±2.914a	13.42±5.826a	15.21±1.454a

Note: The data was the average of triplication ± standard deviation.
Different superscript on the same column indicates significantly at level of α 0,05

Based on the cohesiveness test results (Table 2), the average texture value for the Cohesiveness parameter ranged from 0.348 to 0.503 (gr). The ANOVA test showed that $P < 0.05$, indicating a significant difference among the treatments (P0, P1, P2, and P3) in the texture of skipjack surimi with various concentrations of calcium carbonate. Further Duncan's Multiple Range Test (DMRT) on the Cohesiveness parameter revealed that P0 was significantly different from P3, but P1 and P2 did not show a significant difference compared to P0 and P3. Cohesiveness was a measure of how much a material could change its shape before breaking. In the sensory context, it referred to how much the material could be compressed between the teeth before undergoing damage [31]]. The highest cohesiveness value was found in the control treatment, and the lowest was observed in the treatment with the addition of 1.5% calcium carbonate. Surimi gel with the addition of calcium carbonate showed a slight decrease in cohesiveness, which was not significant. Based on [34], it is explained that the high protein content is directly related to the cohesiveness value of the product. The ability of proteins to bind food components such as water and fat is essential because this ability can directly affect cohesiveness. The closer the cohesiveness value is to 1, the better the product integrity.

Based on the chewiness test results (Table 2), the average texture value for the Chewiness parameter ranged from 13.4163 to 19.7315 (gr). The ANOVA test showed that $P > 0.05$, indicating that there was no significant difference among the treatments (P0, P1, P2, and P3) in the texture of skipjack surimi with various concentrations of calcium carbonate. Chewiness was a parameter that described the level of elasticity or hardness of a food material when chewed. It depicted how much the food material could resist pressure or bite before undergoing a change in shape or breaking [31]. The highest chewiness value was found in the control treatment, and the lowest was observed in the treatment with the addition of 1.0% calcium carbonate. Surimi gel with the addition of calcium carbonate showed a slight decrease in chewiness, which was not significant. Based on [35], the addition of calcium chloride was able to provide an optimum chewiness value.

3.7. Sensory test

The quality testing results of skipjack surimi gel included color, texture, and taste.

3.7.1. Color test. Based on the results of the sensory color test (Figure 4), the average color values ranged from 1.43 to 2.24. The ANOVA test showed that $P < 0.05$, indicating a significant difference among the treatments (P0, P1, P2, and P3) in terms of the color of skipjack surimi with various concentrations of calcium carbonate. Further Duncan's multiple range test (DMRT) showed that the addition of calcium carbonate had a significant effect on the color, where higher concentrations of calcium carbonate led to increased color values in skipjack surimi. The highest color value was obtained in the treatment with 1.5% calcium carbonate addition, while the lowest value was found in the control treatment. From these results, it can be concluded that the addition of calcium carbonate can whiten the skipjack surimi, which initially had a brownish-yellow color, to a brighter yellow color. According to [11], calcium carbonate can effectively brighten fish meat as it is one of the Optical Brightening Agents (OBA).

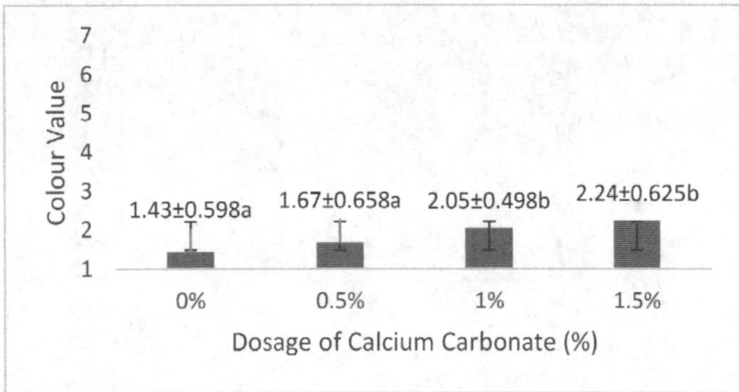

Figure 4. The colour of skipjack surimi with different calcium carbonates concentrations

3.7.2. Texture test. Based on the results of the sensory texture test (Figure 5), the average texture values ranged from 4.86 to 5.43. The ANOVA test showed that P > 0.05, indicating that there is no significant difference among the treatments (P0, P1, P2, and P3) in terms of the texture of skipjack surimi with various concentrations of calcium carbonate. Skipjack surimi without the addition of calcium carbonate had a chewy texture, somewhat compact, and slightly dense. After the addition of calcium carbonate, the texture of skipjack surimi improved to be more chewy, compact, and slightly dense. Calcium carbonate not only helps to brighten the color of surimi, but it also contributes to improving the gel strength [11].

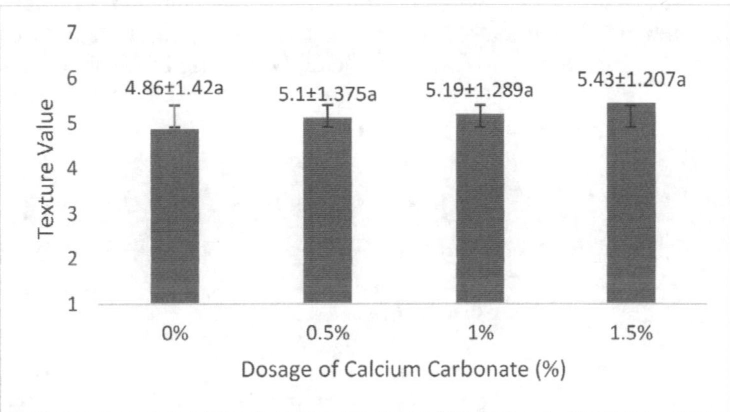

Figure 5. The texture of skipjack surimi with different calcium carbonates concentrations

3.7.3. Aroma test. Based on the results of the sensory aroma test (Figure 6), the average aroma values ranged from 5.90 to 6.38. The ANOVA test showed that P > 0.05, indicating that there is no significant difference among the treatments (P0, P1, P2, and P3) in terms of the aroma of skipjack surimi with various concentrations of calcium carbonate. Free amino acids derived from protein content and free fatty acids from lipid components contribute to the fishy aroma and taste in fish meat. Both skipjack surimi with the addition of calcium carbonate and without calcium carbonate showed similar aromas because calcium carbonate is a compound without any odor [36]. According to [37], the aroma of skipjack fish is not well-liked by people due to its dominant characteristic fishy smell, which can be attributed to compounds such as trimethylamine oxide (TMAO), guanidine, and imidazole derivatives.

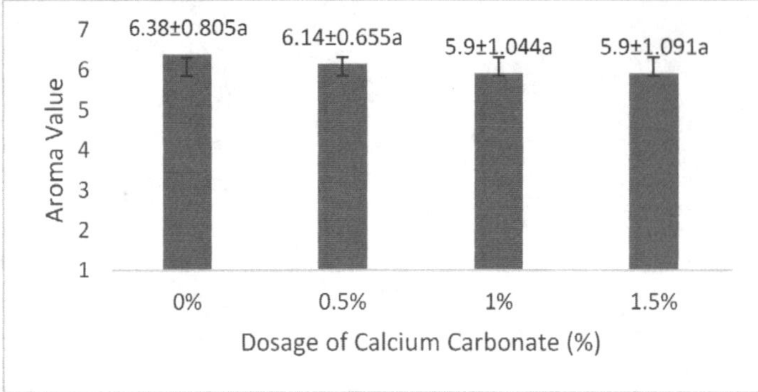

Figure 6. The aroma of skipjack surimi with different calcium carbonates concentrations

3.8. Folding test.

Based on the results of the folding test (Figure 7), the average values ranged from 1.57 to 2.14. The ANOVA test showed that $P > 0.05$, indicating that there is no significant difference among the treatments (P0, P1, P2, and P3) in terms of the folding test values of skipjack surimi with various concentrations of calcium carbonate. In the control skipjack surimi, 5 panelists obtained a value of 1, indicating that the surimi will break completely when folded in half. It can be concluded that the addition of calcium carbonate concentrations affects the folding test values, meaning that higher concentrations of calcium carbonate can increase the folding test value of skipjack surimi, although not significantly. The average assessment of the folding test for skipjack surimi fell under the category of "cracked but still held together when folded in half". The folding test results correspond to the strength of the skipjack surimi gel. Based on research conducted by [38], it is stated that the folding test results are directly related to the texture, especially the gel strength. The better the folding test results (more difficult to crack), the better the quality of the fish gel produced.

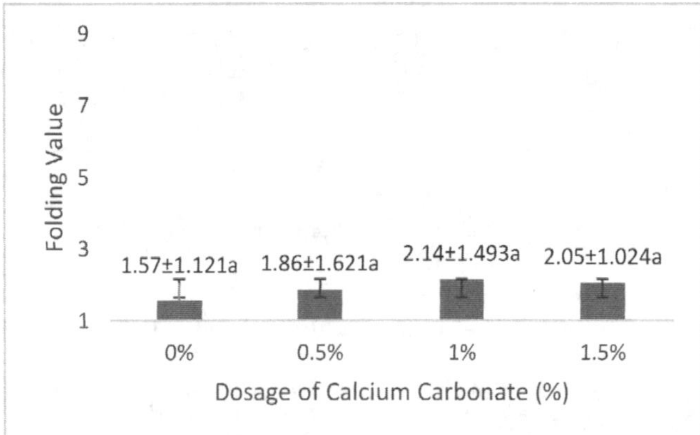

Figure 7. The folding value of skipjack surimi with different calcium carbonates concentrations

3.9. Bite test.

Based on the results of the bite test (Figure 8), the average values ranged from 6.43 to 7.10. The ANOVA test showed that $P > 0.05$, indicating that there is no significant difference among the treatments (P0, P1, P2, and P3) in terms of the bite test values of skipjack surimi with various concentrations of calcium carbonate. From these results, it can be concluded that the addition of calcium carbonate concentrations does not have a significant effect. The average assessment of the bite test for skipjack surimi fell under the category of "moderately strong elasticity". The use of calcium carbonate in the food industry, especially in surimi products, is expected to improve the color and increase the elasticity of surimi products. The addition of binding agents in the production of fishery products aims to improve the elasticity of the final product, bind water, provide color, and form a solid texture [19]. The results of the bite test for skipjack surimi can be categorized as good. This may be due to the good myofibril protein binding in skipjack fish, resulting in a better and more elastic texture of the product.

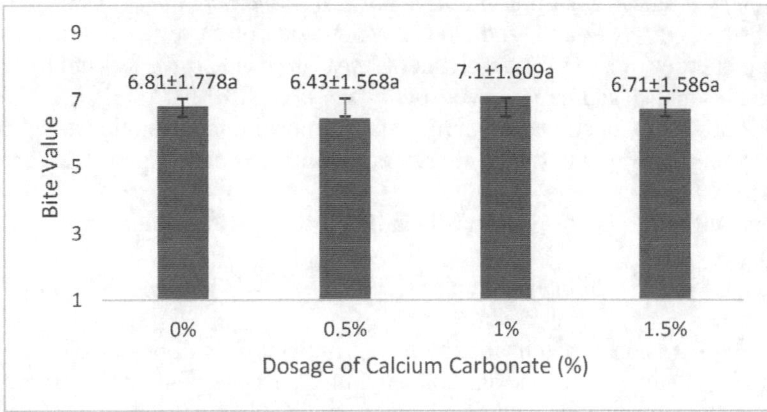

Figure 8. The bite value of skipjack surimi with different calcium carbonates concentrations

4. Conclussion

The addition of the whitening agent $CaCO_3$ did not have a significant effect on skipjack surimi. This study contradicted the initial hypothesis that stated calcium carbonate could whiten and improve the quality of surimi. In this research, there was no significant difference in terms of whitening or quality improvement that could be associated with the addition of $CaCO_3$.

Acknowledgement

The work is supported by collaborative research grants for lecturers and students of the Faculty of Agriculture UGM number: 3162/UN1/FPN/KU/KU.02.05/2023

References

[1] Cando D, Herranz B, Borderías A J, and Moreno H M 2015 Effect of high pressure on reduced sodium chloride surimi gels *J. Food Hydcol.* **51** 176-87
[2] Park J W 2014 *Surimi and surimi seafood: third edition* (New york: CRC Press)
[3] Alatas U, Mardjudo A, Ihsan T and Ekaputra A 2022 Teknologi penangkapan ikan demersal dan aspek ekonomis hasil tangkapan nelayan di Kelurahan Ganti Kecamatan Banawa Kabupaten Donggala, Sulawesi Tengah *J. TROFISH.* **1** 44–50
[4] Noija D, Martasuganda S and Murdiyanto B 2014 Potensi dan tingkat pemanfaatan sumberdaya ikan demersal di Perairan Pulau Ambon Provinsi Maluku *J. Tek. Perik. Dan Kelautan.* **5** 55-64.
[5] Muhtarom A 2017 Analisis kontribusi hasil perikanan laut terhadap kesejahteraan para nelayan dan nasyarakat di Kabupaten Lamongan *J. Penel. Ekon. Dan Akunt.* **2** 265–79
[6] Arfat Y A and Benjakul S 2012 Gelling characteristics of surimi from yellow stripe trevally (*Selaroides leptolepis*) *J. Int. Aq. Res.* **4** 1-13
[7] Statistik B P 2021 *Produksi perikanan tangkap di laut menurut komoditas utama (ton), 2019-2021* (Jakarta: Badan Pusat Statistik)
[8] Inara C 2020 Manfaat asupan gizi ikan laut untuk mencegah penyakit dan menjaga kesehatan tubuh bagi masyarakat pesisir *J. Kalwedo Sci.* **1**, 92–5
[9] Zapata E S, Amensour M, Oliver R, Zaragoza E F, Navarro C, Lopez J F, Sendra E, Sayas E and Alvarez J A 2011 Quality characteristics of dark muscle from Yellowfin tuna Thunnus albacares to its potential application in the food industry *J. Food and Nutr. Sci.* **2** 22-30

[10] Pratama R I, Rostianti I and Kurniawati N 2017 Pemberdayaan masyarakat melalui peningkatan keterampilan produk olahan hasil perikanan di wilayah yang terkena dampak genangan Jatigede Kabupaten Sumedang *J. Pengab. Kepada Masy.* **1** 60-3
[11] Benjakul S, Wonnop V and Yuwathida K 2004 the effect of whitening agents on the gel-forming ability and whiteness of surimi *International J. of Food Sci. and Tech.* **39** 773–81
[12] Sang S, Chen X, Qin Y, Tong L and Ou C 2022 A study on the effects of calcium lactate on the gelling properties of large yellow croaker (*Pseudosciaena crocea*) surimi by low-field nuclear magnetic resonance and raman spectroscopy *J. Foods.* **11** 31-97
[13] Kolanus J P M 2020. Functional properties and chemical composition of dried surimi mackerel (*Scomberomorus* Sp) with different cryoprotectants and drying methods *J. of Physics: Conf. Ser.* **1463** 12-22
[14] Apriyantono A, Fardiaz D, Puspitasari N L and Budijanto S 1989 *Petunjuk Laboratorium Analisis Pangan* (Bogor: IPB Press)
[15] Association Official Analitical Chemistry 2005 *Official Methods of Analysis* (New York: Arlington)
[16] Park J W 1994 Functional protein additives in surimi gels *J. of Food Sci.* **59** 525–27
[17] Saputra E 2018 Pengaruh pencucian dan penyimpanan pada pembuatan surimi dan kamaboko ikan nila (*Oreochromis* sp) *J. of Mar. and Coast. Sci.* **7** 110–9
[18] Standar Nasional Indonesia 2013 *Surimi: 2694:2013* (Jakarta: Standar Nasional Indonesia)
[19] Radityo C T and Darmanto Y S 2014 Pengaruh penambahan egg white powder dengan konsentrasi 3% terhadap kemampuan pembentukan gel surimi dari berbagai jenis ikan *J. Pengolah. Dan Biotek. Has. Perik.* **3** 1–9
[20] Simbolon S E, Onibala H, Pandey E V, Taher N, Mentang F, Dotulong V and Harikedua S D 2021 Kualitas sensori dan mikrobiologi surimi ikan cakalang (*Katsuwonus pelamis* L) yang dipengaruhi oleh waktu pencucian *J. Med. Tek. Has. Perik.* **9** 1–7
[21] Pundoko S S, Onibala H and Agustin A T 2014 Perubahan komposisi zat gizi ikan cakalang (*Katsuwonus pelamis L*) selama proses pengolahan ikan kayu *J. Med. Tek. Has. Perik.* **2** 9-14
[22] Jaya Y A, Suseno T I P and Jati I R A P 2017 pengaruh perbedaan konsentrasi kalsium karbonat (CaCO3) terhadap sifat fisikokimia dan organoleptik sereal sarapan beras hitam-pisang raja *J. Tek. Pang. Dan Gizi.* **17** 43–50
[23] Wawasto A, Santoso J and Nurilmala M 2018 Karakteristik surimi basah dan kering dari ikan baronang (*Siganus* sp.) *J. Pengolah. Has. Perik. Indo.* **21** 367–76
[24] Herlambang A and Martono D H 2008 teknologi pengolahan sampah dan air limbah *J. Air Indo.* **4** 146-60
[25] Zaghbib I, Felix M, Romero A, Arafa S and Hassouna M 2017 Effects of whitening agents and frozen storage on the quality of sardine (*Sardina pilchardus*) surimi: physicochemical and mechanical properties *J. of Aq. Food Prod. Tech.* **26** 29–42
[26] Astuti I 2007 *Penuntun praktikum analisis bahan pakan*. (Surakarta: Jurusan Peternakan Universitas Sebelas Maret)
[27] Astiana I, Lahay A F, Utari S P S D, Farida I, Samanta P N, Budiadnyani I G A and Febrianti D 2023 Karakteristik organoleptik dan nilai gizi biskuit dengan fortifikasi tepung surimi ikan swanggi (*Priacanthus tayenus*) *J. Pengolah. Has. Perik. Indo.* **26** 107-16
[28] Hsu C and Chiang B 2002 Effects of water, oil, starch, calcium carbonate and titanium dioxide on the colour and texture of threadfin and hairtail surimi *gels Inter. J. of Food Sci. & Tech.* **37** 387–93
[29] Han F, Zhao J, Bian Y, Guo J and Chen L 2023 Electro mitigation of calcium carbonate and calcium sulfate scaling in an optimized thermal conductive membrane distillation process. *J. Sep. and Purif. Tech.* **316** 1–10
[30] Kadarini T, Musthofa S Z, Subandiyah S and Priono B 2015 *J. Ris. Akuakul.* **10** 187–97
[31] Szczesniak A S 2002 Texture is a sensory property *J. Food Qual. and Pref.* **13** 215–25
[32] Lesmana S N 2008 Pengaruh penambahan kalsium karbonat sebagai fortifikan kalsium terhadap

sifat fisikokimia dan organoleptik permen jeli susu *J. Tek. Pang. Dan Gizi.* **7** 28–39

[33] Manaf F, Faujan N H, Arifin N and Hanafiah N B 2018 Physicochemical properties and consumer preference of fish burgers produced from black tilapia surimi paste and potato flour *Malay. J. of Sci., Health & Tech.* **1** 25-30

[34] Putranti R T, Anggo A D and Fahmi A S 2020. Pengaruh surimi dari ikan swanggi (*Priacanthus* sp.), ikan kurisi (*Nemipterus* sp.), dan ikan kuniran (*Upeneus* sp.) terhadap karakteristik cumi-cumi analog *J. Ilmu dan Tek. Perik.* **2** 43-53

[35] Laksono U T, Suprihatin S, Nurhayati T and Romli M 2019 Enhancement of textural quality from daggertooth pike conger fish surimi with sodium tripolyphosphate and transglutaminase activator *J. Pengolah. Has. Perik. Indo.* **22** 198-208

[36] Hariyanto A, Sari V K and Pujiastuti C 2020 Kinetika reaksi pembentukan kalsium fosfat dari asam fosfat dan cangkang kerang darah *ChemPro.* **1** 32–8

[37] Srihidayati G and Firdamayanti E 2021 Formulasi dan uji organoleptik otak-otak ikan cakalang (*Katsuwonus pelamis*) dengan berbagai konsentrasi tepung penstabil. *Perbal: J. Pertan. Berkelanj.* 9 123-31

[38] Wibowo T A, Darmanto Y S and Amalia U 2015 *J. Pengolah. Dan Biotek. Has. Perik.* **4** 17–24

IOP Conf. Series: Earth and Environmental Science 1289 (2023) 012028 doi:10.1088/1755-1315/1289/1/012028

Consumer Preference and Nutritional Value of Catfish Nuggets Substituted with Jellyfish

M T P Haji and Ustadi*

Department of Fisheries, Faculty of Agriculture, Gadjah Mada University, Jalan Flora A4, Bulaksumur, Yogyakarta, 55281, Indonesia

Corresponding author mail: ustadi@ugm.ac.id

Abstract. This study aims to determine the effect of jellyfish meat substitution in the production of catfish nugget on the nutrient content and acceptability of the nugget, and to determine the best composition to produce quality nugget. The research design used a completely randomized design (CRD), there were 4 types of treatment with 3 replications. The treatments were the percentage of catfish meat: jellyfish meat including T0 (100:0), T1 (90:10), T2 (80:20) and T3 (70:30). Parameters measured were hedonic organoleptic characteristics (aroma, taste, texture and overall) and nutritive content (moisture, ash, fat and protein content). The data obtained were analyzed using the ANOVA test (Analysis of Variance) and if there was a significant difference then it was continued with the Duncan test. The results showed that the hedonic value of catfish nuggets substituted for jellyfish meat gave a significant difference ($p < 0.05$) in each of the parameters tested, namely aroma, taste, texture and overall. The catfish nugget substituted with 20% jellyfish meat (T2) gave the best sensory quality on aroma (3.77), taste (3.67), texture (3.60) and overall score (3.77). The results of testing the proximate content of catfish nuggets substituted for jellyfish meat showed significantly different results ($p<0.05$) for each parameter. However, all catfish nuggets treated with jellyfish substitution (T1, T2, and T2) as well as treatment without jellyfish substitution (T0) met the quality standards of the nuggets, except for the ash content in the nuggets. It can be concluded that the best formula is in the treatment (T2), the catfish nugget substituted with 20% jellyfish meat with a protein, fat, water and ash content of 10.36%, 2.52%, 13.49% and 5.45%, respectively.

1. Introduction

One of the aquatic products that may be found in virtually all of Indonesia's maritime waters is jellyfish. There are several different types of jellyfish, such as Aurelia Aurita, Cnidaria, Hydrozoa, and Medusozoan [1]. The Kampar Riau River's estuary waters generate 24 tons of dry weight, or 342 tons of fresh weight, of jellyfish each year, which is exported to Malaysia and Singapore [2]. As a kind of medicine, jellyfish are thought to treat conditions like arthritis, high blood pressure, stomach ulcers, soften the skin, and enhance digestive function [3].

Because Indonesia has not done a good job of processing and using jellyfish, more of them are sold abroad. This results from a lack of knowledge about the nutritional value of jellyfish. Because they are concerned about the jellyfish's venom, most people avoid eating jellyfish meat. The tentacles of jellyfish are among the body components that can sting. Jellyfish will have a higher value if processed and cleaned properly. The nutritional value of jellyfish is based on the percentage of dry matter: ash (76.19%), protein (3.49), fat (0.43%) and carbohydrates (19.9%) [4]. Jellyfish can be used as medicine and is believed to cure arthritis, high blood pressure, stomach ulcers, soften the skin, and improve digestive function [3]. Reviewing the benefits of consuming jellyfish makes it possible to process them into a food product.

Nugget is a high-protein side dish created from animal-based products and blended with other ingredients through the giving bread flour and frying process. Animal-based foods, including chicken, beef, fish, and shrimp, are used to make nuggets. Nuggets are currently processed using a wide variety of basic components, including tilapia, potatoes, vegetables, cheese, tofu, mushrooms, chicken, shrimp, catfish, spinach, and carrots. Catfish nuggets are among the most popular fish-based nuggets. choosing catfish because it is relatively inexpensive. Catfish are harvested in large quantities, which lowers their selling price.

There hasn't been a lot of research done on using jellyfish as an ingredient in food products. The use of jellyfish paste in cookies resulted in a larger concentration of jellyfish, which increased consumer approval [5]. According to this study, jellyfish processing is utilized to replace catfish in the production of nuggets, hence increasing the selling value of jellyfish. The purpose of this study is to examine the influence of jellyfish meat substitution in the manufacturing of catfish nugget on nutritious content and acceptability, as well as the ideal composition to produce excellent nugget.

2. Materials and Methods

2.1. Material

This research was conducted from August to November 2022. The production of nuggets was carried out at the Fish Technology Laboratory, Department of Fisheries, Faculty of Agriculture, Gadjah Mada University, Yogyakarta. Proximate analysis was carried out at the Laboratory of Nutrition and Feed Science, Faculty of Animal Husbandry and Agriculture, Diponegoro University, Semarang. The tools used in making catfish nuggets with jellyfish substitution are blenders, basins, spatulas, scales, baking sheets, and steamers. While the ingredients used are catfish meat, jellyfish, tapioca flour, garlic, coriander, sugar, pepper, eggs, onions, salt and water.

2.2. Methods

The process of making nuggets begins with grinding the dried jellyfish which will be substituted into catfish nuggets according to a predetermined treatment. The nugget for each treatment was repeated three times by mixing all the ingredients according to the treatment using a blender. The nugget dough that has been mixed evenly is then placed into a baking dish to be steamed for 20 minutes. The cooked dough is then cooled and cut approximately 1 cm for further processing by dipping it into the beaten egg and breadcrumbs.

The research method that will be used in this study is the experimental method. The design used was a completely randomized design (CRD). There were four treatments and each treatment was repeated three times. The percentage of catfish meat with jellyfish that was tried was:

T0 : catfish (100%) and jellyfish (0%)
T1 : catfish (90%) and jellyfish (10%)
T2 : catfish (80%) and jellyfish (20%)
T3 : catfish (70%) and jellyfish (30%)

The organoleptic test that will be used for the level of acceptance uses the hedonic method. Organoleptic tests were carried out to determine the level of panelists' preference for the jellyfish meat added jellyfish product. This organoleptic test will use 70 untrained panelists by asking them to give a personal assessment of the samples presented. The parameters to be observed were color, aroma, texture, taste and overall rating. The assessment is carried out on a scale of 1 to 5, then it is entered into the scoresheet provided by the researcher. Where the value 1=dislike very much, 2=dislike, 3=neutral, 4=like and 5=like very much.

2.3. Data Analysis
Proximate analysis refers to AOAC 1995 for analysis of water, fat and ash content, while for analysis of protein content refers to SNI 01-2354.4-2006. The data obtained in each treatment and replicates were then carried out by analysis of variance (ANOVA) and Duncan's further test if there was a significant effect at $\alpha = 0.05$ (95% confidence scale).

3. Results and Discussion

3.1. Hedonic Value of Catfish Nugget Substituted Jellyfish Meat
The results of the hedonic test for catfish nuggets substituting jellyfish meat which includes aroma, taste, texture and overall can be seen in table 1.

Table 1. Hedonic Test Results for Catfish Nuggets Substituted Jellyfish Meat

Parameters	Treatment			
	T0	T1	T2	T3
Aroma	3.76^{a}	3.44^{b}	3.77^{a}	3.57^{ab}
Taste	3.27^{b}	3.26^{b}	3.67^{a}	3.46^{ab}
Texture	3.51^{ab}	3.31^{b}	3.60^{a}	3.44^{ab}
Overall	3.59^{ab}	3.51^{b}	3.77^{a}	3.61^{ab}

Note: values followed by different letters (a,b) indicate significantly different ($p<0.05$)

The results of the panelists' assessment of the aroma of catfish nugget substituted with jellyfish ranged on average between 3.44 to 3.77. Based on the ANOVA test in table 1, it shows that the addition of jellyfish makes a significant difference ($p<0.05$). Treatment T1 produced the lowest aroma value (3.44), but was not different from treatment T3 (3.57). The T0 and T2 treatments produced the same strong aroma of 3.76 and 3.77. The addition of 100% catfish at T0 produced a stronger catfish aroma, as did treatment T2 with a combination of catfish:jellyfish (80:20). This research is consistent with the fact that the aroma emitted by catfish nuggets (100%) produces a stronger distinctive catfish odor compared to modified red bean nuggets [6]. The level of aroma preference in treatment T2 showed that the 80:20 combination gave the same aroma as pure catfish nugget. Aroma is a visual image that is often used in organoleptic testing using the sense of smell [7].

Based on the results of the ANOVA analysis of hedonic taste characteristics in table 1, the results show significantly different ($p<0.05$). Treatment T2 produced the highest taste value (3.67), but was not different from treatment T3 (3.46). The taste values in treatments T0, T1 and T3 were not significantly different at 3.27, 3.26 and 3.46. Jellyfish substitution of 20% in this study gave the panelists the most preferred taste. Other research shows that substituting crab fish (80%) using catfish (20%) for crab nugget can produce a higher taste value [8]. This is because the two raw materials used have almost the same aroma organoleptic properties because they are aquatic resources.

Based on the results of the ANOVA analysis of hedonic texture characteristics in table 1, the results show significantly different ($p<0.05$). Treatment T2 produced the highest nugget texture value compared to other treatments at 3.60, but was not different from T0 (3.51) and T3 (3.44). The jellyfish used in the research have a dense and chewy texture, while the flesh of catfish has a soft texture. This can cause differences in the texture of the nugget. The results of other research show that nugget with a combination of crab meat and catfish meat gives a texture that is no different because the raw materials used have the same texture [8]. The addition of 15% flour is harder than 5% when making catfish nugget [9]. The combination of catfish and jellyfish of 80:20 in this research provided the level of texture most liked by the panelists.

Based on the results of the ANOVA analysis of the overall hedonic characteristics in table 1, the results were significantly different ($p <0.05$). The highest overall preference value for catfish nuggets substituted for jellyfish was in the T2 treatment of 3.77. This is because the level of preference for aroma, taste and texture on T2 also produces the highest value.

3.2. *Proximate Content of Catfish Nugget Substituted Jellyfish Meat*

The proximate content of catfish nuggets substituted for jellyfish meat includes moisture content, ash content, crude fat, and crude protein can be seen in table 2.

Table 2. Proximate Test Results for Catfish Nugget Substituted Jellyfish Meat

Parameters	Treatment			
	T0	T1	T2	T3
Moisture content	11.75^{b}	13.28^{a}	13.49^{a}	12.65^{a}
Ash content	5.64^{b}	6.38^{a}	5.45^{b}	5.75^{b}
Crude fat	4.34^{a}	4.32^{a}	2.52^{b}	2.56^{b}
Crude protein	8.76^{d}	9.39^{c}	10.36^{b}	11.04^{a}

Note: values followed by different letters (a,b) indicate significantly different ($p<0.05$)

The water content values in this study met the Indonesian national standards (SNI) for fish natter, which ranged from 11.75 to 13.49. Based on Indonesian national standards, the maximum water content limit in fish nugget is 60% [10]. The results of the ANOVA test for nugget water content in table 2 show significantly different results ($p<0.05$). The water content value of treatment T0 (11.75%) was lower when compared to treatments T1 (13.28%), T2 (13.49%) and T3 (12.65%). This is due to the difference in water content of the main ingredients in making nugget, namely jellyfish and catfish. Based on the results of previous research, the water content of fresh jellyfish reaches 87.50%[11], while the water content for catfish is 75.10%[12]. The higher water content of jellyfish compared to catfish causes an increase in the water content of the nugget product when the jellyfish is substituted.

The production of catfish nuggets which were substituted for jellyfish meat in Table 2 shows significantly different results ($p<0.05$). The ash content of T1 treatment produced a higher value, namely 6.38% compared to T0, T2 and T3. The ash content in this study did not meet the fish natter ash content standard of 2.5% [10]. All treatments showed the highest ash content of 6.38%. The ash content is closely related to the mineral content of the materials used in making nuggets, the higher the mineral content of the raw material will increase the ash content of the nugget [13].

The results of analysis of variance showed that jellyfish substituted for catfish nugget produced significantly different fat content ($p<0.05$). The fat content in all treatments met the quality standards for fish nugget, ranging from 2.52-2.56%. The quality standard for fish nugget is a maximum of 15% [10]. The fat content in treatments T2 (2.52%) and T3 (2.56%) was significantly different ($P<0.05$) from treatments T0 (4.34%) and T1 (4.32). Insignificant differences ($P>0.05$) were seen in treatments T2 and T3, as well as T0 and T1. A higher percentage of jellyfish meat substitute results in a lower fat content value of the nugget product. This is because the fat content of fresh jellyfish is lower at 2.03% [11], compared to the fat content of catfish meat at 4.03% [12]. The results of other research state that the greater the proportion of red beans (60%) or the smaller the proportion of snakehead fish meat used, the higher the fat content of the nuggets produced [13]. This is confirmed by the fact that red beans have a higher fat content of 2.2 g/100 g BDD, while fresh snakehead fish meat is 0.5 g/100 BDD [14].

The results of analysis of variance showed that jellyfish substituted for catfish nugget produced significantly different protein levels ($p<0.05$). The protein content of all treatments met the quality standards for fish nugget, namely more than 5% [10]. The protein content in all treatments was significantly different with treatment T3 producing the highest protein content (11.04%) followed by T2 (10.36%), T1 (9.39%) and T0 (8.76%). A higher percentage of jellyfish substitution results in a higher protein content value. This is because the protein content of jellyfish is higher at 35.4%[15] compared to the protein content in catfish of 17.7%-26.7%[16]. The research results are in line with other research which shows that the more red beans added causes the protein content to be higher [6].

4. Conclusion
The best percentage of catfish substitution with jellyfish was found in treatment T2 with a ratio of 80% catfish meat: 20% jellyfish meat.

References
[1] Landu M 2017 Windows Phone-Based Animal Type Learning Application (Yogyakarta: STMIK AKAKOM).
[2] Rupawan 2018 Fiseries: Research Journal of Aquatic Sciences and Fisheries **7** 12-16.
[3] Murniyati 2009 Indonesian Food Reviews **4** 30-32.
[4] Wakabayashi K, Sato H, Yoshie-Stark Y, Ogushi M, and Tanaka Y 2016 Aquac. Nutr **22** 25–33.
[5] Raposo A, Alice C, Luis A, Amparo G, and Zilda M 2018 J.Sci Food Agric **98** 3972-3981.
[6] Justisia S R W A H and Adi A C 2016 Journal of Indonesian Nutrition Media **11** 106-112.
[7] Setyaningrum H and Saparinto C 2014 *Complete Guide to Agarwood* (Jakarta: Penebar Swadaya).
[8] Ubadillah A and Hersoelistyorini 2010 Journal of Food and Nutrition **1** 45-54.
[9] Tumion F F and Hastuti N D 2017 Journal of Agromix **8** 25-35.
[10] [BSN] The National Standardization Agency 2013 SNI 7758:2013.
[11] Aji D U 2011 BSA Thesis Bogor Agricultural Institute.
[12] Handayani D W and Kartikawati D 2015 UNTAG Semarang Scientific Journal **4** 109–117.
[13] Agusta F K, Ayu D F and Rahmayuni 2020 *Journal of Food Technology* **14** 68-82.
[14] Mahmud M K et al 2018 Table of Indonesian Food Composition 2017. Directorate General of Public Health. Directorate of Community Nutrition. (Ministry of Health of the Republic of Indonesia).
[15] Nurjanah, Jacoeb A M, Nurokhmatunnisa and Pujianti D 2013 Journal of Processing of Indonesian Fishery Products **16** 95-107.
[16] Nurilmala M, Nurjanah, and Utama R H 2009 Journal of Processing of Indonesian Fishery Products **12** 1-16.

IOP Conf. Series: Earth and Environmental Science 1289 (2023) 012029 doi:10.1088/1755-1315/1289/1/012029

Screening of protease-producing lactic acid bacteria from Indonesian fermented fish products and its application in chitin extraction from vannamei shrimp (*Litopenaeus vannamei*) shell

A N Aisyah, M Gavi, and M M P Putra*

Fish Product Technology, Department of Fisheries, Faculty of Agriculture, Universitas Gadjah Mada, Jl. Flora Gedung A4, Bulaksumur Yogyakarta Indonesia 55281

*Corresponding author e-mail: primaputra@ugm.ac.id

Abstract. This study aimed to screen lactic acid bacteria (LAB) from traditional Indonesian fermented fish products, namely *wadi* and *jambal*, with high protease activity for chitin production from *Litopenaeus vannamei* shrimp shells. Twenty-eight colonies were successfully isolated and identified through biochemical tests such as catalase and Gram staining. Seven out of the twenty-eight colonies were identified as lactic acid bacteria based on their Gram-positive and catalase-negative results. These strains from *wadi* samples were named GWB 1.3, GWB 1.7, and GWB 2.5, while those from *jambal* samples were named GMJB 1.3, GMJB 1.6, GMJB 2.6, and GMJB 2.7. The protease activity of these strains was tested using the well-diffusion method on skim milk agar to assess their deproteinized activity. The diameter of the clear zone formed after 24, 48, and 72 hours of incubation was measured. GWB 1.7 and GMJB 2.6 showed the largest clear zone diameter with 21.3 mm and 22.5 mm, respectively, and were selected for the single step-fermentation of shrimp shells. The fermentation process was conducted in 6 days with different inoculum ranging from 0%, 5%, 10%, 15%, and 20% (v/v) as treatments. All treatments, including moisture, ash, and protein, were evaluated for proximate content in the final days. The results indicate that the 20% inoculum showed the highest deproteinization and demineralization for both bacterial strains. GWB 1.7 exhibited a demineralization of 43.7% and deproteinization of 25.2%, while GMJB 2.6 showed a demineralization of 39% and deproteinization of 16.1%. This study demonstrated that lactic acid bacteria from *wadi* and *jambal* samples could produce chitin from shrimp shells, a valuable biopolymer with various applications.

1. Introduction

Indonesia is a major shrimp exporter, with vannamei shrimp being a key contributor. However, 33% of exported vannamei shrimp lack shells, leading to environmental concerns over waste management [1]. Shrimp shell waste contains valuable components that can significantly increase profits for the shrimp processing industry. These components include protein (20%-40%), chitin (15%-40%), calcium carbonate (20%-50%), and other pigments [3]. Chitin has a high market value, reaching USD 42.49 billion in 2020 and is estimated to reach USD 69.3 billion by 2028 [2]. Chitin biopolymer has excellent potential in various fields, including medical, pharmaceutical, cosmetic, food technology, and water treatment industries [3].

The most common method for obtaining chitin from shrimp shells involves a chemical process that uses strong acid for demineralization and a strong base for deproteinization. However, Chemical processes used for chitin extraction from shrimp shells are detrimental to the environment [4], require a significant amount of energy [5], and result in lower-quality chitin [6]. Alternatively, an eco-friendly and safer method is to utilize microorganisms for extraction. Through the biological approach, the deproteinization process of crustacean shells can use proteolytic bacteria, while the demineralization process can involve bacteria that produce acid [7]. Lactic acid bacteria are known as bacteria that can produce lactic acid and, in some species, can also produce protease enzymes [8]. It has been found that lactic acid bacteria can extract chitin from shrimp shells while simultaneously demineralizing and deproteinizing them. The efficiency of this process depends on various factors, such as the duration of fermentation, the species of shrimp, and the carbon source utilized. According to research, lactic acid bacteria can demineralize with 44% to 97.2% efficiency and deproteinize with 54.7% to 94% efficiency [9].

Fermented Indonesian fish products and certain foods like vegetables and dairy contain lactic acid bacteria, which have numerous benefits such as improving health and nutrition, potential as an alternative to antibiotics, and ability in chitin extraction [3, 10, 11]. However, their usefulness in *wadi* and *jambal*, traditional fishery products from Central Kalimantan and Pangandaran, respectively, is yet to be fully explored. *Wadi* has a clay-like texture, salty flavor, and distinct fermented aroma, while *jambal* has a square shape with a crispy, brittle texture like fried bread [12]. This study aims to identify lactic acid bacteria that have the potential to produce protease enzymes for sustainable chitin extraction from *wadi* and *jambal*. It also seeks to establish the most effective percentage of inoculum to achieve optimal deproteinization and demineralization of *Litopenaeus vannamei* shrimp shells.

2. Material and Methods

2.1. Materials

Tools used in this research were microscope (Olympus Model BX51TRF, Olympus Optical Co., Ltd., Tokyo, Japan), pH meter (OHAUS Starter 5000), vortex (Barnstead M37610-33), autoclave (Hi-clave HVE-50 Hirayama), refrigerator -30ºC (Sanyo SR-D180F), incubator (Isuzu SSJ-115), miller (MKS-GH30), moisture analyzers (OHAUS MB120), and spectrophotometer UV-Vis (Genesys 10s UV-Vis). Materials used were fish fermented products, namely *jambal* (Pangandaran, West Java) and *wadi* (Palangka Raya, Central Kalimantan), vannamei shrimp (*Litopenaeus vannamei*), de man rogosa and sharpe broth (MRSB, Sigma Aldrich), de man rogosa and sharpe agar (MRSA, Sigma Aldrich), SM Agar (Skim Milk Agar, HIMEDIA), D-(+)-Glucose monohydrate suitable for microbiology (Sigma Aldrich), K_2HPO_4, yeast extract, 30% glycerol, NaCl 0.85%, 1% $CaCO_3$, alcohol 70%, Gram staining kit (Sigma Aldrich), H_2O_2 3%, and KOH 3%.

2.2. Methods

2.2.1. Isolation of bacteria

The isolation of lactic acid bacteria (LAB) was conducted using the dilution series spread plate technique as described by Lim (2016) [26]. In the initial stage, the *wadi* and *jambal* samples were each dissolved in 225 mL of 0.85% NaCl solution at a ratio of 25 grams per sample. Subsequently, serial dilutions were performed up to 10^{-5} by adding 1 mL of the previous dilution to 9 mL of sterile 0.85% NaCl solution. From each dilution series, 100 µL of bacterial suspension was taken and isolated on MRS agar medium (de-Mann, Rogosa, and Sharpe by Sigma Aldrich) supplemented with 1% $CaCO_3$ (w/v) using the spread plate method. The bacterial cultures were then incubated aerobically at 37°C for 48 hours. LAB isolates were selected based on the formation of clear zones around the colonies, and these colonies were streaked 2-3 times to obtain pure or single isolates. The obtained single isolates were then subjected to further characterization of LAB.

2.2.2. Characterization of lactic acid bacteria
The characterization of LAB was conducted through a series of biochemical tests, including Gram staining, catalase test, and pH test.

2.2.3. Proteolytic activity of isolated lactic acid bacteria
According to Phyu *et al.* (2015) [28], the protease activity testing was done by creating 7 mm wells on skim milk agar, and three replications were performed for each agar plate. A volume of 50 µl of bacterial culture in MRS Broth, with an optical density of 1 at 600 nm, was added to each well using a micropipette. Subsequently, the plates were incubated at 37°C for 24 hours, 48 hours, and 72 hours, and the formation of clear zone around the wells was recorded.

2.2.4. Preparation of sample for chitin extraction
Fresh shrimp samples were subjected to morphological examination utilized for shell extraction and cleaned with running water to remove any adhering impurities. Subsequently, the shrimp shells were heated in boiling water for one hour and rewashed to eliminate any remaining meat residues. The shrimp shells were then dried in an oven at 60°C for 24 hours. Once dried, the shrimp shells were crushed using a miller (MKS-GH30) to a size of 2-5 mm.

2.2.5. Chitin extraction
The fermentation process referred to Sedaghat *et al.* (2017) [5] with some modifications by inoculating 2.5 grams of shrimp shells with 12 mL of fermentation media. For the *wadi* samples, the fermentation media contained 15% (w/v) glucose and inoculum to isolate bacteria, while for the *jambal* samples, the media contained 15% (w/v) glucose, 2% (w/v) yeast extract, and 0.2% (w/v) K_2HPO_4, and inoculum. The fermentation process was conducted in an incubator at 37°C for six days. The independent variable in this research was the variation of inoculum percentage, ranging from 0%, 5%, 10%, 15%, to 20%, to determine the optimal inoculum. After fermentation, the results were filtered, and the obtained solids were dried in an oven at 60°C for 24 hours.

2.2.6. Data analysis
The statistical analysis used was parametric one-way ANOVA with a significance level of 95%, followed by the Tukey post-hoc test to compare differences among the independent variable groups. The independent variable in this study was the variation of the inoculum with five levels ranging from 0%, 5%, 10%, 15%, and 20%. The statistical analysis was performed using R Studio software.

3. Results and Discussion

3.1. Isolation of lactic acid bacteria
In this study, we successfully isolated 28 bacterial isolates from the Indonesian local fermented products, namely *wadi* and *jambal* samples. In the sample taken from the *wadi*, there were nine isolates that had round shapes, clear zones, and yellow-colored colonies. Additionally, five isolates had white-colored colonies. Similarly, in the *jambal* sample, there were eight round-shaped isolates with clear zones and yellow-colored colonies, as well as six isolates with white-colored colonies. Subsequently, the isolates obtained from both samples underwent re-isolation for purification. Furthermore, these 28 single isolates were subjected to biochemical tests to characterize them as lactic acid bacteria (LAB). Lactic acid bacteria (LAB) were identified as Gram-positive (+) bacteria, catalase-negative (-), non-spore-forming, non-motile, and typically displayed a round or rod-shaped morphology [13]. The results of the bacterial isolate characterization are shown in Table 1.

Three bacterial isolates were obtained from the *wadi* samples, while four isolates were identified from the *jambal* samples, exhibiting characteristic features of lactic acid bacteria (LAB) based on biochemical tests. One of the main features of LAB is their ability to produce lactic acid through glucose metabolism, leading to an acidic environment that favors their optimal growth [14]. Table 1 shows that all bacterial isolates could acidify their environment, with the lowest pH value in the suspected LAB isolate GWB 2.5 at 3.79 and the highest in GMJB 1.3 at 5.61. The differences in pH values can be attributed to the specific characteristics of each isolate in terms of its tolerance to acidic conditions. Some LAB strains cannot grow at pH levels below 4, especially homofermentative LAB that require slightly acidic to neutral conditions for glucose fermentation [15].

Table 1. The results of bacterial isolate characterization.

Isolated Strains	Catalase	Gram	Cell Shape	pH	Description
GWB 1.1	-	-	Coccus	4.39	non-LAB
GWB 1.2	+	-	Coccus	3.84	non-LAB
GWB 1.3	-	+	Coccus	4.36	LAB
GWB 1.4	+	-	Coccus	4.35	non-LAB
GWB 1.5	+	-	Bacillus	3.99	non-LAB
GWB 1.6	+	+	Coccus	4.53	non-LAB
GWB 1.7	-	+	Coccus	4.75	LAB
GWB 2.1	+	+	N/D	4.25	non-LAB
GWB 2.2	+	+	N/D	3.94	non-LAB
GWB 2.3	+	+	N/D	3.86	non-LAB
GWB 2.4	+	+	N/D	3.82	non-LAB
GWB 2.5	-	+	Coccus	3.79	LAB
GWB 2.6	+	+	N/D	3.83	non-LAB
GWB 2.7	+	+	N/D	3.86	non-LAB
GMJB 1.1	+	+	Coccus	5.53	non-LAB
GMJB 1.2	+	+	Coccus	5.46	non-LAB
GMJB 1.3	-	+	Coccus	5.61	LAB
GMJB 1.4	-	-	Coccus	5.72	non-LAB
GMJB 1.5	+	-	Coccus	5.22	non-LAB
GMJB 1.6	-	+	Coccus	5.04	LAB
GMJB 1.7	+	-	Coccus	5.22	non-LAB
GMJB 2.1	+	+	Coccus	5.62	non-LAB
GMJB 2.2	+	+	Coccus	5.34	non-LAB
GMJB 2.3	-	-	Coccus	5.42	non-LAB
GMJB 2.4	+	+	Coccus	5.37	non-LAB
GMJB 2.5	+	+	Coccus	5.31	non-LAB
GMJB 2.6	-	+	Coccus	5.44	LAB
GMJB 2.7	-	+	Coccus	5.23	LAB

3.2. Screening for protease activity

The single fermentation of shrimp shells into chitin requires bacterial isolates with specific criteria, namely the ability to produce acid for demineralizing the shrimp shells and protease activity for deproteinization. Protease activity is crucial because the protease enzyme plays a role in deproteinization, breaking down the proteins in the shrimp shells into simpler components. The seven isolated strains identified as LAB through biochemical tests, namely GWB 1.3, GWB 1.7, GWB 2.5, GMJB 1.3, GMJB 1.6, GMJB 2.6, and GMJB 2.7, were subjected to protease activity testing. Using skim milk agar in the protease test aims to assess the bacteria's ability to break peptide bonds in the casein in skim milk. The formation of a clear zone around bacterial colonies determines protease enzyme production. A larger diameter of the zone indicates greater protease activity [16]. Observations of the clear zone formation were recorded after 24 hours, 48 hours, and 72 hours of incubation, as shown in Table 2.

Table 2. Clear Zone Diameters of Selected LAB on Skim Milk Agar Plate

Source	Isolates	Diameter (mm)		
		24 hours	48 hours	72 hours
Wadi	GWB 1.3	18.2 ± 0.14^a	18.9 ± 0.14^a	19.8 ± 0.17^a
Wadi	GWB 1.7	20.3 ± 0.21^a	21.3 ± 0.18^a	21.8 ± 0.15^a
Wadi	GWB 2.5	18.5 ± 0.21^a	19.3 ± 0.22^a	19.3 ± 0.20^a
Jambal	GMJB 1.3	18.8 ± 0.05^a	19.7 ± 0.02^b	20.1 ± 0.02^c
Jambal	GMJB 1.6	19 ± 0.02^a	20.2 ± 0.02^b	20.8 ± 0.03^c
Jambal	GMJB 2.6	19.6 ± 0.05^a	22.2 ± 0.03^a	22.5 ± 0.05^a
Jambal	GMJB 2.7	18.9 ± 0.02^a	20.3 ± 0.02^b	21.6 ± 0.02^b

$^{a-c}$ values with the same superscripts within each column in the *wadi* and *jambal* sections are not significantly different ($P > 0.05$).

The results showed that all bacterial isolates obtained exhibited protease activity, as indicated by forming a clear zone. The diameter of the clear zone formed by the seven bacteria ranged from 18.2 mm to 22.5 mm. These diameter measurements are similar to previous studies on the proteolytic activity of lactic acid bacteria from yogurt products [17] and higher than previous research on *Lactobacillus plantarum* strains [18]. These differences can be attributed to the unique characteristics of each bacterial strain and their specific enzymatic capabilities, resulting in diverse protease activities. This is consistent with the study by Cao *et al.* (2019) [19], which tested the protease activity of 133 lactic acid bacteria and obtained clear zone diameter results ranging from 8.72 mm to 28.51 mm

The results did not show significant differences among the three isolates in the wadi samples. All three isolates exhibited a maximum increase in clear zone diameter of 1 mm over the incubation period. Although there are no significant differences, GWB 1.7 consistently showed the largest diameter at each observation time, with a mean value of 21.13 mm. On the other hand, in the jambal samples, significant differences were observed at the 48-hour and 72-hour observation times. GMJB 2.6 significantly exhibited larger clear zones than the three different bacterial strains. GMJB 2.7 followed with a diameter of 21.6 mm at the 72-hour observation time. Subsequently, the bacterial isolates with the most significant protease activity, as indicated by the size of the clear zone in each sample, were selected for further research on chitin extraction. This study chose GWB 1.7 and GMJB 2.6 as the bacterial strains for fermenting shrimp shells into chitin.

3.3. Chitin Extraction

Analysis of the proximate composition of dried vannamei shrimp (*Litopenaeus vannamei*) shells in flake form before the fermentation step is presented in Table 3.

Table 3. Proximate composition of dried vannamei shrimp shells

Parameter	Content (%)
Moisture	7.82
Ash	30.63
Protein	38.41

Table 4. Proximate composition of fermented vannamei shrimp shells

Isolates	Inoculum	Moisture (%)	Ash (%)	Protein (%)	Demineralization (%)	Deproteinization (%)
GWB 1.7	0%	3.4 ± 0.33^{a}	29.7 ± 0.02^{a}	38.4 ± 0.12^{a}	8.4 ± 0.07^{e}	5.7 ± 0.31^{d}
	5%	2.5 ± 0.15^{b}	23.3 ± 0.08^{b}	34.8 ± 0.05^{b}	28.2 ± 0.25^{d}	14.6 ± 0.13^{c}
	10%	2.9 ± 0.06^{ab}	20.8 ± 0.15^{c}	30.7 ± 0.20^{c}	35.8 ± 0.46^{c}	24.6 ± 0.48^{b}
	15%	2.9 ± 0.31^{ab}	20.2 ± 0.14^{d}	28.7 ± 0.07^{d}	37.8 ± 0.43^{b}	29.4 ± 0.18^{a}
	20%	3.4 ± 0.23^{a}	18.3 ± 0.15^{e}	30.4 ± 0.22^{c}	43.7 ± 0.46^{a}	25.2 ± 0.53^{b}
GMJB 2.6	0%	4.0 ± 0.19^{c}	30.3 ± 0.17^{a}	38.9 ± 0.11^{a}	6.6 ± 0.55^{e}	4.4 ± 0.28^{d}
	5%	4.6 ± 0.06^{bc}	25.1 ± 0.03^{b}	36.2 ± 0.09^{b}	22.6 ± 0.10^{d}	11.2 ± 0.23^{c}
	10%	5.0 ± 0.20^{ab}	23.2 ± 0.11^{c}	34.3 ± 0.11^{c}	28.6 ± 0.35^{c}	15.6 ± 0.26^{b}
	15%	5.6 ± 0.51^{a}	22.1 ± 0.04^{d}	32 ± 0.02^{d}	31.8 ± 0.12^{b}	21.3 ± 0.04^{a}
	20%	5.0 ± 0.51^{ab}	19.8 ± 0.12^{e}	34.1 ± 0.07^{c}	39 ± 0.38^{a}	16.1 ± 0.18^{b}

$^{a\text{-}e}$ values with the different superscripts within each column in both strain-isolated sections are significantly different ($P < 0.05$).
$^{ab \text{ or } bc}$ superscript combination indicates that these groups are not significantly different from each other ($P > 0.05$) but different from groups with different superscripts.

The chitin moisture content obtained is presented in Table 4, showing a reduction of more than 50% for GWB 1.7 and more than 30% for GMJB 2.6 from the dried shrimp shells. This reduction might be attributed to the re-drying process at 60°C for 24 hours after the fermentation stage. The moisture content in chitin is highly dependent on the drying process, packaging, and environmental factors [20]. This is supported by the research of [21], which demonstrated that the drying process influences the moisture content, drying duration, amount of chitin dried, drying area, and the drying method used. ANOVA test on the chitin moisture content based on the inoculum levels in both bacterial strains showed significant differences. In GWB 1.7, there was no significant difference between 0% and 20% inoculum, while a significant difference was observed at 5% inoculum. Different results were found in GMJB 2.6, where 0% inoculum significantly differed from other treatments except for the 5% inoculum. These findings are consistent with previous research, where fermentation of shrimp shells using LAB resulted in a chitin moisture content of 2.11% [22]. Furthermore, according to Indonesian National Standards, the maximum moisture content for chitin is 12%, and the research results indicate that all inoculum treatments meet this standard.

The ash content obtained from both bacterial strains decreases with increasing inoculum levels. This indicates the occurrence of demineralization during the fermentation process of shrimp shells. The results of this study are consistent with previous research, where a higher inoculum amount led to increased lactic acid production and a higher percentage of demineralization [5]. Strain GWB 1.7, with an inoculum of 20%, showed the highest demineralization percentage at 43.7%. This result is higher compared to GMJB 2.6, which had the highest demineralization percentage at 39%. These differences may be attributed to the distinct characteristics of GWB 1.7 and GMJB 2.6 in acid production. However, the demineralization percentage of GWB 1.7 is lower than some references in the literature that reported demineralization reaching 82% at 20% inoculum and a 6-day fermentation process [5]. These variations could be due to differences in experimental conditions, as the fermentation process in this study was conducted without using a shaker for an entire 6-day period and was only done once every 24 hours using a vortex. The lower demineralization could result from shrimp shell sedimentation at the bottom of the fermentation vessel, reducing optimal contact with the acid the bacteria produces and some calcium carbonate not entirely dissolving during the fermentation process [23].

The protein content from the bacterial strains decreased significantly when inoculated with 0% to 15%. This decrease indicates that deproteinization occurred during the shrimp shell fermentation process. At the 20% inoculum level, the percentage of deproteinization decreased, possibly because the high number of bacteria affected the protease enzymes involved in deproteinization [22]. Strain GWB 1.7, with a 15% inoculum, showed the highest deproteinization percentage at 29.4%. This result is higher than strain GMJB 2.6, which had the highest deproteinization percentage at 21.3%. These differences could be attributed to variations in the characteristics of the protease enzymes produced by strain GWB 1.7 and GMJB 2.6. Different fermentation media could also influence the protein content produced by strain GMJB 2.6. GMJB 2.6 uses additional fermentation media containing yeast extract, which is known to have a high protein content of 64.1%, potentially leading to other proteins being processed by bacteria during the fermentation process [24]. However, these deproteinization results are lower than most literature on chitin extraction using biological methods. The lower deproteinization could be associated with some lactic acid bacteria protease enzymes having difficulty accessing the α-chitin structure in the shrimp shells [25]. The use of lactic acid bacteria in single fermentation allows for an inverse relationship between demineralization and deproteinization due to changes in the protease configuration in acidic pH environments, which can affect its activity [23].

4. Conclusion
The highest protease activity of several bacterial strains examined was the GWB 1.7 isolate from *wadi* and the GMJB 2.6 isolate from *jambal*. Further study was conducted on these two bacteria to explore their potential in demineralizing and deproteinizing vannamei shrimp shell chitin. The results were highly encouraging, showing significant success in demineralization and deproteinization. The most promising outcomes were achieved using the GWB 1.7 strain with a 20% inoculum, which produced a demineralization rate of 43.7% and a deproteinization rate of 25.2%. These findings suggest that lactic acid bacteria from *wadi* and *jambal* are promising resources for the chitin extraction industry.

Acknowledgments
This research was part of ANA and MG thesis and funded by Final Project Recognition Grant Universitas Gadjah Mada Number 5075/UN1.P.II/Dit-Lit/PT.01.01/2023 to MMPP.

References
[1] Kementerian Kelautan dan Perikanan 2022 *Statistik Ekspor Hasil Perikanan Tahun 2017-2021* (Jakarta: Ditjen PDSPKP) p12

[2] Vicente F A, Ventura S P M, Passos H, Dias A C R C, Tores-Acosta M A, Novak U and Likozar B 2022 Crustacean waste biorefinery as a sustainable cost-effective business model. *Chem. Eng. J.* **442** 1 -13

[3] Casadidio C D, Peregrina V, Gigliobianco M R, Deng S, Censi R and Martino P D 2019 Chitin and chitosans: characteristics, eco-friendly processes, and applications in cosmetic science. *Mar. Drugs* **17** 1-30

[4] Healy M, Green A and Healy A 2003 Bioprocessing of marine crustacean shell waste. *Acta Biotechnol.* **23** 151-60

[5] Sedaghat F, Yousefzadi M, Toiserkani H and Najafipour S 2017 Bioconversion of shrimp waste *Penaeus merguiensis* using lactic acid fermentation: an alternative procedure for chemical extraction of chitin and chitosan *Int. J. Biol. Macromol.* **104** 883-8.

[6] Muzzarelli R, Tarsi R, Filippini O, Giovanetti E, Biagini G and Varaldo P E 1990 Antimicrobial properties of N-carboxybutyl chitosan *Antimicrob. Agents Chemother.* **34** 2019-23

[7] Jung W J, Jo H, Kuk J H, Kim Y J, Oh K T and Park R D 2007 Production of chitin from red crab shell waste by successive fermentation with *Lactobacillus paracasei* KCTC-3074 and *Serratia marcescens* FS-3 *Carbohydr. Polym.* **68** 746-50.

[8] Duan S, Li L, Zhuang Z and Wu W 2012 Improved production of chitin from shrimp waste by fermentation with epiphytic lactic acid bacteria *Carbohydr. Polym.* **89** 1283-88

[9] Arbia W, Arbia L, Adour L and Amrane A 2013 Chitin extraction from crustacean shells using biological methods – a review *Food Technol. Biotechnol.* **51** 12–25

[10] Gilliland S E 1990 Health and nutritional benefits from lactic acid bacteria *FEMS Microbiol. Lett.* **87** 175-88
[11] Vieco-Saiz N, Belguesmia Y, Raspoet R, Auclair E, Gancel F, Kempf I and Drider D 2019 Benefits and inputs from lactic acid bacteria and their bacteriocins as alternatives to antibiotic growth promoters during food-animal production *Front. Microbiol.* **10** 1-17.
[12] Karyantina M, Anggrahini S, Utami T and Rahayu E S 2021 Karakteristik *jambal* roti ikan manyung (*Arius thalassinus*) dengan suplementasi bakteri asam laktat indigenous *Agrointek J. Tek. Ind. Pertan.* **15** 676-84
[13] Leska A, Nowak A and Motyl I 2022 Isolation and some basic characteristics of lactic acid bacteria from honeybee (*Apis mellifera* L.) environment—a preliminary study *J. Agric.* **12** 1-18
[14] Abedi E and Hashemi S M B 2020 Lactic acid production – producing microorganisms and substrates sources-state of art *Heliyon* **10** 1-32
[15] Wang Y, J Wu, M Lv, Z Shao, M Hungwe, J Wang, X Bai, J Xie, Y Wang and W Geng 2021 Metabolism Characteristics of lactic acid bacteria and the expanding applications in food industry *Front. Bioeng. Biotechnol.* **9** 1-19
[16] Yusmarini, Indrati R, Utama T and Marsono Y 2009 Isolasi dan identifikasi bakteri asam laktat proteolitik dari susu kedelai yang terfermentasi spontan *J. Natur. Indones.* **12** 28-33
[17] Phyu H E, Oo Z K and Aye K N 2015 Screening on proteolytic activity of lactic acid bacteria from various yogurts and fermented milk *Int. J. Adv. Sci. Eng. Technol.* **11** 34-7
[18] Ines E, Maher M, and Mnasser H 2009 Technological and safety properties of *Lactobacillus plantarum* strains isolated from a Tunisian traditional salted meat *Meat Sci.* **81** 203-8
[19] Cao C, Feng M, Sun J, Xu X, and Zhou G 2019 Screening of lactic acid bacteria with high protease activity from fermented sausages and antioxidant activity assessment of its fermented sausages *CyTA J. Food* **17** 347-54
[20] Suhardi 1992 *Buku Monograf Khitin dan Khitosan* (Yogyakarta: PAU UGM)
[21] Saleh M R, Abdillah, Suerman E, Basmal J and Indriati N 1994 Pengaruh suhu, waktu dan konsentrasi pelarut pada ekstraksi kitosan dari limbah pengolahan udang beku terhadap beberapa parameter mutu kitosan *J. Pasca Panen Perikan.* **81** 30-43
[22] Castro R, Guerrero-Legarreta I and Bórquez R 2018 Chitin extraction from *Allopetrolisthes punctatus* crab using lactic fermentation *Biotechnol. Rep.* **20** 1-5
[23] Li J, Song R, Zou X, Wei R and Wang J 2023 Simultaneous preparation of chitin and flavor protein hydrolysates from the by-products of shrimp processing by one-step fermentation with *Lactobacillus fermuntum Molecules* **28** 1-18
[24] Vieira E F, Carvalho J, Pinto E, Cunha S, Almeida A A and Ferreira I M P L V O 2016 Nutritive value, antioxidant activity and phenolic compounds profile of brewer's spent yeast extract *J. Compos. Anal.* **52** 44-51
[25] Tan J S, Abbasiliasi S, Lee C K and Phapugrangkul P 2020 Chitin extraction from shrimp wastes by single step fermentation with *Lactobacillus acidophilus* FTDC3871 using response surface methodology *J. Food Process. Preserv.* **44** 1-11
[26] Lim E S 2016 Inhibitory effect of bacteriocin-producin lactic acid bacteria against histamine-forming bacteria isolated from Myeolchi-jeot *Fish. Aquat. Sci.* **19** 1-10

IOP Conf. Series: Earth and Environmental Science 1289 (2023) 012030 doi:10.1088/1755-1315/1289/1/012030

Ultraviolet irradiation effect on characteristics of protease enzyme from African catfish (*Clarias* sp.) intestines

A Tamaya[1], A Suwanto[2], Hernawan[2], M Anwar[2], C D Poeloengasih[2], B Juligani[2], W Apriyana[2], A Fauzi[2], P Yuliyanto[2], R Suryani[2], Y Khasanah[2], S M Syarifah[1], D J Prasetyo[2*]

[1]Departement of Biotechnology, Faculty of Science and Technology, Universitas 'Aisyiyah Yogyakarta, Indonesia

[2]Research Center for Food Technology and Processing, National Research and Innovation Agency Republic of Indonesia, Yogyakarta, Indonesia

Corresponding author mail: dwi.joko.prasetyo@brin.go.id

Abstract. The goal of this study was to investigate the impact of UV-C irradiation on the activity of crude enzymes derived from African Catfish intestines as well the optimum conditions for subsequent autolysis processes. Three key variables were assessed: UV-C exposure time, pH and temperature. The experiments involved the homogenization of intestines in cold distilled water followed by subjecting the mixture into UV-C treatment for exposure times of 0, 10, 20 and 30 minutes. After centrifugation process, the collected supernatant was utilized as the crude enzyme. At various level of temperature (40, 50, 60 °C) and pH (6, 7, 8) the activity of crude enzyme was assayed with casein as substrate. The findings revealed the increasing of EA as the UV-C exposure time decreased and pH increased, while the highest EA value was observed at temperature of 50°C. Consequently, the optimal conditions were identified as follows: 0 minutes of UV-C exposure time, pH of 8, and temperature of 50°C. Furthermore, UV-Vis absorption, FTIR, and fluorescence spectroscopy of the crude enzyme after UV-C irradiation induction was studied to investigate the its conformational changes. These additional analyses provided valuable insights into the structural alterations of the crude enzyme caused by the UV-C treatment.

1. Introduction

Food loss has become a pressing global issue due to the substantial quantities and varied sources of waste. On the other hand, million people suffer from malnutrition and food security. Both plant and meat-based foods contribute to this waste at various stages, including production, post-harvest handling, processing, distribution, and consumption. The marine and fisheries sector is one of the significant contributors to food loss. For instance, in the South and South-East Asia region, food loss has reached alarming levels, exceeding 30% [1]. In Indonesia specifically, a staggering 35% of seafood waste is generated before consumption [2]. Consequently, there is an urgent need for appropriate handling techniques to address this critical issue.

In Indonesia, African catfish (Clarias sp.) is a popular fish, known for its ease of cultivation and preparation, making it widely available across the region. Typically, African catfish (AC) is directly cooked by frying before consumption, but it can also be processed into various products such as floss, meatballs, nuggets and more. AC processing by-products could reach up to 60-70%, comprising bones, viscera, head and scales [3]. Notably, AC viscera, including the intestines, not only retain nutritional compounds like proteins and fats but also possess protease enzymes [3–5]. This combination of protein and protease enzymes suggests the potential for the material to be converted into protein hydrolysate through the autolysis process [6].

Autolysis process involves utilizing the enzymes present within the substance to convert it into a hydrolyzed product [7]. This process proves to be more economical as it eliminates the need for additional enzymes. Previous research has explored autolysis of marine and fisheries processing by-products, such as shrimp head, rainbow trout waste and jumbo squid viscera, to produce protein hydrolysates [8–10]. To achieve the desired product, proper conditions, especially pH and temperature, must be determined. It is essential to assess the protease enzyme activity (EA) before initiating the autolysis process, as this can minimize the number of experimental trials. For instance, Cao *et al.* (2014) conducted EA assays before processing shrimp head through autolysis, while also employing ultraviolet (UV) irradiation to enhance the protease EA from shrimp head [11]. Their previous study also exploited the ultraviolet (UV) irradiation to enhance EA of protease from head shrimp [8,11]. Their findings contradicted other studies that generally reported enzyme inactivity after UV exposure [12,13]. However, the use of UV irradiation as a pretreatment to retrieve protease from AC intestines has not been explored yet. Therefore, the primary objectives of this research are not only to determine the optimum conditions for EA from AC intestines but also to evaluate the impact of UV irradiation on EA.

2. Materials and methods

2.1. Materials

Alive African catfish (*Clarias* sp.) weighing 7 fish per kilogram were obtained from a fish farmer located in Gunungkidul Regency, Special Region of Yogyakarta, Indonesia. Upon acquisition, the fish were humanely sacrificed and subsequently eviscerated by hand. The focus of this study was on the collection of the digestive tract, specifically the intestines, which were carefully separated for further analysis. The chemicals employed in this research included casein (Sigma), L-tyrosine (Sigma), potassium dihydrogen phosphate/KH_2PO_4 (Merck), hydrochloric acid/HCl (Merck), trichloroacetic acid/TCA (Sigma), sodium hydroxide/NaOH (Merck), distilled water (Waterone), bovine serum albumin/BSA (Himedia) and the BCA Protein Assay Kit (Merck). All chemicals were used in their as-received state without further purifications.

2.2. Ultraviolet irradiation treatment and crude enzyme extraction

The UV irradiation treatment and crude enzyme extraction procedures were carried out based on previous studies by Cao *et al.* (2014) and Dou *et al.* (2023) with some modifications [11,14]. The catfish intestines were diced and chopped using a food chopper for 4 min with the addition of cool pure water (1:3 w/v). Subsequently, the homogenized mixture was placed in a petri dish, and UV irradiation was applied using a Himawari T8 UV-C 20 W lamp. The distance between lamp and sample was fixed at a distance of 20 cm, and the exposure time was varied from 0 to 30 minutes at interval of 10 minutes. After the UV treatment, the mixture underwent centrifugation at speed of 10,000 rpm and temperature of 4°C for 10 minutes. The resulting supernatant, containing the crude enzyme, was collected and stored at a temperature of -20°C until further use.

2.3. Protein content determination

The determination of protein content in the crude enzyme solution was conducted using the BCA Protein Assay Kit from Merck with BSA solution as the standard solution for reference. The protocol provided by the supplier was proceeded.

2.4. Determination of enzyme activity

The method used in this study was based on the approaches described by Klomklao (2010) and Cao *et al.* (2014) with some modifications [3,11]. A phosphate buffer system with a concentration of 0.1 M and pH values of 6, 7 and 8 was employed, and casein served as the substrate for the enzymatic reaction. Three variation of temperature (40, 50, 60°C) were selected for investigation. To begin the reaction, 600 µL of the chosen buffer solution was mixed with 200 µL of a mixture containing 2% (w/v) casein and then heated to the desired temperature. Subsequently, 200 µL of the crude enzyme solution, which had been previously diluted 100 times, was added to the mixture. After allowing the reaction to proceed for 20 minutes, it was halted by adding 200 µL of a 50% (w/v) TCA solution. For the control setup, the same procedure was followed, except for the addition of the enzyme solution, which was done after the TCA solution. The reaction mixture was then subjected to centrifugation at 10,000 rpm and a temperature of 4°C for 10 minutes. The content of tyrosine in the supernatant was determined using the BCA reagent [14]. Activity units (U) were defined as 1 µmol of tyrosine released from casein per minute of the reaction [4]. Specific activity was calculated by dividing the activity units by the protein content of the crude enzyme solution [15].

2.5. Conformational analysis

The crude enzyme from UV irradiated AC intestines at exposure time of 0, 10, 20 and 30 minutes was used as samples to collect of FTIR, UV-Vis and Fluorescene spectra. The protein concentration of crude enzyme extract was diluted to 100 mg/L prior to measurement [14].

2.5.1. Fourier Infra Red Transform (FTIR) spectroscopy. FTIR spectra of crude enzyme extract were measured using Vertex-60 (Bruker) spectrometer in the region of 600-4000 cm^{-1} with resolution of 4 cm^{-1}. The attenuated total reflection (ATR) technique was used by dropping the liquid sample into ATR crystal.

2.5.2. UV-Vis spectrometry. Cary 60 UV-Vis (Agilent) spectrophotometer was used to measure the absorption spectra of crude enzyme extract in the region of 200-800 nm [14].

2.5.3. Fluorescence spectroscopy. Fluorescence spectra of crude enzyme sample was measure using RF-6000 (Shimadzu) at excitation wavelength of 280 nm and the scanning wavelength of 300-600 nm.

2.6. Stastistical analysis

Three way ANOVA technique using CoStat software was used to analyze the result. The variables were UV exposure time (0, 10, 20, 30 minutes), test pH (6, 7, 8) and test temperature (40, 50, 60°C).

3. Results and discussions

3.1. Enzyme activity of crude enzyme

The determination of the significance of each variable was proceeded using Three Way ANOVA and the result was presented in Table 1. The statistical analysis indicated a significant effect not only for all variables (temperature, UV exposure time, pH) but also the their interactions. The R^2 value, reaching 0.9949 further supported the suitability of the three way ANOVA in this case. Moreover, the effect of each variable on the EA of protease enzyme from AC intestines was also studied. The mean value of EA from each variable were statistically compared to identify the optimum condition of each variable and this results were presented in Table 2. Additionally, the trend of enzyme activity (EA) of crude protease enzyme from African catfish (AC) intestines was depicted in Figure 1.

In general, the results demonstrated that EA tend to decrease as ultraviolet (UV) irradiation exposure time increase. This findings aligns with previous study which indicated the protein denaturation and enzyme inactivation due to UV irradiation [12,13,16]. Another study by Kristo *et al.*

(2012) reported structural change of whey protein after UV irradiation [17]. hese structural change and protein denaturation could be the reasons behind the observed decrease in EA of protease enzyme from AC intestines following UV irradiation. However, study by Cao *et al.* (2014) and Dou *et al.* (2023) yielded contradictory result as the EA of head shrimp protease enzyme increased after UV irradiation [11,14]. The difference nature of enzyme sources may be the underlying cause of this contrast result. Thus further in-depth studies are required to investigate the specific reasons that affect the EA of protease enzyme under UV irradiation.

The results presented in Table 2 revealed the highest EA at pH of 8 and temperature of 50 °C. Previous studies also indicated that enzymes extracted from marine and fisheries substance with neutral buffers tend to have optimum EA at alkaline condition (pH of 8-9) and temperature range 45-60°C [3,18,19]. Vannabun *et al.* (2014) reported higher EA of alkaline protease enzyme extracted from giant catfish viscera compared to acid protease [20]. Similar studies on protease enzymes extracted from other surces, such as plants and microbes, have also shown that optimum EA was achieved at pH of 7-9 and temperature of 50°C [21–24]. These findings suggested that alkaline protease likely played a dominant role in the protease enzyme extracted with neutral solvent. However, since this study only explored the effect of temperature and pH on EA from AC intestines within limited range, further investigation is needed, especially to determine the specific optimum pH.

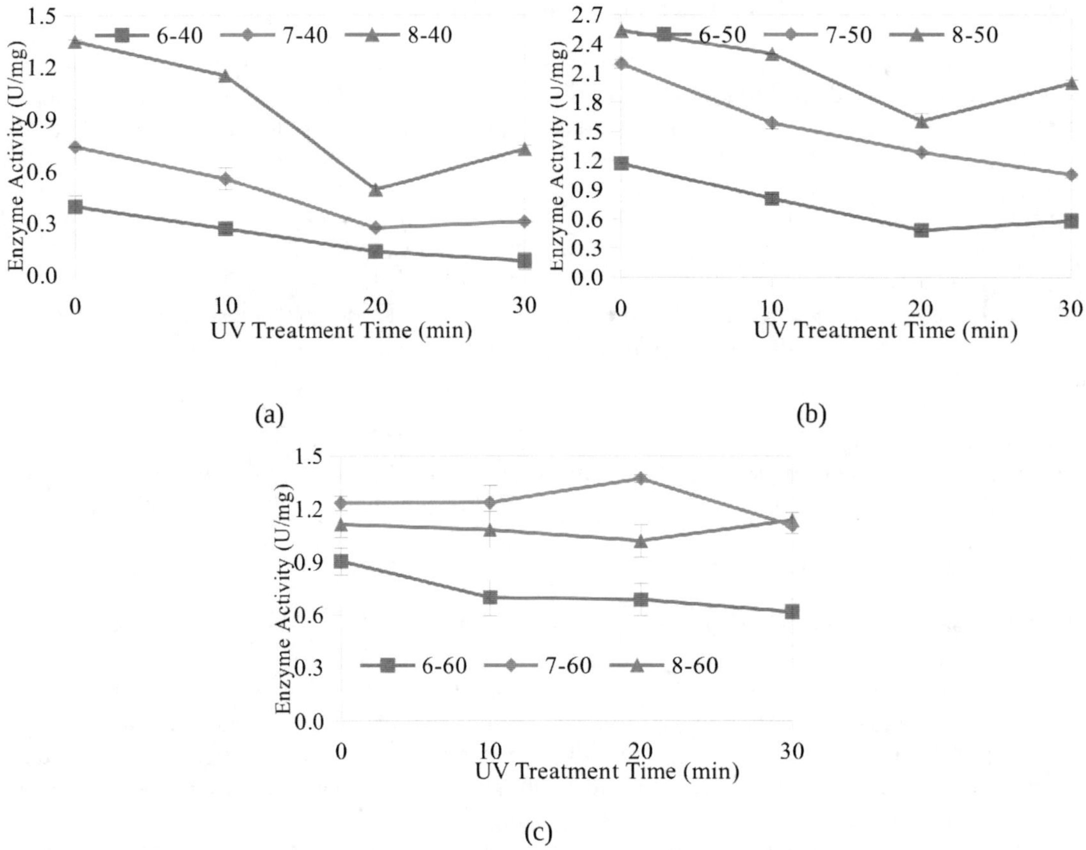

Figure 1. Trend of enzyme activities at temperature of (a) 40; (b) 50; (c) 60 with red, green and blue line indicated pH of 6, 7 and 8 respectively.

IOP Conf. Series: Earth and Environmental Science 1289 (2023) 012030 doi:10.1088/1755-1315/1289/1/012030

Table 1. Three way ANOVA for studying the significance effect of variables on enzyme activity

Source	DF	SS	MS	F	P
Temperature	2	15.5480	7.7740	2871.1593	0.0000
UV-time	3	4.0476	1.3492	498.2995	0.0000
pH	2	12.0316	6.0158	2221.7992	0.0000
Temperature*UV-time	6	1.6878	0.2813	103.8933	0.0000
Temperature*pH	4	3.7767	0.9442	348.7125	0.0000
UV-time*pH	6	0.4942	0.0824	30.4214	0.0000
Temperature*UV-time*pH	12	0.5314	0.0443	16.3550	0.0000
Error	72	0.1949	0.0027		
Model	35	38.1174	1.0891	402.2233	0.0000

Table 2. Effects of different temperature, pH and UV exposure time on enzyme activity

Variable	Level	Enzyme Activity (U/mg)
Temperature (°C)	40	0.5373^{c}
	50	1.4665^{a}
	60	1.0154^{b}
pH	6	0.5668^{z}
	7	1.0773^{y}
	8	1.3751^{x}
UV exposure time (min)	0	1.2928^{p}
	10	1.0745^{q}
	20	0.8435^{r}
	30	0.8147^{s}

Superscripts a-c, x-z and p-s represented significant difference ($P<0.05$) within temperature, pH and UV exposure time, respectively.

3.2. Conformational Analysis

4. The conformational study involved analysis of the alteration of enzyme structure from its FTIR, UV-Vis and fluorescene spectra after UV irradiation pretreatment.

4.1.1. Fourier Infra Red Transform (FTIR) spectroscopy. The FTIR spectra of the enzymes subjected to various UV exposure time was presented in Figure 2. The focus of the observation was focused on the transmittance in the 1600-1700 cm^{-1} region which corresponded to the amide I functional group [14]. Typically peak in the regions of 1600-1640, 1640-1650, 1650-1660 and 1660-1700 cm^{-1} represented structure of β-folding, random coiling, α-helix and β-corner, respectively [25]. In this study, all enzymes exhibited a peak only at region 1640-1650 cm^{-1}, with no substantial changes observed. This finding aligned with previous study by Dou *et al.* (2023) which also reported no significant changes in the FTIR spectra [14].

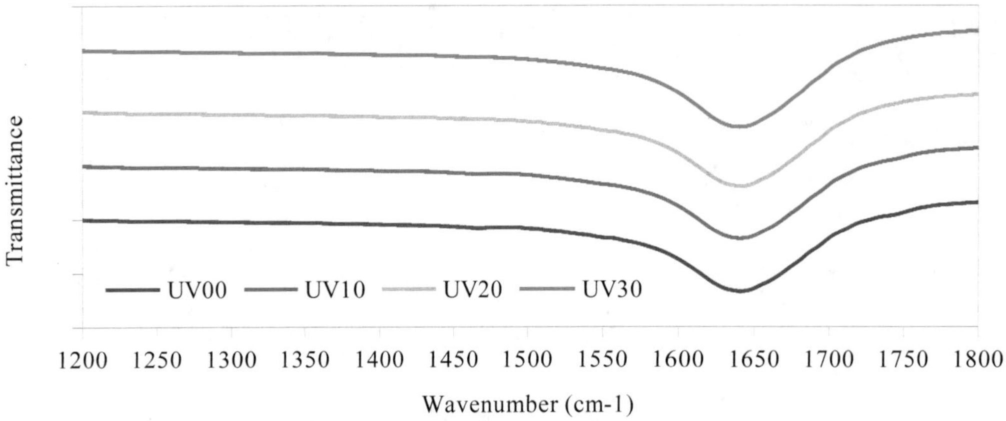

Figure 2. FTIR spectra of enzyme extracted from AC intestines treated by UV exposure time of 0 (UV00), 10 (UV10), 20 (UV20) and 30 minutes (UV30)

4.1.2. UV-Vis spectrometry. UV-Vis spectra of enzymes was recorded and presented in Figure 3. Various peaks appeared in the region of 200-230 nm wavelength, indicating the presence of peptides that absorb in the far UV region (180-230 nm) [26]. The peaks observed in this study were more complex compared to the previous findings by Dou *et al.* (2023) [14] which used purified enzyme, unlike the crude enzyme used in our study [14]. Additionally, the results showed the disappearance of some peaks in this region after UV irradiation, suggesting that certain compounds might have degraded due to UV treatment. Furthermore, a peak in region 240-280 nm was observed in all enzyme spectra, with no significant differences among the enzymes. This peak indicated the presence of aromatic side chain in amino acid mainly phenylalanine, tryptophan and tyrosine [27].

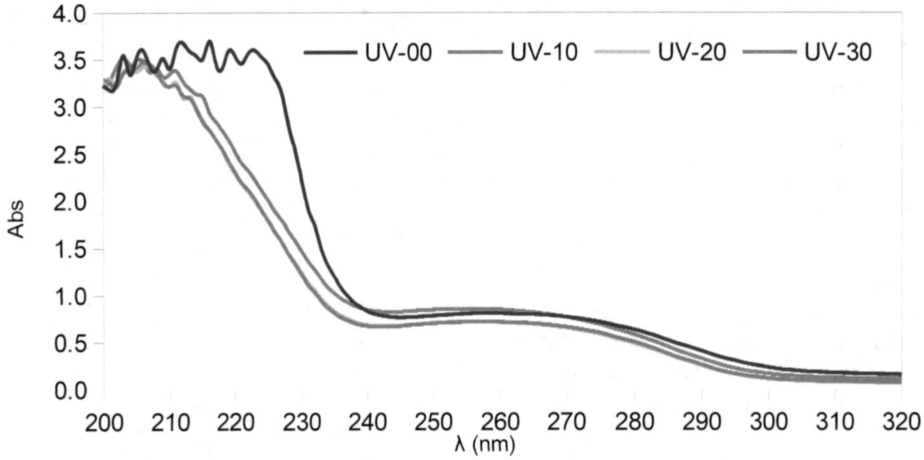

Figure 3. UV-Vis spectra of enzyme extracted from AC intestines treated by UV exposure time of 0 (UV00), 10 (UV10), 20 (UV20) and 30 minutes (UV30)

4.1.3. Fluorescence spectroscopy. Protein generally exhibits intrinsic fluorescene emitted by tryptophan and tyrosine residue at excitation and emission wavelength of 280 and 300-400 nm respectively [14]. In our study, the emission peak of protease enzyme from AC intestines after UV irradiation treatment shifted from 342 to 335 nm and the intensity was increased as UV exposure time increased. Interestingly, this result contrasts with the findings of Dou *et al.* (2023) who reported an escalation of EA after UV irradiation and observed a red shift of the emission peak from 330 to 355 nm [14]. A study by Kristo *et al.* (2012) also demostrated a red shift of the emission peak in whey protein sample after UV treatment, indicating structural change [17]. Considering that UV treatment appears to induce structural changes in proteins, the specific reasons for how this phenomenon affects the EA remain unknown. Thus, further detailed studies are required to uncover the underlying reasons behind this observation.

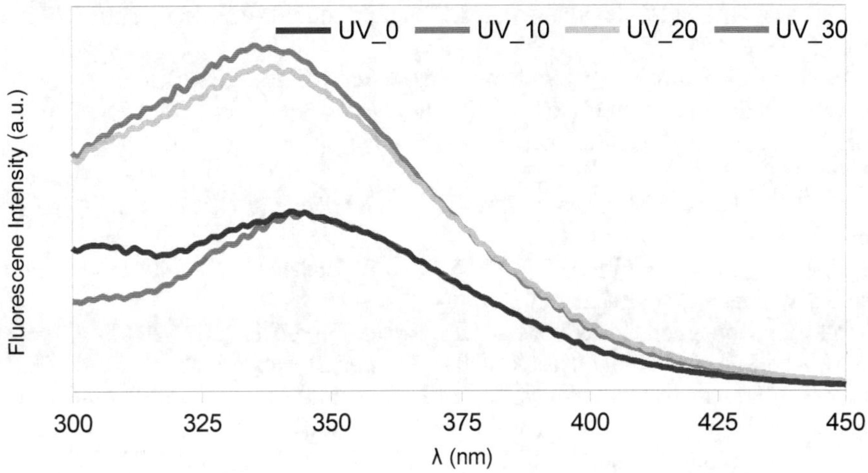

Figure 4. Fluorescence spectra of enzyme extracted from AC intestines treated by UV exposure time of 0 (UV00), 10 (UV10), 20 (UV20) and 30 minutes (UV30)

5. Conclusion

The UV irradiation treatment on AC intestines resulted the decrease in protease EA. The optimum conditions for crude protease enzyme from AC intestines were found to be at UV exposure time of 0 minutes, temperature of 50°C and pH of 8. Conformational change of AC intestines protease enzyme was observed through alterations in their UV-Vis and fluorescence spectra, but no significant change was observed in the FTIR spectra. Further studies are recommended to delve into the detail of UV irradiation effect on EA of AC intestines protease, including investigations with purified enzymes and their characteristics.

Acknowledgment

This work was supported by Research Organization of Agriculture and Food – National Research and Innovation Agency Republic of Indonesia, through research project of "Rumah Program Teknologi Tepat Guna dan Proses 2022". The authors also recognize the facilities, scientific and technical support from Advanced Characterization Laboratories Yogyakarta, National Reseach and Innovation Agency through E-Layanan Sains BRIN.

References

[1] Gustavsson J, Cederberg C and Sonesson U 2011 *Food loss and food waste: Causes and solutions* (Rome)
[2] International Fund for Agricultural Development 2020 Towards zero food waste in Indonesia's fishing communities
[3] Klomklao S, Benjakul S and Kishimura H 2010 *J. Food Biochem.* **34** 711–29
[4] Anati N J, Siti Roha A M and Normah I 2021 *Food Res.* **5** 120–8

[5] Sathivel S, Prinyawiwatkul W, Grimm C C, King J M and Lloyd S 2002 *JAOCS, J. Am. Oil Chem. Soc.* **79** 989–92
[6] Pratama I S, Putra Y, Pangestuti R, Kim S K and Siahaan E A 2022 *Fish. Aquat. Sci.* **25** 357–79
[7] Zamora-Sillero J, Gharsallaoui A and Prentice C 2018 *Mar. Biotechnol.* **20** 118–30
[8] Cao W, Tian S, Wang H, Zhang C and Yuan J 2020 *Food Sci. Nutr.* **8** 170–8
[9] Nikoo M, Regenstein J M, Noori F and Piri Gheshlaghi S 2021 *Lwt* **140** 110702
[10] Arias-Moscoso J L, Maldonado-Arce A, Rouzaud-Sandez O, Márquez-Ríos E, Torres-Arreola W, Santacruz-Ortega H, Gaxiola-Cortés M G and Ezquerra-Brauer J M 2015 *Food Biophys.* **10** 145–54
[11] Cao W, Tan C, Zhan X, Li H and Zhang C 2014 *Food Chem.* **164** 136–41
[12] Lante A, Tinello F and Lomolino G 2013 *Innov. Food Sci. Emerg. Technol.* **17** 130–4
[13] Gabriel M A, Dare E V., Meunier S M, Campbell J L, Sasges M R and Aucoin M G 2019 *Vaccine* **37** 6518–25
[14] Dou B, Wang H, Cao W, Chen Z, Zhu G, Zhang C, Zheng H, Lin H and Gao J 2023 *Int. J. Food Sci. Technol.* 1–10
[15] El-Beltagy A E, El-Adawy T A, Rahma E H and El-Bedawey A A 2004 *Food Chem.* **86** 33–9
[16] Falguera V, Pagán J and Ibarz A 2011 *Lwt* **44** 115–9
[17] Kristo E, Hazizaj A and Corredig M 2012 *J. Agric. Food Chem.* **60** 6204–9
[18] Atta M, Abd El-Hamed S and Keshk S 2017 *Assiut J. Agric. Sci.* **48** 40–54
[19] Bougatef A, Souissi N, Fakhfakh N, Ellouz-Triki Y and Nasri M 2007 *Food Chem.* **102** 343–50
[20] Vannabun A, Ketnawa S, Phongthai S, Benjakul S and Rawdkuen S 2014 *Food Biosci.* **6** 9–16
[21] Park M-H, Walpola B C and Yoon M-H 2013 *African J. Biotechnol.* **12** 1408–18
[22] Reddy N, Deekonda V, Seshagiri S and Reddy R 2022 *Ind. Crop. Prod.* **187** 115403
[23] Abd-Elkhalek A M, Seoudi D M, Ibrahim O A, Abd-Rabou N S and Abd Elazeem E M 2020 *J. Appl. Biotechnol. Reports* **7** 243–50
[24] Kandasamy S, Muthusamy G, Balakrishnan S, Duraisamy S, Thangasamy S, Seralathan K K and Chinnappan S 2016 *3 Biotech* **6**
[25] Ramos Ó L, Reinas I, Silva S I, Fernandes J C, Cerqueira M A, Pereira R N, Vicente A A, Poças M F, Pintado M E and Malcata F X 2013 *Food Hydrocoll.* **30** 110–22
[26] Jamir K and Seshagirirao K 2018 *Int. J. Biol. Macromol.* **106** 277–83
[27] Gu Z, Lai J, Hang J, Zhang C, Wang S, Jiao Y, Liu S and Fang Y 2019 *Int. J. Biol. Macromol.* **128** 603–9

Seasonal effect on the amino acid and fatty acid profiles of *Ulva* spp. collected from Sepanjang Beach, Yogyakarta, Indonesia

CD Poeloengasih*[1], R Pangestuti[1], EA Siahaan[2], Y Putra[2], TB Bardant[3], DJ Prasetyo[1], Hernawan[1], TH Jatmiko[1], Suratno[1], Y Khasanah[1], L Ali[2], DA Christyandari[1], W Widiastuti[4], R Suryani[1], A Windarsih[1], B Kumayanjati[1], N Wahyudin[5]

[1]Research Center for Food Technology and Processing, National Research and Innovation Agency, Gading, Playen, Gunungkidul, Yogyakarta 55861, Indonesia
[2]Research Center for Marine and Land Bio-Industry, National Research and Innovation Agency, North Lombok, West Nusa Tenggara 83352
[3]Research Center for Chemistry, National Research and Innovation Agency
[4]Directorate of Intellectual Property Management, National Research and Innovation Agency, Jakarta 10340
[5]UKM Berkah, Desa Kemadang, Kecamatan Tanjungsari, Kabupaten Gunungkidul

*Corresponding author mail: crescentianadewi@gmail.com

Abstract. In this study, nutritional composition including amino acid and fatty acid profiles of *Ulva* spp. from Sepanjang beach, Yogyakarta, Indonesia collected on different seasons were investigated. The result showed that the dry season had a higher level of total amino acid than the rainy season. The presence of complete essential amino acid (EAA) in the rainy and dry seasons was discovered, accounting for 35.88% and 37.51% of the total amino acid, respectively. Flavour amino acids (Glu, Asp, Ala) were prominent in *Ulva* spp. and reached up to 34.30% of the total amino acid for both seasons. The ratio of essential to non-essential amino acids of *Ulva* spp. from the dry season was greater and substantially different ($p < 0.05$) than that of the rainy season. Meanwhile, the sulfur-containing amino acid concentration during the dry season declined to 2.69% from 6.59% of total amino acid in the rainy. As demonstrated by the assessment of fatty acids, saturated fatty acids (SFAs) were more prevalent in *Ulva* spp. over the dry season, whereas polyunsaturated fatty acids (PUFAs) dominated during the rainy season. *Ulva* spp. harvested in the dry season had higher polyunsaturated fatty acid (PUFA) content than that of the rainy season. Eicosapentaenoic acid (EPA) was only detected in *Ulva* spp. from the rainy season, whereas DHA and AA were not detected in both samples. Generally, the amino acid and fatty acid profiles of *Ulva* spp. correlate with seasonal variations. Overall, the present investigation demonstrates the prospects of under-exploited green seaweed resources from Indonesia to be applied in food industries.

1. Introduction

According to the United Nations, the world's population will reach 9.7 billion by 2050 [1]. Consequently, agriculture will be under more pressure to provide supplies for direct human consumption as well as animal feed because there are more mouths to feed. This condition drives a necessity to find new food sources to feed upcoming populations and fulfill the world's needs. Many terrestrial and marine resources have been proposed as new food sources. From marine ones, macroalgae have been known as promising sources of proteins, polysaccharides, dietary fibers, vitamins, minerals, natural polyphenols, antioxidants, natural pigments, and lipids. Numerous studies on the nutritional value, chemical compositions as well as medicinal value of marine macroalgae have been carried out, including algae from Indonesia. In addition, marine macroalgae have been assessed for their potential as human foods, animal feed, pharmaceutical and nutraceutical uses, and many more. Various marine macroalgae's chemical composition and nutritional properties throughout some regions have been well documented. Previous investigations have found that the nutritional value of algae might vary depending on the species, growth phase, seasons, habitats, geographic location, and environmental variables, including salinity, temperature, intensity of light, and water nutrients [2-4].

In Indonesia, *Ulva* spp. is one of the most potential yet under-exploited green marine macroalgae. Numerous in-depth investigations into the green algae *Ulva* spp. have been conducted intensively. Several authors have reported the chemical composition and nutritional value of green alga *Ulva* collected from Southeast Asia waters including Indonesian waters [5-11]. However, there is a lack of information regarding the amino acids and fatty acids content of the green alga *Ulva* spp. from Gunungkidul waters, Yogyakarta, Indonesia. Notably, up to now, there has yet to be a report on the seasonal variations in the composition of chemicals and nutritional value of green alga *Ulva* spp. collected from Southeast Asia waters, including Indonesia. Therefore, the present study has been carried out to determine the amino acids and fatty acids profiles of the green alga *Ulva* spp. from Gunungkidul waters, Indonesia, and the effect of seasonal variation on their chemistry compositions. To the best of our knowledge, this study is the earliest study on the nutritional composition of *Ulva* spp. harvested in different season. Further, this result contributes to the potential use of *Ulva* spp. from Indonesia as an edible resource and fills the gaps for future research in the development of food-based macroalgae.

2. Materials and methods

2.1. Sample collection

Wild *Ulva* spp. was collected from Sepanjang beach (8°8'12"S 110°34'0"E), Gunungkidul, Indonesia on March 2021 and September 2021 presenting rainy and dry seasons, respectively. Sample collections were collected at a low tide, and the samples were collected based on thallus that measured at least 7 centimeters in length. *Ulva* spp. was gathered from submersed rocks during this study, where the algae are usually abundant during the relevant collecting periods. Fresh *Ulva* spp. was transported to the laboratory in a plastic bag after being rinsed with seawater. When the algae arrived at the laboratory, they were inspected visually and given a second rinsing with filtered seawater. This was done to remove any epiphytes, sand, other biota, and other impurities. A universal oven (Memmert, UF-100, Germany) was used to dry the clean *Ulva* spp. at 45-50 °C after being drained to eliminate excessive water. The dried samples were packed in the zipper plastic bag, stored in a dark container, and analyzed within a week.

2.2 Analysis of amino acid

The chromatographic analysis was employed to analyze the amino acids. Following the protocol provided by Waters [12], an Ultra-Performance Liquid Chromatography (UPLC, Acquity H-class System with PDA Detector, Waters, Massachusetts, USA) was applied to determine the concentration of fifteen amino acids in dried *Ulva* spp., including serine (Ser), aspartic acid (Asp), glutamic acid (Glu), phenylalanine (Phe), isoleucine (Ile), valine (Val), alanine (Ala), glycine (Gly), arginine (Arg), leucine (Leu), lysine (Lys), tyrosine (Tyr), proline (Pro), threonine (Thr), and histidine (His).

Meanwhile, a Liquid Chromatography-Mass Spectrometry (LC-MS, LCMS 8045 with LC Nexera X2, Shimadzu, Kyoto, Japan) was used in assessing cysteine (Cys) and methionine (Met) in dry *Ulva* spp. according to the method by Dahl-Lasses [13]. At the same time, the quantity of tryptophan (Try) was identified through High-Performance Liquid Chromatography (HPLC, Prominence LC-2030, Shimadzu, Kyoto, Japan) according to a method by AOAC 988.15 [14].

The result of the amino acid analysis was then used to analyze its nutritional quality using different parameters, such as total amino acid, essential amino acid, non-essential amino acid, etc.

2.3 Analysis of fatty acid

Gas Chromatography (GC, Clarus 580 & 680, Perkin Elmer, Shelton, USA) connected with a capillary column DB FastFAME and helium as the carrier gas was employed to determine the fatty acid profile of *Ulva* spp. according to the method of AOAC [15]. The injector and detector were set at 240ºC. The fatty acids have been detected by calculating the peak area ratio and expressing it as a proportion of total fatty acids. The result of the fatty acid analysis was then used to assess its nutritional quality using different parameters and was calculated by following formulas:

(1) C18/C20 Polyunsaturated Fatty Acid (PUFA) Ratio:

$$\frac{C18}{C20} = \frac{\Sigma C18\ PUFA}{\Sigma C20\ PUFA} \quad (1)$$

(2) ω6/ω3 Ratio:

$$\frac{\omega 6}{\omega 3} = \frac{\Sigma PUFA\omega 6}{\Sigma PUFA\omega 3} \quad (2)$$

(3) ω3/ω6 Ratio:

$$\frac{\omega 3}{\omega 6} = \frac{\Sigma PUFA\omega 3}{\Sigma PUFA\omega 6} \quad (3)$$

(4) Atherogenicity Index (AI):

$$AI = \frac{[(C12{:}0)+(4 \times C14{:}0)+(C16{:}0)]}{\Sigma MUFA + \Sigma PUFA\omega 6 + \Sigma PUFA\omega 3} \quad (4)$$

(5) Thrombogenicity Index (TI)

$$TI = \frac{C14{:}0+C16{:}0+C18{:}0}{0.5 \times \Sigma\ MUFA + 0.5 \times \Sigma PUFA\omega 6 + 3 \times \Sigma PUFA\omega 3 + \frac{\Sigma PUFA\omega 3}{\Sigma PUFA\omega 6}} \quad (5)$$

(6) Hypocholesterolemic/Hypercholesterolemic (h/H) Fatty Acids

$$h/H = \frac{C18{:}1\omega 9 + C18{:}2\omega 6 + C20{:}4\omega 6 + C18{:}3\omega 3 + C20{:}5\omega 3}{C14{:}0+C16{:}0} \quad (6)$$

(7) PUFAs/SFAs

$$\frac{PUFA}{SFA} = \frac{\Sigma PUFA}{\Sigma SFA} \quad (7)$$

2.4 Statistical analysis

All analytical determinations were performed in duplicate, and the mean value was recorded. The statistical difference between the two groups was examined using a student's t-test. Statistical significance was assigned to differences with p-values less than 0.05. Statistical analyses were performed with Excell Microsoft 365 Apps for enterprise.

3. Results and discussion

The results of this work showed the biochemical and nutritional composition of *Ulva* spp. varies according to the seasons. Figure 1 shows the rainfall profile in Sepanjang Beach, where the algae were cultivated, in the last three years. Collecting samples in March and September were suitable as representative of rainy season and dry season. Seasonal changes including temperature, pH, the salinity of seawater, light intensity, water nutrient, waves, and water currents are responsible for the composition of seaweed [3,16-18]. The findings of the present study concurred with a study by de Melo [4], which found that the chemical composition of *Sargassum vulgare, Crassiphycus corneus*, and *Ulva fasciata* collected from northeast Brazil fluctuated according to the season.

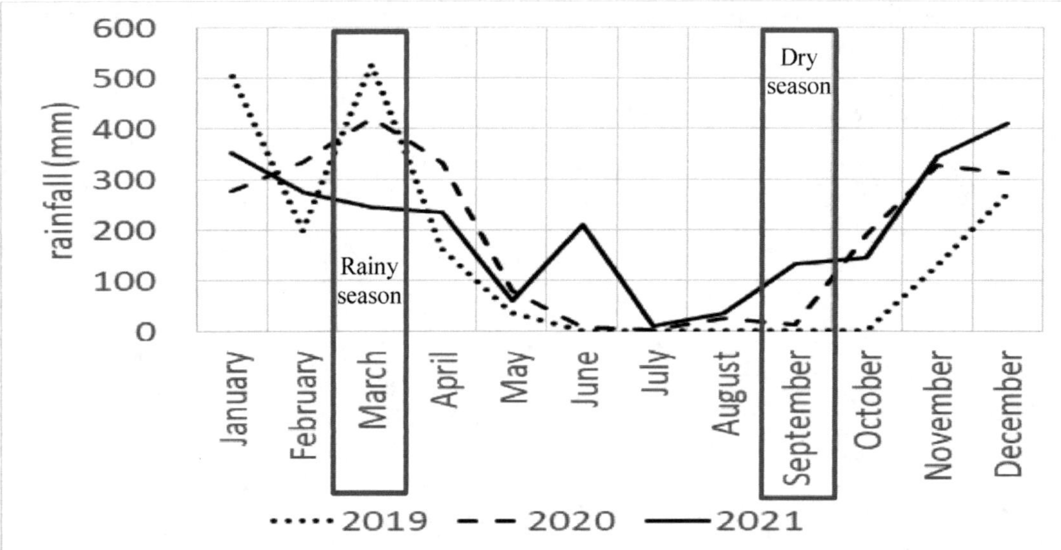

Figure 1. Annual rainfall profile in Sepanjang beach, location of sample collected (rainfall data from BPS Gunungkidul)

The seasonal profile of amino acid composition of *Ulva* spp. is summarised in Figure 2. As illustrated in Figure 2, the composition of total amino acids contained in *Ulva* spp. varied according to the season. A higher level of total amino acid was found in the dry season, which is 4% higher than the total amino acid composition in the rainy season. According to the statistical analysis (Student's t-test, $p < 0.05$), the total amino acid content varied significantly between the rainy and dry seasons. The availability of light and nutrition accessibility could be related to the higher level of total amino acid during the dry season.

As shown in Figure 3, eighteen amino acids were detected in both samples with diverse levels, consisting of nine essential amino acids (EAA) and nine non-essential amino acids (NEAA). The complete profile of EAAs was detected for both rainy and dry season samples. This result is consistent with previous findings described by de Melo et al. [4] that *U. fasciata* contained all essential amino acids. The content of EAAs for rainy and dry seasons accounted for up to 35.88% and 37.51% of the total amino acid, respectively. A student's t-test revealed a significant difference in the level of EAAs between sampling times. However, compared to *Ulva* species from other regions in Southeast Asia, the results found in the present investigation were relatively lower than in earlier reports. The EAAs value of *U. fasciata* collected from the Philippines collected in the rainy season, *Ulva pertusa* from Thailand collected in the rainy season and *Ulva intestinalis* collected in the summer were 42.28% of total amino acid, 41.80% of total amino acid, and 40.15% of total amino acid, respectively [6,11]. These results can be affected by the environment where *Ulva* spp. growth and other environmental parameters. Leucine made up the highest EAA in both samples. *Ulva* spp. harvested in the rainy season was rich in leucine, lysine, phenylalanine, and threonine, whereas *Ulva* spp. collected in the dry season was rich in leucine, lysine, phenylalanine, and valine.

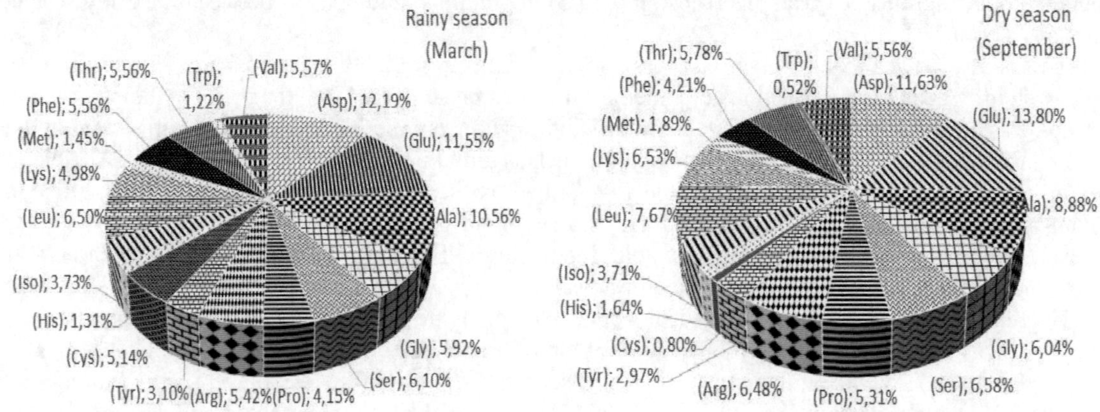

Figure 2. Composition profile of amino acid in *Ulva* spp. harvested in rainy season (left) and dry season (right).

The proportion between EAA and NEAA for rainy and dry seasons was 0.56 and 0.60, respectively. This value implied that the fraction of EAA was lower than that of NEAA. A student's t-test revealed a significant difference in the ratio of EAA/NEAA between sampling times. The ratio of EAA/NEAA could be used to assess the amino acid distribution and seaweed protein quality. The EAA/NEAA ratio determined in this study was lower than that found by Ratana-arporn and Chirapart [5] for *Ulva reticulata* (0.54) harvested from Pattani Bay, Thailand in May. This result is also lower than *U. fasciata* (0.73) harvested from the coastal area of Barangay Lawigan, Philippines in the rainy season in August 2017 [11], and *U. pertusa* (0.72) and *U. intestinalis* (0.67) collected from Pattani Bay, Southern Thailand during the rainy season in December 2007 and summer in April 2008 [6].

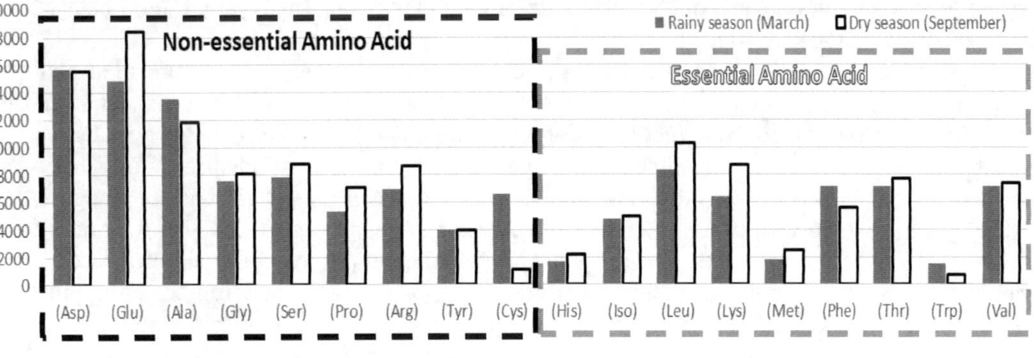

Figure 3. Amino acid profile from *Ulva* spp. harvested in different seasons.

The main amino acids in *Ulva* spp. were aspartic acid, glutamic acid, and alanine reaching up to 34.30% of total amino acid in all samples. These three amino acids contribute significantly to the taste of seaweed. This result is consistent with other works that found glutamic acid to be the most abundant amino acid, followed by aspartic acid in other *Ulva* spp. collected from various locations and seasons [6,19,20]. The flavour and taste of seaweed, including the umami taste, can be attributed to the presence of glutamic and aspartic acid [2,21,22]. However, the potency of aspartic acid to elicit an umami taste is less than that of glutamic acid [23]. Meanwhile, alanine contributes to seaweed's sweet flavour [23]. In this study, a higher level of glutamic acid was found in the dry season, and statistical analysis demonstrated significant variations (t-test, $p < 0.05$) in glutamic acid content between the rainy season and dry season.

This study showed that *Ulva* spp. is a poor source of sulfur-containing amino acids, *i.e.*, methionine and cysteine. As an essential amino acid, methionine must be acquired from the diet. On the other hand, cysteine as a non-essential amino acid could be synthesized from methionine. Both are important in the essential production of intracellular antioxidants, such as glutathione and N-acetyl cysteine [24]. Due to its rarity, the sulfur-containing amino acid is regarded as the amino acid with limited availability in plant-based protein. This study revealed a higher level of sulfuric amino acid was observed in the rainy season representing 6.59% of total amino acid. Interestingly, Benjama and Masniyom [6] reported that methionine and cysteine were not detected in *U. pertusa* and *U. intestinalis* from Thailand.

Branched amino acids, *i.e.*, leucine, isoleucine, and valine are nutritionally crucial compounds since they are essential amino acids. In this study, the sum of branched amino acids was significantly greater ($p < 0.05$) for the dry season in comparison with the rainy season. These results were higher than *Ulva lactuca* [25] obtained from Abu Qir Bay, Egypt but lower than *U. intestinalis* and *U. pertusa* [6] harvested from Pattani Bay Thailand. The branched amino acid plays several important roles, *e.g.*, as a building block for protein and muscle, is involved in regulating blood sugar levels, and reduces fatigue during exercise [26-28].

The results of the fatty acid profile assessment were displayed in Table 1. A total of ten fatty acids were discovered in the rainy season, whereas fourteen fatty acids were detected in dry season. In this study, the total amount of fatty acids during the dry season was substantially greater than during the rainy season. Saturated fatty acids (SFAs) were dominant in the dry season. This result is consistent with other works that SFAs were the primary fatty acids in the green alga *Ulva* collected from the Southeast Asia waters, such as Segara beach, Sanur, Bali, Indonesia; Barangay Lawigan, Philippines, and Tual, Southeast Maluku, Indonesia [7,10,11].

In both seasons, it was observed palmitic acid (C16:0) was the primary SFA. Ten SFAs were identified in the dry season, whereas only four SFAs were detected in the rainy season. In all seasons, only palmitic acid (C16:0) and stearic acid (C18:0) were the only SFAs identified, whereas lauric acid (C12:0) and myristic acid (C14:0) were not identified in the rainy season. Unsaturated fatty acids contents (monounsaturated fatty acids and polyunsaturated fatty acids) also varied significantly between seasonal periods (Student's t-test, $p < 0.05$).

In the monounsaturated fatty acids (MUFAs) profile, only palmitoleic (C16:1) and oleic acid (C18:1 ω9C) were detected in both samples. The dry season sample contained a higher concentration of both fatty acids. *Ulva* spp. exhibited higher levels of polyunsaturated fatty acids (PUFAs) during dry season than that in the rainy season. However, fewer types of fatty acids were identified in the dry season. Linoleic acid/ω3 (C18:3 ω3) was found in both seasons and detected as the prominent omega-3 PUFAs in the dry season. Eicosapentaenoic Acid/EPA (C20:5 ω3), a major omega-3 PUFA, was only found in the rainy season. Meanwhile, DHA and AA were not found in this study. Different results were reported by other authors who identified the fatty acid profiles of *U. fasciata* collected from the Philippines during the rainy season in August 2017 and *U. lactuca* obtained from Tual, Southeast Maluku, Indonesia in June 2018, respectively, where DHA, AA as well as EPA were not detected in both samples [10-11].

In both the rainy and dry seasons, the ω6/ω3 fatty acids concentrations were relatively low with the value of 0.1568 and 0.7531, respectively. Based on the guidelines provided by the World Health Organization's, the ratio of ω6/ω3 fatty acids should not exceed more than 10 in the daily diet [29]. A positive correlation exists between an elevated ratio of ω6/ω3 fatty acids and the prevalence of autoimmune, inflammatory, and allergic diseases [30]. Our finding showed that *Ulva* spp. from Sepanjang beach, Indonesia had a low ω6/ω3 ratio, which suggests that they could be valuable source of ω3 fatty acids.

Table 1. Fatty acid composition of green alga *Ulva* spp. in rainy and dry season

Fatty acid		**Composition (%)**		**Omega (ω) type**
		Rainy	Dry	
Saturated fatty Acid/SFA	Butanoic acid	ND	0.0053	
	Hexanoic acid	ND	ND	
	Caprylic acid	ND	0.0070	
	Capric acid	ND	0.0081	
	Hendecanoic acid	0.0054	ND	
	Dodecanoic acid	ND	0.0767	
	Tridecylic acid	ND	ND	
	Myristic acid	ND	0.0390	
	Pentadecylic acid	ND	0.0055	
	Palmitic acid	0.1223	0.5979	
	Margarinic acid	0.0027	ND	
	Stearophanic acid	0.0028	0.0605	
	Icosanoic acid	ND	ND	
	Heneicosylic acid	ND	ND	
	Docosanoic acid	ND	0.0331	
	Tricosylic acid	ND	ND	
	Carnaubic acid	ND	0.0284	
Monounsaturated Fatty Acid/MUFA	Myristoleic acid	ND	ND	
	Pentadecanoic acid	ND	ND	
	Cis-9-Hexadecenoic acid	0.0078	0.0473	Omega 7
	Heptadecanoic acid	ND	ND	
	ω9C Oleic acid	0.0044	0.2793	Omega 6
	ω9T Oleic acid	ND	ND	Omega 9
	Eicosenoic acid	ND	ND	
	Cis-13-Docosenoic acid	ND	ND	
	ω9 Nervonic acid	ND	ND	Omega 9
Polyunsaturated fatty Acid/PUFA	ω6C Linoleic acid/ω6	0.0183	0.1580	Omega 6
	ω6T Linoleic acid	ND	ND	
	ω6 Linoleic acid/ω6	0.0028	ND	Omega 6
	ω3 Linoleic acid /ω3	0.0557	0.2098	Omega 3
	Icosadienoic Acid	ND	ND	
	ω3 Eicosatrienoic Acid/ω3	ND	ND	Omega 3
	ω6 Eicosatrienoic Acid/ω6	ND	ND	Omega 6
	ω6 Arachidonic Acid (AA)	ND	ND	Omega 6
	ω3 Eicosapentaenoic Acid/ω3 (EPA)	0.0790	ND	Omega 3
	Docosadienoic Acid	ND	ND	
	ω3 Docosaheksaenoic Acid/ω3 (DHA)	ND	ND	Omega 3
	Total SFA	0.1332	0.8614	
	Total UFA	0.1681	0.6944	
	Total MUFA	0.0122	0.3266	
	Total PUFA	0.1559	0.3677	
	Total Oleic Acid	0.0044	0.2793	
	Total Linoleic Acid	0.0183	0.1580	
	Total Linolenic Acid	0.0585	0.2098	
	DHA	ND	ND	
	EPA	0.0790	ND	
	AA	ND	ND	
	ω3 Fatty Acid	0.1348	0.2098	
	ω6 Fatty Acid	0.0211	0.1580	
	ω9 Fatty Acid	0.0044	0.2792	

Table 2. Nutritional quality index of fatty acid from *Ulva* spp. from different seasons

	Rainy season	**Dry season**
C18/C20 PUFA	0.9725	ND
ω6/ω3 Ratio	0.1568	0.7531
ω3/ω6 Ratio	6.3767	1.3278
AI	ND	2.1902
TI	ND	0.3771
h/H ratio	ND	ND
PUFA/SFA	1.1708	0.4269

The nutritional quality indexes of *Ulva* spp. were calculated with its fatty acid profile and the results were presented in Table 2. The ratio of C18/C20 PUFAs was detected only in the rainy season. As determined in this study, the ratio of C18/C20 PUFAs exhibited a lower value than the ratio reported by Magdugo et al. [11] for *U. fasciata* (8.03) collected from the Philippines during the rainy season in August 2017. In both the rainy and dry seasons, the ω6/ω3 fatty acids was relatively low, at 0.1568 and 0.7531, respectively. In this study, atherogenic index (AI) and thrombogenic index (TI) were only identified in the dry season since lauric acid (C12:0) and myristic acid (C14:0) were not identified in *Ulva* spp. collected in the rainy season. AI shows the correlation between the total saturated and the total unsaturated fatty acids. AI and TI correlate with the prevention of coronary artery disease. The lower AI and TI value indicated better protection properties. Meanwhile, h/H could not be calculated since Arachidonic Acid (AA) and Eicosapentaenoic Acid/ω3 (EPA) were not detected in the dry season. However, a high ratio of h/H was desirable as it is correlated with high PUFA content. Since SFA possibly increases the level of blood cholesterol (FAO/WHO, 2008), the FAO/WHO recommending an approximate proportion of PUFA/SFA greater than 0.45.

4. Conclusion

This study presented various compositions of amino acids and fatty acids of green alga *Ulva* spp. harvested in different seasons. The higher total amino acid was found in *Ulva* spp. harvested in the dry season. The complete profile of essential amino acids was detected for the rainy and dry season samples. Flavour amino acid (Glu, Asp, Ala) was found to be the most abundant amino acid on *Ulva* spp. from both seasons. Meanwhile, SFAs were prominent in *Ulva* spp. from the dry season, whereas PUFAs were dominant in *Ulva* spp. from the rainy season. Eicosapentaenoic acid (EPA) was only detected in *Ulva* spp. harvested in the rainy season, whereas DHA and AA were not detected in both samples. Taken together, based on the amino acid and fatty acid profile assessment it can be inferred that *Ulva* spp. exhibit favourable characteristics as a protein and lipids resource; however, further studies on the heavy metals, contaminants, and practical use of *Ulva* spp. as an edible resource are needed.

References

[1] UNDESA 2017 United Nations, Department of Economic and Social Affairs, Population Division 2017. World Population Prospects: The 2017 Revision, Key Findings, and Advance Tables. Working Paper No. ESA/P/WP/248.
[2] Yaich H, Garna H, Besbes S, Paquot M, Blecker C, Attia H 2011 *Food Chem* **128** 895-901.
[3] Gaillard C, Bhatti HS, Novoa-Garrido M, Lind V, Roleda MY, Weisbjerg MR 2018 *Anim Feed Sci Tech* **241** 210–222.
[4] de Melo NSM, Cardoso LG, de Castro Nunes JM, Brito GB, Caires TA, de Saouza CO, Portz L, Druzian JI 2021 *Braz J Bot* **44** 331–344.
[5] Ratana-arporn P and Chirapart A 2006 *Kasetsart J (Nat. Sci.)* **40** (Suppl.) 75 – 83.
[6] Benjama O and Masniyom P 2011 *Songklanakarin J Sci Technol.* **33** (5) 575-583.
[7] Puspawati NM, Suastuti NGAMD, Dewi DAI 2011 *JURNAL KIMIA* **5** (2) 109-116.
[8] Rasyid A 2017 *Trop Life Sci Res* **28** (2) 119–125.

[9] Jatmiko TH, Prasetyo DJ, Poeloengasih CD, Hernawan, Khasanah Y 2018 *2nd International Conference on Natural Products and Bioresource Sciences*. Nutritional Evaluation of Ulva sp. from Sepanjang Coast, Gunungkidul, Indonesia. IOP Conf. Series: Earth and Environmental Science **251** 012011. doi:10.1088/1755-1315/251/1/012011
[10] Pangestuti R, Haq M, Rahmadi P, Chun BS 2021 *Mar Drugs* **19** (10) 578.
[11] Magdugo RP, Terme N, Lang M, Pliego-Cortés H, Marty C, Hurtado AQ, Bedoux G, Bourgougnon N 2020 *Molecules* **25** 2901.
[12] Waters 2012 Acquity UPLC H-Class and H-Class Bio Amino Acid Analysis System Guide. Waters Corporation, USA.
[13] Dahl-Lessen R, van Hecke J, Jørgensen H, Bukh C, Andersen B, and Schjoerring JK 2018 *Plant Methods* **14** 8.
[14] AOAC 1988 AOAC 988.15 Tryptophan in Foods and Food and Feed Ingredients. Official Methods of Analysis of the Association of Official Analytical Chemists, Washington.
[15] AOAC 2000 Official Methods of Analysis of the Association of Official Analytical Chemists. 17th Edition, Washington DC. *Analysis of Oil and Fat*, Chapter 41 26-28.
[16] Handa A, Forbord S, Wang X, Broch OJ, Dahle SW, Størseth TR, Reitan KI, Olsen Y, Skjermo J 2013 *Aquaculture* **414–415** 191–201.
[17] Marinho GS, Holdt SL, Angelidaki I 2015 *J Appl Phycol* **27** 1991–2000.
[18] Schiener P, Black KD, Stanley MS, Green DH 2015 *J Appl Phycol* **27** 363–373.
[19] Fleurence J, Coeur CL, Mabeau S, Maurice M, Landrein A 1995 *J App Phycol* **7** 577 – 582.
[20] Machado M, Machado S, Pimentel FB, Freitas V, Alves RC, Oliveira MBPP 2020 *Foods* **9** (10) 1382-1396.
[21] Paiva L, Lima E, Patarra RF, Neto AI, Baptista J 2014 *Food Chem* **164** 128-135.
[22] Biancarosa I, Espe M, Bruckner CG, Heesch S, Liland N, Waagbø R, Torstensen B and Lock EJ 2017 *J Appl Phycol* **29** 1001–1009.
[23] Mouritsen OG, Duelund L, Petersen MA, Hartmann AL, Frøst MB 2019 *J Appl Phycol* **31** 1213–1232.
[24] Čolović MB, Vasic VM, Djuric DM 2018 *Curr Med Chem* **25** (3) 2018.
[25] Khairy HM, El-Shafay SM 2013 *Oceanologia* **55** (2) 435–452.
[26] Kimball SR, Jefferson LS 2006 *J Nutr* **136** (1) 227S–231S.
[27] Blomstrand E 2006 *J Nutr* **136** (2) 544S – 547S.
[28] Falavigna G, de Araújo Jr JA, Rogero MM, de Oliveira Pires IS, Pedrosa RG, Martins Jr E, de Castro IA, Tirapegui J 2012 *Nutrients* **4** (11) 1767-1780.
[29] Ortiz J, Romero N, Robert P, Araya J, Lopez-Hernández J, Bozzo C, Nvarrete E, Osorio A, Rios A 2005 *Food Chem* **99** 98-104.
[30] DiNicolantonio JJ and O'Keefe J 2021 *Missouri Medicine* **118** (5) 453.

IOP Conf. Series: Earth and Environmental Science 1289 (2023) 012032 doi:10.1088/1755-1315/1289/1/012032

Consumer preference, quality, and financial analysis of tekwan made from snakehead, yellow tail, wolf herring, and spanish mackerels

F Y Pratama and Ustadi*

Fish Product Technology, Department of Fisheries, Faculty of Agriculture, Universitas Gadjah Mada, Jl. Flora Gedung A4, Bulaksumur Yogyakarta Indonesia 55281

*Corresponding authors mail: ustadi@ugm.ac.id

Abstract. Tekwan is a traditional processed fish product from South Sumatra, characterized by its flat shape and small size, resembling fish meatballs. Essentially, tekwan is made using Belida or Snakehead fish, where the fish stocks are running low. Raw materials represent one of the most critical components in a production process. The purpose of this study is to examine the effect of utilizing various fish raw materials on consumer preference, quality, and financial analysis of tekwan. The experimental design employed was a randomized design with 4 treatments and 3 replications. The treatments consisted of snakehead, yellowtail, wolf herring, and Spanish mackerels. The use of different kinds of fish in terms of chemical characteristics shows no apparent differences in fat, water, and ash levels, but significant differences ($p<0.05$) exist for protein levels. In the physical characteristics of the tekwan, a significant difference ($p<0.05$) is observed in the hardness value. In the consumer's preference test, corks and machetes received similar ratings, while yellowtail and mackerel obtained higher preference rates. Based on the financial analysis, snakehead does not seem to be worth the effort, whereas yellowtail, mackerel, and dorab demonstrate business worthiness.

1. Introduction

Today, fast food has become one of the preferred choices for meals due to its ease of serving, convenience, and simple preparation methods. Tekwan can be an effort to preserve traditional food that is easy to serve [1]. Tekwan is a flat-shaped, small-sized dish resembling meatballs and is typically served with shrimp broth [2]. During the tekwan-making process, factors such as the choice of fish type, the freshness of the fish, and the treatment during processing will determine the product's quality [3].

Tekwan was initially made using belida fish (*Notopterus sp.*) or Snakehead fish (*Channa striata*), both of which are native to the Musi River and are known for their white flesh and distinctive savory taste. This was made possible due to the abundance of fish resources during that period [4]. However, today, there are limitations in obtaining raw materials for tekwan, especially with the prohibition of using belida fish for consumption. Additionally, the availability of Snakehead fish cannot meet market demands, leading to increased exploitation [5]. On the other hand, this has driven up the price of Snakehead fish.

To address the aforementioned issues, a search for alternative replacements for Snakehead fish raw materials in tekwan products was conducted. The selected fish needed to have white flesh and meet quality criteria, particularly regarding the texture of the flesh to provide good gelling properties. The gelling properties are essential for fish-based products like tekwan [6]. The most important component in gel formation is actomyosin found in proteins, which, when heated, forms a gel with a mesh-like structure and imparts elasticity to the flesh [7]. Furthermore, differences in the concentration of moisture content in a material can also influence the texture of the product [8]. The substitution of raw materials will also impact production costs and profits to be gained [9]. The use of various types of marine fish such as Wolf Herring, Spanish Mackerels, and Yellow Tail is considered as potential substitutes. The selection of these three types of fish is also based on the availability of fish and more economical fish prices. The choice of these three types of fish is based on the greater availability of fish and more economical fish prices. According to the KKP Statistics, the average fish production from 2014-2018 for Yellow Tail fish was 81,221 tons, and Spanish Mackerels were 188,314 tons, while Snakehead fish was only 55,707 tons [10]. In 2022, in Yogyakarta, the prices for Wolf Herring fish were Rp30,000/kg, Yellow Tail fish Rp50,000/kg, and Spanish Mackerels Rp80,000/kg.

Processing fish-based food products is still limited. Research on the influence of various types of marine fish on tekwan production is expected to increase public interest in consuming fish and can be seen as a business opportunity. Therefore, this study examines the effect of various types of marine fish on the chemical and physical characteristics as well as consumer preference for tekwan. Additionally, it conducts a financial analysis that can change with the substitution of fish raw materials for tekwan production.

2. Research methodology

2.1. Materials
The materials used for the research on tekwan with various types of fish, namely Snakehead fish, Yellow Tail fish, Wolf Herring fish, and Spanish Mackerels, were obtained from the Gudang Lelang Market in the Bumi Waras Subdistrict of Bandar Lampung City. The additional ingredients used included tapioca flour ("cap pak tani"), refined salt, "rose brand" sugar, egg whites, garlic, and water.

The chemicals used for analysis include TCA 5%, distilled water (*aquades*), and concentrated sulfuric acid (H_2SO_4), HNO_3, NaOH-Tio (NaOH 40% + $Na_2S_2O_3$ 5%), Copper complex catalyst, H_3BO_3 5% (has been given the Mr. BCG indicator), Sodium hydroxide NaOH 50%, *phenolphthalein* indicator, boric acid H2O3 2%, HCl (0.1 N), mixed pH indicator (methyl red-blue), and petroleum ether.

2.2. Experimental Design
The experimental design used is Completely Randomized Design with a single factor, which is the use of different types of fish raw materials, with three replications. The treatments applied were as follows:
P0: Snakehead fish meat 250 g
P1: Yellow Tail fish meat 250 g
P2: Wolf Herring fish meat 250 g
P3: Spanish Mackerels fish meat 250 g

Completely Randomized Design (CRD) is the simplest type of experimental design, typically used in experiments with a uniform or homogeneous experimental environment or medium. In this study, CRD was employed because all the conditions for tekwan preparation were the same except for the use of different types of fish raw materials.

2.3. Research Procedure

2.3.1. The Making of Tekwan
Tekwan is made using fresh fish as the raw material, obtained from the Gudang Lelang Lampung market. The acquired fish is filleted, and the yield is calculated. The filleted fish is then weighed, with 250 grams used in accordance with the formulation. The fish meat is subsequently minced using a chopper. Afterward, the minced fish meat is mixed with other additional ingredients to create a tekwan mixture. The tekwan mixture is manually shaped and placed into boiling water. It is then left until the mixture surfaces, indicating that the tekwan is ready to be removed and cooled. The cooled tekwan is vacuum-sealed and stored in a refrigerator. The preparation of the tekwan mixture is carried out for 4 treatments, with each treatment repeated 3 times.

IOP Conf. Series: Earth and Environmental Science 1289 (2023) 012032 doi:10.1088/1755-1315/1289/1/012032

2.3.2. The analysis of moisture content (SNI 2354.2:2015)

In this research, the moisture content of tekwan with various types of fish was tested using a moisture analyzer. The tested tekwan samples, weighing 0.5-0.8 grams, were placed in an aluminum crucible on the moisture analyzer, ensuring even distribution over the entire surface of the crucible. Next, the analyzer was operated at a temperature setting of 105°C until the moisture content value appeared, indicating the completion of the testing process.

2.3.3. The analysis of ash content (SNI 2354.1:2010)

The ash content test is conducted by oxidizing the sample at a temperature of 550°C in a combustion furnace for 8 hours or until obtaining white ash. The weighed ash from the ash content determination is dissolved by adding 25 ml of 10% HCl and boiled for 5 minutes. Afterward, the solution is filtered using ashless filter paper (*Whatman 41*) and washed with distilled water several times until free from chlorides or acid. Then, it is dropped onto pH paper; if the value is 7, it indicates that it is acid-free. Next, dry the filter paper containing the non-soluble ash in the acid in an oven at (100±2)°C. After that, weigh the porcelain crucible until it reaches a constant weight (A g) and insert the dried filter paper containing the non-soluble ash in the acid. Transfer the porcelain crucible with ash to the combustion furnace, then gradually raise the temperature until it reaches 550°C and maintain it for 16-24 hours. Afterwards, remove the porcelain crucible using tongs and place it in a desiccator for about 30 minutes. Then, weigh the weight (c g) immediately after cooling until a constant weight is obtained and calculate it using the formula.

$$\text{Insoluble ash content} = \frac{\text{C-A}}{\text{Sample weight (in grams) x}} \text{ x } 100\%$$

2.3.4. Analysis of protein content (SNI 01-2354.4-2006)

The protein content test was conducted by measuring crude protein using the total nitrogen method. A total of 2 grams of the sample was placed into a 50 ml Kjeldahl flask. Next, 15 ml of concentrated H2SO4 and 3 ml of H2O2 were added, followed by a Kjeldahl tablet, and the mixture was allowed to stand in an acid chamber for 10 minutes. The sample was then destructed for 1 hour until the liquid changed color to clear green. After that, 50 ml of distilled water was added, and distillation was performed. The Kjeldahl flask was connected to the distillation apparatus, and 50 ml of concentrated NaOH was added to obtain 150 ml of distillate collected in an Erlenmeyer flask containing 25 ml of 4% borate acid and 2 drops of pH indicator. The distillation result was then titrated using 0.2 N HCl until the color changed to pink, and the crude protein content was calculated using the formula:

$$\%\text{Crude protein} = \frac{\text{ml HCl sample-ml HCl blanko) N HCl x 14,007 x 6,25}}{\text{Sample weight (gram)x 1000}} \text{ x } 100\%$$

2.3.5. Analysis of fat content (SNI 01-2354.3: 2006)

The fat content was conducted by placing the solvent cup in an oven at 105°C for 1 hour. Afterward, it was cooled in a desiccator for 30 minutes and then weighed. A total of 2 grams of the sample were wrapped in filter paper and placed in an oven at 105°C for 1 hour. The sample was then cooled again in a desiccator for 30 minutes. Next, the sample was placed in a thimble and attached to the Soxhlet extraction apparatus. Then, the condenser is turned on, and the solvent cup is attached to the Soxhlet apparatus. Next, petroleum ether is added to the extraction apparatus, and the extraction process is carried out for 2 hours, followed by rinsing for 1 minute and drying for 5 minutes. Once the process is completed, the solvent cup is removed, and the contents are dried in an oven at 105°C for 1 hour. After drying, the solvent cup is cooled in a desiccator for 30 minutes and weighed. The fat content can then be calculated using the following formula:

$$\%\text{ Fat content } (\textit{wet basis}) = \frac{\text{Final sample weight (grams)}}{\text{Initial sample weight (grams)}} \text{ x } 100\%$$

2.3.6. *Texture analysis*

Texture testing of tekwan was carried out using the Texture Profile Analysis (TPA) method with the Texture Analyzer TXT 32 instrument. The samples were tested following the testing procedure as follows: the tekwan samples were prepared in cube shapes with approximately 3 cm sides, then the data cable of the Texture Analyzer was ensured to be connected to the computer CPU, and the computer was turned on. The sample puncture probe was attached and positioned near the sample. The computer program was operated to run the probe, ensuring that the monitor's values were at zero, then the "start test" menu was selected on the computer, causing the probe to move and puncture the tekwan sample. The test is complete when the probe returns to its original position. The test results will be displayed in the form of graphs and numerical values. The values obtained during the testing include *Hardness* and *Cohesiveness*.

2.3.7. *Consumer Preference Level*

[33] Consumer preference testing falls under the category of affective testing and involves using a large number of untrained panelists. In this study, a total of 80 panelists were utilized. Each panelist was asked to evaluate 4 sensory attributes, which included appearance, texture, aroma, and taste, using a scoring range from 1 to 5, ranging from 'dislike very much' to 'like very much.' After completing the evaluation of one sample, each panelist was required to drink water to neutralize their palate, and they were given a 1-minute break while filling out the hedonic sheets provided.

2.3.8. *Financial analysis*

Financial feasibility aspect aims to understand the funding and cash flow of a business so that the feasibility of the implemented business plan can be assessed [11]. The commonly used and justifiable investment criteria are as follows:

a. *Net Present Value (NPV)* aims to determine the comparison between the net present value of cash flows and the investment over the investment's lifetime. The criterion NPV>0 indicates that a business is feasible to be continued. The formula used for the calculation is:

$$\textit{Net Present Value} = \frac{\text{Cash flow}}{(1+\text{Df})^{\wedge}\text{t}} - \text{Initial investment}$$

Description: Df = dicount factor; t = period

b. *Payback period* aims to determine how long an investment will take to recoup its initial cost [11]. The payback period method is a very straightforward investment assessment. If the payback period is less than a predetermined period, the project is accepted. On the other hand, if it exceeds the predetermined period, the project is rejected. The formula used for the calculation is:

$$\textit{Payback period} = \text{Target year} + \frac{(\text{Initial investment - target year investment})}{(\text{Total final investment - target year investment})} \text{ x 1 year}$$

2.4. *Data analysis*

The chemical test data to be observed includes the analysis of water, protein, fat, and ash content, which will be analyzed using Analysis of Variance (ANOVA) with SPSS 25.0 For Windows software. The purpose of this analysis is to determine the effects of the given treatments and if there are significant results within a 95% confidence interval (significance level $p<0.05$), Duncan's test will be performed. The acceptance level of consumers for tekwan with various types of fish based on appearance, aroma, texture, and taste parameters will also be evaluated using the Kruskal Wallis method, which aims to compare two variables measured from different (independent) samples, where the compared groups are more than two [12]. Then, if there is still a significant effect, the Mann Whitney test will be conducted to determine the differences between two sets of data [14], thus determining the consumer preference for the preferred tekwan.

3. Results and Discussion

3.1. Yield value
The yield value serves the purpose of determining the percentage of the final fish meat produced. Calculating the fish yield is utilized to estimate the amount of the fish's body parts that can be used as food ingredients [14]. Based on the known results, the fish yield values used in the making of tekwan are as follows.

Table 1. The yield value of tekwan using various types of fish

Treatment	Initial weight of fish (grams)	Yield value
Snakehead fish	918	46,30%
Yellow Tail fish	693,5	40,59%
Wolf Herring fish	492,5	43,89%
Spanish Mackerels fish	1.032,5	61,31%

When compared to the initial weight of the fish, the yield weight value decreases due to the removal of a significant portion of the fish through the evisceration and filleting process. Based on the data, it is evident that larger-sized fish have higher yield values. The resulting yield values are incosistent because it is influenced by the type of fish and type of food consumed [15].

3.2. Chemical analysis
The raw material is one of the most crucial components in a production process. Therefore, the selection of alternative raw materials needs to be conducted to meet the requirements for making tekwan and similar products, taking into consideration aspects such as the availability of raw materials, nutritional composition, and economically valuable fish. A comparison of the nutritional content of fish can be seen in Table 2.

Table 2. The average values of chemical analysis of tekwan using various different types of fish

	Treatment			
Analysis	Snakehead	Yellow Tail	Wolf Herring	Spanish Mackerels
Water	61.26 ± 1.090^{a}	59.78 ± 2.152^{a}	62.23 ± 2.525^{a}	60.88 ± 1.812^{a}
Protein	10.84 ± 0.128^{a}	12.20 ± 0.587^{ab}	12.03 ± 0.485^{ab}	13.466 ± 1.271^{b}
Fat	0.786 ± 0.327^{a}	0.894 ± 0.0343^{a}	0.657 ± 0.180^{a}	0.759 ± 0.103^{a}
Ash	0.041 ± 0.014^{a}	0.061 ± 0.004^{a}	0.039 ± 0.023^{a}	0.046 ± 1.812^{a}

3.2.1. Water content
Based on the results of the protein content test in **Table 2**, it can be observed that the average water content of tekwan using different types of fish raw materials ranges from 59.78% to 62.23%. The water content can be influenced by chemical compounds, food components, and temperature [16]. The boiling process during the preparation of tekwan affects the water content in the product. This is consistent with [17], which states that high temperature during boiling causes starch granules to expand, resulting in increased water absorption. The longer the boiling time, the more water is absorbed. Additionally, a higher concentration of fish meat used will lead to an increase in water content. The water content level in fish can vary depending on the fish species, size, and environmental conditions where the fish live [18].

3.2.2. Protein content
Based on the results of the protein content test in **Table 2**, it can be observed that the average protein content of tekwan with different types of fish raw materials ranges from 10.84% to 13.46%. The highest protein content is found in treatment P3, using Spanish Mackerels as the raw material (13.46%), followed by P1 using Yellow Tail fish (12.2%), P2 using Wolf Herring fish (12.03%), and the lowest is in the control treatment P0 using Snakehead fish (10.84%). The protein content in fish tends to increase with the increase in fish size. This is consistent with the raw material yield values [19], which show that Spanish Mackerels

have the highest yield value at 61.31%, followed by other fish species such as Snakehead fish at 46%, Wolf Herring fish at 43%, and Yellow Tail fish at 40.59%. The protein content can also be influenced by various factors such as the type of meat, filler ingredients, and binding materials used [20].

3.2.3. *Fat content*

Based on the results of the protein content test in **Table 2**, it can be observed that the average fat content of tekwan with different types of fish raw materials ranges from 0.65% to 0.89%. If we compare it with other fish gel products such as fish balls, the maximum fat content allowed is 1% [21]. This indicates that the fat content of all tekwan treatments still meets the standard. The higher amount of fish meat used can increase the fat content. The influence of fat content can be affected by dietary habits, maturity, season, and food availability [19]. The low fat content (1%) and the white color of the meat make it suitable for making chewy, delicious, and white pempek and tekwan [22]. The presence of fat content plays a role in forming a compact product and can enhance the taste and aroma [23].

3.2.4. *Ash content*

Based on the results of the protein content test in **Table 2**, it can be observed that the average protein content of tekwan with different types of fish raw materials ranges from 10.84% to 13.46%. Determination of ash content and acid-insoluble ash content aims to assess the quality of a processing and provide an overview of the internal and external mineral content contained in the product. The processing method can also influence the increase and decrease in ash content. Boiling during food processing can lead to a decrease in ash content, while frying can cause an increase in ash content in the food [24]. The varying ash content is also influenced by several factors such as the age of the fish, the type of food consumed, and the fish processing techniques.

3.3. *Texture analysis*

Based on the ANOVA test results, there is a significant influence on the hardness value of tekwan using various types of fish, with scores ranging from 220.89 to 263.70 gf. The Duncan post hoc test showed a significant influence of P2 compared to P0, P1, and P3 [25].

Table 3. Analysis of the texture of tekwan using different types of fish as raw materials.

Parameter	Snakehead	Yellow Tail	Wolf Herring	Spanish Mackerels
Hardness (gf)	263,70±10,636^{a}	260,06±8,811^{a}	220,89±7,768^{b}	249,12±6,645^{a}
Cohesiveness (gf)	0,72±0,034^{a}	0,72±8,0,060^{a}	0,85±0,144^{a}	0,74±0,041^{a}

The high hardness value indicates that the product becomes firmer. Setyowati and Nisa (2014) stated that the hardness parameter in tekwan can be influenced by the moisture content in the ingredients, where higher moisture content in the ingredients leads to lower hardness values. Additionally, it is supported by the opinion of [26] that the high hardness value in kamaboko fish cake is possibly due to the lower content of myofibril protein in the fish.

Based on the ANOVA test results for cohesiveness, the scores ranged from 0.72 to 0.85, indicating no significant influence from any of the treatments [27]. The use of different types of fish did not show any impact on the *cohesiveness* value because each treatment utilized the same concentration of tapioca flour. This is in line with [28], which suggests that a higher *amylpectin* content in the used flour can increase the compactness value of a product.

IOP Conf. Series: Earth and Environmental Science 1289 (2023) 012032 doi:10.1088/1755-1315/1289/1/012032

3.4. The level of consumer preference

Based on the results of treatments P0, P1, P2, and P3 with respective scores of 3.41, 3.78, 3.31, and 3.68, all tekwan treatments exhibit a bright white color. This aligns with the findings of a study [29], where panelists preferred fish sticks with a white color. Upon examining the test results, there is no significant influence from any of the treatments, with the best treatment being P1, which uses Yellow Tail fish with a score of 3.78.

Table 4. The level of consumer preference for tekwan with different fish raw materials

Analysis	Treatment			
	Snakehead	Yellow Tail	Wolf Herring	Spanish Mackerels
Appearance	3.44±0.884^{a}	3.71±0.830^{a}	3.60±0.821^{a}	3.64±0.767^{a}
Aroma	3.43±0.911^{a}	3.65±0.781^{a}	3.38±0.905^{a}	3.68±0.792^{a}
Texture	3.41±0.882^{ac}	3.78±0.616^{b}	3.31±0.922^{c}	3.68±0.897^{ab}
Taste	3.01±0.948^{a}	3.55±0.745^{b}	2.98±0.993^{a}	3.93±0.792^{b}

The aroma of tekwan, which does not show any significant difference, is due to the similar chemical composition present in the fish of each type. Panelists prefer tekwan with a not overly fishy aroma. A stimulating aroma is a good parameter for consumers. The insignificant differences in the aroma resulting from each treatment could be attributed to the use of a 1% concentration of garlic. In line with [30], the aroma of garlic is produced by volatile compounds contained in it, particularly *methyl allyl disulfide*, which gives a spicy and aromatic smell, thereby reducing the fishy aroma from the dish.

Based on the test results, there is a significant influence between P0 and P1, P1 and P2, and P2 and P3, with the best treatment for the texture parameter being P1, using Yellow Tail fish as the raw material with a score of 3.78. Panelists prefer tekwan with a chewy and non-sticky texture. The variation in texture in each treatment can be influenced by various factors, including protein content ratio, fat content, processing temperature, water content, and water activity [31]. Additionally, the texture can be affected by the myofibril structure in the meat and its level of cross-linking and water-binding capacity by proteins [32]. In this study, the observed differences may be attributed to the specific content of actomyosin protein present in each type of fish.

Based on the test results, it is evident that there is a significant influence between P0 and P1, P1 and P2, and P2 and P3. The best treatment is observed in P3, which uses Spanish Mackerels as the raw material with a score of 3.93. Panelists prefer tekwan that has a dominant and savory fish taste [29]. Products with a savory and distinctive fish flavor are favored by consumers.

3.5. Break Event Point

Table 5. The break-even point of tekwan using various different types of fish

Fish	Fixed Cost	Sale/unit	Variable Cost/ unit	BEP Unit
Snakehead	297.287	28.000	24.890	95.5
Yellow Tail	297.287	28.000	18.765	32.1
Wolf Herring	297.287	28.000	11.890	18.4
Spanish Mackerels	297.287	28.000	18.640	31.7

The calculation of tekwan selling price at a selling price of Rp28,000 shows different profits based on the use of raw materials from various types of fish, with the highest to lowest profits being observed in Wolf Herring fish, Spanish Mackerels, Yellow Tail, and Snakehead, respectively.The breakeven points of the company are known from the lowest to the highest, namely Wolf Herring fish, Spanish Mackerels, Yellow Tail, and Snakehead.

3.6. *Net Present Value*

Table 6. The NPV of tekwan using various types of different fish

Fish	*PV Net Benefit*
Snakehead	8.813.270
Yellow Tail	26.814.886
Wolf Herring	34.813.270
Spanish Mackerels	29.928.270

The entire data to be used is in the form of estimations. The estimation plan for discount factors (DF) is set at 10% with the current BRI interest rate at 6%. Discount factor is a value less than 1 used to convert a future value into a present value [11]. Based on the obtained result, the NPV is 29,813,270, which indicates NPV > 0, suggesting that the business is feasible to continue. This is in line with [33] that if the future net cash flows have a higher value than the current investment outlay, then the business is considered feasible to be carried out.

3.7. *Payback Period*

Table 7. The payback period of tekwan using various types of different fish.

Fish	*Payback Period*
Snakehead	3.57
Yellow Tail	2.57
Wolf Herring	2.35
Spanish Mackerels	2.57

The payback period calculation is used to determine how long it takes for the investment value to be recovered. Based on the data used, the objective is to achieve investment payback within a three-year timeframe. Among the invested values, the quickest return is observed in tekwan made from Wolf Herring raw material, followed by Spanish Mackerels and Yellow Tail.

4. Conclusion

The use of different fish raw materials does not affect the physical characteristics; however, it shows differences in the protein content of tekwan in terms of chemical characteristics. The results of consumer preference tests for tekwan with different raw materials showed preferences ranging from neutral to liking. Wolf Herring exhibited results similar to Snakehead fish, whereas tekwan made from Yellow Tail and Spanish Mackerels raw materials were the most preferred among the panelists. From the four types of fish used, it was found that three types of fish, namely Yellow Tail, Spanish Mackerels, and Wolf Herring, were deemed suitable for commercial purposes. The most optimal business feasibility was observed in tekwan made from Wolf Herring as the main ingredient.

Acknowledgement

The work is supported by collaborative research grants for lecturers and studendts of Aqriculture UGM number: 3162/UN1/FPN/KU/KU.02.05/2023.

References

[1] Jayanti D 2018 *Pengembangan Bisnis" Tekwan Instan Khas Palembang" Dengan Pendekatan Matrix Ansoff* (Yogyakarta: UII) hal 3-4

[2] Danar R and Dasir 2017 Studi berbagai jenis bahan pengembang terhadap reabsorpsi tekwan kering ikan gabus *J Edi.* **6** 36-45

[3] Agustini T W and Fronthea S 2003 Pemanfaatan hasil perikanan sebagai produk bernilai tambah (*value-added*) dalam upaya penganekaragaman pangan *J. Food Agroind* **14** 74-81
[4] Suryaningrum T A and Ijah M 2009 Prospek pengembangan usaha pengolahan pempek palembang *J. Squ.* **4** 31-40
[5] Rosadi E and Siti A 2019 Penilaian stok sumberdaya ikan gabus (channa striata,bloch 1793) di kalimantan selatan dengan pendekatan model holistik (equilibrium -non equilibrium state) *J. Fish. Sci.* **9** 64-75
[6] Dewi E N D 2001 *Pengolahan "Fish Jelly" Produk Menggunakan Ikan Berdaging Merah.* (Semarang: PKMT Universitas Diponegoro)
[7] Moniharapon A 2014 Teknologi surimi dan produk olahannya *J. Biam* **10** 16-30
[8] Jannah A 2007 *Pembuatan Gelatin Halal dari Tulang Ikan Bandeng (Chanos chanos Forskal) (sebagai Alternatif Pembuatan Gelatin Halal)* (Malang: LEMLIT UIN Malang)
[9] Faisal F A 2020 *Analisis Biaya Produksi Dalam Menentukan Harga Jual Produk Liquid Zero 57 Study Kasus Pada CV. Cloud Heaven Makassar* (Makassar :Universitas Muhammadiyah Makassar)
[10] Kementerian Kelautan dan Perikanan 2022 *Statistik Ekspor Hasil Perikanan Tahun 2014-2018* (Jakarta: Ditjen PDSPKP)
[11] Hidayati N and Warnara D D 2017 Analisis kelayakan finansial pengembangan kelas alam terbuka kebumian dan lingkungan berkonsep rekreasi dan inspirasi untuk anak di surabaya *J. pros. ISNB* **3** 650-656
[12] Junaidi J 2010 Statistik Uji Kruskal-Wallis. *J. UNJA* **1** 1-5
[13] Sriwidadi T 2011 Penggunaan uji mann-whitney pada analisis pengaruhi pelatihan wiraniaga dalam penjualan produk baru *Binus Biz. Rev.* **2** 751-762
[14] Radityo C T, Darmanto Y S, and Romadhon 2014 Pengaruh penambahan egg white powder dengan konsentrasi 3% terhadap kemampuan pembentukan gel surimi dari berbagai jenis ikan. *JPBHP* **3** 1-9
[15] Heruwati E S Pengolahan ikan secara tradisional prospek peluang dan pengembangan. *J.Litbangper* 92-99
[16] Pratiwi N M, Widiastuti I, and Baehaki 2016 Karakteristik fisiko-kimia dan sensori bakso ikan gabus (Channa striata) dengan penambahan genjer (Limnocharis flava*). J. Teknologi Hasil Perikanan* **5** 178-189
[17] Anggara G, Nopianti R, and Herpandi 2016 Pengaruh suhu dan lama perendaman dalam air dingin pada praperebusan terhadap kualitas bakso ikan patin (Pangasius pangasius) *J. Fish. Tech.* **5** 134-145
[18] Hafiludin H 2015 Analisis kandungan gizi pada ikan bandeng yang berasal dari habitat yang berbeda *J. Mar. Sci. Tech.* **8** 37-43
[19] Nianda T 2008 *Komposisi Protein dan Asam Amino Daging Ikan Gurami (Osphronemus Gouramy) Ada Berbagai Umur Panen* (Bogor: Institut Pertanian Bogor)
[20] Umami M, Rizal ,and Guntoro 2017 Eksperimen sosis sayur hasil olahan dari sayur brokoli (brassica oleracea l.) dan wortel (daucus carota l.) *J. Food Agroind* **4** 73-88
[21] Standar Nasional Indonesia 1995 *SNI 01-3891-1995 Bakso Daging.* (Jakarta: DSNI)
[22] Iljas N 1995 *Peran Teknologi Pangan dalam Upaya Meningkatkan Citra Makanan Tradisional Sumatera Selatan.* (Indralaya: UNSRI)
[23] Riyandi N and Windi A 2010 Diversifikasi dan karakterisasi citarasa bakso ikan tenggiri (scomberomus commerson) dengan penambahan asap cair tempurung kelapa *J.Agroind. Tech.* **3** 1-12
[24] Sundari E, Almasyhuri, and Astuti L 2015 Pengaruh proses pemasakan terhadap komposisi zat gizi bahan pangan sumber protein *J. Med. Litbangkes* **25** 235-242
[25] Haliza W, Kailaku S I, and Yuliani S 2017 Penggunaan Mixture Response Surfa Ce Methodology Pada Optimasi Formula Brownies Berbasis Tepung Talas Banten (Xanthosoma Undipes K. Koch) Sebagai Alternatif Pangan Sumber Serat *J. PP* **9** 96-106

[26] Setyowati W T and Nisa F C 2014 Formulasi biskuit tinggi serat (Kajian proporsi bekatul jagung : tepung terigu dan penambahan baking powder) *J. Food. Agroind.* **2** 224–231
[27] Fadhallah E G, Nurainy F, and Suroso E 2021 Karakteristik sensori, kimia dan fisik pempek dari ikan tenggiri dan ikan kiter pada berbagai formulasi *JPPT* **21** 16-23
[28] Iswara J A, Julianti E, and Nurminah M 2019 Karakteristik tekstur roti manis dari tepung, pati, serat dan pigmen antosianin ubi jalar ungu *J. Food. Agroind.* **7** 12–21
[29] Yanuar V 2020 Analisis tingkat kesukaan konsumen terhadap jenis ikan yang berbeda pada kerupuk stik ikan *J. Zira'ah* **45** 370-377
[30] Meilani F, Purwanti H, and Suharno B 2014 Kandungan Protein, Lemak, Populasi Bakteri dan Sifat Organoleptik Pada Bakso Ikan Rucah Deangan Berbagai Dosis Bawang Putih (*Allium Sativum*) (Semarang: Universitas PGRI Semarang)
[31] Purnomo A H 1995 *Aktivitas Air dan Peranannya dalam Pengawetan Pangan* (Jakarta: UI Press)
[32] Soeparno 2005 *Ilmu dan Teknologi Daging* (Yogyakarta: Gadjah Mada University Press)
[33] Sunyoto D 2013 Analisis Laporan Keuangan untuk Bsinis (teori dan kasus) (Jakarta: CAPS)

IOP Conf. Series: Earth and Environmental Science 1289 (2023) 012033 doi:10.1088/1755-1315/1289/1/012033

Effect of Adding Egg White Powder on Surimi Quality Based on Pangasius and Catfish Flesh

A N F Rahma and Ustadi*

Fish Product Technology, Department of Fisheries, Faculty of Agriculture, Gadjah Mada University, Jl. Flora A4, Bulaksumur, Yogyakarta, 55281, Indonesia.

*Corresponding author e-mail ustadi@ugm.ac.id

Abstract The variety of raw materials from demersal fish in small quantities has resulted in the development of freshwater fish-based surimi processing technology. This study aims to determine the best concentration of egg white powder based on pangasius and catfish on the quality of surimi by physically, chemically, and sensory. The experimental design used was a Completely Randomized Design (CRD) with the addition of egg white powder at concentrations of 0%, 1%, 2%, 3%, and 4%. The surimi quality characteristics tested in this study included texture profile, degree of whiteness, pH, moisture, ash, protein, fat, water holding capacity, sensory and folding tests. The results showed a significant effect ($P < 0.05$) on whiteness, fat, water holding capacity, texture profile, folding and sensory tests. However, it did not affect ($P > 0.05$) on ash, protein, and pH. The results showed that the concentration of 1% egg white powder was able to improve surimi gel based on quality.

1. Introduction

Surimi is a concentrate of myofibrillar protein obtained from pulverized fish meat after being washed in cold water and mixed with cryoprotectant. General raw materials used for making surimi are Alaska Pollock, Pacific whiting, Arrowtooth flounder, Blue whiting, Mackerel, Menhaden, Bigeye snapper, Threadfin bream, Lizardfish, and Croacker [6] [19]. In Indonesia, threadfin bream is often used because it has white flesh, is low in fat, and is cheap. However, when overfishing is carried out, the fish population decreases. The downward trend in cantrang fish production in 2012 - 2019 is shown through reduced catch volumes and dominated by small non-economic fish such as kuniran, swanggi, kurisi, beloso, and others from 45.3 thousand tonnes (2012) to 14.2 thousand tonnes (2019) [12] [25]. Freshwater fish cultivation mostly practiced includes clarias, pangasius hypopthalamus, Oreochromis niloticus, and tilapia. Besides being marketed for restaurants, the abundant cultivation results also need to be processed, so that freshwater fish cultivators can market the fish they produce optimally.

The Indonesian Ministry of Maritime Affairs and Fisheries supports the production of catfish and pangasius cultivation through various technologies. Catfish is one of the production results of the cultivation sector which produces quite high availability. Mixed production of catfish and pangasius expected can be increased surimi products. The addition of food additive that can be added to processed surimi is egg white powder. According to [23], egg white powder can bind protein. Egg white powder not only improves texture but also the quality of surimi gel. Egg white powder can fill the spaces in the fish protein network [15]. Based on [15], added 2.5% egg white powder on surimi was statistically showed effect for pH value. Surimi gel are not only influenced by minced meat quality but also by various additives. Egg white powder is commonly used in surimi manufacturing to modify the textural properties. Egg white powder produced elastic gel as an inhibitor enzyme. Based on [12], inhibitor enzyme aims to inhibit modori stage during thermal process. So that surimi quality stay in good condition.

The purpose of this study was to determine the effect of adding egg white powder with different dosage on surimi quality based on pangasius and catfish flesh.

2. Materials and Methods

2.1 Materials

Primary materials used were catfish and pangasius from a local cultivator. The additional materials were ice, water, salt, Sodium tripolyphosphate, trehalose, and egg white powder. The variable testing materials were distilled water, catalyst N (Na2SO4, CuSO4, Se/TiO2), H2SO4, H3BO3 4% with Mr-BCG indicator, NaOH-Tio (NaOH 40% + Na2S2O3 5%), ice cubes, petroleum ether, and water.

The tools used for producing surimi were digital scales (OHAUS PX523), rinsing containers, filters, rolling pins, molds, freezers, measuring cups, soxhlet, and trays. Tools used for analysis the quality were analytical scales (Deriver Instrument Company AA-200), steam pans, stoves, mercury thermometers, knives, mortar and pestles, erlenmeyer, cutting boards, rulers, polyethylene (PE) plastic, spoons, falcon tube, centrifuge (H.26.F KOKUSAN), pH meter (OHAUS STARTER5000), texture analyzer (OHAUS), moisture meter (Yeasten moisture meter), beaker glass, oven, tissue, filter paper, porcelain cup, desiccator (Buchi), Soxhlet (Buchi Extraction System B-11 brand), chopper (PHILIPS brand), kjeldhal flask (Vepl Scientifica), chromometer, plate, paper cup, panelist sheet [5].

2.2 Research Methods

Pangasius and catfish from local cultivator were deboned directly after gutting and washed with freshwater. The minced meat from both fish washed in cold water with 4°C temperature used ratio 1 : 4. Repeated 3 times. The minced meats mixed with 0.03% trehalose and 0.225% sodium tripolyphosphate. Egg white powder added with a concentration of 1%, 2%, 3%, 4%.

Each surimi sample will be packed into polyethylene bags and saved in low temperature -24. The surimi analyzed for yield, whiteness, profile texture, moisture, ash, fat, protein, measurement of pH (degree of acidity), water holding capacity, sensory and folding tests. All the tests replicated in 3 times. The data were collected and analyzed used a Completely Randomized Design (CRD).

2.3 Texture Profile

Surimi texture profile tests aims to know the gel strength quality. The test carried out by cut into 2.5 cm length. The sample replaced on the universal texture machine. The diameter of the press section used is 25 mm with a pressure load of 0.02 N and a speed of 50 mm/min. The result will be displayed.

2.4 Whiteness

Measurement of whiteness degree use the calorimeter (Minolta CR 400 Chroma-meter). The principle is the measurement of color differences through the reflection of light by the top of the sample. The results showed the value of L* (lightness/darkness), a* (redness/greeness), b* (yellowness/blueness), and whiteness. The value must be calculate with following formula :

$$\textit{Whiteness}\ (\%) = 100 - [(100 - L^*)^2] + (a^2 + b^2)]^{1/2}$$

2.5 Proximate composition

Moisture content carried out used analyzer instrument (Ohaus MB120). Total protein content determined used micro-Kjeldahl analysis method [3]. The fat and ash content tests determined used AOAC method [1].

2.6 Degree of acidity (pH)
The pH measurement test carried out by dissolving the sample with distilled water. Take a sample on it. Buffer solution will calibrated the pH meter.

2.7 *Water Holding Capacity*
Water holding capacity test carried out by putting into a falcon tube. The sample will homogenized used a vortex. It will centrifugation. The supernatant volume calculated by formula [13] :

$$\text{Water Holding Capacity (\%)} : (\text{absorbed volume})/(\text{Total water added}) \times 100\%$$

2.8 Sensory test and Folding test
The sensory test carried out based on [5]. There were seven trained panelists who will evaluate the product properties on the score sheet. Panelist selection carried out using the triangle test.

2.9 Data analysis
The data analysis test used in this study was quantitative data and analyzed used analysis of variance (ANOVA). Data was analyzed used statistical software IBM SPSS 20. Normality test aims to find out the data normally distributed. Homogenity test (Homogenity of Variance) aims to find out the data of two or more groups is homogeneous. The ANOVA test carried out to find the average comparison between data groups. DMRT's (Duncan's Multiple Range Test) used as advance tests.

3. Results

3.1 Yield

Table 1. Surimi yield based on fish species

Type of Fish	Fish weight (g)	Minced meat (g)	Meat yield (%)	Surimi (g)	Surimi yield (%)
Pangasius	4666	1770	38,90	1145,7	24,9
Catfish	3100	1224	34,34	707,63	22,7
Average	3883	1497	36,62	926,66	23,8

The average yield of pangasius meat based on (**Table 1)** was obtained about 38.90%. Highly yield value can causes the fish characteristics include head, bones and patils. Surimi yield on [24] research have value about 22.54%. The yield value decreased after the washing process was carried out from 38.90% to 24.9%. The washing process affects to dissolution of several components in fish meat [10].

The average value of the yield of catfish was 34.34% and was not much different from the yield of catfish from [18] [16], respectively 37.17% and 33.25%. The yield value showed the effective of skinless fillet process. The yield value decreased from 34.34% to 22.7%. The average yield of catfish surimi produced was 22.7% higher than yield value from [16] which had 17.87%. Repeated washing process made the yield value decreased caused losing more chemical compounds such protein, fat, pigment, etc). [9].

The average yield of both surimi was 23.8% after washed 3 times. According to [24], good profit industry started when the surimi has yield value more than 20%. Lower economic value of materials give higher profit when it in commercial products [2].

3.2 Texture profile

Table 2. Texture profile analysis of surimi used egg white powder

Group	Parameter					
	Hardness	*Gumminess*	*Fracture*	*Chewiness*	*Cohesiveness*	*Adhesiveness*
0%	11.07 ± 0.39^{a}	11.85 ± 1.44^{a}	2.25 ± 0.09^{a}	12.02 ± 0.02	0.62 ± 0.34^{a}	0.16 ± 0.02^{ab}
1%	13.95 ± 3.97^{a}	7.44 ± 0.89^{a}	1.92 ± 0.09^{a}	5.53 ± 1.09^{a}	0.26 ± 0.08^{a}	0.06 ± 0.01^{a}
2%	22.7 ± 1.68^{b}	9.99 ± 5.71^{a}	2.39 ± 0.02^{a}	7.97 ± 5.48^{a}	$0.25 \pm 0,008^{a}$	0.25 ± 0.03^{b}
3%	33.84 ± 2.90^{b}	19.52 ± 1.14^{a}	2.48 ± 0.28^{a}	16.41 ± 0.55^{c}	0.48 ± 0.05^{a}	0.69 ± 0.14^{d}
4%	25.54 ± 2.61^{c}	9.57 ± 1.03^{a}	2.05 ± 0.25^{a}	8.33 ± 0.67^{a}	0.41 ± 0.02^{a}	0.38 ± 0.23^{c}

Egg white powder in surimi forms had increased trend by texture parameters. Based on (**Table 2)** can be seen there was a significan effect to hardness and adhesiveness. The analysis also showed there was no significant effect on the gumminess, fracture, chewiness, and cohessiveness parameters ($P > 0.05$).

The analysis showed the addition of egg white powder to surimi had a significant effect on hardness. It was believed able to significantly increase the hardness value because egg white powder can fill the protein spaces inside surimi. The heat-induced value of hardness can caused by increase in water content and the removal of water-soluble proteins to a certain extent [28]. Adhesiveness showed a significant effect on the value of surimi added to egg white powder. A higher adhesiveness value indicates that the texture is softer [28].

Gumminess parameter showed higher level of elasticity than the control was 3% of egg white powder. However, based on ANOVA statistical analysis the addition of egg white powder showed no effect on the gumminess value. Sodium Tripolyphospate and egg white powder known didn't have a significant effect on gel strength on the gumminess. The fracture and cohesiveness values showed that there was no effect given by added egg white powder. Cohesiveness value of the heat-induced gel could be related to increase in myofibrillar protein. However, chewiness as bite aspect showed there was no significant effect. Gel strength can be affected by the presence of thermally induced proteinases and lipids which can disrupt myosin cross-links during the formation of the gel matrix, so that it cannot form a gel [29].

3.3 Whiteness

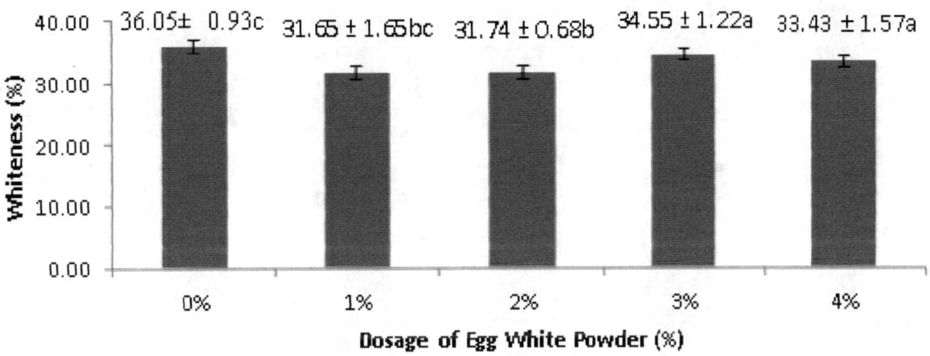

Figure 1. Whiteness analysis of surimi used *egg white powder*

Based on the statistical analysis of DMRT's advanced test, by adding egg white powder into surimi had a significant effect on the degree of whiteness. It can be seen that surimi by adding egg white powder had a whiteness degree of 31.65 - 36.05% (**Figure 1**). The results was directly proportional to the sensory color test results. Sensory test results showed the effect of adding egg white powder. Decreased in the whiteness can be affected by the washing process. Washing process have to optimally because to increase the whiteness value [26]. According to [11], the whiteness of catfish surimi increased after washings in four times from 44.39% to 73.38%. The increase in whiteness of surimi can caused by the wasting of pigment, blood and other impurities during washing process.

3.4 Moisture

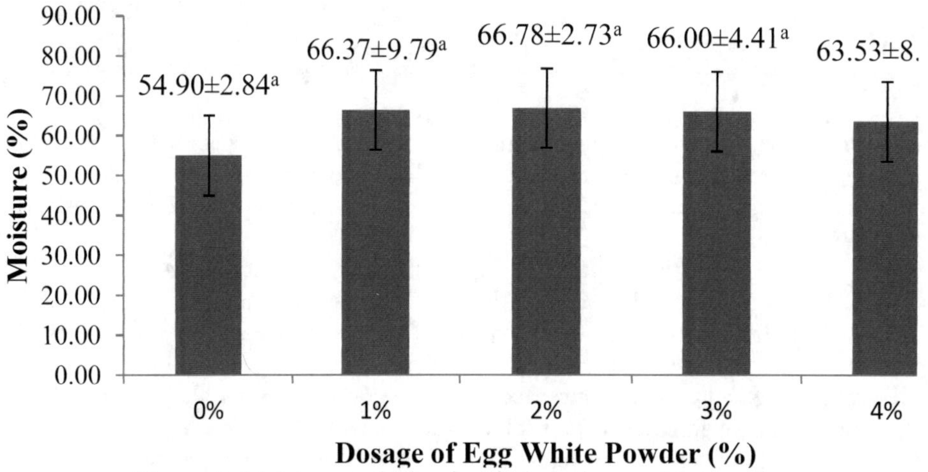

Figure 2. Moisture analysis of surimi used *egg white powder*

Moisture analysis based on (**Figure 2**) showed increased trend with adding egg white powder into surimi. The results of the ANOVA statistical analysis showed the addition of egg white powder concentration did not affect on the moisture of surimi (P>0.05). The results showed that the highest moisture value 45.11% was produced by surimi with egg white powder (0%). Lowest value was 33.22% produced by added of 2% egg white powder concentration.

The water content values produced in this study were different from other research. The water content of Pangasius surimi can be higher due to various factors [8]. The factors are not only caused by raw materials used but also molecular structure in each type of fish. Surimi from catfish flesh with the addition of egg white powder has a decreased trend because it can bind water contained in fish meat [15]. Water content in this research showed increased trend caused the raw material mixture formula having more dominant properties over catfish even though the ratio is the same. However, based on [5], the range moisture produced in this study reached under limit of <80%.

3.5 Protein

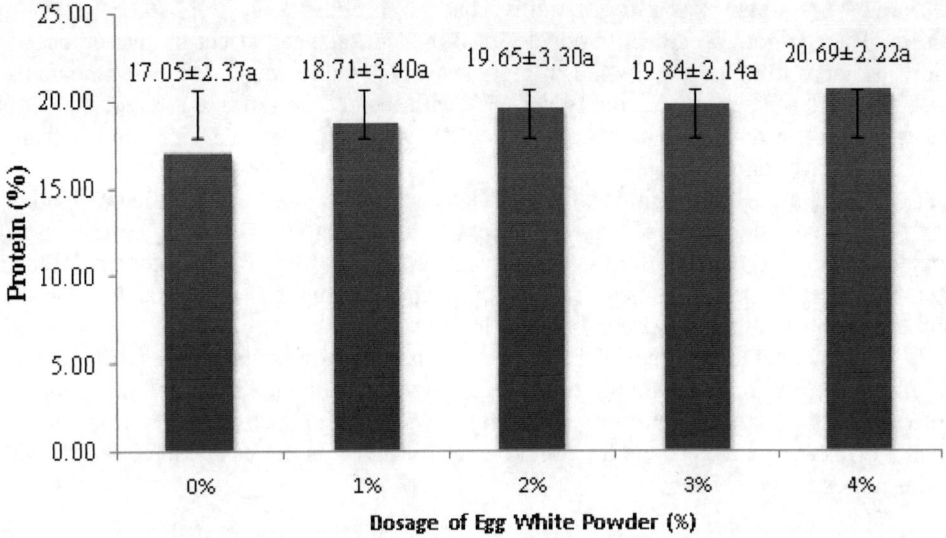

Figure 3. Protein analysis of surimi used egg white powder

The protein content showed on (**Figure 3**) the addition of egg white powder generally forms an increasing trend during highly concentration. The average value of surimi protein content was 17.5% - 20.69% (wb). The highest average protein content was the addition of 4% (wb), and the lowest was 0%. Another research was compared with the protein content of catfish surimi with a washing frequency of three times in the study of [29] which was 11.99% (wb) and [30] 13.07% (wb). Generally the surimi protein content produced in this study was above the surimi quality standard at least 12% based on [5].

Based on ANOVA statistical analysis, the addition of egg white powder as a food additive to surimi had no significant effect ($p > 0.05$) on the protein content of surimi at a significance level of 95%. It can be seen that the treatment of adding egg white powder of 0 - 3% did not affect the protein content. This is presumably because the concentration of added ingredient. Addition of egg white powder was quite low. The results does not show a drastic change [20]. Protein levels greatly affect the quality of surimi gel because in fish protein there is a myofibril protein which important in the gel formation process [9].

3.6 Fat

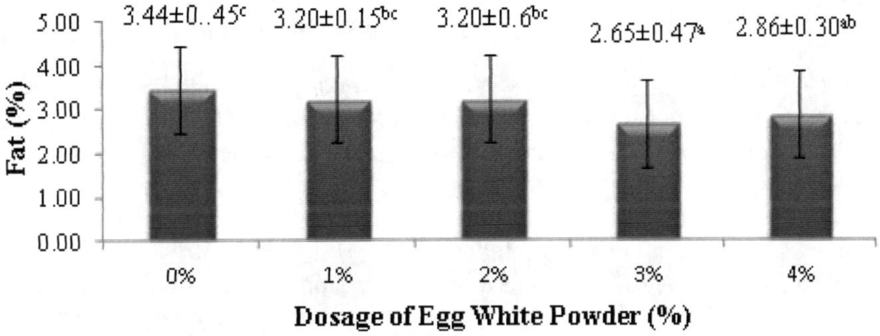

Figure 4. Fat analysis of surimi used *egg white powder*

The fat content results on surimi with the addition of egg white powder showed in (**Figure 4)** generally forms a decreasing trend. Based on ANOVA statistical analysis and DMRT's advanced test, egg white powder as a food additive had a significant effect ($p < 0.05$) on fat content of surimi. The addition of egg white powder of 0 - 4% dosage can affect by fat content of surimi. It showed ranges from 2.65 to 3.44. The highest average value was 3.44% with the addition of 0% dosage. The lowest average value was 2.65% with the addition of 3% dosage. The catfish surimi under fat content obtained about 4.51% - 5.08% (wb) [15]. However, it is higher than compared with 0.82% (wb) fat value.

Based on the composition of protein and fat, catfish used to classified as high protein and low fat. It expected to produce a good gel. Myofibril protein was positively correlated with protein content which is responsible for the gel strength properties. Meanwhile, fat is one of the inhibiting factors for gel formation in meat, so it is important to remove it optimally through the washing process to obtain higher gel strength values [15].

In general, the fat content of surimi in all treatments was above the surimi quality standard based on [5], which is a maximum of 0.55%. The results obtained indicated that all treatments did not reached the limit. In addition, egg white powder can decreased the fat content. Protein will form a matrix that envelops the fat granules, so that the fat is not easily separated from the system and is not extracted [17].

3.7 Ash

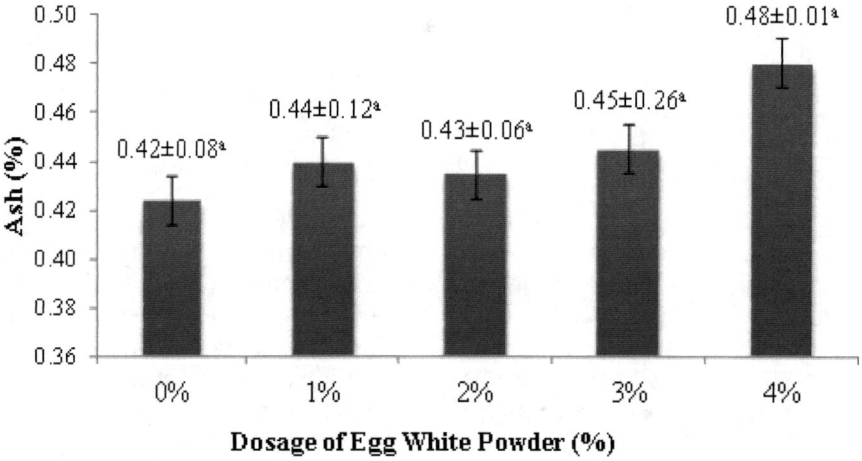

Figure 5. Ash level analysis of surimi used *egg white powder*

Based on (**Figure 5**), the ash value of surimi with added egg white powder generally forms an increasing trend. The average value ranged from 0.42 - 0.48%. The highest average of ash content in surimi (wb) was 4% dosage, and the lowest average ash content was 0% dosage or control. The high value of the ash content could be due to its close relationship with the water content and protein content in a fat-free tissue. Insoluble minerals are associated with fish protein and egg white protein so that the product has a high mineral or ash content [9].

Based on ANOVA statistical analysis, the addition of egg white powder as a food additive to surimi had no significant effect ($p > 0.05$) on ash content. This indicates that the treatment of adding egg white powder with 0-4% dosage did not affect the ash content. This is presumably because the concentration of egg white powder added is quite low so that the ash value is relatively constant [18].

IOP Conf. Series: Earth and Environmental Science 1289 (2023) 012033 doi:10.1088/1755-1315/1289/1/012033

3.8 Measurement of pH (degree of acidity)

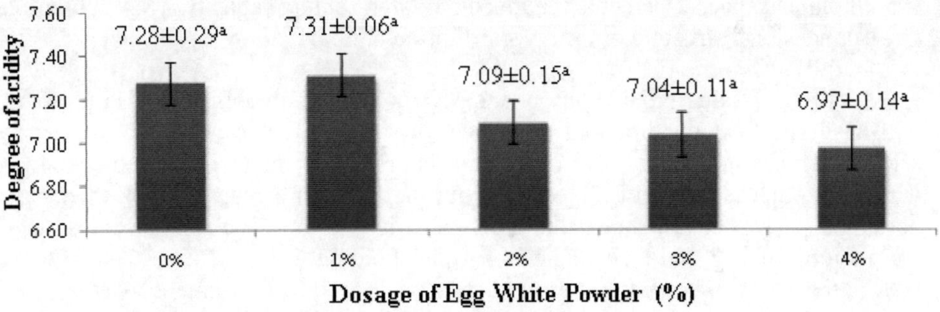

Figure 6. pH analysis of surimi used egg white powder

Surimi pH test results with the addition of egg white powder in (**Figure 6**) showed a downward trend along with the addition of egg white powder concentration. The analysis showed there was no significant effect on surimi ($P < 0.05$). Changes in the pH value can affect the nature of the connective tissue in meat. The pH value can cause denaturation of myofibril proteins. A good pH value for the formation of surimi gel was neutral or slightly alkaline (7.00). A good surimi gel is in the range of 6-7 because the acidic pH is less stable for myofibril protein so that the speed of protein denaturation is faster [15]. If the sol formed is less than 6, a gel that is easily broken will result, whereas at a pH of more than 8, the gel that is formed is not compact.

In general, the pH value of surimi in this study ranged from 6.97 to 7.28. The lowest pH value was in the 4% dosage and highest was 1%. The pH value produced in this study is quite good because the pH of surimi under range 6-7 is the optimum condition for producing a good gel. Based on statistical analysis, the addition of egg white powder to catfish surimi had a significant effect on the surimi pH value ($P > 0.05$).

3.9 Water Holding Capacity

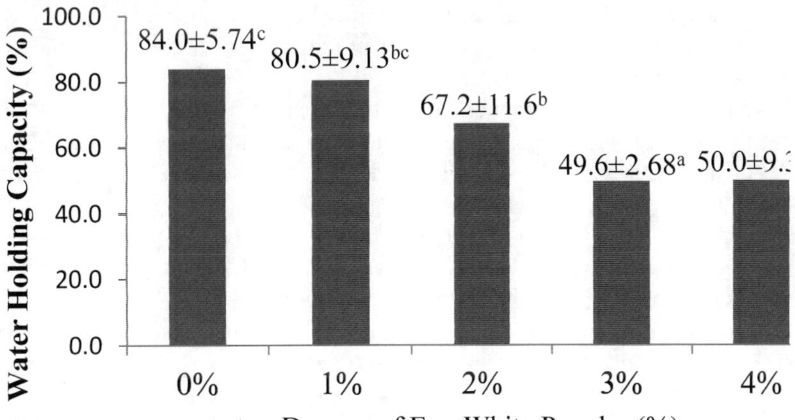

Figure 7. Water Holding Capacity analysis of surimi used egg white powder

The results of the ANOVA statistical analysis and DMRT's advanced test in (**Figure 7**) showed that the addition of egg white powder had a significant effect on the water holding capcity value ($P < 0.05$). Surimi with the addition of egg white powder concentration of 3% and 4% had lower water holding capacity than other treatments. Water holding capacity is strongly influenced by the three-dimensional structure of the myofibril protein.

Myofibril proteins provide an open space for water immobilization, so protein denaturation will have an effect on shrinking the space. In addition, the water holding capacity value also affected by measurement of pH. An increase in pH causes the myofibrils to open more openly, so that more water is trapped [27]. Surimi with the addition of egg white powder has a low water holding capacity because the resulting pH value is also low. The results showed that surimi with the addition of egg white powder had a water holding capacity of 49.63% - 84%. One of the factors that most influences water holding capacity is protein content where protein molecules bind water molecules [8]. Egg white powder can effectively increase the water holding capacity of muscle protein.

3.10 Sensory test

Table 3. sensory analysis of surimi used *egg white powder*

Group	Parameter		
	Odor	Visual	Texture
0%	5.14±0,25^{c}	4.24±0.33^{c}	4.38±0,37^{c}
1%	2.90±0,38^{c}	2.90±0,30^{a}	3.52±0,10^{a}
2%	3.76±0,14^{bc}	3.76±0,30^{b}	4.00±0,14^{b}
3%	4.43±0,17^{a}	4.43±0,00^{c}	4.33±0,17^{c}
4%	4.57±0,9^{ab}	5.19±0,14^{d}	4.76±0,09^{d}

The kind of product that most attracts consumers' attention and gives the impression is color. Color is a very important component to determine the quality or degree of acceptance. Based on the statistical analysis of ANOVA and Duncan's advanced test in (**Table 3**) showed that the addition of egg white powder had an effect on the color of the surimi gel by sensory test ($P < 0.05$). These results are same as the results of color testing using a chromameter which produces significantly different colors. The washing process removes some fat and pigment so that the color of natural surimi is shiny white visually [22].

Odor is a sense that is responded to by the sense of smell. The results showed that the surimi gel with the addition of 4% egg white powder had an odor value that was almost the same as the control surimi gel, which was 4.57. This was the highest value of the resulting surimi gel odor attribute. Surimi gel odor test values ranged from 2.9 to 4.57 (**Table 3**). This range indicates that the surimi gel produced does not smell until it smells strong of fish. Based on [23], revealed that washing process should be able to remove unwanted odors such caused by trimethylamine compounds (one of the main odor-forming compounds in fish).

Texture is the main attribute in determining the functional characteristics of surimi which affects the final quality of surimi-based seafood products [23]. The results of Duncan's advanced test analysis showed that the addition of egg white powder had a significant effect on the texture value of the surimi gel ($P < 0.05$). The values range from 3.52 to 4.76. The value from 1 until 7 represents a texture that is not very rubbery to very rubbery, compact, and very dense. The surimi gel that had the highest texture value was the addition of 4% egg white powder with a value of 4.76. This value indicates that the resulting surimi gel has a chewy, compact, dense to very dense texture. Surimi with the addition of 1% egg white powder had the lowest texture value, which was 3.52 (**Table 3**). This characteristic relates to the softness and hardness of the product.

3.11 Folding test

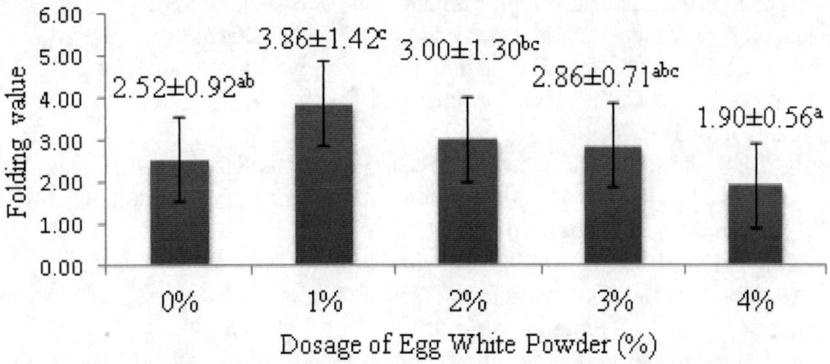

Figure 8. Fold test analysis of surimi used *egg white powder*

The results of the ANOVA statistical analysis and DMRT's further test showed that the addition of egg white powder had an effect on the value of surimi gel folding test ($P<0.05$). The decrease in the value of the folding test can be caused by several factors such as the water content, texture, and the degree of acidity (pH). Excessive increase in water content in surimi gel products can result in a decrease of gel quality and texture [2]. This high value can cause a decrease in the quality of the gel and texture. The folding test results showed lower value.

The folding test is a sensory assessment of gel strength. The results showed that surimi with the addition of egg white powder had a folding test value of 1.90 - 3.86 (**Figure 8**). It value was directly proportional to the strength of the gel. The folding results related to the gel strength. The folding test indicates better elasticity of the surimi gel. According to [23], sucrose and sorbitol have water-binding properties and added to the elasticity of the fish gel paste.

4. Conclusion

Sensory test showed that surimi with 1% treatment had a good odors. The folding test was carried out it showed good results for 1% treatment. Overall, the best treatment is the addition of 1% egg white flour concentration because it has a significant effect in increasing the gel strength of surimi optimally.

Acknowledgement

The work is supported by collaborative research grants for lecturers and students of the Faculty of Agriculture UGM number : 3162/UN1/FPN/KU/KU.02.05/2023

References

[1] [AOAC] Association of Official Analytical Chemists 1995 Official Methods of Analysis.Ed14 Association of Official Analytical Chemists Inc. Washington DC.
[2] Barisetiawan, J 2015 Pengaruh Penambahan Cryoprotectant Kitosan dan Xanthan Gum Terhadap Karakter Fisiko-Kimia Surimi Lele Dumbo (*Clarias gariepinus*) Fakultas Teknologi Pertanian Universitas Brawijaya. Skripsi.
[3] [BSN] Badan Nasional Indonesia 2006 Penentuan Kadar Protein dengan Metode Total Nitrogen Pada Produk Perikanan SNI 01-2354.4-2006. Jakarta Badan standarasi Nasional.

[4] [BSN] Badan Standardisasi Nasional 2006 Surimi Beku–Bagian 1: Spesifikasi. SNI 01 2694.1 2006 Jakarta Badan Standardisasi Nasional (BSN).

[5] [BSN] Badan Standarisasi Nasional 2013 SNI 2694 : 2013 Surimi Jakarta.

[6] Campo,D.L., and Tovar C 2009 The Effect of Egg Albumen on The Viscoelasticity of Crab Sticks Made from *Alaska pollock* and Pacific Whiting Surimi. J. Food Hydrocolloids 23(7):1641-1646.

[7] Dasir, D., Suyatno, S. and Rosmiah, R 2019 Analisis Karakteristik Fisik dan Kimia Surimi Ikan Lele dengan Perlakuan Jenis dan Lama Penyimpanan Dingin. In Seminar Nasional Lahan Suboptimal (pp. 165-171).

[8] Fitriani, R. R 2019 Pengaruh Penambahan Larutan Nanokitosan terhadap Mutu Surimi Ikan Patin (Doctoral dissertation, Universitas Gadjah Mada).

[9] Granada, I.P 2011 Pemanfaatan Surimi Ikan Lele Dumbo (*Clarias gariepinus*) dalam Pembuatan Sosis Rasa Sapi dengan Penambahan Isolat Protein Kedelai. Fakultas Perikanan dan Ilmu Kelautan Institut Pertanian Bogor Skripsi.

[10] Hafluddin 2012 Pengaruh Pencucian dan Penambahan Cryoprotectan pada Karakteristik Surimi Ikan Patin (*Pangasius* sp.) J. Rekayasa 5(1):54-60.

[11] Hassan, Aman., Balange, Amjad., Senapati, Sushree., and Xavier, K.A. Martin 2017 Effect of Different Washing Cycles on the Quality of *Pangasius hypophthalmus* Surimi Fishery Technology Society of Fisheries Technologists (India). 54.

[12] Hikmayani, Y., etc 2017 Alternatif Solusi Bagi Keberlanjutan Industri Surimi di Indonesia. Balai Besar Riset Sosial Ekonomi Kelautan dan Perikanan Gedung Balitbang KP I Lt. 4. Jakarta Utara.

[13] Huda, and T.A. Yang 2012 Technology for Production of Surimi Powder and Potential of Applications Int. Food Research Journal. 19(4): 1313–1323.

[14] Jafarpour, A., Hajiduon, H. A., & Aie, M. R 2012 A Comparative Study on Effect of Egg White, Soy Protein Isolate and Potato Starch on Functional Properties of Common Carp (*Cyprinus carpio*) Surimi Gel *J. Food Processing and Technology*, 3(11).

[15] Kallista, F.A. 2021 Pengaruh Konsentrasi Tepung Putih Telur terhadap Mutu Surimi Lele Dumbo (*Clarias gariepinus*). Fakultas Teknologi Pertanian. Universitas Gadjah Mada. Disertasi.

[16] Listanti, H 2019 Pengaruh Penambahan Larutan Nanokitosan Terhadap Mutu Surimi Ikan Lele. Fakultas Pertanian. Universitas Gadjah Mada. Skripsi.

[17] Nooranisha 2016 Aplikasi Program Linier Pada Optimasi Formulasi Sosis Campuran Tepung Kedelai (*Glycine max* L.) dan Ikan Nila (*Oreochromis niloticus*). Fakultas Teknik Universitas Pasundan. Skripsi.

[18] Nurimala, M., Nurjanah, Utama, R.H. 2009. Kemunduran Mutu Ikan Lele Dumbo (*Clarias gariepinus*) pada Penyimpanan Suhu Chilling dengan Perlakuan Cara Mati. Jurnal Pengolahan Hasil Perikanan Indonesia. 12(1): 1-16.

[19] Perez-Mateos, M., Amato, P. M., & Lanier, T. C. (2004). Gelling Properties of *Atlantic croaker* Surimi Processed by Acid or Alkaline Solubilization. *J. Food Science*, 69,328-333

[20] Prawira, A. 2008. Pengaruh Penambahan Tepung Alginat (Na-Alginat) Terhadap Mutu Kamaboko Berbahan Dasar Surimi Ikan Gabus (*Channa striata*). Fakultas Perikanan dan Ilmu Kelautan.Institut Pertanian Bogor.Skripsi.

[21] Purwadi, S. D., Darmanto, Y.S., dan Wijayanti, I. 2014. Pengaruh Penambahan *Egg White Powder* (EWP) terhadap Kualitas Gel Surimi Beberapa Ikan Air Tawar. *J. Pengolahan dan Bioteknologi Hasil Perikanan*. 3(2): 52-59.

[22] Ramadhan, W., J. Santoso, dan W. Trilaksani. 2014. Pengaruh Defeatting, Frekuensi Pencucian dan Jenis Cryoprotectant Terhadap Mutu Tepung Surimi Ikan Lele Kering Beku. *J. Teknologi dan Industri Pangan* 25(1) : 47-56.
[23] Suryana, M. A., A.B. Patadjai, dan N. Asyik. Pengaruh Pemakain Cryoprotectant Berbeda Terhadap Sensorik dan Kualitas Kimia Surimi dari Ikan Lele Dumbo (*Clarias gariepinus*).*J. Sains dan Teknologi Pangan* 1 (3): 175-183.
[24] Suryaningrum, T.D., Ikasari, D., Syamdidi. 2009. Penambahan Bahan Pembentuk Gel dalam Pembuatan Surimi dari Ikan Patin (*Pangasius hypophthalmus*). *J. Pascapanen dan Bioteknologi Kelautan dan Perikanan*. 4(1):37-47.
[25] Suwarso, S., Taufik, M., and Zamroni, A. (2020). Dinamika Perikanan Cantrang Berbasis di Tegalsari, Tegal: Perubahan Upaya Penangkapan dan Komposisi Hasil Tangkapan Ikan. *J. Penelitian Perikanan Indonesia*, 26(4), 211-220
[26] Wawasto, A., Santoso, J., Nurilmala, M. 2018. Karakteristik Surimi Basah dan Kering dari Ikan Baronang (*Siganus* sp.). *J.PHPI*. 21(2):367-376.
[27] Wijayanti, I., Santoso, J., and Jacoeb, A. M. 2012. Pengaruh Frekuensi Pencucian Terhadap Karakteristik Gel Surimi Ikan Lele Dumbo (*Clarias gariepinus*) *J. Saintek Perikanan* 8(1) : 32-37.
[28] Wiles, J. L., Green, B. W., and Bryant, R. 2004 Texture Profile Analysis and Composition of a Minced Catfish Product *J.Texture Studies*, 35(3),325-337.
[29] Yathavamoorthi, R, Sankar, T.V., Ravishankar, C. 2010. Effect of Ice Storage and Washing on The Protein Constituents and Textural Properties of Surimi from Labeo calbasu (Hamilton,1822). *Indian J. of Fisheries*. 57.
[30] Zuraida, I. 2018. Peran Transglutaminase Indigenous dalam Pembentukan Gel Surimi Lele Dumbo (*Clarias gariepinus*). Fakultas Teknologi Pertanian. Universitas Gadjah Mada. Disertasi.

Valorization of marine by-products using green technology

S Netramai[1], F Saleah[2], C Kawinkamraksa[2], A Wipatanawin[3], T Kijchavengkul[1], F Courtes[4], C K Rosales[4]

[1] School of Bioinnovation and Bio-based Product Intelligence, Faculty of Science, Mahidol University, Nakhon Pathom, 73170, Thailand
[2] School of Bioinnovation and Bio-based Product Intelligence, Faculty of Science, Mahidol University, Bangkok, 10400, Thailand
[3] Department of Biotechnology, Faculty of Science, Mahidol University, Bangkok, 10400, Thailand.
[4] Department of Science and Research, Global Innovation Center, Thai Union Group Public Co., Ltd., S.M. Tower, Bangkok, 10400, Thailand

*Corresponding author email: siriyupa.net@mahidol.ac.th; angkana.wip@mahidol.ac.th

Abstract. In this work, two cases of marine by-product valorization using green technologies were highlighted, *i.e.*, utilization of high-pressure carbon dioxide (HPCD) for demineralization of goldstripe sardinella (*Sardinella gibbosa*) scales and the use of non-specific enzymes for chitosan hydrolysis. After demineralization, type I collagen can be recovered from sardine scales. This work used HPCD process (of up to 10 bar or 1 MPa) to remove minerals from scales instead of the traditional strong acid method. Results from initial phase of the work showed that the process efficiency was significantly affected by pressure applied and treatment time ($p < 0.05$). And the demineralization efficiency reached was significantly lower than that achieved by treatment with hydrochloric acid ($p < 0.05$). In the second study, the chitosanolytic activity of four non-specific enzymes were assessed for chitosan hydrolysis. The results indicated that acid protease and cellulase could be a potential choice to produce chitooligosaccharides (COS) from shrimp shell.

1. Introduction

Typically, approximately 50 to 70% of total weight of marine catches is discarded as by-products, which have little monetary value. Several valuable bio-components, such as, amino acids, enzymes, glucosamine, vitamins, oil, and minerals, can be extracted from marine wastes and, after purification and/or modification process, can be used in food, feed, and cosmetic applications [1]. This work explored the uses of green technologies as parts of recovery processes of bio-components from marine by-products, *i.e.*, extraction of collagen from fish scales and production of chitooligosaccharides from shrimp shell chitosan.

2. High-pressure carbon dioxide process for fish scale demineralization

Collagen can be recovered from many discarded parts of raw fishes, *e.g.*, skin, bone, cartilage, and scales. The production process and yielded collagen type varied, depending on fish species and type of by-product. Generally, the production process includes washing, cutting, chemical pre-treatment, extraction, recovery, and drying steps [2]. Prior to recovering collagen (normally, Type I) from fish scales, the raw material needs to undergo demineralization process to remove minerals which made up to 40 to 45% of total weight [1,3]. This process will improve recovery efficiency and reduce impurities in the final collagen product [2].

The majority of minerals found in fish scales are calcium (Ca) and phosphorus (P). Both minerals typically exist in the form of hydroxyapatite (HA; Figure 1), appearing as opaque white layers [3], which, in acidic solution, will dissolve, releasing calcium, phosphate, and hydroxyl ions (Equation 2) and can be extracted out. The traditional method used to remove the minerals is strong acid treatment, using hydrochloric acid (HCl) or ethylenediaminetetraacetic acid (EDTA). This method is simple and effective but generates a large amount of wastewater. Thus, more environmentally friendly methods, such as biological fermentation, are evaluated as alternative demineralization treatments [4].

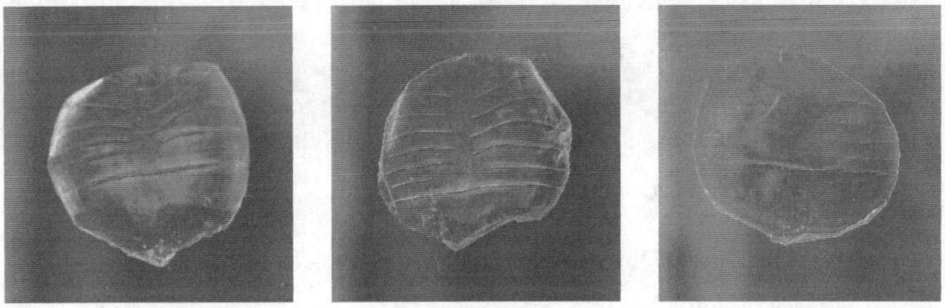
(a) (b) (c)

Figure 1. *Sardinella gibbosa* scales before demineralization (a), and after HCl treatment, with ~12% DM (b) and ~99% DM (c)

$$Ca_{10}(PO_4)_6(OH)_2 \rightleftharpoons 10Ca^{2+}(aq) + 6PO_4^{3-}(aq) + OH^-(aq) \quad \text{(Eq. 1)}$$

This work reported on the first phase of feasibility study of using high-pressure carbon dioxide (HPCD) to temporarily produce carbonic acid (CA; Equation 2 to 4) as demineralization method for sardine scales.

$$CO_2 + H_2O \rightleftharpoons H_2CO_3 \quad \text{(Eq. 2)}$$
$$H_2CO_3 \rightleftharpoons H^+ + HCO_3^- \quad \text{(Eq. 3)}$$
$$HCO_3^- \rightleftharpoons H^+ + CO_3^{2-} \quad \text{(Eq. 4)}$$

2.1. *Materials and methods*

Scales recovered during production of goldstripe sardinella (*Sardinella gibbosa*) were provided by Thai Union Group Public Co., Ltd. (Bangkok, Thailand). Fresh scales were washed using deionized (DI) water, air-dried at 40±2°C, and kept in air-tight container at 25±2°C until use. The dried scales were also determined their ash content (initial ash content; %) [5].

Dried scales were treated using HPCD method under the following conditions: carbon dioxide pressure of 1-10 bar, treatment time of 1-5 h, and scale to water ratio of 5 g of dried scales to 1 L of water. The HPCD treatments were performed in customized high-pressure reactor made from AISI 304 stainless steel, equipped with CO_2 inlet and outlet ports, and pressure gauge (EK Food Tech Ltd. Part., Bangkok, Thailand). The treated scales were determined their ash content (%) [5]. The obtained data was used to calculate demineralization efficiency (DM; %) according to Equation 5.

IOP Conf. Series: Earth and Environmental Science 1289 (2023) 012034 doi:10.1088/1755-1315/1289/1/012034

$$\text{DM (\%)} = \frac{\text{(Initial ash content - Ash content of treated scales)}}{\text{Initial ash content}} \times 100 \qquad \text{(Eq. 5)}$$

All experiments were conducted in triplicates. The data was statistically analyzed using analysis of variance (ANOVA) by using SPSS 18 program (SPSS Inc., Chicago, IL, USA) at the confidence level of 95% ($\alpha = 0.05$) with Tukey's adjustment for comparison of the means.

2.2. *Results and discussion*

Table 1 shows the demineralization efficiency (DM; %) of HPCD under studied conditions. The highest DM achieved under the most severe condition, *i.e.*, 10 bar and 5 h was ~16%. Additionally, physical damage to the treated scales was observed, *i.e.*, the thickness of treated scales appeared to be reduced and fractures, especially around the scale edges, were visible.

Even though demineralization using HPCD method and strong acid method are based on similar principle, demineralization efficiency of traditional method can be as high as ≥98% for fish scales [6]. The differences could partially be because CA is a weak acid, thus providing less hydrogen ions (giving the same concentrations) during the extraction process. Furthermore, at the beginning of the treatments, pH values of the mixture (scales and CA solution) were around 3.5 to 3.9, depending on the pressure applied, and later increased to ~5.5. Towards the end of the treatment, the pH remained at around 5.5, which is a critical pH of HA, and DM did not significantly increase further (data not shown). The presence of phosphate ions could also create a buffering system during the treatment [7].

Table 1. Demineralization efficiency* (DM; %) of high-pressure carbon dioxide treatment for *Sardinella gibbosa* scales

Time (h) \ Pressure (bar)	1.0	5.5	10.0
1	4.62 ± 0.89^{a}	6.22 ± 0.99^{ab}	7.85 ± 0.50^{c}
3	7.98 ± 1.67^{bcd}	9.82 ± 0.93^{d}	13.83 ± 0.35^{e}
5	9.99 ± 2.80^{bcd}	14.21 ± 4.88^{def}	16.13 ± 1.92^{f}

*Values with similar lowercase letter superscript are not statistically different at type I error (α) of 0.05

2.3. *Future works*

Future works on HPCD for fish scale demineralization should include extending the treatment time and preventing saturation of mineral ions, possibly by replacing CA solution periodically during treatment. The quality of recovered collagen from fish scales after HPCD treatment should also be assessed.

3. Non-specific enzyme treatment for shrimp chitosan hydrolysis

Chitosan (CTS) is a polysaccharide which can be commercially produced from crustacean shells, including shrimp shells. Further hydrolysis of CTS can yield chitooligosaccharides (COS), which have degree of polymerization (DP) less than 20. Chitooligosaccharides are water-soluble and reported to have a wide range of biological activities. While COS can be produced using physical, chemical, or enzymatic degradation of either chitin or chitosan, commercial COS are typically prepared by acid or enzyme hydrolysis [8].

Since chemical hydrolysis methods commonly used in chitin or chitosan hydrolysis give unpredictable quality of COS and generate hazardous wastes, enzyme treatments, using either specific or non-specific enzymes, are considered more effective. While using specific enzymes, *i.e.*, chitinases and chitosanases are more effective in controlling COS properties and give higher yield, the enzymes' prices are significantly higher than those of non-specific enzymes [8].

This work explored potential use of four non-specific enzymes, *i.e.*, accellerase, acid protease, cellulase, and lysozyme, for chitosan hydrolysis. The optimal temperature and pH conditions for each enzyme were also identified.

3.1. Materials and methods

Chitosan from shrimp shell, with degree of deacetylation (DD) of 90.15% and average molecular weight of 35 kDa was purchased from Marine Bio Resources Co., Ltd (Samutsakhon, Thailand). Chitosanase (*Streptomyces* sp. C0794; purified) and lysozyme (chicken egg white; purified) were purchased from Merk Ltd. (Bangkok, Thailand). Acid protease (*Aspergillus niger*; crude), accellerase 1000 (*Trichoderma reesei*; crude), cellulase (*Trichoderma reesei*; crude) were purchased from Siam Victory Chemical Co., Ltd. (Bangkok, Thailand).

Chitosan solution (1% w/v) was prepared using method outlined by [9]. Chitosanolytic activity (U/mL) of selected enzymes were assessed using a dinitrosalicylic acid (DNS) method according to Gonçalves *et al.* (2010) [10]. The ratio of enzyme to substrate was fixed at 1:200 (v/v) and treatment time was set at 4 h. The studied range of pH was 3.5 to 6.5 and temperature was 25 to 55°C.

For data analysis, see *Section 1.1.*

3.2. Results and discussion

Based on chitosanolytic activities of all five enzymes listed in Table 2, acid protease and cellulase demonstrated strong potential as enzymes for chitosan hydrolysis.

Several studies reported high chitosanolytic activity for cellulase and attributed this to the structural similarity between cellulose and chitin/chitosan [8,11,12].

No work on chitin/chitosan degradation has been documented on acid protease. Acid protease refers to a water-soluble group of protease enzymes, with high efficiency at low pH. They are used in food and beverage industries to digest peptide bonds in proteins to improve their solubility, digestibility, and dispersibility [13]. On the other hand, crude proteases from *Portunus segnis*, *Erwinia chrysanthemi*, *Bacillus cereus* SV1, *Bacillus mojavensis* A21, and *Scorpaena scrofa* were shown to have significant chitosanolytic activity for shrimp shell hydrolysis [11], but the mechanisms of their chitin/chitosan degradation are still not clear. It also should be noted that most commercial non-specific enzymes derived from microorganisms are crude enzymes (not purified), thus, it is possible that the demonstrated chitosanolytic activity might be the activity of other enzymes in the mixture [14].

Table 2. Chitosanolytic activity (U/mL) of chitosanase and non-specific enzymes obtained at their optimal pH and temperature

Enzyme	**pH**	**Temperature (°C)**	**Chitosanolytic activity* (U/mL)**
Chitosanase	5.5	37	0.689 ± 0.033
Accellerase	5.0	37	0.515 ± 0.075
Acid protease	3.5	37	0.840 ± 0.043
Cellulase	6.0	52	0.649 ± 0.022
Lysozyme	6.2	25	0.037 ± 0.002

*One unit (U) is the amount of enzyme that released 1 μmol of reducing sugar per minute. Data are shown as mean ± SD (n=3).

3.3. Future works

Since chitosan solubility increases in acidic conditions [9], utilization of acid protease will be beneficial for COS production. Additional research on the use of acid protease in chitosan hydrolysis should be done, especially on the possible hydrolysis mechanism and quality of resulting hydrolysates. Furthermore, Li *et al.* (2007) reported that chitosan with higher DD and higher molecular weight showed lower affinity for protease, leading to slower rate of hydrolysis [15]. Thus, the effects of chitosan properties and enzyme to substrate ratios should also be considered.

References

[1] Kim S-K and Venkatesan J 2014 Introduction to seafood processing by-products. In: Kim S-K, editor *Seafood Processing By-Products: Trends and Applications* New York (USA): Springer Science+Business Media doi: 10.1007/978-1-4614-9590-1_1

[2] Jafari H, Lista A, Siekapen MM, Ghaffari-Bohlouli P, Nie L, Alimoradi H and Shavandi A 2020 Fish collagen: Extraction, characterization, and applications for biomaterials engineering *Polym.* **12** doi: 10.3390/polym12102230.

[3] Ikoma T, Kobayashi H, Tanaka J, Walsh D and Mann S 2003 Microstructure, mechanical, and biomimetic properties of fish scales from *Pagrus major J. Struct. Biol.* **142** 327-33.

[4] Yang H, Gözaydın G, Nasaruddin RR, Har JRG, Chen X, Wang X and Yan N 2019 Toward the shell biorefinery: Processing crustacean shell waste using hot water and carbonic acid *ACS Sustain. Chem. Eng.* **7** 5532-42 doi: 10.1021/acssuschemeng.8b06853

[5] AOAC 2018 *Official Methods of Analysis* Washington, DC (USA): Association of Official Analytical Chemists, Inc

[6] Bellali F, Kharroubi M, Hmimid F, Loutfi M, Bourhim N and Correspondence F 2016 Response surface methodology optimization of demineralization from sardine (*Sardina pilchardus*) scale *J. Entomol. Zool. Stud.* **554** 554-58

[7] Palaghias G 1985 The role of phosphate and carbonic acid-bicarbonate buffers in the corrosion processes of the oral cavity *Dent. Mater.* **1** 139-44 doi: 10.1016/S0109-5641(85)80006-3

[8] Liaqat F and Eltem R 2018 Chitooligosaccharides and their biological activities: A comprehensive review *Carbohydr. Polym.* **184** 243-59 doi: 10.1016/j.carbpol.2017.12.067

[9] Li M, Han J, Xue Y, Dai Y, Liu J, Gan L, Xie R and Long M 2019 Hydrogen peroxide pretreatment efficiently assisting enzymatic hydrolysis of chitosan at high concentration for chitooligosaccharides *Polym. Degrad. Stab.* **164** 177-86

[10] Gonçalves C, Rodriguez-Jasso RM, Gomes N, Teixeira JA and Belo I 2010 Adaptation of dinitrosalicylic acid method to microtiter plates *Anal. Methods.* **12** 2046-48

[11] Kaczmarek MB, Struszczyk-Swita K, Li X, Szczęsna-Antczak M and Daroch M 2019 Enzymatic modifications of chitin, chitosan, and chitooligosaccharides *Front. Bioeng. Biotechnol.* **7** doi: 10.3389/fbioe.2019.00243

[12] Poshina DN, Raik SV, Poshin AN and Skorik YA 2018 Accessibility of chitin and chitosan in enzymatic hydrolysis: A review. *Polym. Degrad. Stab.* **156** 269-78 doi: 10.1016/j.polymdegradstab.2018.09.005

[13] Raveendran S, Parameswaran B, Ummalyma SB, Abraham A, Mathew AK, Madhavan A, Rebello A and Pandey A 2018 Applications of microbial enzymes in food industry *Food Technol. Biotechnol.* **56** 16-30 doi: 10.17113/ftb.56.01.18.5491

[14] Miguez N, Kidibule P, Santos-Moriano P, Ballesteros AO, Fernandez-Lobato M and Plou FJ 2021 Enzymatic synthesis and characterization of different families of chitooligosaccharides and their bioactive properties. *Appl. Sci.* **11** doi: 10.3390/app11073212

[15] Li J, Du Y and Liang H 2007 Influence of molecular parameters on the degradation of chitosan by a commercial enzyme *Polym. Degrad. Stab.* **92** 515-24 doi: 10.1016/j.polymdegradstab.2006.04.028

Acknowledgement

This research has received funding from the NSRF via the Program Management Unit for Human Resources & Institutional Development, Research and Innovation [grant number: MOU-CO-2564-13555-TH]. Thanks also to the cosupport of scholarshop from Graduate Program in Science Innovation (SCIN), Mahidol University and to Thai Union Group Public Co., Ltd. for the collaboration and supports of research facilities.

Characteristics of ready to cook product from dried pempek with freezing treatment before drying

A S Fahmi* and Romadhon

Department of Fisheries Product Technology, Faculty of Fisheries and Marine Science, Universitas Diponegoro, Semarang, Indonesia 50275

Corresponding author mail: suhaeli.fahmi@live.undip.ac.id

Abstract. Pempek is a popular ready to cook fish cake in Indonesia with a high level of moisture content. At room temperature, pempek can only stored for a few hours. There are several variants of pempek available in the market in the form of dried pempek which can be stored at room temperature for a long time. These dried products need to be rehydrated before cooking.The drying technology of dried pempek will affect the characteristics of the ready-to-cook product after the rehydration process. This study aims to determine the effect of freezing before drying on the characteristics of ready to cook rehydrated dried pempek (curly shape pempek). The experimental treatments tested were freezing treatment and non-freezing treatment. Each treatment was repeated in 3 replications of experiment. Freezing process were carried out in a household freezer for 24 hours and the drying process were carried out in an electric oven at 45°C for 12 hours. The results showed that after rehydration, dried pempek could returned to its original characteristics like before drying. Freezing treatment before drying gave a higher rehydration ratio, better texture and lighter appearance than without freezing treatment.

1. Introduction

Pempek is one of popular fisheries product in Indonesia, a traditional cuisine from Palembang, capital city of South Sumatra Province Republic of Indonesia. Pempek very popular and widely known by people outside Palembang. Pempek well known as an adaptation food from China like fish cake. Fish cake in China was made by wheat flour however pempek was made by starch flour [1]. Pempek usually consume after being fried and served with cuko and cube sliced cucumber. Cuko is a sour sweet sauce with dark color made from palm sugar, ebi (dried small prawn) and condiments.

There are several various types of pempek namely lenjer, kapal selam, adaan, keriting. Ready to cook pempek usually available in the form of boiled pempek. Although boiled pempek is ready to be directly eat, pempek usually fried or grilled before consumed [2]. Boiled pempek has relatively high moisture content 60-67% [3]. Shelf life of boiled pempek in room temperature is only one day [4,5]. The short shelf life of pempek caused by high nutrition contents which had degradation by microbes into volatile components such as trimethylamine, H_2S, CO_2, ammonium that caused stink and rancidity smells [5]. To extend shelf life of pempek, several studies have reported some research on dried pempek.

Due to drying process, only several types of pempek with small size like tekwan and keriting were decent to be dried. Dried pempek not only have a longer shelf life but also lighter in weight. With longer shelf life and lighter in weight, dried pempek can be distributed to distant places.

Drying of boiled pempek can be done directly or can be done after boiled pempek is being frozen first before drying. Freezing before drying can be applied for queueing pempek for drying process due the capacity of drying process. Freezing of pempek before drying will also affect the characteristics of pempek. Compare with sun drying and oven drying, freeze drying is known as the best drying process to keep the shape and the color of product but freeze drying needs a special and an expensive machine. Freezing with regular freezer can be applied in small enterprises to improve the quality of the product and for queueing products storage.

Dried pempek need to be rehydrated to return into its original shape before cooked. After rehydration, dried product will change in moisture content, volume and texture [6]. The processes involved in drying pempek will affect the characteristics of rehydrated product. This research was aimed to study the effects of freezing before drying to the characteristics of ready to eat rehydrated dried pempek keriting (pempek with curly shape).

2. Methods

2.1 Pempek Ingredients

The ingredients used in pempek with 100 g barracuda fish meat were: 100 g ground fresh barracuda fish meat (*Sphyraena* sp), 100 g tapioca flour, 4 g salt, 10 g sugar, 4 g flavour enhancer, 2 g pepper, 10 g garlic, 5 g candlenut, and 140 g chilled water; all materials were purchased from local market.

2.2 Experimental Procedure

The process of making pempek referred to Alhanasir et al.[7], started with mixing ground barracuda fish meat, tapioca flour, salt, sugar, flavour enhancer, pepper, garlic, candlenut and chilled water to become a dough, then the dough shaped into pempek keriting (curly pempek) with diameter 3 cm and boiled for 5 minutes. After these processes, boiled pempek were ready.

To produce dried pempek there are two treatments applied namely: without freezing/ directly dried (DD) and with freezing (FD). For sampel DD boiled pempek were directly dried in electrical oven with temperature 45°C for 12 hours. For sample FD boiled pempek were frozen in household freezer for 24 hours before dried in electrical oven with temperature 45°C for 12 hours.

2.3 Product Characteristic Assessments

Characteristics assessments for dried pempek were: moisture content, water activity, dimension lightness and microstructure image. Characteristics assessments for boiled pempek and rehydrated dried pempek were moisture content, dimension, rehydration ratio, texture profile analysis, lightness and sensory.

3. Result and Discussion

3.1 Moisture Content, Water Activity and Lightness of Dried Pempek

Table 1 shows that freezing treatment before drying made the lower moisture content and water activity on dried pempek. Average moisture content of dried pempek with freezing treatment before drying was 7.00% while directly dried pempek was 8.20%. Irsalina et al reported that dried fish skin pempek could achieve 6.02% water content after freezing treatment at -17°C for 24 hours and dried with oven 60°C for 24 hours [8].

Average water activity of dried pempek with freezing treatment was 0.44 while directly dried

IOP Conf. Series: Earth and Environmental Science 1289 (2023) 012035 doi:10.1088/1755-1315/1289/1/012035

pempek was 0.46. The water activity of dried pempek was lower than that of duck meatball in [9]. These results showed that the value of duck meatball with the sun drying method was 0,6083 and that with the mechanical drying method was 0,6175.

The colour of dried pempek with freezing treatment was looks lighter than directly dried pempek. Freezing treatment affect colour of dried pempek.

Table 1. Moisture Content and Water Activity of Dried Pempek

Treatment	Moisture content	Aw	Lightness
DD	8.20±0.72	0.46±0.005	36.69±0.91
FD	7.00±1.10	0.44±0.018	37.00±1.10

Note: data were average of three replications of treatments ± standard deviation

3.2. Microstructure of Dried Pempek

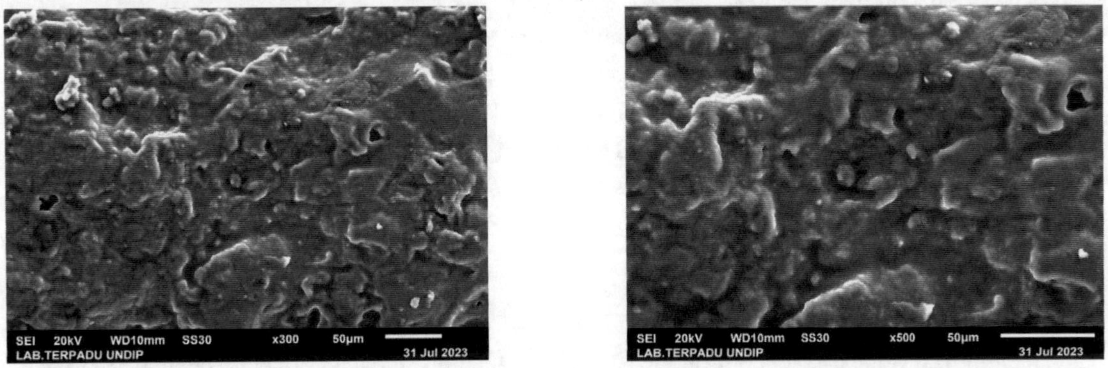

Figure 1. Microstructure image of dried pempek without freezing treatment before drying

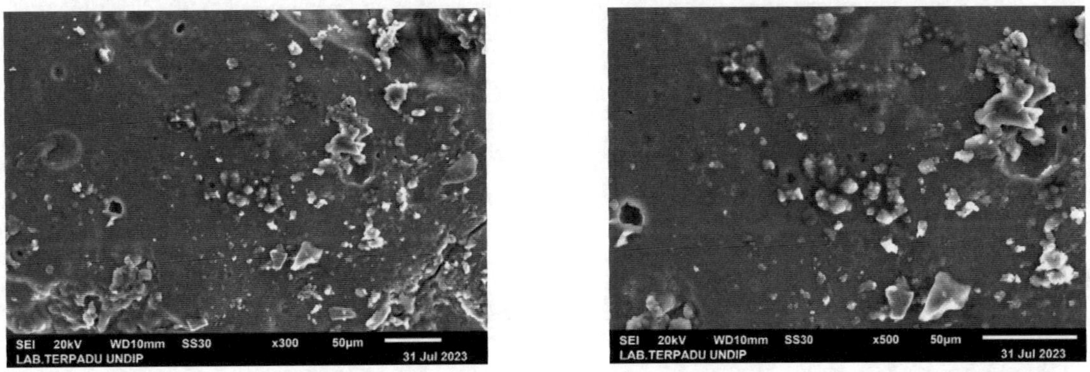

Figure 2. Microstructure image of dried pempek with freezing treatment before drying

Dried pempek with freezing treatment (FD) showed more pores than dried pempek that directly dried without freezing treatment (DD). Pores in dried FD were also distributed more evenly with relatively uniform size. It showed that freezing treatment affected more porous dried product. These results was similar with products dried with freeze drying. [10] reported that freeze drying could maintain microstructure of the product and gave lower shrinkage of product while oven drying cause collapse in microstructure and higher shrinkage. Using freeze drier, the shrinkage obtained is very low, the dried product has a very porous structure and good rehydration ability [11]. The lower freezing temperature and higher freezing rate in freeze drying increased nucleation rate of ice, which determines the number of ice crystals. Consequently, the number of small ice crystals increases after freezing, and causes smaller pores [12].

3.3. Moisture Content of Boiled and Rehydrated Pempek

Table 2. Moisture Content of Pempek

Treatment	Moisture content (%)
Boiled	66.76±2.04
Boiled Rehydrated DD	64.90±6.95
Boiled Rehydrated FD	69.50±3.51

Note: data were average of three replications of treatments ± standard deviation

Moisture content of all ready to cook samples were still met [13] for pempek (max 70%). Moisture content of ready to cook FD (69.50%) was higher than DD (64.90%). It showed that during rehydration process and boiling, water could easily enter the cavities in FD dan DD. These results related to the image of microstructure in Figure 1. There were more pores in dried pempek FD so water could easily enter the cavities or pores of the products.

3.4 Lightness

The lightness value is colour appearance that shows the light and dark effects. The increase in the lightness value shows that the colour appearance of the product is lighter. Table 3 shows that the lightness value of ready to cook pempek with freezing treatment before drying was the highest compared with the other. This result was related with freezing process before drying and higher moisture content in FD. This result is similar to the colour of freeze dried product. The freeze-dried method resulted in less discoloration than the cabinet-dried method, which was caused by ice sublimation, reduced oxygen availability, and polifenoloksidase enzyme deactivation. Therefore, low temperature restrained enzymatic browning to stabilize the color [14]. Related results also found on dried cherry laurel and kiwi fruit reported by [15], freeze drying affected lightness value of dried cherry laurel and kiwi fruit where freeze dried cherry laurel and kiwi fruit had a higher lightness than fresh cherry laurel and kiwi fruit.

Table 3. Lightness of Pempek

Treatment	Lightness
Boiled	40.78±1.38
Boiled Rehydrated DD	41.88±1.85
Boiled Rehydrated FD	44.47±0.79

Note: data were average of three replications of treatments ± standard deviation

3.5 Rehydration Ratio

Dried pempek need to be rehydrated before it becomes ready to cook pempek. The rehydration characteristics of dried products are shown in the rehydration ratio form [16]. Rehydration ratio indicates the ability of a product to reabsorb water. The higher rehydration ratio, the greater ability of the product to reabsorb water. Rehydration ratio of dried pempek is presented in table 4.

Table 4. Rehydration Ratio of Pempek

Treatment	Rehydration Ratio (%)
DD	112.93±24.55
FD	156.59±9.19

Note: data were average of three replications of treatments ± standard deviation

Table 4 shows that dried pempek with freezing treatment before drying (FD) gave higher rehydration ratio that dried pempek without freezing treatment before drying (DD). These results showed that freezing treatment made the product (dried pempek) absorb more water. Without freezing treatment, average rehydration ratio was 112.93% and with freezing treatment average rehydration ratio was 156.59%. Dried pempek with FD and DD treatment required rehydration time for approximately 2 hours (120 minutes) until it was become ready to cook product. This rehydration time was shorter than rehydration time of several products from pempek reported in previous research. The rehydration time of dried fish skin pempek was 141.33–171.33 minutes [8]. Rehydration time of dried tekwan is usually done for 4 to 12 hours. The longer time of rehydration process allows the product to be contaminated by bacteria, which can affect the quality of the product.

3.6 Sensory analysis

Table 5. Sensory Value of Ready to Cook Pempek

Treatment	Appearance	Aroma	Flavor	Texture
Boiled	8.60±0.82	8.90±0.45	8.50±0.89	8.20±1.01
DD	8.10±1.02	7.80±1.01	7.50±0.89	7.80±1.01
FD	8.40±0.94	8.10±1.02	7.50±0.89	8.20±1.01

Note: data were average of 15 panelist ± standard deviation

Boiled pempek, the undried one showed the higher value in appearance, aroma and flavour. After rehydrated and boiled, directly dried pempek (DD) showed the lowest value in appearance, aroma and texture. Dried pempek with freezing treatment (FD) showed higher sensory value compare to the directly dried pempek in appearance, aroma and texture. Texture is an important physical atribut that driving the consumer acceptance. Table 6 showed that freezing treatment could maintain the texture quality of dried pempek after rehydrated.

3.7 Dimensions

Diameters and heights of pempek are presented in table 6 and table 7. The result showed that diameter of rehydrated FD was the same with the diameter of the sample before dried. Diameter of rehydrated DD was lower than diameter of the sample before dried. These results related with rehydration ratio in table 4 where freezing treatments gave higher rehydration ration than without freezing treatment.

Table 6. Dimension of Samples DD

Samples	Diameter (cm)	Height (cm)
Boiled	3.2	1.0
Dried DD	2.7	0.6
Rehydrated DD	2.9	0.9
Boiled rehydrated DD	2.9	0.9

Table 7. Dimension of Samples FD

Samples	Diameter (cm)	Height (cm)
Boiled	3.3	1.0
Dried FD	2.5	0.4
Rehydrated FD	3.3	0.9
Boiled rehydrated FD	3.3	0.9

4. Conclusion

After rehydration in the water, dried pempek with curly shape could returned to its original shape as before drying. Freezing treatment before drying of pempek will produce ready to cook pempek with greater rehydration ratio, better texture and lighter appearance compare with dried pempek without freezing treatment before drying.

References

[1] Surya R, Destifen W and Nugroho D 2023 Pempek: traditional fishcake dish from South Sumatra, Indonesia *Canrea Journal: Food Technology, Nutritions, and Culinary Journal*. 57-76

[2] Isnawijayani I, Rozalena A and Caropeboka RM 2021 The meaning of longing for the tradition of "Ngirup Cuko Pempek" Palembang as friendship communication in the Covid-19 pandemic era *Int J Commun Service Engagem.* **2(4)** 99–111

[3] Supriadi A, Saputra D, Priyanto G, Baehaki A and Pambayun R 2020 The profile of pempek as a determining factor of quality, originality and ethnicity In *Journal of Physics: Conference Series.* **1485(1)** p 012032

[4] Moulia MN, Syarief R, Suyatma NE, Iriani ES and Kusumaningrum HD 2019 Application of bionanocomposite edible coating on pempek during storage at room temperature *Jurnal Teknologi dan Industri Pangan.* **30(1)** 11-19

[5] Karneta R and Gultom NF. The development strategy of packaged pempek industry. *Int J Bus Manag.* **12(8)** 227–33

[6] Muhandri T, Yulianti SD and Herliyana EN 2017 Karakteristik pengeringan jamur tiram (*Pleurotus ostreatus var. florida*) menggunakan pengering tipe fluidized bed dryer *Agritech: Jurnal Fakultas Teknologi Pertanian UGM.* **37(4)** 420-427

[7] Alhannasir, Rejo A, Saputra D and Priyanto G 2017 Physical characteristics: rehydration, porosity diameter, and colors of instant pempek out of treatment with freeze drying pressure *Food Science and Quality Management.* **67** 64-70

[8] Irsalina R, Pratama F and Widowati T 2018 Physical and chemical properties of dried fish skin-pempek with the addition of red and black rice flours *International Journal of Recent Scientific Research.* **9** 27853-27856

[9] Bhaskar ON, Biswas S, Bano S and Bhattacharya D 2020 Effect of drying methods on value added duck meat balls *International Journal of Current Microbiology and Applied Sciences.* **9(3)** 1534-1539

[10] Baeghbali V, Ngadi M and Niakousari M 2020 Effects of ultrasound and infrared assisted conductive hydro-drying, freeze-drying and oven drying on physicochemical properties of okra slices *Innovative Food Science & Emerging Technologies*. **63** 102313

[11] Aksoy A, Karasu S, Akcicek A and Kayacan S 2019 Effects of different drying methods on drying kinetics, microstructure, color, and the rehydration ratio of minced meat *Foods*. **8(6)** 216

[12] Thuy NM, Tuyen NTM, Thanh NV and Tai NV 2020 Evaluation of freeze-drying conditions on the process kinetics and physicochemical properties of purple shallot *Food Res*. **4** 1630-1636

[13] National Standardization Agency of Indonesia (Badan Standardisasi Nasional) 2019 *SNI 7661-2019: Pempek* (Badan Standardisasi Nasional, Jakarta)

[14] Shams R, Singh J, Dash KK and Dar AH 2022 Comparative study of freeze drying and cabinet drying of button mushroom *Applied Food Research*. **2(1)** 100084

[15] Gümüşay Ö A and Yalçın M Y 2019 Effects of freeze-drying process on antioxidant and some physical properties of cherry laurel and kiwi fruits *Akademik Gıda*. **17(1)** 9-15

[16] Phahom T, Juntharat N, Premsuttarat P, Paosunthia Y and Roudaut G 2019 Evaluation of desorption isotherms, drying characteristics and rehydration properties of crab stick by-product *Heat and Mass Transfer*. **57** 1039-1052

The characteristics and consumer acceptance of jelly drink fortified with *Ulva lactuca* fatty acid microemulsion

D R S Kusumajati and S A Budhiyanti*

Department of Fisheries, Faculty of Agriculture, Universitas Gadjah Mada, Jalan Flora, Fisheries Building A4, Bulaksumur, Yogyakarta 55281, Indonesia

*Corresponding Author : sitabudhiyanti@ugm.ac.id

Abstract. This study aims to determine the fortification of *Ulva lactuca* fatty acid microemulsion in jelly drinks and the characteristics and consumer acceptance of jelly drinks. The stages in this research were extraction of *U. lactuca* fatty acid, production of *U. lactuca* fatty acid microemulsion, fortification of *U. lactuca* fatty acid microemulsion in jelly drinks, and physical, chemical, and sensory analysis. Extraction of *U. lactuca* fatty acid was carried out using hexane and ethanol solvents with a ratio of 1:19 (v/v) at 70°C for 3 hours. Production of *U. lactuca* fatty acid microemulsion was done using three types of surfactants: Tween 80, Tween 20 and Span 80 in a ratio of 87:7.5:5.5(v/v) with 80 percent water content and 200 ppm of *U. lactuca* fatty acid concentration. Fortification of microemulsion in the jelly drink was carried out with several ratios: 0:10; 0.25 : 9.75; 0.5:9.5; 0.75:9.25; and 1:9. The characteristics of the jelly drink were determined by chemical and physical testing. The chemical analysis consists of pH, antioxidant content, and fatty acid composition. Physical analysis consists of syneresis, viscosity, and color. The sensory analysis consists of a preference level for color, aroma, taste, and texture. The results showed that the best fortification was microemulsion and jelly with a ratio of 0,25:9,75, with a significant effect on syneresis, viscosity, antioxidant content, color, and sensory. The results were 2.6 mg/min, 11.847 cP, 15.56%, a change in color, and a decrease in the level of preference, but they did not have a significant effect on pH.

1. Introduction

Seaweed is a marine biological resource that has the potential to be developed because it has high economic value in the food and non-food industries. *Ulva lactuca* is one of the green seaweeds that is abundant in the coastal and marine areas of Indonesia [1]. The abundant availability of *U. lactuca* in Gunungkidul is indicated by the density of *U. lactuca* species higher than the density of green seaweed Enteromorpha intestinalis at Ngandong Beach, Gunungkidul, amounted to 21.25 ind/m^2 and 6.00 ind/m^2, and at Drini Beach Gunungkidul amounted to 76.39 ind/m^2 and 21.25 ind/m^2 [2]. *U. lactuca* has nutrients in the form of 11.53% water content, 2.94% ash content, 5.17% fat, 17.43% protein, and 62.93% carbohydrates [3]. The people around the coast use *U. lactuca* only as fresh due to a lack of knowledge, and technology [4]. People have not used other content in *U. lactuca* such as its fatty acid content. Although the fatty acid content in *U. lactuca* is low, it has benefits for human health. Unsaturated fatty acids from green macroalgae are important to preventing various diseases such as narrowing of blood vessels, diabetes, inflammation, cardiovascular disorders, and nervous system disorders [5][6].

The availability of fatty acids in *U. lactuca* provides benefits for the body so that they can be fortified into food products. However, *U. lactuca* fatty acid is insoluble in water, so it requires a carrier system to be applied in a hydrophilic system, one of which is in the form of a microemulsion. Microemulsion is an isotropic dispersion consisting of an oil phase and a water phase, which are stabilized by surfactant molecules in the interface layer with the help of a cosurfactant with a droplet size <100 nm [7]. Microemulsions have thermodynamically stable properties, a transparent appearance, low viscosity, and can be used as a delivery system [8]. Fatty acids can be fortified in food products using a microemulsion system, as done by [9], [10], and [11] who respectively apply microemulsion to lemon juice, green tea, and jelly drinks.

Jelly drinks are soft drinks that have elastic properties and a soft gel consistency, making them easy to drink and able to quench thirst as well as overcome hunger [12]. Jelly drinks are liked by all groups, including children, adolescents, and adults [13]. According to [14], 12% of the 1200 respondents included commercial jelly drinks among their favorite soft drinks for adults and teenagers. The fat content in the jelly drink of flower leaves is 0% [15], while [11] shows that commercial jelly drinks have saturated fatty acids and unsaturated fatty acids, respectively 464.45 ± 0.004 µg/mg, and 524.85 ± 0.006 µg/mg. The addition of *U. lactuca* fatty acids to jelly drinks can enrich their nutritional value.

The application of *U. lactuca* fatty acid microemulsion into jelly drinks [11] This study showed that the microemulsion applied to the jelly drink was stable in the dilution, incubation, centrifugation, and heating treatments, but the solvent used in the extraction process did not meet food-grade standards. Besides, testing related to the physical, chemical, and consumer acceptance of jelly drinks fortified with fatty acid microemulsion has not been carried out. This study focused on determining the microemulsion fortification formulation of *U. lactuca* fatty acids that were extracted using food-grade solvents in jelly drinks and knowing their effects on physical, chemical, and consumer acceptance.

2. Materials and Methods

2.1. Materials and equipments

The materials used in this study were *Ulva lactuca* was taken from Trenggole Beach Gunungkidul in November 2019, hexane (Merck, Germany), ethanol (Merck, Germany), the internal standard of nanodecanoic acid (C19:0) (Sigma-Aldrich), nitrogen gas (CV.Perkasa), 1.1-diphenyl-2-picrylhydrazil (DPPH) (Merck, USA), Aquades, Na_2SO_4 (Merck, Germany), surfactant Tween 20 (Merck, Germany), Tween 80 (Merck, Germany), and Span 80 (Sigma), jelly powder (Semesta), lychee essence (Koepoe-Koepoe), natrium benzoate (Koepoe-Koepoe), sugar (Rose Brand), citric acid (Cap Sirsat), pH 7 buffer. The equipments used were oven (Rikakikai CO. LTD Japan), analytical scales (Driver Instrument Company), vortex (Maxi Mix II, US), centrifuge (Kokusan Corporation, Japan), water-bath shaker (Shibata, Japan), GCMS (Gas Chromatography-Mass Spectrometry) (Shimadzu, Japan), pH meter (Hanna), hot plate stirrer (Velp AREC), micropipette (Socorex), viscometer, sealing machine (Super Sealer), MB120 Moisture analyzer OHAUS.

2.2. Sample Preparation

Green algae samples were taken in Trenggole Beach, Gunungkidul. Green algae brought to the laboratory using a cooler. Green algae samples were washed using running water to remove impurities. Samples were then identified and packed in plastic and then stored in the freezer (-18°C) until the research was carried out.

2.3. Macroalgae Extraction

Macroalgae extraction was carried out using the LRC method (Lepage and Roy) referring to the research of [16] and [17]. The extraction stage required 25 grams of algae (fresh), finely chopped and then extracted by adding 250 ml of a solution of hexane and ethanol (1:19 v/v). The sample was heated at 70°C at 120 rpm for 3 hours in a water shaker, then the sample was cooled. The samples were added with 50 ml of distilled water and 100 ml of hexane. The sample was homogenized by vortex and centrifuged at a speed of 3.000 rpm for 7 minutes. The organic phase is collected using a syringe, Na_2SO_4 is added to absorb the distilled water. The fatty acids were evaporated with nitrogen then 2000 μl of hexane was added, and analyzed by GC-MS.

2.4. Preparation Microemulsion
The microemulsion was prepared by mixing 10 ml of surfactant, water, and *U. lactuca* fatty acid. The microemulsion was formed by mixing hydrophilic surfactant (Tween 20, and Tween 80) and hydrophobic surfactant (Span 80) respectively, with a ratio of 7.5: 87: 5.5 (% v/v) without a carrier oil. 200 ppm of *U. lactuca* fatty acids was put into the microemulsion system, and the addition of 80% water drop by drop at a temperature of 70°C ± 5°C then vortexed. This formulation is following the results of a stable microemulsion in the o/w microemulsion system from [18].

2.5. Production of Jelly Drinks and Addition of Microemulsion
The production of jelly drinks refers to [19] with modifications. The making of jelly drink begins with 5 grams of jelly powder mixed with 750 ml of water into a pot, then heated to a boil. Next, add 63 grams of sugar, 1 gram of citric acid, 0.7 ml of lychee essence, and 0.25 grams of natrium benzoate, then stir. After boiling, the solution is removed and allowed to stand until it reaches a temperature of 70°C, then a microemulsion that has been made according to the ratio is added and then stirred with a spoon. The addition of microemulsion in jelly drinks was carried out based on the formulation of the ratio between microemulsion and jelly drinks. The ratio of microemulsion and jelly drink used was 0:10; 0.25: 9.75; 0.5: 9.5; 0.75: 9.25, and 1: 9 (v/v).

2.6. Test Parameters
Extracted *U. lactuca* fatty acid sample was tested in the form of water content, and fatty acid analysis using GCMS. The microemulsion of *U. lactuca* fatty acid tested for pH, Turbidity refers to [20], and analysis of fatty acids using GCMS. The jelly drink with the addition of *U. lactuca* fatty acid microemulsion was to physical, chemical, and sensory tests then the best samples were taken. Physical testing such as the syneresis test refers to [21], viscosity using a Brokfield viscometer and color. Chemical testing such as an antioxidant test that refers to [22] and pH test. The sensory test such as a hedonic test, then the best sample is selected. The best sample was carried out by the proximate test in the form of water content, ash content, protein content, and fat content according to [23], fiber test according to [24], carbohydrates were calculated by difference, and fatty acid analysis using GCMS.

2.7. Data Analysis
This study used a one-factor randomized block design (CRD), with variation of the microemulsion ratio of *U. lactuca* fatty acids in jelly drinks with 2 replications. The data were analyzed for normality and homogeneity, if not normally distributed then tested with the Kruskal Wallis test and further tested with Mann Whitney and pair comparison at the 95% confidence level. Normally distributed data were then analyzed using ANOVA (one-way analysis of variance) at a 95% confidence level. If a significantly different effect was obtained, it was followed by the DMRT test at the 95% confidence level. Statistical analysis of data was processed using the SPSS version 20 application.

3. Result and Discussion
3.1. Fatty Acid Microemulsion
Microemulsions were tested by physical testing such as turbidity, and chemical testing such as acidity and fatty acid analysis. The turbidity analysis of the microemulsion was 0.81%. The microemulsion with a transparent appearance and a turbidity index of less than 1% was defined as a microemulsion that was not cloudy, transparent, or stable, so the microemulsion in this study was stable and can be further applied to jelly drinks [20]. The microemulsion and the respective surfactants were chemically tested in terms of their degree of acidity. The results of acidity degree (pH) on the surfactants Tween 20, Tween

80, and Span 80 were 6, 7.6, and 6, respectively, while the degree of acidity in the surfactant mixture with a ratio of Tween 20, Tween 80, and Span 80 was 7.5: 87: 5.5 (%v/v) was 7.9. This result was higher when compared to the research of Suciati & Patricia (2012)[25], which combined the Tween 80 and Span 80 (3:1) surfactants with a pH of 7.02. The pH test results of the fatty acid microemulsion showed a result of 6.4. These results were not different from the research conducted by[26], who made a microemulsion with a ratio of Tween 80: Span 80 (9:1) 30.6%, 8.2% sesame oil, and 61.2% water with a pH of 6.4, and a Tween 80: Span 80 formulation (8:1) 28.9%, 7.1% sesame oil, and 64.1% water with a pH of 6.42. The acidity degree of the microemulsion showed results close to neutral pH (7). These results indicated that the microemulsion had a degree of acidity that was not too acidic or too alkaline, so that when fortified in food, it did not change the pH of food products too much.

The fatty acid content of *U. lactuca* extract in 0 ppm, 200 ppm microemulsion, control jelly drink, and microemulsion jelly drink is shown in Table 1. The total fatty acid concentration in the extraction of *U. lactuca* fatty acid using the LRC method (Lepage and Roy) amounted to 44.96%. The results in this study had a higher total fatty acid content compared to the research of Saraswati (2020)[17] which extracted *U. lactuca* fatty acids with the same solvent, namely hexane and ethanol solvents, with a total fatty acid content of 7.67%. Total fatty acids consist of 3,91% saturated fatty acids such as palmitic acid and 3.76% unsaturated fatty acids such as oleic acid. These different results could be caused by differences in the methylation method when testing fatty acids using GCMS. The PUFA/SFA ratio in the extraction of *U. lactuca* fatty acids by the LRC method was 0,10. The results showed that it still meets the limits recommended by[27] stated that a maximum PUFA/SFA ratio of 1-1.5 can prevent the risk of CVD. Ratios that exceed the recommended maximum limits can increase total cholesterol and low-density lipoprotein (LDL) cholesterol in the blood, which has a negative effect on health [28].

The results of fatty acids analysis in the 0 ppm *U. lactuca* fatty acid microemulsion showed that it did not contain fatty acids. Meanwhile, [29] showed that the surfactants used in the production of Tween 20 and Tween 80 microemulsions contained various fatty acids. Microemulsion with the addition of *U. lactuca* fatty acid at 200 ppm contains a total fatty acid content of 99.93%. The fatty acids contained in the 200 ppm *U. lactuca* fatty acid microemulsion were sourced from added fatty acids and surfactants. Fatty acids on the 200 ppm *U. lactuca* fatty acid microemulsion showed the same fatty acid content as [29] such as capric acid, linoleic acid, palmitic acid, and oleic acid.

Total fatty acids contained in the control jelly drink and the microemulsion jelly drink were 90.72% and 97.15%, respectively. The fatty acid content in the control jelly drink was not different from the results of [11] research, which showed that the control jelly drink without the addition of microemulsions contained 100% total fatty acids with various fatty acids including butyric acid, lauric acid, tetradecanoic acid, palmitic acid, palmitoleic acid, octadecanoic acid, linoleic acid, oleic acid, and heneicosanoanic acid. The results of this study, microemulsion jelly drink, which had a higher fatty acid content than the control jelly drink, was due to the addition of 200 ppm fatty acid microemulsion in the jelly drink.

3.2. Physical and Chemical Analysis of Jelly drink
A physical analysis is an analysis carried out on the properties of jelly drinks that can be felt by the five senses, such as color and texture. The physical analysis of the jelly drink includes analysis for syneresis, viscosity, and color. Chemical analysis on jelly drinks including acidity and antioxidant tests, these results can be seen in Figure 1.

3.2.1 Syneresis of Jelly Drink.
The syneresis value of microemulsion jelly drinks was shown in Figure 1 (A). The syneresis value of the microemulsion jelly drink ranged from 2.45±0.15 to 3.01±0.02 mg/min. This result was not different from the syneresis in commercial jelly drinks, which was 2.83 mg/min. The syneresis in this study is also not different from the research of [21] which stated that drinks added with starfruit have a syneresis of 2.26 mg/min. The syneresis value of jelly drinks tended to increase ($P < 0.05$) with the addition of *U. lactuca* fatty acid microemulsion. This was due to the 80% water content in the microemulsion used in

this study, so the more microemulsion that was added to the jelly drink, the more water was added. According to [21], the more water was added, causing the carrageenan in jelly drinks to bind too much water so that the tissue formed in the jelly drink was no longer able to hold water, the more syneresis occured. Syneresis also utilized some hydrocolloid compounds as gelling agents. If the hydrocolloid compound absorbed too much water, the gel that was formed was brittle and there was syneresis [30]. When the addition of hydrocolloid compounds formed a strong double helix structure, it could bind more water, and the water molecules in the gel do not easily escape, so syneresis can be reduced [31]. For the production of drinks, it was necessary to formulate an optimal ratio between the number of hydrocolloid compounds and the amount of air added so that syneresis could be reduced.

Table 1. Total fatty acids in *U. lactuca* extraction, microemulsion 0 ppm, 200 ppm, control jelly drink, and microemulsion jelly drink.

No	Number of Lipid	Common Name	Fatty Acid Extraction	Microemulsions		Jelly Drink	
				0 ppm	200 ppm	Control	Microemulsions
1	C16:0	Palmitic Acid	25.61	n.d	6.85	7.88	2.53
2	C24:0	Lignoceric Acid	3.94	n.d	n.d	n.d	n.d
3	C10:0	Capric Acid	n.d	n.d	1.55	n.d	n.d
4	C16:1	Palmitoleic Acid	6.46	n.d	0.78	n.d	n.d
5	C16:3	Roughanic Acid	0.67	n.d	n.d	n.d	n.d
6	C22:1	Erucic Acid	n.d	n.d	0.25	n.d	5.95
7	C18:1	Oleic Acid	5.94	n.d	85.08	4.38	n.d
8	C18:2	Linoleic Acid	2.34	n.d	5.42	2.66	n.d
9	C18:0	Stearic Acid	n.d	n.d	n.d	75.8	88.27
10	C14:0	Myristic Acid	n.d	n.d	n.d	n.d	0.4
Total Fatty Acid			44.96	n.d	99.93	90.72	97.15
Saturated Fatty Acid (SFA)			29.55	n.d	8.4	83.68	91.2
Unsaturated Fatty Acid (UFA)			15.41	n.d	91.53	7.04	5.95
Mono Unsaturated Fatty Acid (MUFA)			12.4	n.d	86.11	4.38	5.95
Poly Unsaturated Fatty Acid (PUFA)			3.01	n.d	5.42	2.66	n.d
PUFA/SFA			0.10	n.d	0.65	0.031	0

Note: n.d : not detected

3.2.2. Viscosity of Jelly Drink.

The viscosity of the microemulsion jelly drink can be seen in Figure 1 (B). These results show a significant difference in the range of 10.147±123.0 cP -11.847±284.6 cP. The results of this study had a higher viscosity when compared to the research of [53], with the viscosity of the jelly drink the addition of ginger emprit was 743 cP. The viscosity of control jelly drinks and jelly drinks with the addition of microemulsion had a higher viscosity compared to commercial jelly drinks is 2.175±0.00 cP due to the different materials used and production methods. The viscosity of the jelly drink decreased as the number of *U. lactuca* fatty acid microemulsions was added to the jelly drink. The addition of a microemulsion is the same as adding water to a jelly drink because the water content in the microemulsion is 80%. The results obtained were in accordance with the opinion of [21] which showed that jelly drinks with the addition of starfruit have a viscosity value that tends to decrease with the large proportion of water added. This is influenced by the ability of hydrocolloid compounds to form gels. The more hydrocolloid compounds that are added, the more water will be bound, the low water content in the solution causes the viscosity to increase.

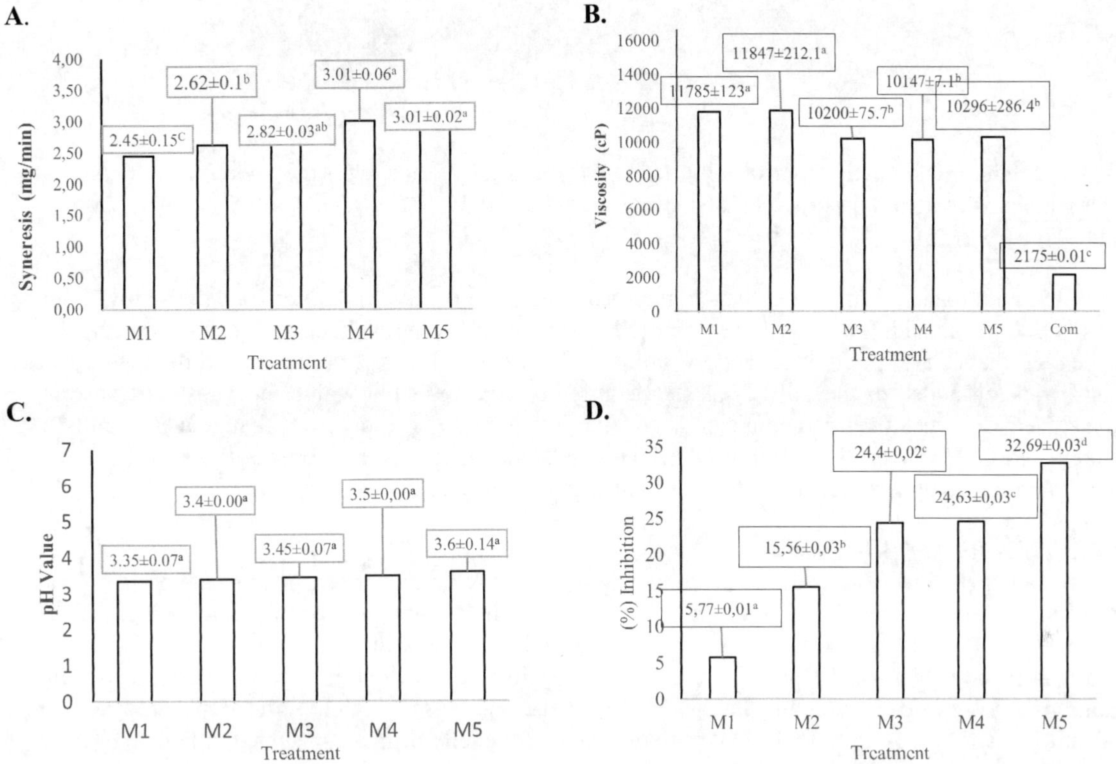

Figure 1. Physical, and chemical testing of jelly drinks (A) syneresis; (B) viscosity; (C) pH; (D) % inhibition

Note : control jelly drink (M1), jelly drink, the addition of *U. lactuca* fatty acid microemulsion with a ratio (microemulsion: jelly drink) of 0.25: 9.75 (M2), 0. 5: 9.5 (M3), 0.75: 9.25 (M4), 1: 9 (M5), Com (Commercial Product). Differences in letters indicate significant differences (P <0.05).

3.2.3. Color of Jelly Drink

Color is the visual element that affects the appearance of a product. Color can affect a person's perception of the flavor of a product [32]. The addition of *U. lactuca* fatty acid microemulsion to the jelly drink has a effect on the color of the jelly drink. Color changes in microemulsion jelly drinks can be seen in Figure 2.

The figure 2 was showing the color change in the jelly drink, which was increasing the green color on the addition of the microemulsion. This indicated that the more fatty acid microemulsions were added, the stronger the green color level. This was because the *U. lactuca* seaweed extract added to the jelly drink had a green color. According to [3], *U. lactuca* contained natural pigments that gave a green color, namely pigments from the chlorophyll a and b groups.

Figure 2. The image of the control jelly drink (M1) and jelly drink with the addition of *U. lactuca* microemulsion. with a ratio (microemulsion: jelly drink) of 0.25: 9.75 (M2), 0.5: 9.5 (M3), 0 , 75: 9.25 (M4), and 1: 9 (M5).

The addition of *U. lactuca* fatty acid microemulsion to the jelly drink, increased the yellow color. This is due to the addition of surfactants in jelly drinks. According to [29], surfactants Span 80 and Tween 80 have a characteristic yellow color and Tween 20 has a yellowish-white color. The microemulsion added to the jelly drink has a turbidity level of <1% which indicates a transparent appearance. So, when the microemulsion is added to the jelly drink, it does not change the turbidity in the jelly drink. However, the addition of a microemulsion can change the color of the jelly drink, the more microemulsion is added to the jelly drink, the more green, and yellowish it will be.

3.2.4. Acidity (pH) of Jelly Drink.
The average pH value of jelly drinks with the addition of microemulsions in various comparisons was not significantly different (P>0.05). The pH measurement results in the jelly drink ranged from 3.35±0.07 to 3.6±0.14. These results indicate that the pH of the jelly drink in this study is classified as acidic when compared to the pH of the commercial jelly drink, which is 5.5. Jelly drinks with the addition of fatty acid microemulsions have a pH range that is still included in the pH threshold in foodstuffs, namely 2.5-7.5 [33]. The average pH value of the jelly drink is presented in Figure 1(C).

The pH value in the jelly drink sample tends to increase with the added fatty acid microemulsions. This is because the microemulsion has a more alkaline pH than the jelly drink sample. The degree of acidity (pH) in the 200 ppm *U. lactuca* fatty acid microemulsion showed a result of 6.4. In addition, adding a microemulsion is the same as adding water to a jelly drink because the water content in the microemulsion is 80%. According to [21], an increase in the amount of water added causes the [H +] ion which comes from organic acids to also dilute, so that the [H +] ion that forms acid will decrease and the pH of the material will increase.

3.3. Antioxidant Activity of Jelly Drink.
The antioxidant activity in jelly drinks can be seen in Figure 1(D). The antioxidant activity in jelly drinks showed 5.77±0.01 - 32.69±0.03 % inhibition. The antioxidant activity in jelly drinks can be derived from citric acid and natrium benzoate which is added to jelly drinks. Natrium benzoate can act as an antioxidant because it has a phenolic group that appears on the phenyl ring in benzoate compounds and their derivatives [34]. According to [35], citric acid added to food can increase, and stabilize antioxidant activity.

The antioxidant activity was further increased by the addition of *U. lactuca* fatty acid microemulsion. This is because the results of the extraction of *U. lactuca* fatty acids contain antioxidant compounds in the form of phenolic compounds, namely 2,6-Di-tert-butyl-4-methyl phenol (BHT) is 0.49%. The compound 2,6-Di-tert-butyl-4-methyl phenol ($C1_5H_{24}O$) in the fatty acid *U. lactuca* has a phenolic group and can be synthesized endogenously by some phytoplankton under conditions of oxidative stress [36]. The 2,6-Di-tert-butyl-4-methyl phenol molecule is an antioxidant because it can donate hydrogen atoms to free radicals at the initiation stage of lipid oxidation, then delocalized electrons in the benzene ring. BHT or 2,6-Di-tert-butyl-4-methyl phenol is a lipophilic bio-active organic and a phenol derivative that is added to many food ingredients as an antioxidant [37].

3.4. Sensory Analysis (Hedonic) of Jelly Drinks
Fortified products need to attend to the nutrients added, but also need to pay attention to consumer acceptance of sensory tests. The preference test (organoleptic) was carried out based on the panelists' level of preference using the hedonic method [38]. The parameters tested are the parameters of taste, aroma, texture, color, and the whole can be seen in Figure 3.

3.4.1. Color.
Color is a visual element that can affect the appearance of a product. A person's perception of the flavor in the product can be influenced by the color and outer appearance of a product [32]. The average value of color preference in control jelly drinks and microemulsion jelly drinks is presented in Figure 3(A). The average value of preference for color parameters in jelly drinks ranges from 2.52±0.95 -3.92±0.94, which indicates the quality is not like to like. The average value of preference color parameters tends to decrease with the magnitude of the microemulsion ratio of fatty acids added. The color produced in jelly drinks with the addition of *U. lactuca* fatty acid microemulsion is a yellowish-green color. The more fatty acid microemulsions that were added, the stronger the green, and yellow colors were produced. This is due to the addition of *U. lactuca* extract. According to [3], *U. lactuca* contains chlorophyll a and b which can give a green color. Also, the added surfactant gives it a yellow color. According to [29], surfactants span 80 and Tween 80 have a characteristic yellow color and Tween 20 has a yellowish-white color.

Based on the hedonic test results of the color parameters in Figure 2, the average value produced by the M2 microemulsion jelly drink has a high hedonic value or has a preferred quality compared to M3, M4, and M5 microemulsion jelly drinks. The control sample (M1) is the sample that is preferred by the panelists because of its clear yellowish color, while the jelly drink with the most microemulsion addition (M5) shows the sample that is the least preferred because it has a dark green color. The results of research by [39] showed that the higher the concentration of Spirulina in jelly drinks, the green color of the jelly drink is so that it can reduce the level of preference for the color parameter.

3.4.2. Aroma.
The aroma is one of the parameters in determining the quality of a product and is an important parameter in product assessment [40]. The average value of preference for aroma parameters in jelly drinks is around 2.35±1.01 – 3.93±0.82 which indicates dislike to like. The results of the average value of the jelly drink aroma parameters are presented in Figure 3(B). The resulting average value tended to decrease with the increase in the microemulsion of fatty acids in the jelly drink. Panelists did not like the aroma of jelly drinks with the addition of microemulsions due to the unpleasant aroma. According to [29], surfactants Tween 80, Tween 20, and Span 80 have a distinctive aromatic surfactant. The more microemulsion added, the stronger the distinctive aroma. Based on the results of the hedonic test on aroma parameters, it shows M2 jelly drink has a higher or preferred hedonic average value compared to M3, M4, and M5 jelly drinks.

3.4.3. Taste.
Taste is a parameter that requires a sense of taste, namely the tongue. Taste is a component that determines whether a product is good to eat or not. If a product has a bad taste, the product will not be accepted by consumers even though the color and aroma are good [40]. The average value of the taste preference for jelly drinks with the addition of fatty acid microemulsion is presented in Figure 3(C). The hedonic test average value of taste parameters in jelly drinks tended to decrease ($P < 0,05$) with increasing microemulsions added to jelly drinks. This is because there is a bitter taste in jelly drinks with the addition of a microemulsion. According to [29], surfactants Tween 80, Tween 20, and Span 80

in microemulsions have a bitter taste, so the more the amount of fatty acid microemulsion is added, the stronger the bitter taste. The average value of the taste parameters in the jelly drink which was added with fatty acid microemulsion ranged from 1,87±1,02 to 4,07±0,86. This value shows very dislike to like. The hedonic test results on taste parameters showed that the M2 microemulsion jelly drink had a higher or preferred hedonic average value compared to the microemulsion jelly drinks M3, M4, and M5.

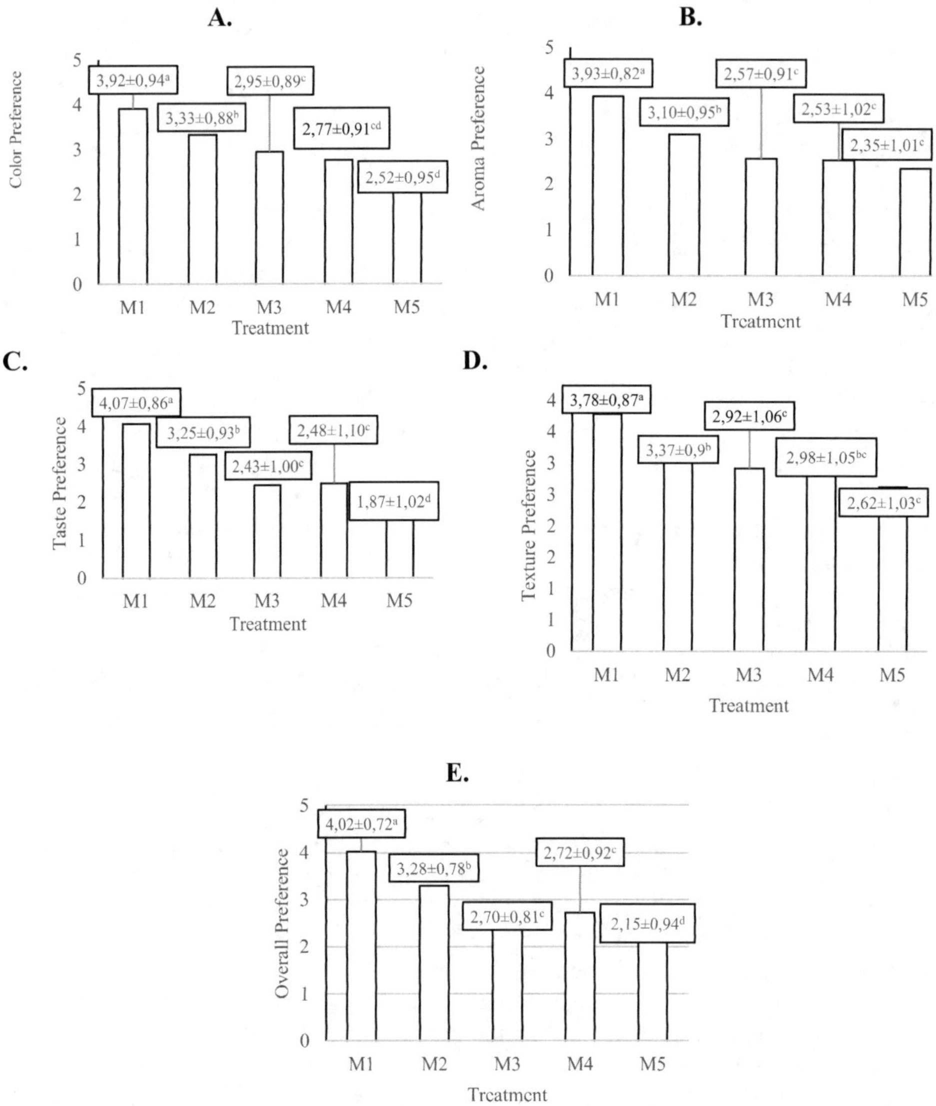

Figure 3. Sensory value on jelly drinks, (A) Color; (B) Aroma; (C) Taste; (D) Texture; (E) Overall preference.

Note: control jelly drink (M1), jelly drink with the addition of *U. lactuca* microemulsion, with a ratio (microemulsion: jelly drink) of 0.25: 9.75 (M2), 0.5: 9.5 (M3), 0 , 75: 9.25 (M4), and 1: 9 (M5).

Different letters indicate a significantly among the samples (P <0,05).\

3.4.4. Texture.
The texture is one of the most important parameters in consuming jelly drinks so that in the mouth you can feel the texture of the jelly. Every consumer has a different way of feeling the texture of food in the mouth. The average value of the texture parameter preference for jelly drinks is around 2.62±1.03 - 3.78±0.87. The average value of texture parameter preference in jelly drinks is presented in Figure 3(D). These results indicate that the average value of preference in the M2 microemulsion jelly drink texture parameter has a higher or preferred hedonic average value compared to the microemulsion jelly drinks M3, M4, and M5. The more microemulsion added to the jelly drink causes the viscosity to decrease, and the syneresis increases so that it gives a watery texture, the results of viscosity and syneresis can be seen in Figure 1 (A), and (B). This causes a decrease in the preference level of the jelly drink texture parameter due to the watery texture. The research results of [41] showed that the Rosella-Soursop jelly drink has a watery jelly texture, and crumbles easily so it is not liked by panelists.

3.4.5. Overall.
Overall preference is a combined overall assessment of the sensory attributes of the jelly drink. The average results of the overall assessment attributes of the jelly drink with the addition of fatty acid microemulsions in various concentrations are presented in Figure 3(E). The resulting overall average value ranged from 2.5±0.94 to 4.02±0.72. This value shows the results of dislike to like. The overall preference test results tended to decrease with the increase in microemulsion added to the jelly drink. The addition of fatty acid microemulsion had a significant effect on the overall attributes of the jelly drink ($P < 0.05$). The average range of the overall attributes of the M2 jelly drink tends to have a higher or preferred hedonic value compared to the jelly drinks M3, M4, and M5. This indicated that overall M2 microemulsion jelly drink products were more acceptable to the panelists than M3, M4, and M5 microemulsion jelly drinks.

3.5. Chemical Characteristics of Control and Best Jelly Drink
Chemical characteristics testing was carried out on M1 control jelly drink and M2 microemulsion jelly drink as the best samples. The M2 microemulsion jelly drink sample was the best in terms of the syneresis and hedonic test results. The syneresis test results on the best samples have a low value compared to other jelly drink samples. Based on hedonic testing, M2 microemulsion jelly drinks were more acceptable to the panelists than M3, M4, and M5 microemulsion jelly drinks. Chemical characteristics testing is carried out such as testing fat content, moisture content, ash content, protein content, total food fiber content, carbohydrates, and energy. The results of proximate testing on control jelly drink samples and jelly drinks fortified with *U. lactuca* fatty acid can be seen in Table 2.

Table 2. Proximate analysis for control jelly drinks, jelly drinks fortified with *U. lactuca* fatty acid microemulsion, and commercial jelly drinks [52].

Proximate	Control (M1) (% dry basis)	Addition of microemulsion with ratio 0,25:9,75 (M2) (% dry basis)	Commercial
Water	90.59%	90.94%	94.68%
Ash	0.19%	0.27%	-
Protein	0.17%	0.26%	-
Fat	0.037%	0.06 %	-
Dietary Fiber	0.38%	0.56 %	2.57 g/180 ml (1.43%)
Carbohydrates	8.5%	7.92 %	11 g/ 180 ml (6.1%)
Energy	34.267 cal/100 g	32.133 cal /100 g	45000 cal/180 ml

Fat is a compound that is included in the lipid group consisting of carbon, and hydrogen [42]. The results of the fat content test showed that the M2 microemulsion jelly drink had a fat content of 0.06%. This result is higher when compared to the M1 control sample of 0.037%. This could be due to the microemulsion fortification of *U. lactuca* fatty acids that provide additional fatty acids in the jelly drink. The fatty acids in the *U. lactuca* extract produced in this study contained saturated fatty acids, monounsaturated fatty acids, and polyunsaturated fatty acids, 29.55%, 12.4%, and 3.01%, respectively. M1 and M2 jelly drinks contain fat, while commercial jelly drinks on the market did not contain fat. The fat content in M1 and M2 jelly drinks can be sourced from the essence used. One of the ingredients listed on the label of the essence of the Koepoe-Koepoe brand is vegetable oil, which can be a source of fat in jelly drinks. The jelly drink in this study had a higher fat content when compared to the results of [43] which showed that jelly drinks with the addition of 0.025% *Spirulina platensis* carotenoid nanocapsules had a fat content of 0.04%. The more fat contained in jelly drinks, the easier the fat oxidation process will be [44].

Ash content testing is one of the proximate tests on food that is used to evaluate the nutritional value of food. Ash is an inorganic residue from the combustion or oxidation process of the organic components of food. The ash content of a food ingredient shows the total minerals contained in the foodstuff [40]. The jelly drinks M1 and M2 in this study had an ash content of 0.19% and 0.27%, respectively. The results of the ash content in this study were not different from the hantap leaf jelly drink in [45] which had an ash content of 0,18% and the research of [46] which showed that the pineapple mixed jelly drink had an ash content of 0.17%. However, it is different from commercial jelly drinks on the market which does not contain ash content. The ash content in the M2 microemulsion jelly drink and M1 control jelly drink can be sourced from the hydrocolloid material used. The ash can be contained in hydrocolloid compounds [39].

The protein content obtained in the M1 control jelly drink and M2 microemulsion jelly drink is 0.17% and 0.26%, respectively. These results are lower when compared to the research of [43] which shows that jelly drinks with the addition of 0.025% *Spirulina platensis* carotenoid nanocapsules have a protein content of 0.09%. The protein levels found in M1 and M2 jelly drinks can be caused by the main ingredients in the production of jelly drinks, namely jelly powder whose chemical composition is a hydrocolloid compound. The hydrocolloid compound that is often used in making jellies is carrageenan [47]. Carrageenan extracted from *Eucheuma cottoni* seaweed using KOH solvent has a protein content of 0.8% [48].

Dietary fiber is a part of plants that can be consumed and is composed of carbohydrates that are resistant to digestion and absorption in the human small intestine and undergo partial or complete fermentation in the large intestine [49]. The results of the food fiber test on the M1 control jelly drink sample is 0.38% which consisted of 0.35% insoluble food fiber and 0.05% soluble food fiber. The M2 microemulsion jelly drink contains a total food fiber content of 0.56% consisting of 0.51 insoluble food fiber and 0.05% soluble food fiber. According to [48], carrageenan produced from *Eucheuma cottonii* seaweed contains 4.12% crude fiber. The dietary fiber contained in jelly drinks can be sourced from hydrocolloid materials for making jelly, one of which is carrageenan. The dietary fiber in both control and microemulsion jelly drinks has a relatively low fiber content when compared to commercial jelly drinks and other fortified jelly drinks. The content of dietary fiber in commercial jelly drinks on the market is 2.57g/180 ml (1.42%). The research of [45] showed that jelly drinks with the addition of raw leaves have a dietary fiber content of 2.08%. This is because in the raw leaf jelly drink there is a source of fiber other than carrageenan, which is the addition of raw leaves. However, the insoluble fiber in control jelly drinks and microemulsion jelly drinks has a higher value when compared to the [43] which shows insoluble fiber content in jelly drinks with the addition of 0.025% *Spirulina platensis* carotenoid nanocapsules by 0.12%. The high fiber content in jelly drinks can benefit the body's metabolism [50].

Energy is one of the metabolic products of carbohydrates, protein, and fat. Energy is needed to support various bodily functions, including respiration, circulation, physical work, and protein synthesis. Excess energy in the body will be stored in the form of glycogen as a short-term energy reserve and stored in the form of fat as a long-term reserve [51]. According to the Food and Drug Administration

(2016), the energy that must be fulfilled every day for adults is 2150 kcal. The energy of the control jelly drink and the microemulsion jelly drink is 34.267 cal/100 g and 32.133 cal/100 g, respectively. The energy produced by jelly drinks in this study shows a value that is not much different when compared to the research of [43] which adds 0.025% *Spirulina platensis* carotenoid nanocapsules to jelly drinks, which can contribute energy of 34.39 cal/100g. However, the results in this study are lower when compared to commercial jelly drinks which can contribute energy of 45000cal/180ml. This can be caused by the use of sugars and hydrocolloid compounds which can affect the number of calories in jelly drinks. According to [14], one of the calories contained in jelly drinks comes from the sugar used. Jelly drinks in this study can be used as an alternative to drinks that delay hunger during the hours between eating the main menu.

4. Conclusions

The best formulation of fortified *U. lactuca* fatty acid microemulsion in jelly drinks is in the ratio 0.25: 9.75 because it is the most preferred, and accepted by the panelists, and has a syneresis of 2.6 mg/minute, a pH of 3.4, a viscosity of 11.847 cP, and antioxidant content of 15.56%. The addition of *U. lactuca* fatty acid microemulsion to jelly drinks affects physical parameters (syneresis, viscosity, color), chemical (anti-oxidant content), and sensory level of jelly drink preference (P <0.05), but does not affect chemical parameters in the form of pH.

Acknowledgements

The authors would like to thank Universitas Gadjah Mada with Final Project Recognition Program, letter of assignment number 2488/UN1.P.III/DIT-LIT/PT/2020.

References

[1] Mahmud I, Pertiwi., N. R. Azis., dan D. N. Reviana 2014 *Jurnal Pekan Ilmiah Mahasiswa*

[2] Safia, R.N. 2013 *Jenis dan Sebaran Makroalga di Zona Intertidal Pantai Ngadong dan Drini Kabupaten Gunung Kidul* Skripsi Fakultas Pertanian. Universitas Gadjah Mada

[3] Da Costaa J. F,W. Merdekawatib and F.R. Otu 2018 *Journal of Food Technology and Nutrition* **17**(1): 1-17.

[4] Praphesti, L.A. 2017 *Karakterisasi Produk Inori dari Rumput Laut Lokal U. lactuca dan Glacilaria sp.1-33* Skripsi Institut Pertanian Bogor

[5] Ginneken, V.J.T.V, Johannes, P.F.G H., Willem de, V., Herman van, K., and Willem, A.B 2011 *Journal Lipids in Health, and Disease* **10**: 104

[6] Suparmi dan Sahri, A. 2009 *Mengenal Potensi Rumput Laut: Kajian Pemanfaatan Sumber Daya Rumput Laut dari Aspek Industri dan Kesehatan* Tesis Universitas Diponegoro. Semarang

[7] McClements, D.J., and J. Rao. 2011 *Critical Reviews n Food Science, and Nutrition* **51**: 285-330

[8] Flanagan, J., and Singh H 2006 *Journal Food Science Nutrition* **4**: 221-237

[9] Sary, F.M 2019 *Formulasi Mikroemulsi Asam Lemak Makroalga Ulva lactuca serta Aplikasinya pada Minuman Sari Buah Lemon* Skripsi Fakultas pertanian. Universitas Gadjah Mada

[10] Mufidah, H 2019 *Formulasi Mikroemulsi Asam Lemak Makroalga Ulva lactuca serta Aplikasinya pada Minuman Teh Hijau* Skripsi Fakultas pertanian. Universitas Gadjah Mada

[11] Safitri, S.S 2019 *Formulasi Mikroemulsi Asam Lemak Makroalga Ulva lactuca serta Aplikasinya pada Minuman Jeli* Skripsi Fakultas pertanian. Universitas Gadjah Mada

[12] Novelina, N. Nazir and M.R.Adrian 2016 *Journal Agriculture, and Agricultural Science Procedia* **9**: 328 – 334

[13] Masluha, D 2013 *Formulasi Jelly Drink Berbasis Rumput Laut (Eucheuma cottonii) dan Spirulina platensis* Skripsi Departemen Teknologi Hasil Perairan Fakultas Perikanan dan Ilmu Kelautan. Institut Pertanian Bogor

[14] Febriyani N.M.P, S. Hardinsyah, dan D. Briawan 2012 *Jurnal Gizi dan Pangan* **7**(1): 35-42

[15] Ratnasari, A.D 2018 *Analisis Nilai Gizi dan Daya Terima Minuman Jeli Daun Kembang Bulan*

(Thitonia diversifolia) dengan Flavor Sirsak dan Stroberi sebagai Minuman Fungsional Anti Hiperglikemik Tesis Universitas Esa Unggul
[16] Kumari P, C.R.K. Reddy and B. Jha 2011 *Journal Analytical Biochemistry* **415**: 134-144
[17] Saraswati, D 2020 *Pengaruh Volume, Kecepatam Pengadukan, dan Lama Ekstraksi pada Produksi Asam Lemak Ulva lactuca* Skripsi Fakultas Pertanian Universitas Gadjah Mada
[18] Ningrum, Y.P 2020 *Variasi Jenis dan Konsentrasi Surfaktan untuk Formulasi Mikroemulsi Asam Lemak Ulva lactuca* Skripsi Fakultas Pertanian. Universitas Gadjah Mada
[19] Mujiati 2010 *Pengaruh Suhu Pengolahan Terhadap Kadar B-Karoten Minuman Jeli Spirulina platenisis* Skripsi Fakultas Pertanian. Universitas Gadjah Mada
[20] Cho Y.H, Km, S., Bae, E.K., Mok, C.K., and Park, J 2008 *Journal Food Science* **73**(3): E115-E121
[21] Agustin, F.dan W.D.R. Putri 2014 *Jurnal Pangan dan Agroindustri* **2**(3): 1:9
[22] Zarranappa, Vagdevi,M.H. Lokesh, M.R., and Gowdarshivannavar,B.C 2012 *International Journal of Chem Tech Research* **4**(4): 1527-1534
[23] AOAC 2005 . Official Method of Analysis. Arlington AOAC International
[24] AOAC 1995 Official Methods of Analysis of Association of Official Analytical Chemist. AOAC International Virginia USA
[25] Suciati, T., and L. Praticia 2012 *Jurnal Acta Pharmaceutica Indonesia* **23**(3):90-94
[26] Ghorbanzadeha M, N. Farhadiana, S. Golmohammadzadeh., M. Karimic, and M. Ebrahimi 2019 *Journal Colloids, and Surfaces B: Biointerfaces* **179**: 393–404
[27] Kang M. J, M. S. Shin., J. N. Park, and S. S. Lee 2005 *British Journal of Nutrition* **94**: 526-532
[28] Sartika, R 2008 *Pengaruh Asam Lemak Jenuh, Tidak Jenuh dan Asam Lemak Trans terhadap Kesehatan* Skripsi Universitas Indonesia
[29] Rowe R.C, P.J. Sheskey and M.E. Quinn 2009 *Handbook of Pharmaceutical Excipients* (Eds) Sixth Edition Pharmaceutical Press. London
[30] Muriana, E 2013 *Sifat Fisikokimia dan Organoleptik Jelly Drink Buah Naga Merah (Hylocereus polyrhizus) dengan Variasi Konsentrasi Karagenan* Skripsi Fakultas Teknologi Pertanian UKWMS Surabaya
[31] Achayadi, N.S., Y. Taufik, dan S. Selviana. 2016 *Pengaruh Konsentrasi Karagenan Dan Gula Pasir Terhadap Karakteristik Minuman Jeli Black Mulberry* (*Morus nigra* L.) Skripsi Universitas Pasundan
[32] Garber LL, Hyatt EM and Starr RG 2000 *Journal of Marketing Theory, and Practice* **8**(4): 61-72.
[33] McClements, D. J. 2016 CRC Press Taylor & Francis Group. Francis
[34] Muchtadi, D. 2004 *Jurnal Gizi Mindo* **3** (7): 4-6.
[35] Primurdia, E.G dan Kusnadi, J. 2014 *Jurnal Pangan dan Agroindustri* **2**(3) :98-109
[36] Babu, B and Wu J.T. 2008 *Journal Phycol* **44**: 1447–1454
[37] Aprilia V, H. Saebani, S.K.L. Bhima, dan A. Ismail 2018 *Jurnal Kedokteran Diponegoro* **7**(2): 1154-1165
[38] Soekarto, S. T 1985 *Penilaian Organoleptik (untuk Industri Pangan dan Hasil Pertanian)* Bharata Karya Aksara Jakarta
[39] Trilaksani W, I. Setyaningsih dan D. Masluha 2015 *JPHPI* **18**(1): 74-82
[40] Winarno, F. G 2008 *Ilmu Pangan dan Gizi* Gramedia Pustaka Utama Jakarta
[41] Gani, T.F 2014 *Jurnal Teknologi Pangan dan Gizi* **13**(2): 87-93.
[42] Hartono, A 2006 *Terapi Gizi dan Diet Rumah Sakit* Penerbit Buku Kedokteran EGC Jakarta
[43] Prawitasari .R.R 2020 *Pengaruh Penambahan Nanokapsul Karotenoid Spirulina platensis Terhadap Karakteristik dan Tingkat Penerimaan Konsumen pada Minuman Jeli* Skripsi Fakultas pertanian. Universitas Gadjah Mada
[44] Yustitie, N dan N.Rustanti 2016 *Journal of Nutrition Collage* **5**(4): 307-31.
[45] Pamungkas A 2014 *Jurnal Gizi Pangan* **9**(3): 195-202
[46] Ekafitri R, R. Kumalasari dan D. Desnilasari 2016 *Jurnal Penelitian Pascapanen Pertanian*

13(3): 115-124.

[47] William, P.A and Phillips. 2000. *Introduction to Food Hydrocolloid In: Handbook Of Hydrocoloid* (Eds). Woodhead Publishing Limited, and CRC Press LLC. P.19

[48] Ega L, C.G.C. Lopulalan dan F. Meiyasa 2016 *Jurnal Aplikasi Teknologi Pangan* **5** (2): 38-44

[49] American Association of Cereal Chemists (AACC). *The Definition of Dietary Fiber*. 2001. 46 (2): 112-126. Cfw AACC report

[50] Pranajaya, D 2007 *Pendugaan Sisa Umur Simpan Minuman Jeli di Pasaran* Skripsi Fakultas Teknologi Pertanian. Institut Pertanian Bogor

[51] IOM. Institute of Medicine. 2002. *Dietary Reference Intake for Energy, Carbohydrate, Fiber, Fat, Fatty Acids, Cholesterol, Protein, and Amino Acids. A Report of the Panel on Macronutrients, Subcommittees on Upper Reference Levels of Nutrients, and Interpretation, and Uses of Dietary Reference Intakes, and the Standing Committee on the Scientific Evaluation of Dietary Reference Intakes*. National Academies Press. Washington

[52] Merita. 2011. *Konsumsi Mie, Susu, dan Minuman Ringan Terhadap Kecukupan Gizi pada Mahasiswi dengan Status Gizi Normal dan Kegemukan*. Skripsi Departemen Gizi Masyarakat Fakultas Ekologi Manusia. IPB. Bogor

[53] Firdaus .A.N., B. Kunarto., dan E.Y. Sani 2018 *Jurnal Teknologi Hasil Pertanian* **2**: 1-9

IOP Conf. Series: Earth and Environmental Science 1289 (2023) 012037 doi:10.1088/1755-1315/1289/1/012037

High temperature short time blanching enhances the antioxidative properties of sea grapes (*Caulerpa racemosa)* powder

E Windrayani, S A Budhiyanti, and N Ekantari
Department of Fisheries, Faculty of Agriculture, Universitas Gadjah Mada. Jl. Flora, A4 2.01 Building, Bulaksumur, Yogyakarta 55281 Indonesia

Corresponding author mail: emiwindrayani@gmail.com

Abstract. Green alga sea grapes (*Caulerpa racemosa) were* consumed by Asian and Pacific Island people. It contains antioxidant compounds, such as phenols and caulerpin (CAU) that can potentially prevent disease. These sea grapes need to be blanched before drying to minimize oxidation damage. This research aims to understand the effect of temperature during hot water blanching (HWB) on the antioxidant activities, TPC, CAU, color, and moisture content of sea grapes powder. HWB temperatures used 70, 80, 90, and 100 °C at 20 seconds. The HWB had a more significant increase in antioxidant activity than control of sea grapes powder. The 100 °C had the highest value on ABTS, FRAP, and TPC, respectively 13.18 ± 0.13 (µg TE/g), 1.52 ± 0.03 (mg Ferro E/g), 0.96 ± 0,01(mg GAE/100 g). It is increasing its value because enzyme deactivation is better in high temperatures. Short time with high-temperature blanching can prevent the loss of bioactive compounds. CAU has specific aromatic rings, -NH, C=O, and -O-; all samples have peaks at these frequencies. Blanching has an intense green color (-a*) because it can increase color quality but has no significant result in moisture content. This study suggested blanching sea grapes can be done at 100°C for 20 seconds.

1. Introduction

Caulerpa known as sea grapes is a species of green algae (Chlorophyta), 97 species are found in Indo-Pacific seas [1]. *Caulerpa racemosa*, *Caulerpa racemosa var turbinata*, and *Caulerpa lentillifera* are all widespread in Indonesia [2]. The Java Sea (Jepara) [3]; Kuala Langsa Aceh [4]; and Kei, Tual Maluku Tenggara [5] are among the Indonesian places that are promoting sea grapes cultivation.

Sea grapes are consumed raw or as a side dish because it has high fiber and low in fat [2,6,7], and also have high antioxidant activity [8]. Polyphenols (phenols, flavonoids), caulerpin (CAU), and sulfated polysaccharides are a few sea grapes compounds having antioxidant activity [6,9–11]. These compounds promote health through antioxidant, antiinflammatory, anticancer [12,13], antiinflammatory-cardiovascular [8], antineuroinflammator [14], and antiobesity [15]. Sea grapes positive effects have the potential to be applied as functional and nutraceutical foods [16,17].

Seaweed, especially sea grapes, is a perishable product with a moisture content of about 90%. However, drying reduces antioxidant activity [18,19], which is assumed to be related to oxidation damage [20]. Blanching protects antioxidant compounds from oxidation, enhances color and texture, and eliminates off taste [21]. Several studies have shown that blanching combined with drying increases antioxidant activity and TPC, of seaweed tea [22]; *Saccharina latissima* [23]; and peony flowers [24].

Hot water blanching (HWB) is a typical procedure involving immersing or dipping the components in water as a medium. Temperatures range from 70-100 °C for a short time because it can inactivate enzymes by denaturing the structure [21,25]. The temperature and duration of blanching can have an impact on the final product [26,27]. High-temperature short-time (HTST) use in blanching has been studied [28].

Using high temperatures with a short time HWB is recommended for seaweed. *Caulerpa lentillifera* HWB at 95 °C for 20 seconds provided the highest sensory score and no coliform bacteria growth [29]. HWB of *Undaria pinnatifida* 20 seconds turned the brown color to greenish, however the longer duration, the weaker the green hue and the lower amounts of metabolites such as amino acids [30]. HWB *Sargassum* at 98.3 °C for 30 seconds enhanced fucoxanthin content compared to blanching in salt water, microwave, and steam [31]. Previous research has not investigated HWB sea grapes (*C. racemosa)* on antioxidative properties. This study investigates how HWB temperature affects the antioxidative properties of sea grapes.

2. Method

2.1. Sample preparation

Sea grapes *(C. racemosa*) were harvested from the Jepara Brackish Water Cultivation Fisheries Center (BBPBAP) and transported in styrofoam wrapped in banana leaves. Fresh samples are stored in perforated baskets covered with banana leaves at room temperature for four days after delivery without washing with fresh water. Fresh samples were washed three times in fresh water with a ratio of 1:2 before blanching. Clean sea grapes were weighed 500 mg for each treatment.

HWB was performed in distilled water for 20 seconds, starting after the samples were entirely submerged in the medium using household kitchen equipment. The temperature treatments were carried out at 70, 80, 90, and 100 °C using a thermocouple to monitor the temperature (±0.5 °C). After that, the samples were drained and dried for 24 hours in an oven drier at 50 °C [32]. One gram of the dry sample was taken out to determine the moisture content, and the rest was blended and sieved at 60 mesh.

A 0.5 g of sea grapes powder was macerated in 10 mL of ethanol for 24 hours at room temperature in the dark. The supernatant was collected after centrifuging at 3000 rpm for 15 minutes. The liquid extract (50 mg/mL) was kept in the refrigerator (0-3°C) for the following tests.

2.2. ABTS Assay

The method of ABTS (2,2′-azinobis (3-ethylbenzothiazoline-6-sulfonic acid)) -(NH4)2) was developed by [33]. In the dark, stock solutions 7 mM ABTS and 2.4 mM potassium persulfate (1:1) were reacted at room temperature for 12-16 hours. Reagents were obtained from 0.75 mL stock solution with 60 mL ethanol to get an absorbance of 0.700. A 75 µL liquid extract from 20 mg/mL dilution reacted with 1.425 µL ABTS+ at room temperature for 2 hours in the dark, with 734 nm detected. The % inhibition was calculated using the distilled water blank = ((blank absorbance - sample absorbance) *100)/ blank absorbance. Using Trolox as a reference, the curve equation y=0,1279x-0,1381, R^2 = 0,9985 was derived. The reducing power is expressed in µg Trolox equivalent (TE)/g of dry basis (DB).

2.3. FRAP Assay

The FRAP method refers to [34]. A FRAP reagent was made by combining 300 mM acetate buffer, 10 mM TPTZ, 40 mM HCl, and 20 mM FeCl3.6H2O using a 10:1:1 (v/v) ratio. The liquid extract sample was diluted to 10 mg/mL, and 120 µL was taken out to be reacted with 900 µL reagent in the dark for 2 hours at room temperature. Determination of FRAP using UV-Vis spectrophotometer at 595 nm. With a concentration of 10-100 ppm of FeSO4.7H2O as a reference, the calibration curve equation y = 0.0083x + 0.1239, R^2 = 0.9963 was achieved, with values reported in mg FerroE/g dry basis (DB).

2.4Determination Total Phenolic Content (TPC)

The TPC method refers to the approach adopted by [35]. The liquid extract was diluted to a concentration of 5 mg/mL. In a 10 mL volumetric flask, 400 µL of liquid extract and standard (gallic acid) were combined for 5 minutes with 0.4 mL of Folin Ciocalteu reagent. 4 mL Na2CO3 (7%) and aquabidest until its limit was added to the mixture. The mixture was then incubated in the dark for one hour. Absorbance was measured at 765 nm. The calibration curve for gallic acid 0.1-0.7 mg/100 mL was $y = 1.2436x + 0.0243$, $R^2 = 0.9968$. TPC content was expressed in milligrams of GAE per gram of (mg GAE/g) dry basis (DB).

2.5. Caulerpin compound

Caulerpin (CAU) was determined based on the absorption of functional groups using the Shimadzu IRspirit Fourier Transform-Infra Red (FT-IR) devices. [36] used functional group absorption with a frequency of 4000 - 4500 cm^{-1}, scan duration of 6, resolution of 4, and scan speed of 0.2 cm/s.

2.6. Moisture Content

Moisture content is determined using the thermogravimetric method (AOAC, 2007), where the sample is heated at 105-110°C until the sample weight is constant. Sample 0.5 g was placed in the moisture content tool (moisture analyzer, Ohaus MB120), and results were presented in %.

2.7. Color analysis

The color of sea grapes powder was evaluated using the Konica Minolta cr portable Colorimeter NH310 chroma meter, with the experimental parameters being L* (Lightness), a*(redness/greenness), dan b* (yellowness/blueness).

2.8. Data Analysis

All data is the mean of three replicates and was analyzed with one-way ANOVA and Duncan's posthoc (5% confidence level). Statistics were analyzed using SPSS Version 26. Duncan's test outcomes indicated that the results were statistically significant when the confidence level was $p < 0.05$.

3. Results and Discussion

3.1. Antioxidant Activity

Table 1. shows the results of antioxidant activity (ABTS and FRAP) and Total Phenolic Content.

Table 1. ABTS, FRAP, and TPC sea grapes powder

Temperature	Antioxidant Activity		
(°C)	ABTS (µg TE/g) DB	FRAP (mg FerroE/g) DB	TPC (mg GAE/g) DB
Control	9.11 ± 0.2^{a}	0.54 ± 0.05^{a}	0.74 ± 0.02^{a}
70	11.96 ± 0.27^{b}	1.16 ± 0.04^{b}	0.87 ± 0.019^{b}
80	12.44 ± 0.13^{c}	1.19 ± 0.04^{b}	0.88 ± 0.013^{b}
90	12.58 ± 0.09^{c}	1.31 ± 0.02^{c}	0.92 ± 0.005^{c}
100	13.18 ± 0.13^{d}	1.52 ± 0.03^{d}	0.96 ± 0.005^{d}

Different letters above the different columns indicate a significant difference ($p < 0.05$). Data were presented as mean ± SD ($n = 3$).

The ABTS values varied from 9.11 to 13.18 µg TE/g dry weight, as shown in Table 1. According to statistical analyses, there was a significant difference ($p<0.05$) between the treatments with the highest value, the HWB 100 °C treatment with 13.18 µg TE/g DB. The lowest value was 9.11 µg TE/g DB for sea grapes without blanching.

Based on the statistical data, the antioxidant activity of the FRAP method was significantly different ($p < 0.05$) between the treatments presented in Table 1. The highest value was in the 100 °C treatment, followed by 90 °C, 80 °C, and 70 °C, and the lowest in the control treatment, with values of 1.51, 1.31, 1.19, 1.16, and 0.53 mg Ferro/g DB respectively.

The results show that short-term HWB at high temperatures enhances antioxidant activity. First, HWB sea grapes are considered to improve antioxidant compounds, including FRAP values, as observed in oven-dried pomegranate with blanching at 100 °C for 60 seconds [37]. Second, inactivation oxidator enzymes such as peroxidase and polyphenol oxidase activity reduce the oxidation of compounds [38]. Peroxidase reaction with H_2O_2 produces phytotoxic free radicals that cause breakdown compounds affecting undesired changes in food quality such as color, odour, flavour, and nutritional loss [39]. The enzyme catalyzes the oxidation of the functional OH group of monohydroxyl phenols to o-dihydroxy phenols (catechol, dopamine, adrenaline) and the dehydrogenation of o-dihydroxy phenols to o-quinones [40]. Inactivation of peroxidase and polyphenol oxidase activity more efficiently in blanching at high temperatures [41]. Third, the cell wall's permeability improves, which maximizes extraction [42,43]. The highest HWB temperature, 100 °C for a short period (20 seconds), increases antioxidant activity. High temperatures for long-term potentially decrease the value of antioxidant activity because its compounds are suspected to dissolve in the HWB medium.

3.2. TPC

Based on Table 1, the TPC sea grapes powder is significantly different ($p < 0.05$). The control value is 0.74 mg GAE/g DB, whereas the four temperature treatments have values of 0.87, 0.88, 0.93, and 0.96 mg GAE/g DB, respectively. The results indicate that short-term blanching at high temperatures results in better outcomes than long-term blanching. Decreasing 58.82% TPC value of soaking Caulerpa at 90 °C for 5 minutes compared to fresh [44]. TPC increases with the ability to extract carotenoids and phenolics from the matrix structure during blanching [37]. High temperatures in a short time have also been shown to increase extraction by breaking covalent bonds and breaking down insoluble phenolic bonds with proteins [24,45,46].

3.3. Caulerpin (CAU)

Based on peak frequency (Fig. 1) and similarity to the CAU standard (Table 2), all treatments indicated the presence of CAU functional groups. This resemblance may be seen in the classic CAU groups, notably N-H, C-H, C=O, -C=C-, C-O-C, and the aromatic ring appearing [36].

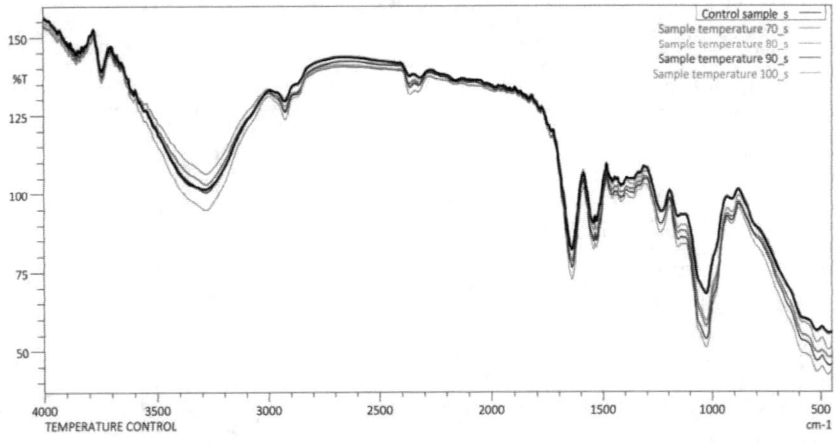

Figure 1. FT-IR spectrum of sea grapes powder

Table 2. IR frequencies of sea grapes powder and CAU standards

	[36]	Control	Temperature (°C)			
			70	80	90	100
N-H/O-H	3322	3289	3279	3279	3279	3279
C-H	2958	2924	2923	2923	2924	2923
C=O	1678	1638	1638	1636	1638	1638
-NH		1540	1538	1538	1538	1540
-C=C-	1442	1457	1417	1417	1417	1417
C-O	1330	1357	1331	1331	1331	1331
C-O-C	1155	1152	1151	1152	1152	1152
	1049	1026	1025	1025	1025	1025
aromatic ring	937	908	910	909	910	909
	602	517	580	582	580	582

The -OH and -NH functional groups are in the 3000-3600 cm^{-1} frequency range. The absorption of secondary amino groups at a frequency of 1525-1541 cm^{-1} supports the amino group (-NH). The absorption peak shows the ether group (C-O-C) at 1151-1165 cm^{-1} [47]. IR frequency's aromatic character includes 3030, 1631, 1613, 1582, 1488, and 1445 cm^{-1}; aromatic ring system (730 and 920 cm^{-1}); amino (3380 and 1585 cm^{-1}); and carbonyl functions at 1684, 1631, 1613 cm cm^{-1} [48,49]. Caulerpin is found in the genus Caulerpa, specifically *Caulerpa racemosa* and *Caulerpa var. clavifera* [9]; *C. macrophysa, C. racemosa var. laetevirens,* and *C. lentillifera* [2]; *C. cupressoides, C. paspaloides, C. polifera, C. setulariodes* [50] The benefits of CAU include antioxidants, anti-tuberculosis activity against Mycobacterium tuberculosis strain H37Rv [51], Anti-microbial activity against human food pathogenic bacteria and antiinflammatory activity against the murine macrophage cell line, RAW 264.7 [2]; anticancer [52]; and antiinflammatory [53,54].

3.4. Color Profile

Table 3 shows the results of the color profile of sea grapes powder. Values of control in l*, a*, and b* were 50.57, -1.28, and 8,22, respectively. The statistical analysis revealed that the higher the temperature, the lower L* and b* decreased, but -a* increased (p 0.05). The -a* parameter indicates that high temperatures enhance the intensity of the greenish color. The same effect appears in the Wakame and Mekabu research, with the 20 second HWB improving the greenish color [30]. Water-blanching green beans produce an increasingly deep green hue [55]. The improved green color quality can be related to the inactivation of the chlorophyllase enzyme during the blanching process [39].

Table 3. Color profile of sea grapes powder and dry sea grapes moisture content

Temperature (°C)	Color			Moisture content (%)
	L*	a*	b*	
control	50.56 ± 0.42^{b}	-1.28 ± 0.05^{c}	8.22 ± 0.27^{c}	4.92 ± 0.51^{a}
70	51.27 ± 0.13^{c}	-1.44 ± 0.06^{b}	8.23 ± 0.15^{c}	5.06 ± 0.22^{a}
80	51.32 ± 0.60^{c}	-1.47 ± 0.24^{ab}	8.10 ± 0.24^{bc}	4.72 ± 0.17^{a}
90	51.18 ± 0.73^{c}	-1.49 ± 0.15^{ab}	8.00 ± 0.15^{b}	4.72 ± 0.35^{a}
100	49.32 ± 0.45^{a}	-1.59 ± 0.09^{a}	7.02 ± 0.18^{a}	5.05 ± 0.39^{a}

Different letters above the different columns indicate a significant difference ($p < 0.05$). Data were presented as mean ± SD ($n = 3$).

3.5. Moisture content

The moisture content of dried sea grapes is shown in Table 3. According to the statistical results, the moisture content ranged from 4.72% to 5.06%, with samples without HWB reaching 4.92%. The oven was dried for 24 hours, which is thought to have evaporated all bound water, resulting in an equilibrium moisture content in both the control and the HWB treatment. Blanching reduces drying time by increasing the permeability of the cell walls, allowing water to evaporate quickly [43].

4. Conclusion

Blanching at high temperatures (100°C) for a short time (20 seconds) increased values for ABTS, FRAP, TPC, and green color intensity (-a*) in sea grapes powder.

References

[1] Guiry M D and Guiry G M 2020 Genus: Caulerpa taxonomy browser. *AlgaeBase version 4.2. World-wide electronic publication, National University of Ireland, Galway, Ireland.*

[2] Nagappan T and Vairappan C S 2014 Nutritional and bioactive properties of three edible species of green algae, genus Caulerpa (Caulerpaceae) *J Appl Phycol* **26** 1019–27

[3] Cahyanurani A B and Ummah R 2020 Study of Water Quality in Sea Grape Cultivation (Caulerpa racemosa) in Brackish Water Aquaculture Institute (BBPBAP) Jepara *Samakia: Jurnal Ilmu Perikanan* **11** 58–65

[4] Muliani S, Rosmiati and Isma M F 2021 Analysis Of The Suitability of The Quality of The Water of Kuala Langsa For The Cultivation of Sea Grapes (Caulerpa Recemosa) Reviewed By Gis *Jurnal Ilmiah Samudra Akuatika* **IV** 66–75

[5] Tapotubun A M 2018 Komposisi Kimia Rumput Laut (Caulerpa lentillifera) dari Perairan Kei Maluku dengan Metode Pengeringan Berbeda *J Pengolah Has Perikan Indones* **21** 13

[6] Kumar M, Gupta V, Kumari P, Reddy C R K and Jha B 2011 Assessment of nutrient composition and antioxidant potential of Caulerpaceae seaweeds *Journal of Food Composition and Analysis* **24** 270–8

[7] Tapotubun A M, Matrutty T E A A, Riry J, Tapotubun E J, Fransina E G, Mailoa M N, Riry W A, Setha B and Rieuwpassa F 2020 Seaweed Caulerpa sp position as functional food *IOP Conference Series: Earth and Environmental Science* vol 517 (Institute of Physics Publishing)

[8] Matanjun P, Mohamed S, Mustapha N M, Muhammad K and Ming C H 2008 Antioxidant activities and phenolics content of eight species of seaweeds from north Borneo *J Appl Phycol* **20** 367–73

[9] Aguilar-Santos G 1970 Caulerpin, a New Red Pigment from Green Algae of the Genus Caulerpa" *J. Chem. SOC. (C)*

[10] Tanna B, Choudhary B and Mishra A 2018 Metabolite profiling, antioxidant, scavenging and anti-proliferative activities of selected tropical green seaweeds reveal the nutraceutical potential of Caulerpa spp. *Algal Res* **36** 96–105

[11] Yang P, Liu D Q, Liang T J, Li J, Zhang H Y, Liu A H, Guo Y W and Mao S C 2015 Bioactive constituents from the green alga Caulerpa racemosa *Bioorg Med Chem* **23** 38–45

[12] Lee J C, Hou M F, Huang H W, Chang F R, Yeh C C, Tang J Y and Chang H W 2013 Marine algal natural products with antioxidative, anti-inflammatory, and anticancer properties *Cancer Cell Int* **13**

[13] Tanna B, Yadav S and Mishra A 2020 Anti-proliferative and ROS-inhibitory activities reveal the anticancer potential of Caulerpa species *Mol Biol Rep* **47** 7403–11

[14] Pangestuti R and Kim S K 2011 Neuroprotective effects of marine algae *Mar Drugs* **9** 803–18

[15] Lange K W, Hauser J, Nakamura Y and Kanaya S 2015 Dietary seaweeds and obesity *Food Science and Human Wellness* **4** 87–96

[16] Fithriani D 2015 Opportunities And Challenges for Developing Caulerpa Racemosa as Functional Foods *KnE Life Sciences* **2** 85

[17] Rushdi M I, Abdel-Rahman I A M, Attia E Z, Abdelraheem W M, Saber H, Madkour H A, Amin E, Hassan H M and Abdelmohsen U R 2020 A review on the diversity, chemical and pharmacological potential of the green algae genus Caulerpa *South African Journal of Botany* **132** 226–41

[18] Gupta S, Cox S and Abu-Ghannam N 2011 Effect of different drying temperatures on the moisture and phytochemical constituents of edible Irish brown seaweed *LWT* **44** 1266–72

[19] Jiménez-Escrig A, Jiménez-Jiménez I, Pulido R and Saura-Calixto F 2001 Antioxidant activity of fresh and processed edible seaweeds *J Sci Food Agric* **81** 530–4

[20] Lim Y Y and Murtijaya J 2007 Antioxidant properties of *Phyllanthus amarus* extracts as affected by different drying methods *LWT* **40** 1664–9

[21] De Corcuera J I R, Cavalieri R P and Powers J R 2004 Blanching of Foods *Encyclopedia of Agricultural, Food, and Biological Engineering*

[22] Sinurat E and Suryaningrum Th D 2019 Aktivitas Antioksidan dan Sifat Sensori Teh Rumput Laut Sargassum sp. Berdasarkan Variasi Lama Perendaman *JPHPI* **22** 581–8

[23] Nielsen C W, Holdt S L, Sloth J J, Marinho G S, Sæther M, Funderud J and Rustad T 2020 Reducing the high iodine content of *Saccharina latissima* and improving the profile of other valuable compounds by water blanching *Foods* **9**

[24] Xu H, Wu M, Zhang X, Wang B, Wang S, Zheng Z, Li D and Wang F 2022 Application of blanching pretreatment in herbaceous peony (*Paeonia lactiflora Pall.*) flower processing: Improved drying efficiency, enriched volatile profile and increased phytochemical content *Ind Crops Prod* **188**

[25] Guo Y, Wu B, Guo X, Liu D, Wu P, Ma H and Pan Z 2021 Ultrasonication and thermosonication blanching treatments of carrot at varying frequencies: Effects on peroxidase inactivation mechanisms and quality characterization evaluation *Food Chem* **343**

[26] Gupta S, Lakshmi J and Prakash J 2008 Effect of Different Blanching Treatments on Asorbic Acid Retention in Gereen Leavy Vegetables *Natural Product Radiance* **7** 111–6

[27] Dey P, Kundu A, Kumar A, Gupta M, Lee B M, Bhakta T, Dash S and Kim H S 2020 Analysis of alkaloids (indole alkaloids, isoquinoline alkaloids, tropane alkaloids) *Recent Advances in Natural Products Analysis* (Elsevier) pp 505–67

[28] Liu F, Wang Y, Li R, Bi X and Liao X 2014 Effects of high hydrostatic pressure and high-temperature short time on antioxidant activity, antioxidant compounds and color of mango nectars *Innovative Food Science and Emerging Technologies* **21** 35–43

[29] Minh N P, Thi Yen Nhi T, Kim Tuyen L, Hung Phi T, Tan Khoa D and Quoc Thuan T 2019 Technical Factors Affecting Seagrape (*Caulerpa lentillifera*) Production By Cultivation And Its Stability By Post-Harvest Treatment *J. Pharm. Sci. & Res.* **11** 783–6

[30] Hamid S S, Wakayama M, Ashino Y, Kadowaki R, Soga T and Tomita M 2020 Effect of blanching on the concentration of metabolites in two parts of Undaria pinnatifida, Wakame (leaf) and Mekabu (sporophyll) *Algal Res* **47**

[31] Nie J, Chen D, Lu Y and Dai Z 2021 Effects of various blanching methods on fucoxanthin degradation kinetics, antioxidant activity, pigment composition, and sensory quality of Sargassum fusiforme *LWT* **143** 1–9

[32] Windrayani E and Ekantari N 2021 The Fortification Effects of Sea Grapes (*Caulerpa racemosa*) Powder on Color and Sensory of Hakau Dim Sum Wrappers *IOP Conference Series: Earth and Environmental Science 919* vol 919 (IOP Publishing Ltd)

[33] Blanc N, Hauchard D, Audibert L and Ar Gall E 2011 Radical-scavenging capacity of phenol fractions in the brown seaweed Ascophyllum nodosum: An electrochemical approach *Talanta* **84** 513–8

[34] Nursyamsi F N and Husni A 2021 Soaking Time in Lime Solution Increases the Antioxidant Activity, Antidiabetic Activity, and Consumer Acceptance Level of Sargassum Polycystum Seaweed Tea *Journal of Hunan University (Natural Sciences)* **48** 276–85

[35] Vasanthi C, Appa Rao V, Narendra Babu R, Sriram P and Karunakaran R 2020 In-vitro antioxidant activities of aqueous and alcoholic extracts of Sargassum species—Indian brown seaweed *J Food Process Preserv* **44**
[36] Pusecker K, Laatsch H, Helmke E and Weylandb H 1997 Dihydrophencomycin Methyl Ester, a New Phenazine Derivative from a Marine Streptomycete *The Journal of Antibiotics* **50** 479–83
[37] Adetoro A O, Opara U L and Fawole O A 2021 Effect of blanching on enzyme inactivation, physicochemical attributes and antioxidant capacity of hot-air dried pomegranate (*Punica granatum* l.) arils (cv. wonderful) *Processes* **9** 1–18
[38] Wang Y, Sun J, Ma D, Li X, Gao X, Miao J and Gao W 2019 Improving the contents of the active components and bioactivities of *Chrysanthemum morifolium Ramat.*: The effects of drying methods *Food Biosci* **29** 9–16
[39] Mónica, Monsalves M and Bifani V 1998 Chlorophyllase Inactivation as a Measure of Blanching Efficacy and Colour Retention of Artichokes (*Cynara scolymus L.*) *Lebensm Wiss Technol* **31** 50–6
[40] Singh B, Suri K, Shevkani K, Kaur A, Kaur A and Singh N 2018 Enzymatic browning of fruit and vegetables: A review *Enzymes in Food Technology: Improvements and Innovations* (Springer Singapore) pp 73–8
[41] Chhe C, Imaizumi T, Tanaka F and Uchino T 2018 Effects of hot-water blanching on the biological and physicochemical properties of sweet potato slices *Engineering in Agriculture, Environment and Food* **11** 19–24
[42] Deylami Mahsa Z, Abdul Rahman R, Tan C P, Bakar J and Olusegun L 2016 Effect of blanching on enzyme activity, color changes, anthocyanin stability and extractability of mangosteen pericarp: A kinetic study *J Food Eng* **178** 12–9
[43] Deng L Z, Pan Z, Mujumdar A S, Zhao J H, Zheng Z A, Gao Z J and Xiao H W 2019 High-humidity hot air impingement blanching (HHAIB) enhances drying quality of apricots by inactivating the enzymes, reducing drying time and altering cellular structure *Food Control* **96** 104–11
[44] Nurjanah N, Jacoeb A M, Asmara D A and Hidayat T 2019 Phenol Component of Fresh and Boiled Sea Grapes (*Caulerpa* sp.) from Tual, Maluku *Food ScienTech Journal* **1** 31
[45] Lachos-Perez D, Baseggio A M, Mayanga-Torres P C, Maróstica M R, Rostagno M A, Martínez J and Forster-Carneiro T 2018 Subcritical water extraction of flavanones from defatted orange peel *Journal of Supercritical Fluids* **138** 7–16
[46] Park J S, Han J M, Surendhiran D and Chun B S 2022 Physicochemical and biofunctional properties of *Sargassum thunbergii* extracts obtained from subcritical water extraction and conventional solvent extraction *Journal of Supercritical Fluids* **182**
[47] Januar H, Wikanta T and Hastarini E 2004 Hubungan antara Musim dengan Kadar Caulerpin dalam *Caulerpa racemosa Jumal Penelitian Perikanan lndonesia* **10**
[48] Parvez M, Ara J, Sultana V, Qasim R and Uddin Ahmad V 2000 Caulerpin *Acta Crystallogr* **C56** 96–7
[49] Kamal C and Sethuraman M G 2012 Caulerpin-A bis-indole alkaloid as a green inhibitor for the corrosion of mild steel in 1 M HCl solution from the marine alga *Caulerpa racemosa Ind Eng Chem Res* **51** 10399–407
[50] Vest S E, Dawes C J and Romeo J T 1983 Distribution of Caulerpin Caulerpicin in Eight Species of the Green Alga Caulerpa *Botanica Marina* **XXVI** 313–6
[51] Chay C C I, Cansino R G, Espitia Pinzón C I, Torres-Ochoa R O and Martínez R 2014 Synthesis and anti-tuberculosis activity of the marine natural product caulerpin and its analogues *Mar Drugs* **12** 1757–72
[52] Abdelrheem D A, Abd El-Mageed H R, Mohamed H S, Rahman A A, Elsayed K N M and Ahmed S A 2021 Bis-indole alkaloid caulerpin from a new source *Sargassum platycarpum*:

isolation, characterization, in vitro anticancer activity, binding with nucleobases by DFT calculations and MD simulation *J Biomol Struct Dyn* **39** 5137–47

[53] Lucena A M M, Souza C R M, Jales J T, Guedes P M M, De Miranda G E C, de Moura A M A, Araújo-Júnior J X, Nascimento G J, Scortecci K C, Santos B V O and Souto J T 2018 The bisindole alkaloid caulerpin, from seaweeds of the genus Caulerpa, attenuated colon damage in murine colitis model *Mar Drugs* **16**

[54] De Souza É T, De Lira D P, De Queiroz A C, Da Silva D J C, De Aquino A B, Campessato Mella E A, Lorenzo V P, De Miranda G E C, De Araújo-Júnior J X, De Oliveira Chaves M C, Barbosa-Filho J M, De Athayde-Filho P F, De Oliveira Santos B V and Alexandre-Moreira M S 2009 The antinociceptive and anti-inflammatory activities of caulerpin, a bisindole alkaloid isolated from seaweeds of the genus Caulerpa *Mar Drugs* **7** 689–704

[55] Muftugil' N 1986 Effect of Different Types of Blanching on the Color and the Ascorbic Acid and Chlorophyll Contents of Green Beans *Journal of Food Processing and Preservation* **10** 69–69

IOP Conf. Series: Earth and Environmental Science 1289 (2023) 012038 doi:10.1088/1755-1315/1289/1/012038

Effect of carboxymethyl cellulose addition on the characteristic of chitosan-based bioplastic

L N Putranti and P S Nugraheni*
Department of Fisheries, Faculty of Agriculture, Universitas Gadjah Mada, Jl. Flora Gd. A4, Bulaksumur, Yogyakarta, Indonesia

Corresponding author: nugrahenips@ugm.ac.id

Abstract. Bioplastic is an alternative solution in solving the problem of the plastic waste accumulation. Chitosan is a marine product waste from the crustacean group that can be used as a raw material for bioplastic because of its good film-forming ability. However, the bioplastic from chitosan has the property of being easily broken, so another material is added, namely carboxymethyl cellulose (CMC), that increase the bond's compactness. This study aims to determine the effect of the addition of CMC on the characteristics of bioplastic made from 2% chitosan and 0.5% glycerol with a volume ratio of 10:1 chitosan:CMC. The experimental design used was a completely randomized design (CRD) with the addition of CMC (0%, 0.25%, 0.5%, 0.75% and 1%). The characteristics of the bioplastic tested in this study included film thickness, density, water content, solubility, tensile strength, elongation, antibacterial activity, and functional group analysis. The addition of 0.5% CMC resulted the bioplastic with a film thickness of 0.072 mm, the density of 0.836 g/cm3, the water content of 16.99%, the solubility of 40.5%, the tensile strength of 12.83 MPa, the elongation of 41.76%, inhibition zone of the antibacterial activity of 2.83 mm and a new group is formed in film that was made.

1. Introduction

Plastic is a crucial artificial material that covers contemporary civilization. Plastic is well recognized for having distinctive qualities including as lightweight, flexible, impervious to water, heat, and electricity, as well as being simple to produce. Plastics may be made from petroleum-based polymers, and the phrase fossil-based plastics was coined as a result of the successful downstream processing of these materials [1]. However, it is challenging to recycle and degrade this plastic substance. Thrown-away plastic products in the environment can contribute to environmental pollution [2].

One alternative effort that can be done to reduce the use of plastic is to develop eco-friendly plastics or bioplastic. Bioplastic is a kind of plastic with all components of the raw materials that are renewable, able to return to nature and able to be decomposed by the activity of decomposing microorganisms. The raw materials that are widely used in the manufacture of bioplastic are biopolymers. Biopolymers are natural biodegradable polymers that are accumulated by microorganisms [3]. One of the biopolymers that has been widely used as a raw material for making bioplastic is chitosan [4].

Chitosan is a polysaccharide produced from chemical deacetylation of chitin that can be found in the exoskeleton of crustacea, insect's cuticles, algae and in the cell wall of fungi [5]. The use of chitosan in the manufacture of bioplastic is based on the ability of chitosan in good film-forming, non-toxic, biocompatible and biodegradable. In addition, chitosan has antibacterial properties that can increase its usefulness as a bioplastic to protect food [6]. However, to be used as a bioplastic material, chitosan needs to be combined with other materials because bioplastic made from chitosan alone are still brittle and fragile, due to its stiff polymer backbone with large amounts of intermolecular and intramolecular hydrogen bonds [4]. One of the materials that can be used to improve the quality of bioplastic so that the bonds become more compact is from polysaccharides, namely cellulose. The previous research on bioplastic from chitosan and rice husk cellulose, where the addition of cellulose to chitosan increased the tensile strength and elongation values. Another type of cellulose that can be used as an additive in the manufacture of bioplastic because of its gel-forming, elastic, edible and renewable properties is carboxymethyl cellulose (CMC) [7]. Carboxymethyl cellulose derived from cellulose can be used to increase chemical crosslinking so that it can improve the properties of film products and can be biodegradable [8]. Chitosan has good affinity with cellulose because of the similarity in its basic chain structure [9]. The basic chemical chain structure is almost the same, making the bonds formed between chitosan and cellulose stronger so that the membrane structure formed is stronger and not easily torn [6].

Based on the characteristics of the chitosan and CMC materials, this research aims to develop bioplastics sourced from chitosan and CMC by varying the addition of carboxymethyl cellulose in order to obtain bioplastic that are in accordance with bioplastic standards, then bioplastic from a mixture of chitosan and CMC are expected to be able to help overcome the problem plastic waste that pollutes the environment.

2. Materials and Methods

2.1. Materials

Chitosan powder (Bio Chitosan Indonesia), glycerol, carboxymethyl cellulose (CMC, food grade stabilizer), distilled water, glacial acetic acid (Merck) 1%, pure bacterial culture of *Escherichia coli* (from the Laboratory of Agricultural Microbiology UGM), nutrient agar (Merck KGaA 64271 Darmstadt), nutrient broth (Merck KGaA 4271 Darmstadt) and silica gel.

2.2. Optimization of the ratio of the volume of chitosan and CMC solution

The volume ratio of the chitosan and CMC solutions was optimized based on [10], which put 1 mL of CMC solution into 50 mL of chitosan solution at a ratio of 50:1. Further research focused on the production of bioplastics utilizing chitosan and CMC solutions at ratios of 5:1 and 10:1. Each optimization ratio uses the same amount of the same materials 2% chitosan, 0.25% CMC, and 0.25% glycerol to make bioplastic solutions.

2.3. Optimization of glycerol concentration

The initial reference to the use of glycerol concentration was made in the research of [11], which produced bioplastic from cellulose solid waste derived from seaweed *Eucheuma cottonii*, chitosan, CMC, and tapioca using glycerol concentrations of 0%, 0.25%, 0.5%, 0.75%, and 1%. These films had optimal values that were close to *Standar Nasional Indonesia* (SNI) at a concentration of 0.25%. This led to the optimization of glycerol concentration being carried out in this investigation utilizing the same concentration variations, namely 0%, 0.25%, 0.5%, 0.75%, and 1%, each of which was mixed with 2% chitosan solution and 0.25% CMC.

2.4. Optimization of CMC concentration in chitosan bioplastic

2.4.1. Preparation of 2% chitosan solution

Chitosan (2 gram) is dissolved in 1% glacial acetic acid with continuous stirring by heating to a temperature 60^0C until the solution is homogenous.

2.4.2. Preparation of carboxymethyl cellulose (CMC) solution

Carboxymethyl cellulose (CMC) (0.25 gram, 0.5 gram, 0.75 gram, and 1 gram) is dissolved in 100mL aquadest with continuous stirring at room temperature until the solution is homogenous.

2.4.3. Bioplastic production

Bioplastic is produced by adapting earlier research techniques [11]. While stirring continuously for 15 minutes, a homogeneous 2% chitosan solution at a temperature of 60°C was added with 0.5% glycerol. The CMC solution was then combined with chitosan in the following amounts: 0%, 0.25%, 0.5%, 0.75%, and 1%. CMC can be heated to a homogeneous state for 30 minutes at a temperature between 60 and 65°C. The homogenous solution is filtered and let to stand for 24 hours. The solution volume (mL) is then calculated by dividing the volume (mL) by the mold area (cm). For 3.5 hours, bioplastic is formed in an oven set to a temperature of 80°C.

2.5. Bioplastic analysis and characterization

2.5.1. Bioplastic thickness

The measurement procedure based on [12] which was carried out by measuring the thickness of the film sample using a Microcal messmer thickness gauge with an accuracy of 0.0001 mm in five different places. The thickness value is measured from the average of five film thickness measurements.

2.5.2. Density

The film sample testing procedure was carried out by modifying [13] by means of a 4x2 cm film sample being weighed using a digital scale and recorded as the mass of the film in grams. After that, the sample was placed in a measuring cup containing 5 mL of distilled water for ± 1 minute then the difference in the initial and final volumes was recorded in the measuring cup. Density is calculated using the equation:

$$\rho = m/v$$

Information:
ρ = density (g/cm^3)
m = mass of film (g)
v = difference between final volume and initial volume (cm^3 or mL)

2.5.3. Solubility

The procedure for testing the solubility of biodegradable plastics or bioplastic in water is carried out by modifying those of [2, 14] by means of a 2x2 cm film sample placed in an oven at a temperature of 80°C for 10 minutes and then weighed first as the weight of dry bioplastic. After that, the sample was placed in a plastic container and immersed in 10 mL of distilled water for 24 hours. Samples that did not dissolve in distilled water were removed and dried in an oven for 15 minutes at a temperature of 80°C. Then the weight of the dry sample was reweighed as the final weight after immersion. The percentage solubility can be calculated using the following formula:

$$\text{Percent Solubility} = (a-b)/a \times 100\%$$

Information:
a = initial weight (g)
b = final weight (g)

2.5.4. Water content

The film sample testing procedure was carried out using a moisture content analyzer. A sample of 0.5 grams was placed on a pallet provided in the tool then closed and waited a few minutes until the drying process was complete. Then record the results of the moisture content that can be seen on the moisture content analyzer screen.

2.5.5. Tensile strength

The tensile strength testing procedure refers to the American Standard Testing Method ASTM D882-18 (2018) using Zwick & Roel Retroline tC II for Z020 TN. Sheets of film size 7x15 cm with humidity (RH) 50±10% for 40 hours. The instron tool is mounted on an initial grip separation of 50 mm, a crosshead speed of 50 mm/minute and a loadcell of 50 kg, and is measured using the formula:

$$\tau = (F\ max)/A$$

Information:
τ = tensile strength (Mpa)
F max = tensile strength (N)
A = cross-sectional area (mm2)

2.5.6. Elongation

The elongation test procedure refers to the American Standard Testing Method ASTM D882 (2018) using Zwick & Roel Retroline tC II for Z020 TN. Sheets of film size 7x15 cm with humidity (RH) 50±10% for 40 hours. Instron equipment is set at a crosshead speed of 50 mm/min, initial grip separation of 50 mm, and loadcell of 50 kg. The calculation is carried out when the film is broken or torn and is calculated by the formula:

$$\text{Percent Elongation} = (l1\text{-}lo)/lo$$

Information:
lo = initial length
l1 = length after break up

2.5.7. Antibacterial activity

The film sample testing procedure was carried out by modifying [15] which was carried out using the paper disc diffusion method on *Escherichia coli* bacteria. Escherichia coli bacteria that have been grown in NB are measured for turbidity OD to determine the estimated value of OD using a spectrophotometer. 0.1 mL of the bacterial suspension was put into a petri dish which already contained solid NA and then leveled with a drigalsky. In the media, a paper disk with a diameter of 6 mm was inserted which had been dipped in a bioplastic solution with treatments a1, a2, a3, a4, and a5. After that, it was incubated for 24 hours at 37°C and the diameter of the inhibition zone formed was measured. The diameter of the inhibition zone was calculated between the difference in the diameter of the clear zone (including the diameter of the paper disk) and the diameter of the paper disk formed in mm using a ruler.

2.5.8. Functional group analysis

The analytical procedure was carried out using a wavelength of 4000-400 nm. The film sample used with a minimum diameter of 2 mm is inserted into the set holder, then the appropriate spectrum is searched. The results obtained are in the form of a diffractogram of the relationship between wave number and intensity. The FTIR spectra were recorded using a spectrophotometer at room temperature.

3. Results and Discussion

3.1. Optimization of the ratio of the volume of chitosan and CMC solution

Table 1. The results of the optimization of the volume ratio of chitosan and CMC solution

Chitosan : CMC ratio	Solution homogeneity	Viscosity
50:1	There are no clots	The solution is dilute similar with chitosan solution without addition of CMC
10:1	There is a little clot	The solution is thicker
5:1	There are many clots, large clumps	The remaining solution is dilute

Based on Table 1., the most appropriate use of the volume ratio is 10:1 for the ratio of the volumes of chitosan and CMC solution due to it showed just a little clot formed and the viscosity was very suitable to support film formation.

3.2. Optimization of glycerol concentration

Table 2. The results of the optimization of the glycerol concentration

Glycerol concentration	Firmness	Elasticity
0,25%	The film is not mushy	Easy to remove from the mold, and when folded and pulled easily torn and brittle
0,5%	The film is slightly mushy	Easy to remove from the mold, and when folded and stretched it does not tear easily and is slightly brittle
0,75%	The film is a bit mushy	A bit difficult to remove from the mold and when folded and pulled easily brittle
1,00%	The film is mushy	Somewhat difficult to remove from the mold and when folded and stretched becomes brittle

Based on Table 2., the most appropriate use of glycerol is 0.5% glycerol because it has better elasticity. The result of Table 1 and Table 2 were used to carry out the next step experiment.

IOP Conf. Series: Earth and Environmental Science 1289 (2023) 012038 doi:10.1088/1755-1315/1289/1/012038

3.3. Optimization of CMC concentration in chitosan bioplastic

3.3.1. The effect of CMC Addition on Chitosan Bioplastic Thickness

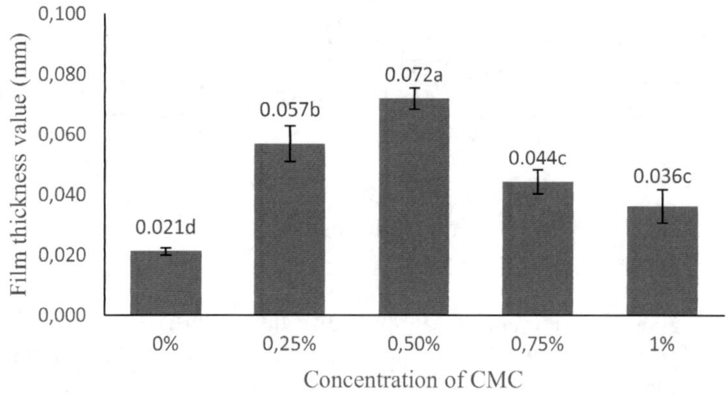

Figure 1. Effect of CMC addition on chitosan bioplastic thickness

The effect of CMC addition on chitosan bioplastic thickness is presented in Figure 1. The addition of CMC in chitosan bioplastic had a significant effect ($p<0.05$) on the film thickness at a significance level of 95%. This shows that the addition of CMC with a concentration of 0%-1% affects the thickness of the bioplastic produced. The film thickness that exceeds the standard can affect the organoleptic of the product, but if it is too thin it indicates that the film is easily damaged. Bioplastic that is too thick or exceeds the standard will produce a darker color of the bioplastic. If this is applied to food products that you want to highlight, it will have a negative impact. CMC is a hygroscopic material so that the more concentration of CMC is added, the more water will be absorbed and bound to the material. The hydroxyl group owned by CMC able to bind water that is free from solution, emulsion, or suspension as water hydrate so that when added to the emulsion solution it can become thick and affect the thickness of the bioplastic [16]. This shows that the more concentration of CMC added will tend to increase the thickness of the bioplastic. As the film thickness increases (maximum 0.25 mm), the retained ability will also improve so that the shelf life of packaged products will be longer [17].

The decrease in the average value of bioplastic thickness with the addition of 0.75% and 1% CMC is not in line with [16] that states that the addition of CMC concentration will tend to increase the value of plastic thickness. That is probably because at the time of dissolved chitosan with the addition of CMC 0.75% and 1% it is not homogeneous and does not bind perfectly so that the bonds formed are in the form of lumps or undergo agglomeration. This is evidenced by the presence of many lumps in the resulting solution. The number of lumps will be wasted when filtering is carried out which indicates that the solution has experienced a lot of biomass loss. This is supported by [18] which states that at high concentrations CMC is difficult to spread evenly so that it can cause clots to form in the solution.

Based on the Japanese Industrial Standard (JIS), the standard value for the maximum thickness of bioplastic is 0.25 mm. Thick bioplastic produced in all treatments met the standard. The thickness of the bioplastic obtained in this study was not much different from the thickness of the bioplastic made from chitosan and canna starch in the research of [14] which is 0.035-0.045 mm and the thickness of bioplastic made from chitosan and cassava starch in the research of [19] which is 0.0175-0.0755 mm.

3.3.2. The effect of CMC Addition on Chitosan Bioplastic Density

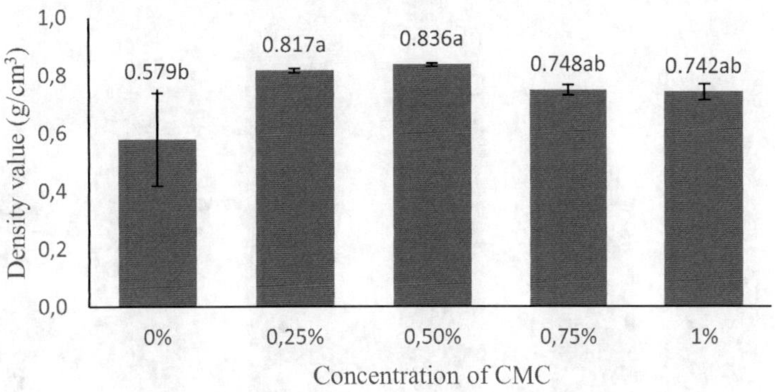

Figure 2. Effect of CMC addition on chitosan bioplastic density

The effect of CMC addition on chitosan bioplastic density is presented in Figure 2. The addition of CMC in chitosan bioplastic had a significant effect ($p<0.05$) on the density value at the 95% significance level. This shows that the addition of CMC with a concentration of 0% up to 1% affects the density of the resulted bioplastic. The denser the bioplastic film produced, the less the number of cavities in the bioplastic film. The effect of the addition of CMC with an increase in the density number because CMC able to fill the empty space in the bioplastic film which makes the density better [1].

The decrease density that occurs when the addition of CMC 0.75% and 1% can be caused by the manufacture of chitosan bioplastic solution with the addition of CMC 0.75% and 1% not bound properly so that the form or agglomeration is formed. This can be seen from the results of the solution before it was carried out there were many images indicating that CMC was not able to fill the empty space in the bioplastic film. This is supported by [18] which states that a high concentration of CMC is difficult to spread evenly so that it can lead to the formation of a solution.

Based on the Indonesian National Standard (SNI) 7188.7:2016 about the mechanical properties of bioplastic, the standard set for the density of bioplastics is 0.95 g/cm3. The average density value of chitosan bioplastic that is close to the standard is chitosan bioplastic with the addition of 0.5% CMC, which is 0.836 g/cm3. The density results obtained in this study were not much different from the density of bioplastic based on chitosan, CMC and sorbitol in the research of [12] which is about 1 g/cm3 and in the research of [13] which makes bioplastics with the basic ingredients of chitosan and sorghum stem cellulose, which is 0.915 gr/cm3.

3.3.3. The effect of CMC Addition on Chitosan Bioplastic Water Content

The effect of CMC addition on chitosan bioplastic water content is presented in Figure 3. Based on statistical analysis (ANOVA), the addition of CMC in chitosan bioplastics had a significant effect ($p<0.05$) on the water content value at a significance level of 95%. This shows that the addition of CMC with a concentration of 0% up to1% has effect on the water content of the bioplastic produced. Moisture content affects shelf life because it is related to the possible activity of microorganisms in bioplastic during storage. The lower the water content value indicates that the bioplastic film produced is better because it able to maintain the product to be packaged longer [21]. The addition of CMC concentration from 0%-0.5% increased due to the addition of CMC which was able to bind high water, so that the more concentration of CMC that was added and able to bind other materials, the greater the amount of water bound to the film and would increase the water content [22].

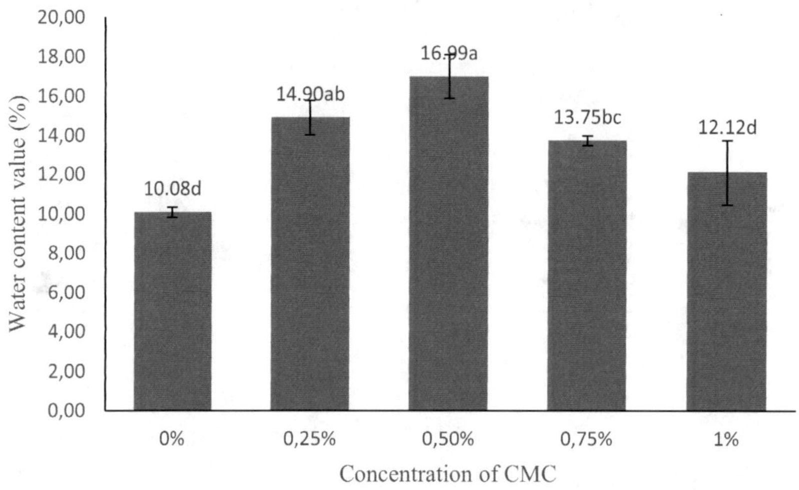

Figure 3. Effect of CMC addition on chitosan bioplastic water content

The decrease that occurred in the addition of 0.75% and 1% CMC could be caused because during the manufacture of chitosan bioplastic solution with the addition of 0.75% and 1% CMC it was not homogeneous and did not bind perfectly so that the bonds formed were in the form of lumps or agglomeration. This can be seen in the solution before filtering there are many lumps so that chitosan dominates the hydrophobic nature of the film formed. This is supported by [18] which states that at high concentrations CMC is difficult to spread evenly so that it can cause clots to form in the solution.

According to the Food and Agriculture Organization of the United Nations (FAO), the maximum moisture content value is 12%. The average value of water content in the study that met the standard was only in the treatment using 0% CMC. However, overall the average water content obtained in this study was slightly lower than the water content value of the chitosan and starch-based films in the research of [23] which ranged from 16.48-23.96% and research by [17] which made films based on chitosan and yellow sweet potato starch ranging from 21.77-26.37%.

3.3.4. The effect of CMC Addition on Chitosan Bioplastic Solubility

The effect of CMC addition on chitosan bioplastic solubility is presented in Figure 4. Based on statistical analysis (ANOVA), the addition of CMC in chitosan bioplastics had a significant effect ($p<0.05$) on the solubility value at a significance level of 95%. This shows that the addition of CMC with a concentration of 0% up to 1% affects the solubility of the resulting bioplastic. The solubility of bioplastic is an important factor in packaging materials which is influenced by hydrophilic and hydrophobic components. Hydrophilic components are components that like water or are soluble in water while hydrophobic are components that like fat or are insoluble in water. The higher the hydrophilic nature of a material, the higher the solubility and the higher the hydrophobic value of a material, the lower the solubility [24]. Solubility parameters are used to determine the ability of bioplastic to be degraded in aqueous media. The higher the solubility value of a film, the higher its biodegradability and the faster the film is degraded due to the presence of hydrophilic components in water and soil. However, if the solubility value is higher, the film's resistance to water will be lower [25].

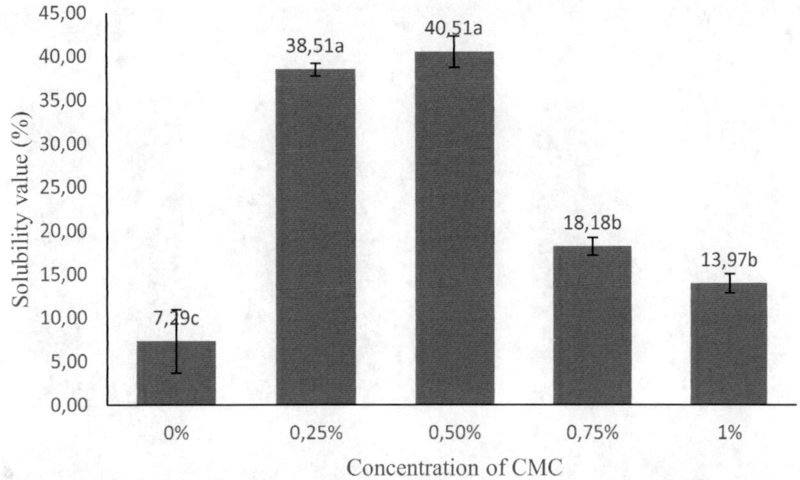

Figure 4. Effect of CMC addition on chitosan bioplastic solubility

The decrease that occurred in the addition of 0.75% and 1% CMC could be the fact that during the manufacture of chitosan bioplastic solution with the addition of 0.75% and 1% CMC it was not homogeneous and did not bind perfectly so that the bonds formed were in the form of lumps or agglomerations. This can be seen in the solution before being filtered there are many lumps. This is supported by [18] which states that at high concentrations CMC is difficult to spread evenly so that it can cause clots to form in the solution.

The solubility results obtained in this study are lower than the solubility values of chitosan and carrageenan-based films produced in the research of [26] which ranged from 63.5-77% and in the research of [11] which made films based on chitosan and cellulose waste from seaweed, ranging from 54.4-98.4%.

3.3.5. The effect of CMC Addition on The Diameter of The Chitosan Bioplastic Inhibition Zone

The effect of CMC addition on chitosan bioplastic inhibition zone is presented in Figure 5. The average diameter of the inhibition zone obtained in this study when categorized in the diameter of the inhibition zone is based on Table 3. All of them are included in the category of 5 mm diameter with weak inhibitory strength. The clear zone is an indication of an inhibitory response to the growth of bacteria from antibacterial compounds. The number of bacteria required for the sensitivity test is 10^5-10^8 cfu/mL. The clear zone area shows the effectiveness of bioplastic in inhibiting bacteria. The wider the clear zone, the better the ability of bioplastic to inhibit bacteria [27].

Based on the Kruskal-Wallis analysis, the addition of CMC to chitosan bioplastic had a significant effect ($p<0.05$) on the diameter of the inhibition zone at a significance level of 95%. This shows that the addition of CMC treatment with a concentration of 0%-1% affects the diameter of the bioplastic inhibition zone produced in *Escherichia coli* bacteria.

The addition of CMC will fill the plastic composition so that the role of chitosan is not too dominant which causes a decrease in the antibacterial power of chitosan [28]. The average results of the inhibition zone against *Escherichia coli* bacteria obtained in this study were not much different when compared to the inhibition zone of chitosan-based bioplastic in the previous research [29] which is 1.83 mm.

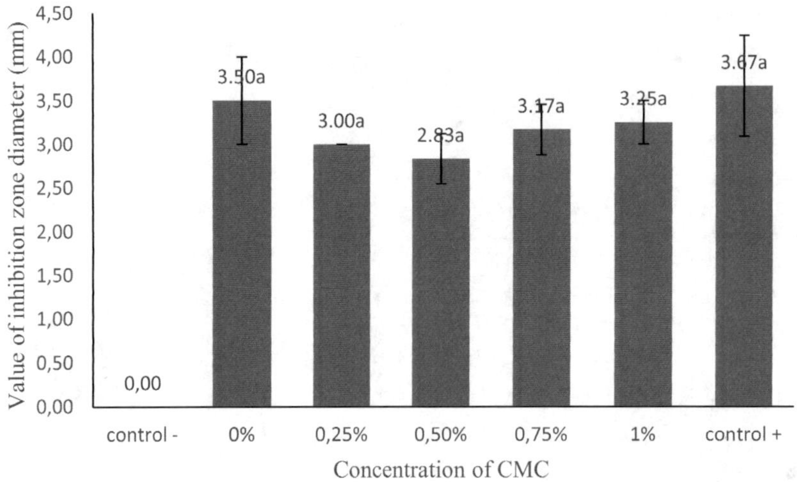

Figure 5. Effect of CMC addition on the diameter of the chitosan bioplastic inhibition zone

Table 3. Inhibition zone category

Diameter	Inhibitory power
≤ 5 mm	Weak
6-10 mm	Medium
11-20 mm	Strong
≥ 21 mm	Very strong

3.3.6. The Effect of CMC Addition on Chitosan Bioplastic Functional Group Analysis

Table 4. The results of the FTIR analysis of the main bioplastic constituent groups and their comparisons

Component	Wavenumber (cm^{-1})				
	O-H	N-H	C-O	C-H	C=N
Glycerol (Ningsih *et al.*, 2019)	3427,51			2881,65	
CMC	3278,29		1020,36	2918,87	
Chitosan+CMC 0%	3265,11	1027,15		1408,41	
Chitosan+CMC0,25%	3260,45	1027,67	1151,89	2881,00	2341,77 2359,83
Chitosan+CMC 0,5%	3256,41	1026,46	1151,79	2880,01	2360,05
Chitosan+CMC0,75%	3247,92	1026,05		1406,9	
Chitosan+CMC 1%	3263,95	1027,20		1408,11	

The results of the FTIR analysis of the main groups of bioplastic that compose and their comparisons is presented in Table 4. Chitosan bioplastic with the addition of CMC has a main functional group that is unique to chitosan, namely the presence of an N-H group. The treatment that showed the presence of a typical CMC functional group, namely the presence of the main functional group OH, CO and CH groups [24] and showed the formation of a new functional group, namely C = N was a treatment with the addition of 0.25% and 0.5%. According to [30], the presence of amide and ester functional groups in bioplastic film samples causes plastic to be degraded and can be said to be environmentally friendly plastic. In the treatment with the addition of 0.75% and 3% CMC, no new functional groups were formed. This can happen because the manufacture of chitosan bioplastic solution with the addition of 0.75% and 1% CMC is not perfectly homogeneous and it is possible that the chitosan solution and CMC solution do not bond perfectly and cause agglomeration. This is supported by [18] which states that at high concentrations CMC is difficult to spread evenly so that it can cause clumps to form. In addition, this shows that the bioplastic formed are the result of physical mixing.

3.3.7. The Effect of CMC Addition on Chitosan Bioplastic Tensile Strength

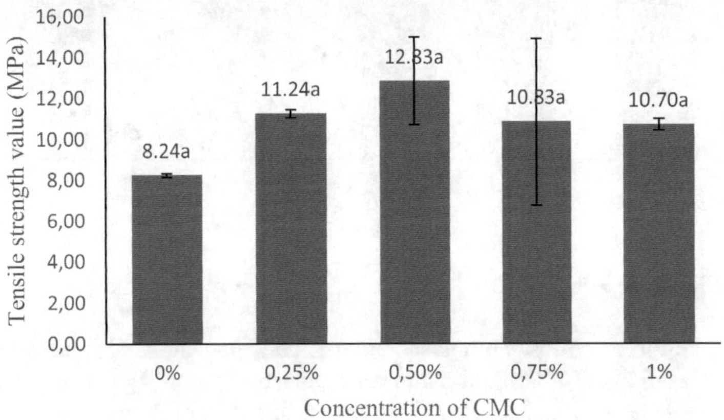

Figure 6. Effect of CMC addition on chitosan bioplastic tensile strength

The effect of CMC addition on chitosan bioplastic tensile strength is presented in Figure 6. Based on statistical analysis (ANOVA), the addition of CMC in chitosan bioplastic had no significant effect ($p>0.05$) on the tensile strength value at the 95% significance level. This shows that the addition of CMC with a concentration of 0%-1% does not affect the tensile strength of the resulting bioplastic and it is possible that the concentration of plasticizer is glycerol which affects the tensile strength of the film. In this research, the concentration of glycerol used in all treatments was 0.5% so that the tensile strength values between treatments were not significantly different. This is supported by previous research [11] that the factor that affects the tensile strength of the film is the concentration of glycerol. The more glycerol added, the lower the tensile strength, but if too little glycerol is added, the resulting bioplastic will easily crack or be less elastic [31]. According to [23] the higher the tensile strength value is influenced by the higher filler content. This is because the greater the filler content, the more bonds there are in bioplastics so that the chemical bonds will be stronger and harder to break because it requires a lot of energy to break them.

The average tensile strength value produced in this study did not meet SNI. The average value of the tensile strength produced and close to SNI is in the treatment with the addition of 0.5% CMC concentration, which is 12.83 MPa. The results obtained have a higher tensile strength value compared to bioplastic made from starch and CMC in the research of [24] which ranged from 0.2894 to 0.4739 MPa and almost the same as bioplastic made from chitosan and cellulose stems of sorghum in the research of [13] which is around 11.64 MPa.

3.3.8. The Effect of CMC Addition on Chitosan Bioplastic Elongation

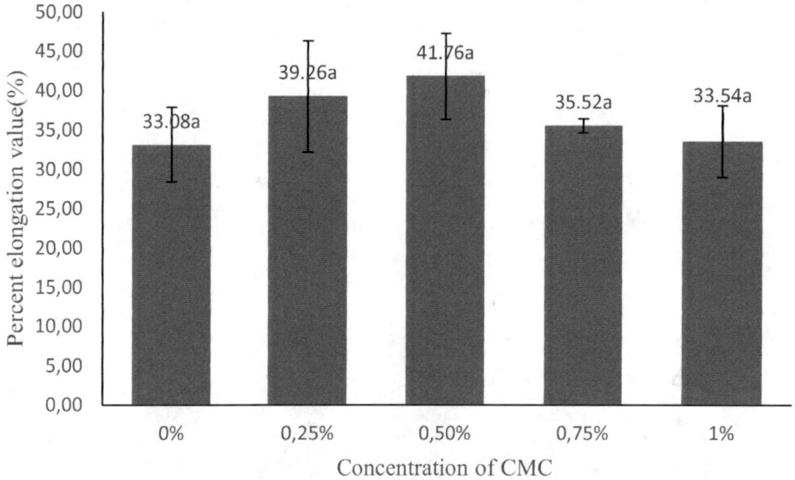

Figure 7. Effect of CMC addition on chitosan bioplastic elongation

The effect of CMC addition on chitosan bioplastic elongation is presented in Figure 7. Based on statistical analysis (ANOVA), the addition of CMC in chitosan bioplastic had no significant effect ($p>0.05$) on the elongation value at the 95% significance level. This shows that the addition of CMC with a concentration of 0%-1% does not affect the elongation value of the resulting bioplastic. This is supported by the research of [32] that what affects the elongation value is the addition of plasticizers that can increase the volume of bioplastic so that a wider space is formed to increase the segmental motion of polymer molecules in bioplastic. Plasticizer molecules can interfere with the compactness of the filler, reduce intermolecular interactions and increase the mobility of the polymer, so that with the addition of a plasticizer hydrogen bonds can be broken. Plasticizers can increase the volume of bioplastic so that a wider space is formed to increase the segmental length of polymer molecule motion in bioplastic thereby increasing the elongation value.

Based on the Indonesian National Standard (SNI) 7188.7:2016 regarding the mechanical properties of bioplastic, the standard set for the percent elongation of bioplastic is 21-220%. The average value of the percent elongation of bioplastic produced in all treatments has met SNI. The average elongation result obtained in this study was higher than that of bioplastic made from chitosan and sorghum stem cellulose in the study of [20] which is around 10.98% and bioplastic made from chitosan and starch in the research of [14] which is around 1.86-3.7%.

3.4. Discussion

The addition of CMC was able to increase the bioplastic properties of chitosan on film thickness, density, water content, and solubility. This shows that the resulting bioplastic has good characteristics as packaging that is in direct contact with food in terms of film thickness and density. If it is seen from the value of the water content and solubility of bioplastic that are higher with the addition of CMC, it is assumed that the bioplastic produced are more suitable for foods with low water content, for example in powdered foods and have the potential to dissolve immediately. For example, this bioplastic can be used as a wrapper for fine spices found in instant foods, such as instant noodles and instant porridge. When the instant food is cooked, you can immediately pour hot water into it and later the bioplastic will also dissolve with the food dissolved in hot water. This can be an alternative solution to reduce plastic waste from wrapping powdered spices in previous instant foods.

Based on the results of the study, the tensile strength value obtained did not meet the Indonesian National Standard (SNI), so it was necessary to re-optimize. Tensile strength describes the ability of the film structure to withstand loads without being damaged which is influenced by the addition of fillers and plasticizers in the process of making bioplastic [32]. The use of plasticizer, namely glycerol in this study, is quite good when viewed from the resulting elongation value. Therefore, optimization was carried out to increase the tensile strength by adding other hydrocolloid materials that have the potential to strengthen the bond between chitosan, CMC, and glycerol. Hydrocolloids that can be used as crosslinkers are compounds that have low molecular weights to weaken polymer stiffness while increasing its flexibility [25]. An example of a low molecular weight hydrocolloid is an oligosaccharide. Oligosaccharides also have short polymer chains so that the hydrogen bonds that cause film stiffness are only slightly.

4. Conclusions

The addition of 0.25% and 0.5% CMC concentrations will form new groups resulting in an increase in the value of bioplastic thickness, density, water content, and solubility, but will decrease the value of the diameter of the inhibition zone against bacteria and give the same results in tensile strength and percent elongation. The best formulation that produces bioplastic with values close to the standard is the use of 2% chitosan and 0.5% carboxymethyl cellulose (CMC) with a film thickness of 0.072 mm, a density of 0.8364 g/cm3 and an elongation value that meets SNI is 41.76%.

5. Acknowledgment

The research was financially supported by Universitas Gadjah Mada through Final Project Recognition Program Grant (RTA) with grant number 3143/UN1.P.III/DIT-LIT/PT/2021. Our appreciation to Atika Dyah Kusumastuti and Fiskina Zulfa Kamala who helped take part in this research.

References

[1] Dang B, Xuan-Thanh B, Duyen P H T, Huu H N, Long D N, Thi-Khanh-Dieu H, Phuong-Thao N, Hai H N, Thi-Kim-Quyen V, Chitsan L, Kun Y A L, Sunita V 2022 Current application of algae derivatives for bioplastic production: A review *Bioresource Technology* 347 pp 1-16

[2] Fortunati E, Luzi F, Yang W, Kenny J M, Torre L, Puglia D 2018 Chapter 4- bio-based nanomaterials for food packaging *Elsevier* pp 71–110

[3] Maarif L, Ulia F and Endaruji S 2020 Bioplastic biodegradation based on ganyong umbi states with addition of sorbitol and cmc (*carboxymethyl cellulose*) in soil media *Journal of Chemistry* **3** pp 429-435

[4] Jiang S, Congde Q, Runpeng L, Qinze L, Jing X, Jinshui Y 2023 Structure and properties of citric acid cross-linked chitosan/poly(vinyl alcohol) composite films for foof packaging applications *Carbohydrate Polymers* 312 pp 120842

[5] Aranaz I, Andres R A, Maria C C, Concepcion A, Begona E, Angeles H C, Niuris A 2021 Chitosan: An overview of its properties and applications *Polymers* 13 19 pp 3256

[6] Florez M, Esther G, Patricia C, Manuel V 2022 Chitosan for food packaging: Recent advances in active and intelligent films *Food Hydrocolloids* 124 pp 1-16

[7] Rahman S, Saif H, Ashis S N, Sunghyun N, Aneek K K, Shameem A, Muhammad J A S, Mohammad B A 2021 *Polymers* 13 pp 1-48

[8] Syuhada M, Sintia A and Endaruji S 2020 The Effect of Cassava Peel Starch Addition to Bioplastic Biodegradation Based on Chitosan on Soil and River Water Media *Journal of Biology, Medicine, and Natural Product Chesmistry* **9** 1 pp 7-13

[9] Ningsih E, Dahlena A and Sunardi 2019 Effects of carboxymethyl cellulose addition on the characteristics of bioplastic from nagara sweet potatoes (*Ipomea batatas L.*) starch *Indonesian Journal of Chemical* **7** 1 pp 77-85

[10] Nugroho A, Basito and Katri R B 2013 Kajian pembuatan edible film tapioka dengan penambahan pektin beberapa jenis kulit pisang terhadap karakteristik fisik dan mekanik *Jurnal Teknosains Pangan* **2** 1 pp 1-12

[11] Harumarani S, Widodo F and Romadhon 2015 The effect of different glycerol concentration on characteristic of composite *Eucheuma cottonii* semi refined carrageenan edible film and beeswax *Journal of Fishery Products Processing and Biotechnology* **5** 1 pp 101-105

[12] Sari D, Syelvia P and Bahruddin 2019 Pembuatan bioplastik berbasi pati sagu dengan modifikator asam sitrat dan filler *carboxy methyl cellulose* (CMC) *Jurnal Online Mahasiswa Fakultas Teknik* **6** 1 pp 1-6

[13] Arizal V D. Yuli, Edwin A, Lia L, and Heri U 2017 Application *Eucheuma cottonii* in bioplastic synthesis based on sorghum with plasticizers glycerol *Proceedings of the National Seminar on Industry Riset* pp 32-39

[14] Oktaviani C, Khairat and Bahruddin 2019 Modifikasi bioplastik berbasis pati sagu dengan asam sitrat dan filler *microcrystalline cellulose* (MCC) *Jurnal Online Mahasiswa Fakultas Teknik* **6** 1 pp 1-5

[15] Katili S, Harsunu B T and Irawan S 2013 Effect of concentration of glycerol plasticizer and composition of chitosan in solvent on physical properties of chitosan edible film Journal of Technology 6 1 pp 29-38

[16] Krisnadi R, Yuni H and Kartika U 2019 Effect of plasticizer type on characteristics of biodegradable plastics from rice bran *Proceedings of the National Seminar on Applied Science and Technology* pp 125-130

[17] Martinez D G, Barneto G and Partal M 2011 Modelling of pyrolysis and combustion

[18] Umar 2017 *Pengaruh Penambahan CMC, Kitosan, dan CMC dengan Kitosan terhadap Efisiensi Mikroenkapsulasi Ekstrak Kasar Daun Avicennia alba dengan Metode Freeze Drying* (Malang: Fakultas Perikanan dan Ilmu Kelautan Universitas Brawijaya)

[19] Supeni G, Cahyaningtyas A and Fitriana A 2015 Karakteristik sifat fisik dan mekanik penambahan kitosan pada edible film karagenan dan tapioka termodifikasi *Jurnal Kimia Kemasan* **37** 2 103-110

[20] Rahman S, Saif H, Ashis S N, Sunghyun N, Aneek K K, Shameem A, Muhammad J A S, Mohammad B A 2021 Recent Developments of Carboxymethyl Cellulose *Polymers* 13(8) pp 1345

[21] Handayani, P A and Hesmita W 2015 Bidegradable plastic films from waste durian (*Durio Zibethinus Murr*) seeds *Journal of Renewable Natural Materials* **4** 1 pp 21-26

[22] Gontard N, Guilbert, and Cuq J L 1992 Edible wheat gluten films: influence of main process variables on film properties using response surface methodology *Journal of Food Science* **57** pp 190-195

[23] Nurmala N A, Eko B S and Fransisca W M 2018 Synthesis of chitosan from crab shells with beeswax composite and its application as an edible coating on strawberries *Indonesian Journal of Chemical Science* **7** 3 pp 278-284

[24] Nishiyama M 1995 Biodegradable plastics derived from cellulose and chitosan *American Chemical Society* **7** 1 pp 113-123

[25] Krisna D 2011 *Pengaruh Regelatinisasi dan Modifikasi Hidrotermal terhadap Sifat Fisik pada Pembuatan Edible Film dari Pati Kacang Merah (Vigna Angularis Sp.)* (Semarang: Magister Teknik Kimia Universitas Diponegoro)

[26] Fara M 2018 Pembuatan dan Uji Kemampuan Membran Komposit Kitosan-Selulosa terhadap Penurunan Kadar Logam Kromium (Cr) pada Industri Penyamakan Kulit (Malang: Fakultas Teknologi Pertanian Universitas Brawijaya)

[27] Apriyanti A, Widhi M, and Warlan S 2013 Study of physico-mechanical and antibacterial properties of glycerol modified chitosan plastics *Indonesian Journal of Chemical Science* **2** 2 pp 148-153

[28] Surjowardojo P, Tri E and Gabriel R 2015 Daya hambat deok kulit apel manalagi (*Malus sylvesters Mill.*) terhadap pertumbuhan *Staphylococcus aureus* dan *Pseudomonas sp.* penyebab mastitis pada sapi perah *Jurnal Ternak Tropika* **16** 2 pp 40-48

[29] Katrin D, Nora I and Berlian S 2015 Test of antibacterial activity of malek leaf extract (*Litsea garciae Vidal*) against *Staphylococcus aureus* and *Escherichia coli* bacteria *Journal of Chemical and Packaging* **4** 1 pp 7-12

[30] Gozali T, Willy P and Iqbal R 2020 Effect of cmc concentration and glycerol concentration on edible packaging characteristics of instant coffee from mung bean starch (*Vigna radiate L.*) *Pasundan Food Technology Journal* **7** 1 pp 1-9

[31] Sari D P and Abdiani I M 2015 Pemanfaatan kulit udang dan cangkang kepiting sebagai bahan baku kitosan *Jurnal Harpodon Borneo* **8** 2 pp 142-147

[32] Sinaga R F, Gita M, Hendra S and Rosdanelli H 2014 Effect of glycerol addition on tensile strength and elongation properties of bioplastics from taro tuber starch *Journal of Chemical Engineering* **3** 2 pp 19-24

[33] Saputro A and Linggar A 2017 Synthesis and characterization of biopastic from chitosan-ganyong starch (Canna edulis) *Journal of Chemistry and Chemistry Education* 2 1 pp 13-21

IOP Conf. Series: Earth and Environmental Science 1289 (2023) 012039 doi:10.1088/1755-1315/1289/1/012039

Effect of carrageenan addition on the characteristic of chitosan-based bioplastic

E Favian and P S Nugraheni*

Department of Fisheries, Faculty of Agriculture, Univesitas Gadjah Mada, Jl Flora, Bulaksumur, Yogyakarta. Indonesia

Corresponding author: nugrahenips@ugm.ac.id

Abstract. Synthetic plastic is a product that widely used in daily life. However, the use of plastic has a negative impact that is very difficult to decompose by nature, so it is necessary to find other alternatives by switching to bioplastics. Chitosan is a polysaccharide that can be used as raw material for bioplastics because of its good plastic-forming ability. On the other hand, it has poor elasticity so it is necessary to add carrageenan which can improve the mechanical properties of bioplastics. The purpose of this research is to determine the characteristics of bioplastics made from a mixture of chitosan and carrageenan. The concentration of chitosan used was 2%, glycerol was 0.5%, and carrageenan concentrations were varied from 0%, 0.25%, 0.5%, 0.75%, and 1%. Characteristic tests carried out were film thickness, density, moisture content, solubility, tensile strength, elongation, FTIR, and antibacterial activity. The result showed that the addition of 0.5% of carrageenan resulted in better bioplastic with film thickness of 0.06 mm, density of 0.85 g/cm^3, water content of 17.27%, solubility of 36.89%, the tensile strength of 13.72 mPa, the elongation of 46.03%, inhibition zone of the antibacterial activity of 2.66 mm and a new group is formed in the FTIR test.

1. Introduction

Synthetic plastic is a product that functions as a grocery bag, food wrapper, and daily necessities. The use of high amounts of plastic in various sectors makes synthetic plastic production widespread, this is due to the strong and stable nature of synthetic plastics. The main raw materials used in the manufacture of synthetic plastics come from fossil fuel reserves [1]. The process of production, consumption and disposal of synthetic plastics produces high carbon emissions that contribute to climate change which causes the earth's conditions to heat up [2,3]. Due to the many problems that arise from the use of synthetic plastics, it is necessary to look for other alternatives that are environmentally friendly but still have strong and stable properties. One of the alternatives offered is to switch to using bioplastics.

Bioplastics are plastics derived from renewable natural materials, which are environmentally friendly because they can be decomposed by microorganisms as well as by moisture and radiation from the sun, into compounds that are environmentally friendly and safe for human health [4]. Raw materials that are widely used in the manufacture of bioplastics usually utilize natural polymers. Chitosan is a natural polymer that can be used as a raw material for making bioplastics. Chitosan can form a good film because the resulting film is not easily soluble in water so that it can be used as a constituent of bioplastics [5].

Chitosan is produced through the deacetylation of chitin from crustacean waste [6]. Chitosan is environmentally friendly because it can be naturally degraded by the environment and has an amine group so that it has antibacterial properties which are added value as a bioplastic making material. Bioplastics made from chitosan have good properties, but they still have the disadvantage that they tend to be brittle so they have low tensile strength and elasticity. The shortcomings of bioplastic made from chitosan can be overcome by adding a crosslinker that can improve the mechanical properties of chitosan bioplastic, one of which is carrageenan derived from red seaweed.

A carrageenan is a group of linear polysaccharides derived from the extraction of red seaweed species [7]. Carrageenan can increase ionic and chemical cross-linking so that it can improve the mechanical properties of bioplastics [8]. This study aims to determine the characteristics of chitosan-based bioplastics that added various concentrations of carrageenan, which produces bioplastics according to bioplastic quality standards. Glycerol was used as a platicizer in this study. The resulting bioplastic is expected to replace the use of plastic food packaging with low water content.

2. Materials and Methods

2.1. Materials

Chitosan powder (Bio Chitosan Indonesia), glycerol, carrageenan powder (RVN-M from CV Carrageenan Indonesia Semarang), distilled water, glacial acetic acid (Merck) 1%, pure bacterial culture of *Escherichia coli* (from the Laboratory of Agricultural Microbiology UGM), nutrient agar (Merck KGaA 64271 Darmstadt), nutrient broth (Merck KGaA 4271 Darmstadt) and silica gel.

2.2. Optimization of the ratio of the volume of chitosan and carrageenan solution

Chitosan (2 g in 100 mL) was dissolved in 1% of acetic acid. Carrageenan was also prepared with a concentration of 0.5%. Chitosan and carrageenan solution was mixing in two ratios, namely 5:1 and 10:1 (v/v). The mixing solution was heated on a hot stirrer using a rotational power of 600 rpm until it reached a temperature of 60ºC. Then 0.5% of glycerol was added as a plasticizer. The final solution was observed for the color clarity and the homogeneous solution was compared by following whether any lumps form.

2.3. Optimization of glycerol concentration

Preparation of mixing chitosan solution and carrageenan solution is as described in 2.1 with a chitosan: carrageenan ratio of 100: 10 (v:v). The glycerol concentration used is 0.25%; 0.5%; 0.75% and 1%. Observations of bioplastic included appearance, elasticity, and solubility in water.

2.4. Optimization of concentration of carrageenan solution

The preparation was conducted by dissolved 2 g of chitosan powder into 100 ml of 1% acetic acid. The solution was heated on a hot stirrer using a rotational power of 600 rpm until it reached a temperature of 60ºC. Glycerol 0.5% was added to the chitosan solution after the solution temperature reached 60ºC. After 15 minutes of mixing, pour the carrageenan solution with different concentrations, 0.25%; 0.5%; 0.75%; and 1% with chitosan: carrageenan ratio of 10:1 (100 mL chitosan solution and 10 mL carrageenan solution).

The bioplastic solution was stirred for 30 minutes using a hot plate stirrer and the temperature was kept at 60ºC. The bioplastic solution was left in the solution for 24 hours at room temperature, this serves to remove air bubbles in the solution. The solution was poured into a mold and continue with oven processing. The oven process is carried out at a temperature of 70º-75ºC for 3-4 hours. The bioplastic removed from the mold and cooled at room temperature. This bioplastic was used to determine the characteristics covered thickness, density, water content, solubility, tensile strength, elongation, antibacterial activity and functional grup analysis.

2.5. Bioplastic characterization

2.5.1. Thickness measurement

The thickness of the bioplastic obtained was measured using a screw micrometer with an accuracy of 0.001 mm at five different point of bioplastic. The thickness value is measured from the average of the five thickness measurements.

2.5.2. Density test

Density in bioplastic serves as a marker of the quality of the mechanical properties of bioplastics. This test is carried out using a digital balance to obtain the weight of the bioplastic sample with a size of 4x2, then the sample is placed in a measuring cup filled with 5 ml of distilled water for ±1 minute, then calculated the difference in volume using the following equation:

$$\rho = m/v$$

information :
ρ = density (g/cm^3)
m = film mass (g)
v = difference between final volume and initial volume (cm^3 or mL)

2.5.3. Water content test
Testing the moisture content of bioplastics using the OHAUS MB120 Moisture Analyzer. The test is carried out by placing the sample above the pen handle as much as 0.5 g. Then the pen handle is placed on the sample chamber, make sure the pen handle is in the available hollow. Furthermore, setting the tool according to the needs of the analysis. After that, a display will appear indicating the ongoing drying process. The display on the screen will show the method name, temperature and drying time, current temperature, and moisture content.

2.5.4. Solubility test
Bioplastic sheets measuring 3 x 3 cm were weighed first to determine the dry weight, then put into a container containing distilled water and then allowed to stand for 24 hours. After 24 hours, dry at room temperature and then weighed. The percentage of solubility is calculated using the following formula:

$$\text{Percent Solubility} = (a\text{-}b)/a \times 100\%$$

Information:
a = initial weight (g)
b = final weight (g)

2.5.5. Tensile strength test
The test refers to the American Standard Testing Method (ASTM) D882 (2013) using the MPY testing machine. Bioplastic sheet size 7 x 15 cm with humidity (RH) 50% for 40 hours. The instron tool is installed on the initial grip separation of 50 mm, crosshead speed 50 mm/minute and loadcell 50 kg, the measurement uses the following formula:

$$\tau = (F\ max)/A$$

Information:
τ = tensile strength (Mpa)
F max = tensile strength (N)
A = cross-sectional area (mm2)

2.5.6. Elongation test
The elongation test refers to ASTM D882 (2013) using a MPY testing machine. The film sample used was 25 x 5 cm and was conditioned at 50% humidity (RH) for 48 hours. The instron equipment used is set at a crosshead speed of 5 mm/minute, initial grip separation of 50 mm, and loadcell of 50 kg. The calculation is carried out when the film is broken and is calculated by the following formula:

$$\text{Percent Elongation} = (l1\text{-}lo)/lo$$

Information:
lo = initial length
l1 = length after break up

IOP Conf. Series: Earth and Environmental Science 1289 (2023) 012039 doi:10.1088/1755-1315/1289/1/012039

2.5.7. Antibacterial activity

The method carried out by the paper disc diffusion method on *Eschericia coli* bacteria. The diameter of the inhibition zone was measured after incubation for 24 hours at 37°C. Observations were made by measuring the clear zone formed around the paper disk that had been dipped in a solution of chitosan and carrageenan bioplastics. The planting medium used is Nutrient Agar (NA), while the Nutrient Broth (NB) solution is used for bacterial growth media. 1 ose of bacteria was added to 10 mL NB and stirred using a shaker incubator for 24 hours at a speed of 150 rpm. After 24 hours, 1 mL of bacteria were grown in NA medium for 30 minutes. Place it on a plate and incubate for 24 hours at 37°C, then the zone of inhibition in mm diameter is calculated.

2.5.8. Function group analysis

This analysis is used to determine the functional groups contained in the resulting bioplastic film. The analysis is carried out using wave numbers between 400-4000 nm. The results obtained in the form of a diffractogram of the relationship between wavelength and intensity. The spectrum was recorded using an infrared spectrophotometer.

3. Results and Discussion

3.1. Optimization chitosan and carrageenan solution volume ratio

Table 1. Results of optimization of chitosan and carrageenan mixing ratio

Chitosan : carrageenan ratio	Solution homogeneity	Color clarity
5:1	There are many clots, large clumps	Murky color
10:1	There are fewer clots, small clots	Relatively clear color

Optimization is done to obtain a solution that produces fewer clots because it indicates the solution is homogeneous. Based on Table 1, the optimization ratio obtained using the mixing ratio of chitosan and carrageenan 10:1 showed bioplastic with better mechanical properties on homogeneity and color clarity. This volume ratio was used in the next step experiment.

3.2. Optimization of glycerol concentration

Table 2. Results of glycerol concentration optimization

Glycerol concentration	Appearance	Elasticity	Water solubility
0,25%	Clear color	Broken when folded in half	Slow soluble in water
0,5%	Yellowish color	Does not break when folded in half	Slow soluble in water
0,75%	Yellowish color	Does not break when folded in half	Slow soluble in water
1,00%	Yellowish color	Does not break when folded in half	Quickly soluble in water

Based on Table 2., glycerol concentration of 0.5% resulted better mechanical properties. Research of Hidayati et al [9] showed that a lower concentration of glycerol (0.25%) when applied in bioplastic that made from chitosan, CMC, tapioca, and glycerol. Differences in ingredients affect the effectiveness of the use of glycerol concentrations.

3.3. Optimization of carrageenan concentration in chitosan bioplastic

3.3.1 Effect of carrageenan concentration to analysis of chitosan bioplastic functional groups

Functional groups were observed using FTIR Method. Bioplastic chitosan with the addition of carrageenan concentrations of 0.25% and 0.50% formed bonds between chitosan and carrageenan. This result is evidenced by the presence of C=N group at these concentrations. The addition of carrageenan concentrations of 0.75% and 1% did not form a bond between chitosan and carrageenan probably due to in high concentration of carrageenan, there were aglomeration and when it was filtered before tested the clots that contain unbonded carrageenan were separated (Table 3).

Table 3. Results of the analysis of the function cluster

Component	Wave Number (cm^{-1})				
	-H (3200-3500)	N-H (1000-1350)	C-O (1000-1300)	C-H (2850-3000)	C=N (±2260-2340)
Carrageenan	3359,44		1023,02	2915,68	
Chitosan+Carrageenan 0%	3265,11	1027,15			
Chitosan+Carrageenan 0,25%	3259,45	1027,98	1152,00	2925,12	2341,78 2359,69
Chitosan+Carrageenan 0,5%	3259,28	1026,48	1151,36	2876,70	2359,82
Chitosan+Carrageenan 0,75%	3247,92	1026,05			
Chitosan+Carrageenan 1%	3263,95	1027,20			

3.3.2. Effect of variations carrageenan concentration on thickness of chitosan bioplastic

The thickness of chitosan bioplastics with the addition of carrageenans at concentrations of 0.00%, 0.25%, 0.75%, and 1.00% can be seen in Figure 1.

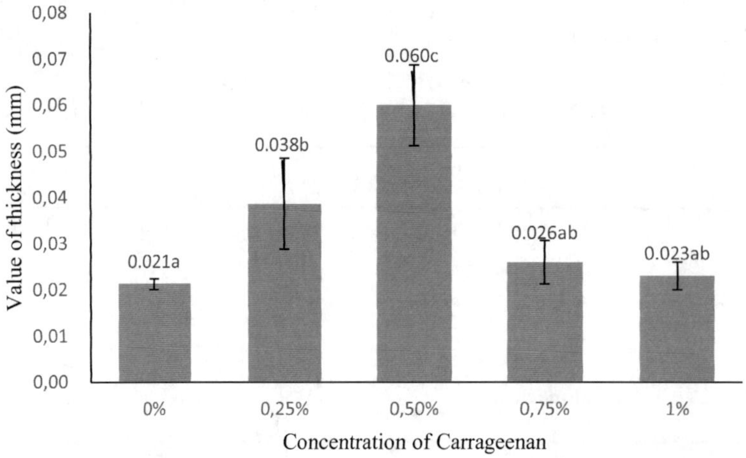

Figure 1. The effect of variations in carrageenen concentration on the thickness of chitosan bioplastics

Based on Figure 1. The thickness of the bioplastic mixture of chitosan and carrageenan increased from a concentration of 0% to 0.5%, then decreased at a concentration of 0.75% to 1%. The highest thickness bioplastic was obtained at the addition of 0.5% carrageenan concentration, which was 0.06 mm. All solutions were molded using a volume of 10 mL on 8x3x3cm concrete. The thickness of the bioplastic tends to increase with the addition of carrageenan concentration, the increase in film thickness is due to an increase in dissolved solids in the film along with the added carrageenan concentration. The thickness of the bioplastic is strongly influenced by the concentration of dissolved solids in the solution that forms the bioplastic. The number of carrageenan polymers is increasing in the bioplastic matrix, where each carrageenan monomer is connected to intermolecular hydrogen which causes cavities to form in the bioplastic matrix, so that the bioplastic becomes thicker.

The addition of carrageenan at concentrations of 0.75% and 1% caused the thickness of the bioplastic to decrease. The clumps formed in the chitosan solution with the addition of carrageenan produced when the mixing process is carried out can cause a decrease in film thickness [10]. The clumps formed are carrageenan which undergoes a gel formation process due to not reacting with glycerol to form cross-links. Carrageenan has hydroxyl (OH) and sulfate groups which cause it to be hydrophilic [11]. Carrageenan is a hydrocolloid compound that is able to absorb water and form a gel. The non-interacting glycerol causes the carrageenan gel to form lumps. The interaction between filler and matrix becomes weak [12]. The clumps will separate during the filtering process so that the dissolved solids of carrageenan cannot fill the film and there is no increase in dissolved solids in the film along with the addition of 0.75% and 1% carrageenan concentrations. The results at concentrations of 0.75% and 1% were not significantly different with 0% addition of carrageenan. The results obtained were supported by functional group analysis at concentrations of 0.75% and 1% carrageenan, there was no carrageenan group, so there was a decrease in the concentration of 0.75% and 1% carrageenan addition.

The results obtained have met the standard used, namely JIS (Japanese International Standard) with a thickness of <0.25 mm. Bioplastics with a sufficiently high thickness can provide better protection for packaged products, and will increase the tensile strength of bioplastics [8]. The results of this study are not much different from the results of the study of [7], bioplastic carrageenan with a concentration of 1% produces a film thickness of 0.041 mm and also does not differ much from the results study [13], making bioplastic from carrageenan with a concentration of 0.4% produces a thickness of 0.4%. 0.036mm film. Research belonging to [7] and [13] only used carrageenan without the addition of chitosan. Carrageenan solution mixed in chitosan solution will be more difficult to homogenize than when mixed with distilled water, this is because the chitosan solution has a higher viscosity than distilled water.

3.3.3 Effect of variations carrageenan concentration on density of chitosan bioplastic

The average value of chitosan bioplastic density with the addition of carrageenan at concentrations of 0%, 0.25%, 0.5%, 0.75%, and 1% can be seen in Figure 2.

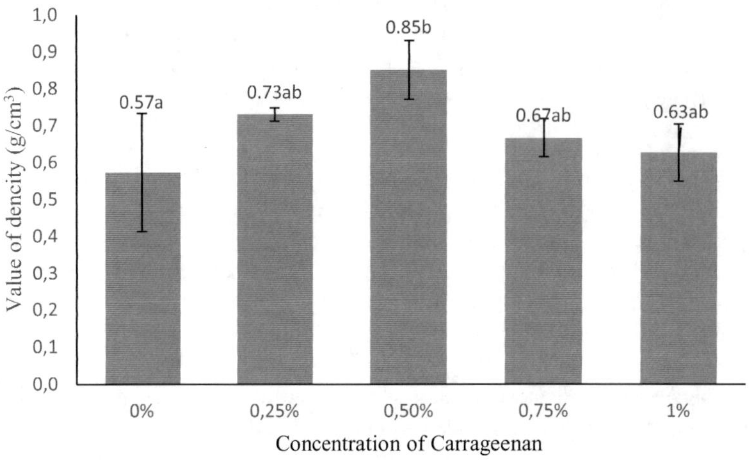

Figure 2. The effect of variations in carrageenen concentration on density of chitosan bioplastics

Based on Figure 2, the density values of chitosan and carrageenan bioplastics increased from a concentration of 0% to 0.5%, then decreased at a concentration of 0.75% to 1%. The highest bioplastic density was obtained at the addition of 0.5% carrageenan concentration, which was 0.85 g/cm3. The density value of bioplastics tends to increase along with the addition of carrageenan concentration, the increase in the value of film density is because along with the addition of carrageenan concentration, the cavities or pores in bioplastics will be increasingly filled with carrageenan so as to make the density value of bioplastics better [14]. The density value in the resulting bioplastic matrix causes the cross-linking between polymers to increase [7]. The increase in the density value is also influenced by the properties of carrageenan, namely the function of colloid compounds as suspenders and thickeners, as well as the interaction between the constituent materials of bioplastics [15]. Bioplastics that have a high density value, the mechanical properties will be better in maintaining the packaged product, the denser the bioplastic produced, the fewer cavities or pores of the bioplastic produced [16].

The addition of carrageenan concentrations of 0.75% and 1%, the density value of bioplastics decreased significantly, this was influenced by the addition of carrageenan concentrations and could be caused at concentrations of 0.75% and 1% to form lumps during the stirring process, due to the increasing concentration of carrageenan [10]. The clumps formed as a result of the bond between chitosan and carrageenan are not filtered, so that at concentrations of 0.75% and 1% the interaction between the constituent materials of bioplastic becomes weak and causes the density value in the matrix to be low. The fewer hydroxyl groups (-OH) resulted the lower the density value [16]. Heriyanto et al [17] said that the gel strength of carrageenan is strongly influenced by the concentration of the solvent.

The density results obtained do not meet the standard used, namely SNI 7188.7:2016 with a density value of 0.95 g/cm3. The density of bioplastics will be related to the permeability of bioplastics to air so that less dense conditions will cause bioplastics to be less able to protect food. There needs to be an effort to increase the density by adding crosslinkers derived from oligosaccharides that have a large number of hydrogen bonds, so that they can absorb water at high humidity [6]. The density value of carrageenan that is closest to SNI 7188.7:2016 is the addition of 0.5% concentration is 0.85 g/cm3. The results of this study are still higher than the Darni et al study result [16] which made bioplastics from sorghum and Eucheuma spinosium at a ratio of 8.5:1.5 resulted in a density of 0.125 g/cm3.

3.3.4. Effect of variations carrageenan concentration on water content of chitosan bioplastic

The water content of chitosan bioplastics with the addition of carrageenans at concentrations of 0%, 0.5%, 0.25%, 0.75%, and 1% can be seen in Figure 3.

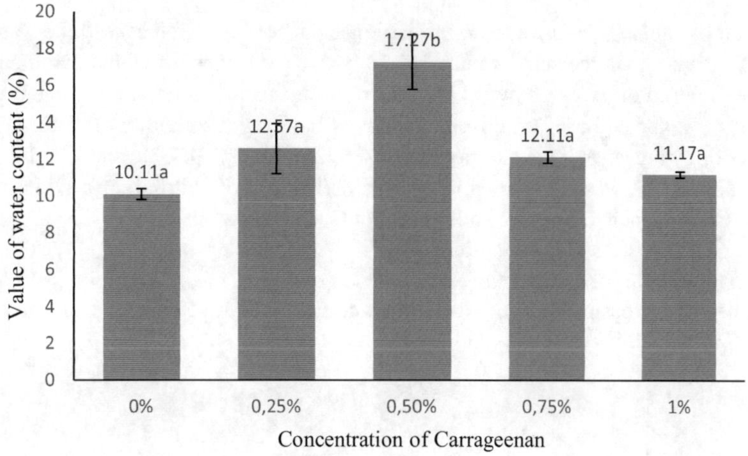

Figure 3. The effect of variations in carrageenan concentration on the water content of chitosan bioplastics

Based on Figure 3. The results showed that the water content of chitosan:carrageenan bioplastic increased from a concentration of 0% to 0.5%, then decreased at a concentration of 0.75% to 1%. The highest water content of bioplastic was obtained at the addition of 0.5% carrageenan concentration, which was 17.27%. The value of the water content of chitosan bioplastic with the addition of carrageenan increased from the addition of 0% carrageenan concentration to the addition of 0.5% carrageenan concentration, then decreased significantly at the addition of 0.75% and 1% carrageenan concentrations. The value of the water content of bioplastics tends to increase along with the addition of carrageenan concentration, the increase in the value of water content is due to the fact that carrageenan has a large number of hydroxyl groups (-OH) so that the higher the concentration of carrageenan added, the greater the water absorption, this causes lower water resistance and water absorption process occurs [18].

Another influencing factor is that the carrageenan crosslinker has properties that can bind to water (hydrophilic) so that the water content possessed by each material can cause an increase in water content as the concentration of carrageenan increases. The increase in water content is caused by the greater molecular weight so that it can cause the gap between molecules to get bigger and water molecules can be inserted [19]. The water content of bioplastics has a function in maintaining the stability of the packaged product, low water content is expected not to contribute water to the product which can have an impact on product damage [8]. The amount of water content contained in carrageenan can affect the growth of microorganisms, where the higher the water content, the growth of microorganisms, fungi, and bacteria will increase [21]. Bioplastic has a water content that is not too high so as not to contribute water to the packaged product [8].

The addition of carrageenan concentration of 0.75% and 1% of water content in bioplastics decreased significantly. This can be due to the formation of more lumps when the mixing process is carried out [10]. The clumps formed due to the bond between chitosan and carrageenan are not filtered, so at a concentration of 0.75% and 1% the interaction between the constituent materials of bioplastics becomes weak, as a result, during the oven process, the thickness becomes thinner, the water content at the concentration of the addition of carrageenan is reduced. 0.75% and 1% became less because the amount of carrageenan which tends to be hydrophilic is less. Chitosan is hydrophobic so it is difficult to dissolve in water. The water content of chitosan bioplastics

The results obtained with the addition of carrageenan concentration caused the water content value to be slightly higher than the standard used, namely the FAO standard with a maximum water content value of 12%. Based on research by [8] bioplastic carrageenan 1%-3% water content produced ranges from 17.14% to 20.86%. Based on this, the results of the overall solubility of chitosan bioplastics with the addition of carrageenan concentrations of 0%, 0.25%, 0.5%, 0.75%, and 1% did not differ much from the study, which was better than research [8]. The use of chitosan affects the water content of bioplastics. Chitosan is hydrophobic so it tends not to absorb water.

3.3.5. Effect of variations carrageenan concentration on solubility of chitosan bioplastic
The solubility of chitosan bioplastics with the addition of caragenants at concentrations of 0.00%, 0.25%, 0.75%, and 1.00% can be seen in Figure 4.

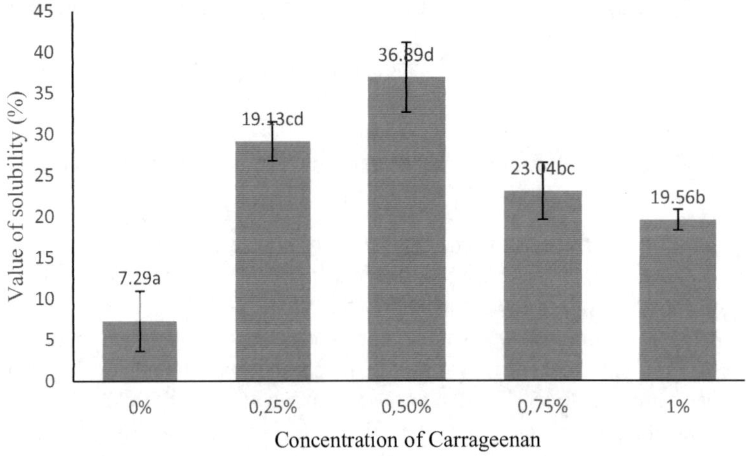

Figure 4. The effect of variations in carrageenan concentration on the solubility of chitosan bioplastics

Based on Figure 4. The results obtained that the solubility value of chitosan and carrageenan bioplastics increased from a concentration of 0% to 0.5%, then decreased at a concentration of 0.75% to 1%. The highest bioplastic solubility was obtained at the addition of 0.5% carrageenan concentration, which was 36.89%. The solubility value of bioplastics tends to increase with the addition of carrageenan concentration, the increase in the solubility value of the film is because carrageenan has properties that can bind to water (hydrophilic) so that when immersed in water it will absorb water from the outside. The solubility of carrageenan in water is influenced by other factors such as storage temperature, type of carrageenan, pH, and other solutes [6]. The increase in the solubility value is also influenced by the properties of carrageenan, namely the function of colloid compounds as suspenders and thickeners, as well as the interaction between bioplastic constituents [15]. The solubility value of bioplastics shows that these bioplastics are easily degraded in nature so that they can be used as primary packaging for ready-to-eat food that has low water activity [8]. The higher the solubility value, the easier it will be to dissolve in water, so it can be said that the quality is not good as a liquid product packaging [9]. Bioplastics that have a high solubility value are more suitable for ready-to-eat food products because

IOP Conf. Series: Earth and Environmental Science 1289 (2023) 012039 doi:10.1088/1755-1315/1289/1/012039

they can dissolve when consumed [22]. The solubility of a bioplastic is closely related to its biodegradability and water content.

The addition of carrageenan concentrations of 0.75% and 1% of the solubility value of bioplastics decreased which was not too significant, this could be due to concentrations above 0.5% forming more clumps when the mixing process was carried out [10]. The clumps formed due to the bond between chitosan and carrageenan were not filtered, so that at a concentration of 0.75% and 1% the interaction between the ingredients of the bioplastic was weak. The amount of carrageenan concentration contained in the solution is reduced so that the hydrophilic properties of bioplastics become weak and tend to be hydrophobic. The nature of carrageenan which tends to be hydrophilic causes higher water absorption, while the hydrophobic chitosan bioplastic solution causes bioplastics to be difficult to dissolve in water [7]

Research by Dwimayasanti and Bayu [6], using 2% chitosan and 1-2.5% carrageenan with 2% glycerol resulted in the solubility values obtained ranging from 63%, 73%, 62%, and 77%. The solubility of bioplastics does not always increase with increasing carrageenan concentration. The results of the solubility of 2gr/100mL chitosan bioplastic with 0.5% glycerol and the addition of 0%-1% carrageenan concentration were better than the three previous studies. Low solubility is an important requirement when used as food packaging in general [8].

3.3.6. Effect of variations carrageenan concentration on antibacterial activity of chitosan bioplastic

The mean value of the growth inhibition zone for the antibacterial activity of chitosan bioplastic with the addition of carrageenan at concentrations of 0%, 0.25%, 0.5%, 0.75%, and 1% using a negative control in the form of NB solution and a positive control in the form of a only chitosan solution without glycerol addition can be seen in Figure 5.

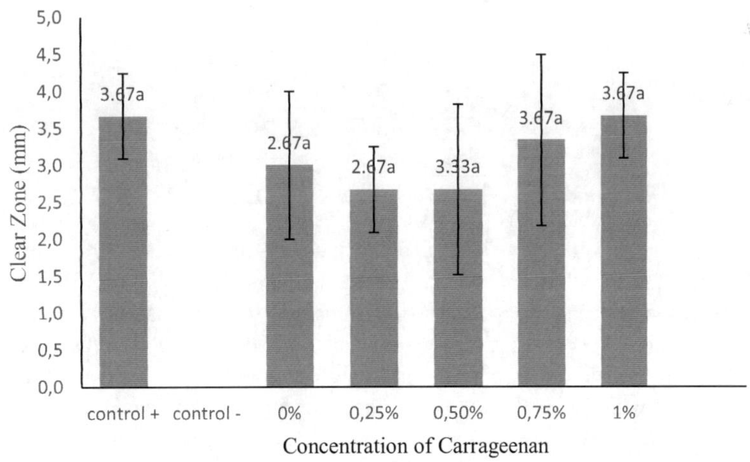

Figure 5. The effect of variations in carrageenen concentration on the antibacterial properties of chitosan bioplastics

Based on Figure 5. the results showed that an inhibitory zone of antibacterial activity was formed in the positive control solution, namely chitosan solution without the addition of glycerol, then in the control solution - namely the NB solution, no bacterial inhibition zone was formed. Chitosan bioplastic with the addition of carrageenan formed inhibition zones at all concentrations of carrageenan addition. The addition of carrageenan concentrations of 0%, 0.25%, and 0.5% inhibition zones experienced an insignificant decrease from the positive control used but at a concentration of 0.75% the inhibition zone formed was close to the positive control used while at 1% the addition of carrageenan The inhibition zone formed was the same as the positive control. The negative control solution did not form an inhibition zone because NB solution was the most common medium used to grow most bacteria [22].

Chitosan is a chelating agent that can bind metal ions so that it can inhibit toxin production and microbial growth. Chitosan also has the ability to inhibit enzymes and bind water [23]. The inhibitory value of the antibacterial activity of bioplastics tends to decrease with the addition of 0.25% and 0.5% carrageenan concentrations because the carrageenan crosslinker has properties that can bind to water (hydrophilic) and is negatively charged so that at that concentration the chitosan solution which has an amine group interacts with the charge. negative for carrageenan. Chitosan solution that has been bound to carrageenan molecules will be entangled in the matrix so that the antibacterial activity tends to decrease [24]. Another cause of the decrease in the inhibition zone formed is the water content possessed by each ingredient which can cause an increase in water content as the concentration of carrageenan is given. The increase in water content causes the gap between the molecules to get bigger and water molecules can be inserted, thereby reducing the ability of chitosan bioplastics to bind water and inhibit the growth of microorganisms [8]

The addition of carrageenan concentrations above 0.5% inhibition zone of antibacterial activity increased again, this was due to the presence of clumps between chitosan and carrageenan so that they formed lumps when the mixing process was carried out. The clumps formed will separate during the filtering process so that the chitosan solution can bind to bacteria more optimally. The category of bacterial inhibition zone can be seen in Table 4.

Table 4. Categories of bacterial inhibition zone diameter [25]

Diameter	Inhibitory power
≤ 5 mm	Weak
6-10 mm	Medium
11-20 mm	Strong
≥ 21 mm	Very strong

The average diameter of the inhibition zone obtained in this study when categorized in the diameter of the inhibition zone is in Table 6. The results of bioplastics obtained by the addition of carrageenan concentrations of 0%, 0.25%, 0.5%, 0.75%, and 1% included into the category of diameter 5 mm, namely the strength of the inhibition is weak. The results obtained on chitosan bioplastics with the addition of carrageenan concentrations of 0%-1% compared to previous research, [23] the results obtained on the inhibition of E.coli bacteria ranged from 0.139-0.625 mm. The results of the chitosan bioplastic research with the addition of carrageenan concentrations were better than previous studies.

Chitosan bioplastic with the addition of carrageenan concentration was significantly different between treatments in the negative control solution, so it can be said that chitosan was proven to have antibacterial properties, but the comparison of the positive control chitosan solution to the chitosan bioplastic solution with the addition of carrageenan concentration had no effect on the ability of chitosan bioplastics in inhibiting antibacterial activity.

3.3.7. The effect of variations in carrageenan concentration on the tensile strength of chitosan bioplastics

The tensile strength of chitosan bioplastic with the addition of carrageenan at concentrations of 0%, 0.25%, 0.5%, 0.75%, and 1% is shown in Figure 6. Based on Figure 6, the results showed that the tensile strength of chitosan and carrageenan bioplastics increased from a concentration of 0% to 0.5%, then decreased at a concentration of 0.75% to 1%. The highest tensile strength of bioplastic was obtained at the addition of 0.5% carrageenan concentration, which was 12.11 MPa. Chitosan bioplastic with the addition of carrageenan was not significantly different from the addition of the other four concentrations of carrageenan. This could be due to the effect of using the concentration of glycerol. In the research of Dwimayasanti and Bayu [6], the use of 2% glycerol concentration just resulted in significantly different bioplastics between treatments. This is because the glycerol molecules enter the polymer matrix and turn it into an irregular flexible structure. Hidayati et al [9] stated that the factor that affects the tensile strength of the film is the concentration of glycerol.

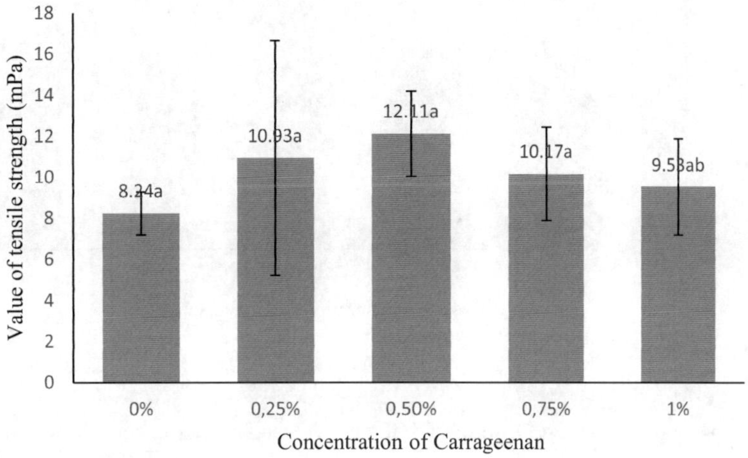

Figure 6. The effect of variations in carrageenan concentration on the tensile strength of chitosan bioplastics

Carrageenan has a lot of hydroxyl (OH) groups so that the higher the concentration of carrageenan added, the greater the water absorption, this produces a gel matrix that can increase the tensile strength of bioplastics [26]. The increase in the tensile strength value is due to the increase in the intermolecular bonds that make up the bioplastic so that a more compact bioplastic is obtained [8]. Another influencing factor is that carrageenan as a mixing biopolymer can increase the molecular interaction between carrageenan and glycerol in the matrix, so that a greater force is needed to break the bond [7]. The addition of carrageenan concentrations above 0.5% of the tensile strength value in bioplastics experienced a significant decrease, this could be due to the influence of the use of plasticizers, where glycerol molecules in the polymer matrix through hydrogen bonds might disrupt the polymer structure and then turn it into an irregular flexible structure and is said to be a restructuring in the polymer matrix with increased chain mobility thereby reducing resistance to stress [27]. Bioplastic which has a high tensile strength value is one of the requirements as a good packaging material because it is considered to be able to protect the product more optimally [8].

The results obtained do not meet the standard used, namely SNI 7188.7:2016 with a tensile strength value of 24.7-3022 mPa. Efforts need to be made to increase the tensile strength by adding crosslinkers from the oligosaccharide group with short chains. The research by [6] using 2% chitosan and 1-2.5% carrageenan with 2% glycerol yielded tensile strength values of 56.20%, 48.84%, 117.35%, and 53.73% respectively. The tensile strength value produced by this study is still low when compared to previous studies.

3.3.8. The effect of variations in the concentration of chitosan and carrageenan on the elasticity of bioplastics

The elasticity of chitosan bioplastic with the addition of carrageenan at concentrations of 0%, 0.25%, 0.5%, 0.75%, and 1% can be seen in Figure 7.

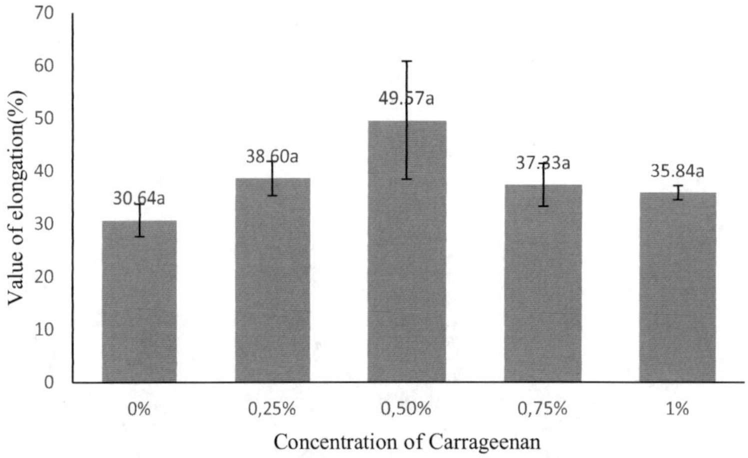

Figure 7. Effect of variation of carrageenan concentration on chitosan bioplastic elasticity

Based on Figure 7. the results showed that the chitosan and carrageenan bioplastics increased from a concentration of 0% to 0.5%, then decreased at a concentration of 0.75% to 1%. The highest bioplastic slackness was obtained at the addition of 0.5% carrageenan concentration, which was 46.03%. Chitosan bioplastics with the addition of carrageenan concentrations were not significantly different between treatments, so it can be said that the addition of carrageenan concentrations had no effect on the ability of chitosan bioplastics to affect the increase in elongation of chitosan bioplastics. This can be caused by the influence of the type and concentration of plasticizer used because it can affect the ability of its function in the bioplastic network which in turn will determine the distribution and interaction in the carrageenan network. Glycerol can cause the elongation value of bioplastics to vary due to its ability to affect the interaction between water molecules with chitosan and carrageenan. In addition, glycerol is hygroscopic which can affect the strength between macromolecules [6].

The results obtained with the addition of 0%-1% carrageenan concentration have met the standard used, namely the SNI 7188.7:2016 standard with an elongation value ranging from 21%-220%. Based on the research of [6] using 2% chitosan and 1-2.5% carrageenan with 2% glycerol resulted in slack values of 52%, 61%, 17.5%, and 27% respectively. The fatigue value produced by this study is still low when compared to previous studies, but it has met the SNI standard.

4. Conclusion

Based on the results of overall bioplastic parameter, the use of carrageenan 0% to 0.5% resulted in the average value of film thickness, density, water content, and solubility increasing and then decreasing again with the use of concentrations of 0.75% and 1%. It is still necessary to do further research in order to increase the tensile strength by adding other hydrocolloid polymers to increase the tensile strength of bioplastics up to meet the standart.

5. Acknowledgement
This research was able to be carried out with financial support from the Final Project Recognition Program Grant (RTA) with grant number 3143/UN1.P.III/DIT-LIT/PT/2021 from Universitas Gadjah Mada. Our thanks also go to Atika Dyah Kusumastuti and Fiskina Zulfa Kamala who helped with the data analysis.

References

[1] Gade R, Tulasi M S, Bhai V A 2013 Seaweed: A novel biomaterial *International Journal of Pharmacy and Pharmaceutical Sciences* **5** 2: 40-44.

[2] Coniwanti P, Laila L, Alfira M 2014 Biodegradable plastic films from corn starch with the addition of chitosan and glycerol as plasticizers *Journal of Chemical Engineering* **4** 20 pp 22-30

[3] Costa, L P M, D M V Miranda, A C C de Oliveira, L Falcon, M S S Pimenta, I G Bessa, S J Wouters, M H S Andrade, J S Pinto 2021 Capture and Reuse of Carbon Dioxide (CO2) for a Plastics Circular Economy: A Review *Processes* **9** 759: 1-79. https://doi.org/10.3390/pr9050759

[4] Katili S, Harsunu B T, Irawan S 2013 Effect of concentration of glycerol plasticizer and composition of chitosan in solvent on physical properties of chitosan edible film *Journal of Technology* **6** 1 pp 29-38

[5] Viorica N, Olteanu S M, Spriroiu M F, Pincu E, Meltzer V 2011 Strach/chitosan film forming hydrogel *Revue Rumaine de Chimie* **56 8 pp** 827-832

[6] Dwimayasanti R, Bayu K 2019 Characterization of layer-by-layer assembly of carrageenan and chitosan-based edible film *Jurnal Pascapanen dan Bioteknologi Kelautan dan Perikanan* **14** 2 pp 141-150

[7] Maryuni A E, Mangiwa S 2018 Karakterisasi bioplastik dari karaginan dari rumput laut merah asal Kabupaten Biak yang dibuat dengan metode blending menggunakan pemlastis sorbitol *Jurnal Avogadro* **2 1 pp** 9-16

[8] Rusli A, Metusalach, Salengke, Tahir M M 2017 Karakterisasi edible film karagenan dengan pemlastis gliserol *Jurnal Pengolahan Hasil Perikanan Indonesia* **20 2 pp** 219-229

[9] Hidayati S, Zulferiyenni, Satyajaya W 2019 Optimization of biodegradable film from cellulose of seaweed solid waste Eucheuma cottonii with addition of glycerol, chitosan, CMC, and tapioca *Indonesian Journal of Fishery Products* **22** 2 pp 340-354

[10] Nafiyanto, I 2019 Pembuatan plastik biodegradable dari limbah bonggol pisang kepok dengan plasticizer gliserol dari minyak jelantah dan komposit kitosan dari limbah cangkang bekicot (*Achatina fullica*) *Integrated lab Journal* ISSN 2339-0905

[11] Supeni G, Cahyanintyas A A, Fitrina A 2015 Karakterisasi sifat fisik dan mekanik penambahan kitosan pada edible film karagenan dan tapioka termodifikasi *Jurnal Kimia Dan Kemasan* **37 2 pp** 103-110.

[12] Lusiana S W, Putri D, Nurazizah I Z, Bahruddin 2019 Bioplastic properties of sago-PVA strarch with glycerol and sorbitol plasticizers *Journal of Phyics* 1-8

[13] Handito, D 2011 Pengaruh konsentrasi karaginan terhadap sifat fisik dan mekanik bioplastik *Jurnal Agroteksos* **21 pp**

[14] Arizal V D. Yuli, Edwin A, Lia L, Heri U 2017 Application *Eucheuma cottonii* in bioplastic synthesis based on sorghum with plasticizers glycerol *Proceedings of the National Seminar on Industry Riset* pp 32-39

[15] Galus S, Lenart A 2013 Development and characterization of composite edible films based on sodium alginate and aectin *Journal of Food Engineering* **115 5 pp** 59- 465

[16] Darni Y, Lia L, Muhammad H, Sarkowi, Dita S 2019 Peningkatan kuat tarik bioplastik dengan *filler microfibrillated cellulose* dari batang sorgum *Jurnal Teknik Kimia Indonesia* **18** 2 pp 37-41

[17] Heriyanto H, Indar K, Denni KS 2018 The effect of temperature and time of extraction on the quality of Semi Refined Carrageenan (SRC) *MATEC Web of Conferences* 154, 01034

[18] Aritonang D H, Hartiati A, Harujuwono B A 2020 Karakteristik komposit bioplastik pada variasi rasio pati ubi talas belitung (*Xanthosoma sagittifolium*) dan karagenan *Jurnal Rekayasa dan Manjamen Agroindustri* **8** 3 : 348-359

[19] Sitompul A J W S, Zubaidah E 2017 Pengaruh jenis dan konsentrasi plasticizer terhadap sifat fisik edible film kolang kaling (*Arenga pinnata*) *Jurnal Pangan dan Agroindustri* **5 1 pp** 13- 25.

[20] Fransiska, D, Giyatmi, Irianto H E, M. Darmawan, Melanie S 2018 Karakteristik film k-karaginan dengan penambahan plasticizer polietilen glikol *JPB Kelautan dan Perikanan* **13** 1 pp 13-20.

[21] Pitak N, Rakshit S 2011 Physical and antimicrobial properties of banana flour/chitosan biodegradable and self sealing films used for preserving fresh cut vegetables *LWT-Food Science and Technology* **44 pp** 2310–2315

[22] Thohari N M, Pestariati, Istanto W 2019 Pemanfaatan tepung kacang hijau (*Vigna radiata L.*) sebagai media alternatif NA (*Nutrient Agar*) untuk pertumbuhan bakteri *Eschericia coli Analisis Kesahatan Sains* **8 2 pp** 725-737

[23] Winarti C, Miskiyah, Widaningrum 2012 Teknologi produksi dan aplikasi pengemas *edible* antimikroba berbasis pati *J. Litbang Pert.* **31 3 pp** 85-93

[24] Surjowardojo P, Tri E, Gabriel R 2015 Daya hambat deok kulit apel manalagi (*Malus sylvesters Mill.*) terhadap pertumbuhan *Staphylococcus aureus* dan *Pseudomonas sp*. penyebab mastitis pada sapi perah *Jurnal Ternak Tropika* **16** 2 pp 40-44

[25] Sorour M A, Abdou E S 2014 Preparation and characterization of starch/carrageenan edible films *International Food Research Journal* **21 1 pp** 189-193

[26] Afifah N E, Sholichah N, Indriati, Darmajana D A 2018 Pengaruh kombinasi plasticizer terhadap karakteristik edible film dari karagenan dan lilin lebah *Biopropal Industri* **9** 1 pp 49-60.

[27] Gontard N, Guilbert, and Cuq J L 1992 Edible wheat gluten films: influence of main process variables on film properties using response surface methodology *Journal of Food Science* **57** pp 190-195.

IOP Conf. Series: Earth and Environmental Science 1289 (2023) 012040 doi:10.1088/1755-1315/1289/1/012040

Cinnamon and amylum mucilage alter the physicochemical, sensory, and antioxidant activity of *Arthrospira platensis* granules

A W Elmaghriba, M D Candraningsih and N Ekantari*
Laboratorium of Fish Processing Technology, Fisheries Department, Faculty of Agriculture Universitas Gadjah Mada, Yogyakarta, Bulaksumur, Yogyakarta
*corresponding author email: nurfitri@ugm.ac.id

Abstract. This study was done to find out how creating *Arthrospira platensis* granules using the wet granulation method and a blend of cinnamon powder and amylum mucilage affected their physical, physicochemical, and antioxidant properties. *Arthrospira platensis* is granulated by the wet granulation method using cinnamon, amylum mucilage, and stevia sugar as sweeteners. The 3x3 factorial design was used, with factors for amylum mucilage of 2.5%, 5%, and 7.5% w/w and cinnamon powder of 0.5%, 1%, and 1.5% w/w. Granules were tested for their physical characteristics (moisture content and pH), physicochemical characteristics (water holding capacity, oil holding capacity, solubility, foam ability, emulsion activity, and stability), and antioxidant activity using the DPPH-RSA and FRAP procedures. The results demonstrated that all granules had moisture contents that met the required standard of less than 7%. *Arthrospira platensis* dry biomass can be processed into granules, which can raise the OHC but reduce the WHC, solubility, foam ability, emulsion activity, emulsion stability, and antioxidant ability. Amylum mucilage affected the solubility of *Arthrospira platensis* granules, while cinnamon powder had an impact on the pH level, solubility, and activity of the emulsion. Consumer preferences and buying intention of granules are higher than dry biomass *Arthrospira platensis*. The combination of 1.5% cinnamon powder and 5% amylum mucilage is the optimal treatment based on pH, solubility, and emulsion activity, making it appropriate for fortified in food product.
Keywords: granulated, *Arthrospira platensis*, cinnamon, amylum mucilage, stevia sugar.

1. Introduction

Arthrospira platensis is a species of cyanobacteria microalga that contains nutrients such protein, essential fatty acids, vitamins, minerals, and antioxidants in the form of phycocyanins [1]. Additionally, compared to calcium-rich milk or calcium carbonate, *A. platensis* calcium content has a higher bioavailability [2]. Capsules, pills, and dry biomass (powder) are the commercially available forms of *A. platensis*. *A. platensis* is typically used in culinary products as dry biomass (powder). According to [3], *A. platensis* powder is utilized as a food additive for instant drink mixes, sauces, soups, chocolate bars, noodles, cookies, bread, and biscuits. Added of 10% and 15% concentrations of *A. platensis* reduced consumer preference for the aroma and flavor of chocolate bars. The addition of *Arthrospira platensis* powder causes a change in the color of the product to green, the smell becomes fishy and tastes bitter [4], It is necessary to make Arthrospira preparations other than powder form, namely granule form, which is anticipated not to cause off-flavor when added to food products. In addition, the granule form is typically more physically stable, i.e., denser so that when mixed in the product it does not change the color of the product and is resistant to oxidation. Granulation is used because of the presence of a binder that can retain the bioactive components of *A. platensis*. Processing of *A. platensis* into granule form can be done to increase food diversity. The granulation method used is wet granulation because the resulting granules are more compact than dry granulation. The wet granulation method entails adding a binder to the powder in the granulator during the granulation process, allowing particle agglomeration to take place [5]. This method is used to increase the size of powder particles. The result of the particle agglomeration process is an increase in the granules strength and hardness, making them more difficult to crush.

The binder used in this granulation process is amylum mucilage. Amylum mucilage was selected for this investigation because it has the benefit of having high gelatinization characteristics and is anticipated to be able to bind green colors and active compounds from *Arthrospira platensis*. Tapioca starch is a possible filler component in the wet granulation process. The binder used is made from amylum tapioca starch because it has a higher percentage of amylopectin content than other starch sources [6]. The physical and physicochemical characteristics of granules are affected using amylum mucilage as a binder to create granule texture.

At the time of making granules, a masking agent can be added. Research by [7] states that noodle products, flakes, cookies added to *Arthrospira platensis* powder produce green products and the addition of cinnamon to these products can mask the off-flavor of Arthrospira. Based on research by [8], the addition of cinnamon to *Arthrospira platensis* ice cream can mask the bitter taste and aftertaste of Arthrospira. The masking agent used in this study was cinnamon, because cinnamon has a sweet taste and contains sweet aromatic compounds and volatile compounds such as cinnamaldehyde which have been shown to cover the off-flavor taste in *Arthrospira platensis*. An intermediate product with promise as a functional additive is called granules. *Arthrospira platensis* granules can be made as an option to create preparations of *Arthrospira platensis* that do not impart off odors when added to food products but still preserve the antioxidant and nutrient content of *Arthrospira platensis*.

Therefore, it is necessary to test the physical, physicochemical, and sensory characteristics of the granules. It is hoped that this research can increase consumer preference for *Arthrospira platensis* preparations, so that the granules can be added to commercial food products and can be widely applied to the food industry.

2. Material and Method

2.1 *Material*

The materials used in this study included *Arthrospira platensis* powder (PT. Transpangan Spirulindo), tapioca starch (Rose Brand), cinnamon, stevia sugar (Stevigrow®), and mineral water (Aqua).

2.2 *Method*

A 3x3 Completely Randomized Design (CRD) with two factors—cinnamon concentration (0.5%; 1%; 1.5%) and amylum mucilage concentration (2.5%; 5%; 7.5%)—was employed in this study. Table 1 displays the combination of formulas for the *Arthrospira platensis* granules.

Table 1. Experimental designed (combine formula of *Arthrospira platensis* granules)

Independent variabel	Amylum mucilage		
	M1 (2.5%)	M2 (5%)	M3 (7.5%)
cinnamon (C1) 0,5%	C1M1 (0.5%:2.5%)	C1M2 (0.5%:5%)	C1M3 (0.5%:7.5%)
cinnamon (C2) 1%	C2M1 (1%:2.5%)	C2M2 (1%:5%)	C2M3 (1%:7.5%)
cinnamon (C3) 1,5%	C3M1 (1.5%:2.5%)	C3M2 (1.5%:5%)	C3M3 (1.5%:7.5%)

2.2.1 *Formulation of Arthrospira* Granules

The treatment of the *Arthrospira platensis* granule formulation used was a combination of cinnamon and amylum mucilage shown in Table 2. The binder used was amylum mucilage with concentrations of M1 (2.5%), M2 (5%), and M3 (7.5%). The concentration of mucilage was chosen based on a study by [9]. The concentration of cinnamon powder used was C1 (0.5%), C2 (1%) and C3 (1.5%). Concentration of cinnamon powder based on research by [7] using cinnamon powder above 2.5% can leave a strong cinnamon aroma, thereby reducing consumer preferences. Cinnamon powder is used as a masking agent for the fishy smell of *Arthrospira platensis*. The same concentration of *Arthrospira platensis*, which is equal to 10% according to research by [3], added *Arthrospira platensis* to lozenges as much as 10% and has the highest antioxidant content, namely 28.67%.

The sweetener used is stevia sugar. The concentration of stevia sugar used is 4%, according to research by [10] stevia sugar of 4% is the most preferred concentration of panelists because it can mask the bitter taste. The filler material used is tapioca flour with a ratio of tapioca flour and water of 5:3. The addition of 50% tapioca flour to make the granules more voluminous [11]. The addition of 30% water based on the results of preliminary trials to obtain the consistency of a granular dough form. Amylum mucilage is prepared by dissolving tapioca starch in hot water with a ratio of 10% tapioca flour and 90% hot water, then stirring until homogeneous [12].

Table 2. The formulation of granules Arthrospira

Ingredients	C1M1	C1M2	C1M3	C2M1	C2M2	C2M3	C3M1	C3M2	C3M3
Arthrospira platensis	10%	10%	10%	10%	10%	10%	10%	10%	10%
Cinnamon powder	0.5%	0.5%	0.5%	1%	1%	1%	1.5%	1.5%	1.5%
Amylum mucilage	2.5%	5%	7.5%	2.5%	5%	7.5%	2.5%	5%	7.5%
Stevia	4%	4%	4%	4%	4%	4%	4%	4%	4%
Starch	51.9%	50.3%	48.7%	51.6%	50%	48.4%	51.25%	49.7%	48.1%
Water	31.1%	30.2%	29.25%	30.9%	30%	29.1%	30.75%	29.8%	28.9%

2.2.2. *Preparation of Arthrospira* Granules

Arthrospira powder, amylum mucilage, cinnamon powder, stevia sugar, tapioca, and water are mixed until smooth. The ready wet dough was sieved through an 8-mesh sieve and dried in an oven at 55°C for 24 hours. Furthermore, the dry granules were sifted through a 14-mesh sieve.

2.2.3. *Moisture content determination*

Testing the moisture content of *A. platensis* granules was carried out using a tool, namely a Moisture Analyzer (OHAUS, MB 120) based on the principle of thermogravimetry.

2.2.4. *Degree of Acidity (pH) determined.*

The working principle of pH testing is to measure the hydrogen ion activity of the sample potentiometrically using a pH meter. Sample solution concentration of 10% (gram/mL)

2.2.5. *Water Holding Capacity determined.*

Water Holding Capacity (WHC) test is to determine the water holding capacity of the sample by comparing the amount of water bound in 1 gram of sample. Water Holding Capacity (WHC) is calculated by finding the ratio between the mass of centrifuged pellets and the mass of the sample used [13].

2.2.6. *Oil Holding Capacity determined.*

Oil Holding Capacity (OHC) test is to determine the binding capacity of the oil by the sample by comparing the amount of oil bound in 1 gram of the sample. The oil used is olive oil. Oil Holding Capacity (WHC) is calculated by finding the ratio between the mass of centrifuged pellets and the mass of the sample used [13].

2.2.7. *Solubility determined*

The solubility test is to calculate the percentage of the number of samples dissolved using filter paper. Calculation of the solubility value is by calculating the percentage ratio between the dissolved sample mass and the initial sample mass.

2.2.8. Foam Ability

Foam ability or the ability to form foam is to measure the amount of foam formed (mL) in a sample solution that has been homogenized using a standardized blender or vortex. Foam ability test refers to research by [14] with modifications to the volume of distilled water used.

2.2.9. *Emulsion Activity*

The principle of emulsion activity testing is to determine the ability of the emulsifier to form an emulsion between oil and water which is indicated as the surface area per unit protein mass unit (m2/gram) [15]. The emulsion activity test was carried out by preparing a sample solution with a concentration of 1% (gram/mL). The sample solution and corn oil were mixed in a ratio of 3:1, then vortexed for 2 minutes. Measurement of emulsion activity begins at minute 0, or when the emulsion has just been formed by the spectrophotometric method. The cuvette contains 20 µL of emulsion and 2 mL of 0.1% SDS (Sodium Dodecyl Sulphate) solution, then it is measured using a UV-Vis Spectrophotometer at a wavelength of 500 nm.

2.2.10. *Emulsion Stability*

Emulsion stability testing is to determine the ability of the emulsion formed between oil and water to remain in a dispersed state. The process of testing the stability of the emulsion begins with making a 1% concentration sample solution. Sample solution with corn oil mixed with a ratio of 3:1, then vortexed for 2 minutes. After that, it was heated using a hot plate stirrer at 80°C. The cuvette contains 20 µL of emulsion and 2 mL of 0.1% SDS (Sodium Dodecyl Sulphate) solution, then it is measured using a UV-Vis Spectrophotometer at a wavelength of 500 nm. Absorbance measurements were made every 20 minutes for 1 hour [16].

2.2.11. *Antioxidant Activity*

Activity of the granule Arthrospira antioxidant were evaluated using DPPH-RSA and FRAP method. Testing the antioxidant activity using the DPPH-RSA (2,2-diphenyl-1-picrylhydrazyl) method consisted of 2 stages, namely preparing the solution, and reading the absorption. The solutions required for the DPPH antioxidant activity test include sample solutions which include *A. platensis* granules, *A. platensis* powder, vitamin C and BHT (Butylated hydroxytoluene) and DPPH reagent solutions. *A. platensis* powder was carried out by a process of equalizing the concentration of *A. platensis* contained in the granules, namely 10%. The IC-50 value is represented for the antioxidant capacity. The stages of testing the antioxidant activity of the FRAP method are preparing the solution and reading the absorbance. Solutions needed for the FRAP method antioxidant test include acetate buffer solution pH 3.6, 10 mM/mL TPTZ (2,4,6 Tripyridyl-s-triazine) solution, 20 mM/L $FeCl_3$ solution, FRAP reagent solution, $FeSO_4$.7H2O standard solution, and sample solution. *A. platensis* powder was carried out by a process of equalizing the concentration of *A. platensis* contained in the granules, namely 10%. Absorbance values and concentrations of standard solutions were curves using Microsoft Excel to obtain the equation y=ax+b. After that, the sample absorption value is entered as a y value to find the x value, namely the sample concentration needed to reduce Fe^{3+} to Fe^{2+} [17].

2.2.12. *Consumer preference using hedonic test*

The consumer acceptance level test uses the hedonic test with 105 untrained panelists. The weight of the sample tested was 0.3 grams. The hedonic test rating scale from a score of 1-5 (1=dislike very much; 2=dislike; 3=rather like, 4=like, and 5=like very much, followed by questions related to buying interest in the samples tested.

3. Results and Discussion

3.1. *The effect of the addition of cinnamon and amylum mucilage on the moisture content and pH of Arthrospira granules*

The granule sample of *A. platensis* moisture content values varied from 3.86 to 5.12%, which is within the acceptable range of [18], which is less than 7%. The shelf life increases as the water content value decreases. The moisture content of the granule samples and the dry biomass of *A. platensis* did not differ significantly. The moisture content of the granules was unaffected by the addition of cinnamon powder and amylum mucilage in a range of concentrations. The pH range of the *A. platensis* granule samples was 6.43 to 6.67, which is within the acceptable range for granule pH of 7. *Arthrospira platensis* dry biomass had a pH of 7.13, which was higher than the pH of the granules. It is believed that the usage of tapioca as a filler, which has a pH of 5.8 and is added to the formula up to 50%, can cause the pH of *A. platensis* granules to fall.

The usage of cinnamon powder in various quantities may result in different treatment combinations. The pH of cinnamon powder is 8.5 [19]. Comparing 0.5% of cinnamon powder to 1% and 1.5% revealed substantial differences. The pH of the granules rises with the amount of cinnamon powder present. It was clear that the treatment combinations C3M1 and C3M2 had the best pH since they were close to neutral pH. There were variances in the treatment combinations of *A. platensis* granules with respect to pH values.

Table 3. The effect of the addition of cinnamon and amylum mucilage on the moisture content and pH of Arthrospira granules

Treatments	Moisture (%)	pH
C1M1	5.03 ± 1.1^a	6.43 ± 0.12^a
C1M2	4.22 ± 1.2^a	6.47 ± 0.12^{ab}
C1M3	4.05 ± 0.8^a	6.43 ± 0.06^a
C2M1	4.37 ± 1.0^a	6.57 ± 0.06^{bc}
C2M2	4.12 ± 1.2^a	6.57 ± 0.06^{bc}
C2M3	3.86 ± 1.6^a	6.60 ± 0.00^c
C3M1	5.12 ± 0.7^a	6.67 ± 0.06^c
C3M2	4.38 ± 0.4^a	6.67 ± 0.06^c
C3M3	4.31 ± 0.4^a	6.60 ± 0.00^c
Dry biomassa *A. platensis*	4.63 ± 0.5^a	7.13 ± 0.06^d

The number that is followed by different letters in the same column indicate a significance between treatments ($p<0.05$)

3.2. *The effect of adding cinnamon and amylum mucilage on physicochemical properties of Arthrospira platensis granules*

3.2.1. *The capacity of granules Arthrospira on WHC, OHC, Solubity and Foam ability*

Table 4 shows the outcomes of the WHC, OHC, and solubility tests. Significant variations were seen between the WHC values of the *A. platensis* dry biomass and the granules. The difference arises from the folding of the polymer during granule formation, which results in an obstruction from the binder that prevents the -OH groups in amylose and amylopectin from binding water. Additionally, *A. platensis* dry biomass has a greater WHC value than granules because the hydrophilic group of the dry biomass has a better ability than the granules. The water retention capacity was unaffected by the different amylum mucilage and cinnamon powder concentrations. High WHC compounds can be utilized to prevent other materials, like agar, from synthesizing [20]. Granules of *A. platensis* did not have a higher capacity to bind water than *A. platensis* dry biomass, according to the WHC value. As a result, it is probable that when added to food products, *A. platensis* granules will not be able to slow down the syneresis process. However, it is anticipated that *A. platensis* granules with lower WHC than dry biomass will be able to preserve the bulk density of the solid substance.

Table 4. The effect of adding cinnamon and amylum mucilage to the ability of water holding capacity, oil holding capacity, solubility and foam ability of Arthrospira granules

Treatments	WHC	OHC	Solubility (%)	*Foam Ability* (ml)
C1M1	1.71 ± 0.05^{a}	2.71 ± 0.02^{a}	11.24 ± 2.70^{a}	24.67 ± 2.08^{a}
C1M2	1.80 ± 0.09^{a}	2.61 ± 0.17^{a}	13.48 ± 1.00^{ab}	24.00 ± 1.00^{a}
C1M3	1.74 ± 0.02^{a}	2.49 ± 0.06^{a}	11.44 ± 1.33^{a}	24.00 ± 2.65^{a}
C2M1	1.79 ± 0.08^{a}	2.53 ± 0.09^{a}	12.30 ± 3.26^{a}	24.00 ± 1.00^{a}
C2M2	1.78 ± 0.02^{a}	2.37 ± 0.27^{a}	12.69 ± 1.36^{a}	25.00 ± 2.65^{a}
C2M3	1.80 ± 0.10^{a}	2.45 ± 0.11^{a}	15.88 ± 4.44^{abc}	24.00 ± 2.65^{a}
C3M1	1.81 ± 0.09^{a}	2.49 ± 0.38^{a}	12.14 ± 1.11^{a}	22.67 ± 2.08^{a}
C3M2	1.78 ± 0.02^{a}	2.63 ± 0.09^{a}	17.99 ± 1.54^{c}	25.00 ± 3.61^{a}
C3M3	1.77 ± 0.02^{a}	2.54 ± 0.07^{a}	14.89 ± 3.96^{bc}	23.00 ± 2.00^{a}
Dry biomassa *A. platensis*	3.73 ± 0.04^{b}	1.83 ± 0.11^{b}	28.13 ± 5.22^{d}	30.00 ± 3.61^{b}

The number that is followed by different letters in the same column indicate a significance between treatments ($p < 0{,}05$)

The ability of the granules to absorb oil is better than the dry biomass of *A. platensis* because the OHC value of the granules is higher. This is because tapioca flour, a filler with strong oil-holding ability, was added [21]. The presence of fat and protein has an impact on an object's capacity to bind oil. High protein levels will promote oil absorption, and fat will build a hydrophobic layer [22]. In 100 grams of flour, tapioca has 3.39% fat and 0.59% protein [23]. The value of oil retaining capacity was unaffected by the concentrations of cinnamon powder and amylum mucilage used in the granule manufacturing process. For the manufacturing of food products related to the viscosity of meals like dough or soup, WHC and OHC tests are conducted [24].

The C3M2 combination has the best solubility value, as can be shown by looking at the solubility value of *A. platensis* granules, since it has the highest value among the other combinations. Cinnamon powder and amylum mucilage were added, and this had the desired effect of producing the differences in the mixtures. In comparison to the 1.5% concentration shown in Figure 4.2, the addition of 0.5% cinnamon powder concentration was significantly different. Because cinnamon is hygroscopic, a growing number of its granules can bind water, increasing its solubility value [25]. Because starch has hygroscopic qualities, the addition of 2.5% amylum mucilage concentration was significantly different from the 7.5% concentration [26].

The solubility of *A. platensis* dry biomass is greater than the solubility of granules, because during the process of making granules it is possible for protein denaturation to occur which causes changes in protein conformation. During denaturation the hydrogen bonds and hydrophobic bonds are broken, so that the solubility of the protein decreases, and other biological activities decrease [27]. In addition, there is a steric hindrance by the binder which inhibits the –OH group in the amylose and amylopectin contained in tapioca, so that it cannot bind with water. The solubility of granules value ranges from 11.24 to -17.99% (w/w), and this value denotes that more than 80% (w/w) of the granules total weight will precipitate.

The results of the foam ability test indicated that the volume of foam formed by the granules was lower than the dry biomass of *A. platensis*. Cinnamon powder factor and amylum mucilage had no effect on the ability of the granules to form foam. This is possible because the concentration of *A. platensis* in the granule is only 10%, so that the interaction between the protein and the foam that is formed is reduced in the granule. Based on the solubility value, *A. platensis* granules are not suitable for application in beverage products. As a result, *A. platensis* granules should be added to foods that are either dry or have minimal water, such cookies. Due to the existence of granules that do not dissolve in the dough, *A. platensis* can be used to improve the appearance of cookies.

3.2.2. *The effect of adding cinnamon and amylum mucilage on emulsion activity and stability of Arthrospira platensis Granules*

Arthrospira platensis granules can be utilized as emulsifiers in food products that contain water and oil based on the findings of assessing the activity and stability of the emulsion. Table 5 shows the activity and stability parameters of the emulsion made from *A. platensis* dry biomass granules. The gelatin fraction of tilapia emulsion activity ranged from 15.16 to 39.17 m2/gram, while the average activity of *A. platensis* emulsion granules was 11.15 m2/gram [28]. High emulsion activity values indicate that a substance is suitable for use as an emulsifier. Due to its higher protein content than granules, the dry biomass emulsion of *A. platensis* exhibits increased activity and stability. According to [29], the protein in microalgae controls functional characteristics such as gel strength, emulsion activity, emulsion stability, water activity, water retaining capacity, and sensory. If the protein is very hydrophobic and can be bound to oil, this can lower the surface tension and result in the formation of an emulsion [30].

The emulsion activity value for the C3M3 combination is the highest. Due to the eugenol compound high hydrophobicity, the concentration of cinnamon powder added caused an increase in emulsion activity ($p< 0.05$). According to [31], eugenol dissolves more readily in non-polar solvents than it does in water. The granule emulsion activity was unaffected by the addition of amylum mucilage. The C3M2 combination offers the best emulsion stability for granules, as can be shown in Table 4. There are discrepancies in the results of emulsion stability in the treatment combinations.

Table 5. *The effect of adding cinnamon and amylum mucilage on emulsion activity and stability of Arthrospira platensis Granules*

Treatments	Emulsion activity	Emulsion stability
C1M1	10.27±4.9^{ab}	9.12±0.54^{ab}
C1M2	9.54±1.84^{ab}	9.09±0.14^{ab}
C1M3	8.82±1.61^{a}	9.43±0.20^{ab}
C2M1	10.50±2.74^{ab}	9.38±0.40^{ab}
C2M2	11.30±3.51^{ab}	9.28±0.76^{ab}
C2M3	11.61±1.23^{ab}	9.28±0.27^{ab}
C3M1	11.38±1.08^{ab}	8.88±0.68^{ab}
C3M2	12.81±2.01^{ab}	9.57±0.74^{b}
C3M3	14.15±1.32^{b}	8.48±0.38^{a}
Dry biomassa *A. platensis*	18.87±3.89^{c}	12.17±0.65^{c}

The number that is followed by different letters in the same column indicate a significance between treatments ($p<0,05$)

3.3. *The effect of adding cinnamon and amylum mucilage on emulsion activity and stability of Arthrospira platensis Granules*

Table 5 shows the outcomes of assessing the antioxidant activity of granules using the DPPH and FRAP methods. The dry weight of the sample is used to calculate the DPPH value of the *A. platensis* dry biomass and granules. [32] states that antioxidant activity is categorized as extremely strong if the IC-50 is less than 50 ppm, strong if the IC-50 is between 50 and 100 ppm, and moderate if the IC-50 is between 150 and 200 ppm. The antioxidant activity of *A. platensis* granules was rated as being very poor since the IC-50 values from samples of *A. platensis* granules ranged from 788.976 to 874.061 ppm. However, [33] study found that the antioxidant activity of *A. platensis* granules was higher than that of

A. platensis lozenges in this investigation, at 2310.90 ppm. The dry biomass of *A. platensis* had an IC-50 value of 71.956 ppm, indicating that it has a strong level of antioxidant activity. Because of the heat treatment involved in granule production, which has the potential to lower antioxidant activity, the IC-50 values of *A. platensis* granules and dry biomass differ from one another. Additionally, because a high solubility value shows the number of soluble antioxidant components, greater antioxidant activity is indicated, the solubility value difference between the dry biomass and the granules of *A. platensis* can impact the IC-50 value. The standard samples of vitamin C and BHT had IC-50 values of 8.3 and 13.93 ppm, which qualified them as exceptionally potent antioxidants. In comparison to *A. platensis* granule samples, the IC-50 value of *A. platensis* dry biomass against vitamin C and BHT was relatively good. The IC-50 value was unaffected by the amylum mucilage or the cinnamon powder component.

Table 6. Antioxidant activity of Arthrospira granules using the DPPH-RSA and FRAP methods

Treatments	DPPH-RSA	FRAP	
	IC-50 (ppm)	Concentration (ppm)	FRAP (μM/gram)
C1M1	816.40[a]	500	207.69[a]
C1M2	836.51[a]	500	212.20[a]
C1M3	854.93[a]	500	214.98[a]
C2M1	874.06[a]	500	210.01[a]
C2M2	842.69[a]	500	208.10[a]
C2M3	853.11[a]	500	227.89[a]
C3M1	788.98[a]	500	224.98[a]
C3M2	860.91[a]	500	223.49[a]
C3M3	848.73[a]	500	217.10[a]
Dry biomassa *A. platensis*	71.96[b]	250	139.14[b]
Vitamin C	8.30±1.81	100	70.23±2.53
BHT	13.93±0.50	100	93.77±1.91

The number that followed by different letters in the same column indicate a significance between treatments ($p<0.05$)

Granules and dry biomass of *A. platensis* had substantially varied FRAP values. At 100 ppm, vitamin C and BHT showed FRAP values of 70.23 and 93.77 μM/gram, respectively. The FRAP value was unaffected by the cinnamon powder component or amylum mucilage. The dry biomass of *A. platensis* and cinnamon powder served as the granules primary sources of antioxidants. The addition of cinnamon powder should have an impact on the FRAP value of *A. platensis* granules, as study by [34] found that cinnamon bark extract includes an antioxidant component in the form of cinnamaldehyde up to 68.65%. However, because the amount of cinnamon powder applied in this study was so small, only between 0.5 and 1.5% the granule FRAP value was unaffected. The FRAP value between granules compared to its dry biomass suggests that Arthrospira antioxidant capacity is classed as a secondary antioxidant when DPPH-RSA and FRAP methods are evaluated.

3.4. *The effect of adding cinnamon and amylum mucilage on sensory of Arthrospira platensis Granules*

The effect of cinnamon and amylum mucilage on the preference consumer (appearance, aroma, taste and texture can be seen on Figure 1,2,3 and 4 respectively. Based on Figure 1, the C2M1 formula (1% cinnamon:2.5% amylum mucilage) has the highest preference rating for the appearance attribute at 3.94. *Arthrospira platensis* dry biomass had the lowest level of preference (2.86). According to the results of the Kruskal Wallis test (data not shown), the level of preference for the appearance attribute varied considerably between the *Arthrospira platensis* granule and dry biomass. Due to the presence of water, tapioca starch, and amylum mucilage, which serves as an entrapping agent, the green color of dry biomass of *Arthrospira platensis* tends to be darker than that of granules, resulting in a lower level of consumer acceptance of the appearance.

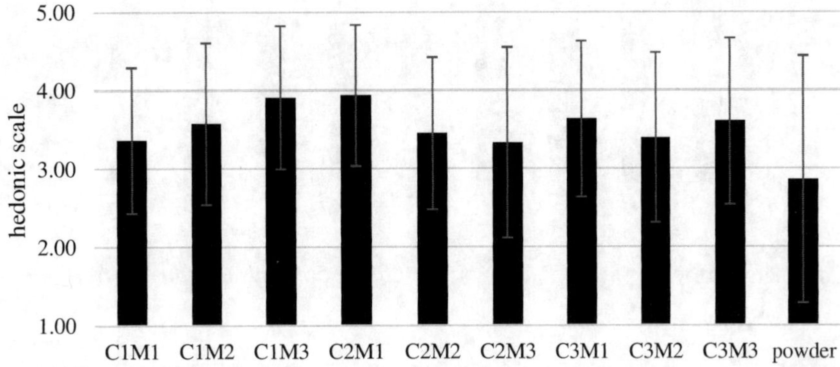

Figure 1. Consumer preference (hedonic test) on the appearance of granule Arthrospira platensis (N=105)

According to [1], the high mineral concentration of *Arthrospira platensis* causes its odour to generally have a fishy scent. Besides that, *Arthrospira platensis* has 54 volatile compounds and 23 volatile carbonyl compounds which cause a strong fishy odor [35]. Because cinnamon includes cinnamaldehyde and eugenol, two compounds that give spices their characteristic flavor, the inclusion of cinnamon to the granule formulation is anticipated to mask the fishy smell and bitter taste of *Arthrospira platensis*. The preference level for the granule fragrance is larger than the dry biomass of *Arthrospira platensis*, according to Figure 2. Because of the dry biomass of *Arthrospira platensis* very strong fishy odor brought on by its high mineral content, consumer approval of its scent is lower than that of granules [1]. The addition of cinnamon as a masking agent of 0.5%-1.5% in granules can cover the fishy smell of Arthrospira. Taste is the factor that influences customers choice regarding whether to accept a food product. In this investigation, the flavour of the granules was determined through sucking. Taste is the factor that influences customers choices regarding whether or not to accept a food product. In this investigation, the flavour of the granules was determined through sucking. Based on Figure 5, the granules taste preferences are higher than the dry biomass of *Arthrospira platensis*. *Arthrospira platensis* dry biomass has a bitter taste because it contains alkaloids and tannin compounds [36], whereas granules go through an agglomeration process and contain stevia sugar, which can mask the taste. As a result, consumers are less likely to accept the taste of granules than they are of dry biomass. Due to the addition of mixed substances, it became bitter [37]. Cinnamon is added as a masking agent to hide Arthrospira off flavor [7], along with other substances to hide *Arthrospira platensis* inadequacies.

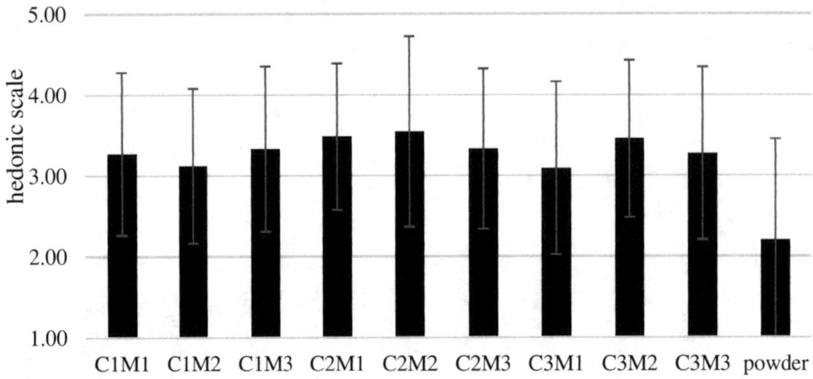

Figure 2. Consumer preference (hedonic test) on the aroma of granule *Arthrospira platensis* (N=105)

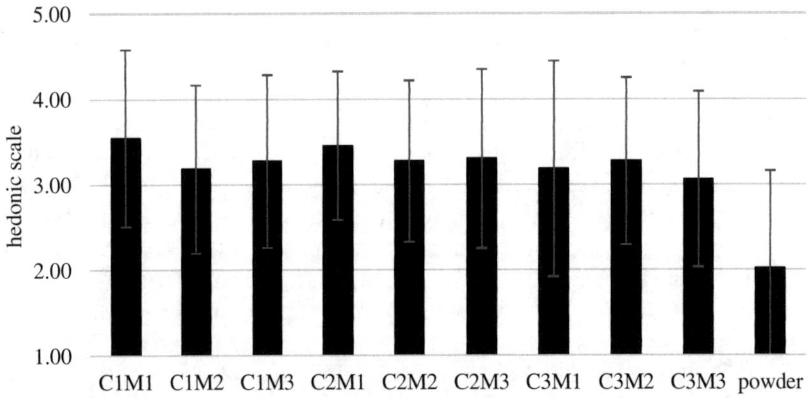

Figure 3. Consumer preference (hedonic test) on the taste of granule *Arthrospira platensis* (N=105)

Texture is a product of physical properties that are observed with the mouth when eaten or touched with the fingers [38]. In this study, texture observations were made by assessing the hardness of the granules by touching them with a finger. Based on Figure 4, the preference level of granule texture is higher than the dry biomass of *Arthrospira platensis*. It was found that the addition of amylum mucilage acts as an entrapping agent that can strengthen the bond between the components of the material [39]. The addition of amylum mucilage can also increase the compatibility of the granule texture, so it is hoped that when it is added to food products it does not dissolve easily and does not change the color of the food product. The level of consumer acceptance of the texture of dry biomass of *Arthrospira platensis* is lower than granules because the texture of dry biomass of *Arthrospira platensis* is powdery and smooth, while granules are dense and textured.

The C1M1, C2M1, and C2M2 alternate formulations are among those that panelist favour based on the amount of customer desire for granules. Similar results were also obtained when the physico-chemical characteristics were tested.

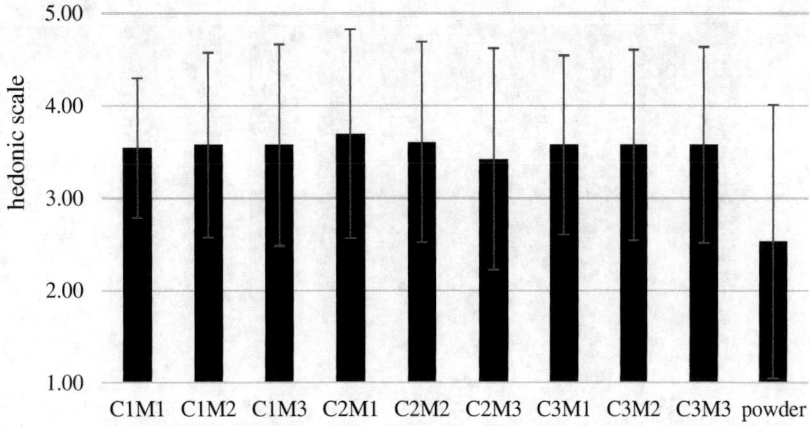

Figure 4. Consumer preference (hedonic test) on the texture of granule *Arthrospira platensis* (N=105)

3.5. *Buying intention of granules compare to dry biomass Arthrsopira*

Comparing *Arthrospira platensis* dry biomass to Arthrospira granules, a significant increase in purchasing intention was detected. C1M1, C2M1, and C2M2 were the samples chosen for the hedonic test, 81.82% of consumers in the C1M1 sample expressed a desire to purchase, whereas 18.18% did not. The C2M1 sample has a buying interest of up to 78.79% and a buying disinterest of up to 78.79%. The C2M2 sample is unwilling to purchase up to 18.18% and is only interested in purchasing up to 81.82%. In contrast, in the dry biomass samples of *Arthrospira platensis*, up to 71.21% of respondents did not want to buy, while up to 28.79% did want to buy. The off-flavor that the panelists perceived when they tasted the dry biomass of *Arthrospira platensis* is the cause of the low degree of consumer purchase interest. The level of consumer preference for a food product might drop as a result of off-flavor in items, which can be bad for producers [40].

The panelists most popular justification for purchasing, at 46.8%, was the food nutritional value. The nutritional value of a product is one factor people consider when choosing a food item because it is good for the human body. Since taste and aroma are the primary factors in choosing food products, a whopping 43.8% of respondents cited their interest in purchasing the product as a result of these factors. As many as 4.4% of the panelists selected this rationale, but there were fewer respondents because *Arthrospira platensis* granules were a novel product that had not yet gained widespread public awareness. The choice of panelists for the least preferred cause for purchasing interest was the product price, which might have increased by up to 3%, as opposed to other factors like product utilization and *Arthrospira platensis* granules' qualities. Customers do not purchase things for the sake of price; nonetheless, a high price may encourage customers to compare it to other similar products, while a low price may prompt them to question the product's quality [41].

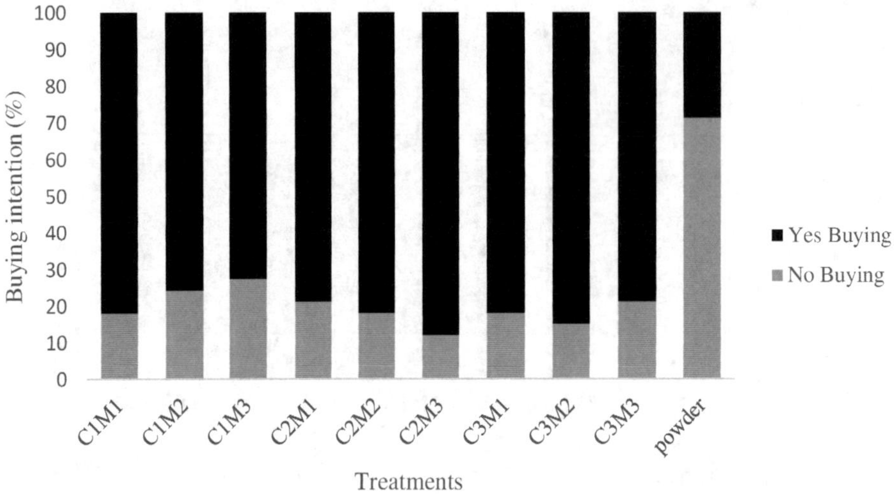

Figure 5. The Percentage of Buying Intention of Granules Arthrospira

4. Conclusion

Arthrospira granule preparation with amylum mucilage binder and cinnamon masking agent results in granule water content that meets dry product standards, granule pH ranges from pH 7, and can increase granule ability to bind water and oil, though granule activity and stability become lower than dry biomass. The granules that are created can be strengthened by raising the concentration of amylum mucilage, and the addition of cinnamon can serve as a masking agent, boosting the level of consumer preference and consumer buying intention for granules that are greater than dry biomass. There are different granule formulae that can be created based on sensory and physicochemical characteristics, including C1M1, C2M1, and C2M2. The Temporary Dominance of Sensation test can be used to further optimize the granules and decide which of the three possibilities provides a better sensation, particularly when it comes to masking the distinctive flavor and odor of Arthrospira.

5. Acknowledgements

Authors would like to thank Universitas Gadjah Mada for financial funding through Rekognisi Tugas Akhir (RTA~Final Task Recognition) No: 2488/UN1.P.III/DIT-LIT/PT/2020.

6. References

[1] Christwardana M, Nur M M A and Hadiyanto 2013 Spirulina platensis: Potensinya sebagai Bahan Pangan Fungsional *Jurnal Aplikasi Teknologi Pangan* **2** 1–4

[2] Ekantari N, Harmayani E, Pranoto Y and Marsono Y 2017 Calcium of Spirulina platensis Has Higher Bioavailability than Those of Calcium Carbonate and High-Calcium Milk in Sprague Dawley Rats Fed with Vitamin D-Deficient Diet *Pak J Nutr* **16** 179–88

[3] Utomo M T S and Prabakusumo A 2009 Formulasi pembuatan tablet hisap berbahan dasar mikroalga Spirulina platensis sebagai sumber antioksidan alami. *Jurnal Sains MIPA* **15** 67–176

[4] Negara H P, Iwan Y B, Lalena and Ekantari N 2014 Pengkayaan betakaroten pada cokelat batang dengan penambahan Spirulina platensis *Jurnal Perikanan* **16** 17–28

[5] De Simone V, Caccavo D, Lamberti G, d'Amore M and Barba A A 2018 Wet-granulation process: phenomenological analysis and process parameters optimization *Powder Technol* **340** 411–9

[6] Parker R 2003 *Introduction to Food Science* (United States of America: Delmar)
[7] Fitriya W and Alfionita K 2019 The Capability of Cinnamon as an Off-Flavor Masking Agent for Spirulina platensis enriched Food Product *Jurnal Perikanan Universitas Gadjah Mada* **20** 95
[8] Listyowati T 2017 *Kayu Manis sebagai Agen Masking Aftertaste pada Es Krim Spirulina platensis.* Skripsi (Universitas Gadjah Mada)
[9] Sheth B B, Bandelin F J and Shangraw R F 1980 Compressed Tablet *Pharmaceutical Dosage Forms, Tablets, Volume I* (New York: Marcel Dekker Inc)
[10] Lynatra C, Wardiyah and Elisya Y 2018 Formulation of effervescent tablet of temulawak extract (Curcuma xanthorrhiza Roxb.) *SANITAS: Jurnal Teknologi dan Seni Kesehatan.* **9** 72–82
[11] Mulyadi M D, Astuti I Y and Dhiani B A 2011 Formulasi granul instan jus kelopak bunga rosela (Hibiscus sabdariffa L) dengan variasi konsentrasi povidone sebagai bahan pengikat serta kontrol kualitasnya. *Pharmacy* **8** 29–41
[12] Susanti G A M , Jemmy G N, Prasetia A and Putra D 2015 *Pengaruh variasi kadar mucilago pati singkong 13%, 15% dan 17% sebagai pengikat terhadap sifat fisik tablet vitamin B kompleks untuk anjing*
[13] Onyango C, Henle T and Ziems A 2004 Effect of extrusion variables on fermented maize-finger millet blend in the production of uji *Lebensmittel Wissenchaft und Technologie* **37** 409–15
[14] Makri E, Papalamprou E and Doxastakis G 2005 Study of functional properties of seed storage proteins from indigenous European legume crops (lupin, pea, broad bean) in admixture with polysaccharides *Food Hydrocoll* **19** 583–94
[15] Mulvihill D M 1997 Production, Functional Properties, and Utilization of Milk Protein Products *Advanced Dairy Chemistry-1: Proteins* ed P F Fox (London: Elsevier Applied Science)
[16] Pearce K N and Kinsella J E 1978 Emulsifying properties of proteins: evaluation of a turbidimetric technique *J Agric Food Chem* **26** 716–23
[17] Clarke G, Ting K N, Wiart C and Fry J 2013 High correlation of 2,2-diphenyl-1-picrylhydrazyl (DPPH) radical scavenging, ferric reducing activity potential and total phenolics content indicates redundancy in use of all three assays to screen for antioxidant activity of extracts of plants from the malaysian rainforest *Antioxidants* **2**
[18] Voight R 1994 *Buku Pelajaran Teknologi Farmasi* (Yogyakarta: Gadjah Mada University Press)
[19] Barnett A 1996 *Examining food technology* (Heinemann Education Publishers)
[20] Rahmawansyah Y 2006 *Pengembangan produk minuman cincau hitam (Mesona palustris) dalam kemasan cup polipropilen di PT. FITS Mandiri Bogor* Skripsi (Bogor: Institut Pertanian Bogor)
[21] Giyarto, Hafidoh I S and Windrati W S 2016 Sifat fungsional tepung bumbu hasil formulasi dengan penggunaan tepung koro kratok *Prosiding Seminar Nasional Apta* pp 227–30
[22] Sedjati S, Yudiati E and Suryono 2012 . Profil Pigmen Polar dan Non Polar Mikroalga Laut Spirulina sp dan Potensinya sebaagai Pewarna Alami *Jurnal Ilmu Kelautan* **17** 176–81
[23] Soediaoetomo A J *Gizi dan Profesi untuk Mahasiswa* (Jakarta: Dian Rakyat)
[24] Adebowale K O and Lawal O S 2004 Comparative study of the functional properties of bambarra groundnut (Voandzeia subterranean), jack bean (Canavalia ensiformis) and mucuna bean (Mucuna pruriens) flours *Food Research International* **37** 355–65
[25] Hamidah S, Burhanudini V and Istikowati W T 2009 Kajian sifat-sifat dasar kayu manis sebagai pertimbangan pemanfaatan limbah pemanenan kulit kayu manis (Cinnamomum burmanii, Blume). *Jurnal Hutan Tropis Borneo* **10** 210–23
[26] Krocta J M and Mulder C L C D 1997 Edible and biodegradable polymer filmschallenges and opportunities (A Scientific Status Summary). *Food Technol* **51** 61–74
[27] Estiasih T 2016 *Kimia dan Fisik Pangan* (Jakarta: Bumi Aksara)
[28] Suryanti, Marseno D W, Indrati R and Irianto H E 2017 Karakteristik emulsi beberapa fraksi gelatin dari kulit ikan nila (Oreochromis niloticus) *Jurnal Pendidikan Biologi Kelautan dan Perikanan* **12** 43–54

[29] Iqbal M, Ma'aruf W F and Sumardianto 2016 Pengaruh penambahan mikroalga Spirulina platensis dan mikroalga Skeletonema costatum terhadap kualitas sosis ikan bandeng (Chanos chanos Frosk). **5** 56–63

[30] Pratiwi A R and Hartayanie L 2012 *Pengembangan Produk Pangan dari Mikroalga Laut S. platensis Berdasarkan Sifat Fungsional dan Molekular Proteinnya* (Semarang)

[31] Kong X, Liu X, Li J and Yang Y 2014 Advances in pharmacological research of eugenol *Intertional Journal of Complementary and Alternative Medicine* **1** 8–11

[32] Molyneux P 2004 The user of the stable free radicals diphenylpicryl-hydrazyl (DPPH) for estimating antioxidant activity *J. Science and Technology* **26** 211–9

[33] Wulandari D A 2013 *Formulasi tablet hisap Spirulina platensis sebagai suplemen makanan* Skripsi (Bogor: Institut Pertanian Bogor.)

[34] Ekaprasada M T, Nurdin H, Ibrahim S and Dachriyanus 2009 Chemical composition and antibacterial activity of the essential oil of the Toona sureni (Blume) Merr. *Jurnal Ris. Kim* **3** 90–5

[35] Aguero J, Lora J, Estrada K, Concepcion F, Nunez A, Rodriguez A and Pino J A 2003 Volatile Components of a Commercial Sample of the Blue-Green Algae Spirulina platensis *Journal of Essential Oil Research* **15** 114–7

[36] Tietze H W 2004 *Spirulina micro food macro blessing* ed Harald W (Australia: Tietze Publishing)

[37] Saputra J S E, Agustini T W and Dewi E N 2014 Pengaruh penambahan biomassa serbuk Spirulina platensis terhadap sifat fisik, kimia, dan sensori pada tablet hisap (lozenges). *JPHPI* **17** 281–91

[38] Kartika B, Hastuti P and Supartono W 1988 *Pedomen uji inderawi bahan pangan* (Yogyakarta: PAU pangan dan gizi UGM)

[39] Putri A P, Ridwan M, Darmawan T A, Darusman F and Gadri A 2017 Calcium modified edible Canna Canna edulis L starch for controlled released matrix *IOP Conf Ser Mater Sci Eng* **223** 1–8

[40] Kilcast D 2003 Sensory Analytical Methods in Detecting Taints and Off-Flavours in Food *Taints and off-flavours in food* ed B Baigrie (Florida: CRC Press) pp 5–30

[41] Utami R P and Saputra H 2017 Pengaruh harga dan kualitas produk terhadap minat beli sayuran organik di pasar sambas Medan *Jurnal Niagawan* **6** 44–53

IOP Conf. Series: Earth and Environmental Science 1289 (2023) 012041 doi:10.1088/1755-1315/1289/1/012041

Physical, chemical and sensory properties of cookies added with *Arthrospira platensis* granules and stevia sugar

S Raharjo and N Ekantari*

Laboratorium of Fish Processing Technology, Fisheries Department, Faculty of Agriculture Universitas Gadjah Mada, Yogyakarta, Bulaksumur, Yogyakarta

*corresponding author email: nurfitri@ugm.ac.id

Abstract. Cookies can be supplemented with the protein, fatty acids, vitamins, and minerals present in *Arthrospira platensis*, however adding the powder alters the product's flavor, color, and aroma. Granulating the powder, which has the benefit of not being easily soluble in the dough, is one technique to reduce the fishy smell and flavor and boost consumer acceptance. Stevia sugar is used in this study to make low-calorie cookies in place of sucrose sugar. This study conduct to ascertain how the inclusion of Arthrospira granules and stevia sugar affected sensory qualities and levels of customer acceptance. The study 2x4 totally random design included stevia sugar replacement (30%; 50%) and granule addition (3%; 6%; 9% (w/w dough)) with 0% serving as the control. The results showed that stevia sugar affected swelling volume and that granules affected granule appearance, color, and flavor. A follow-up test was carried out because, with sensory averages of like-really like, the G1S1 (3% granules: 30% stevia) and G1S2 (3% granules: 50% stevia) treatments were not significantly different based on hedonics. Although the ash content was not considerably different, the water content and overall sugar content were. In the depth investigation revealed that 3% of granules can be added to arthrospira to lessen its fishy flavor. The Time Intensity results show that G1S1 treatment produces the best *Arthrospira platensis* granula cookies that can be utilized as an alternative functional food, whereas G1S2 treatment produces a very bitter aftertaste.

Keywords: *Arthrospira platensis*, granules, stevia sugar, time intensity, and cookies

1. Introduction

Arthrospira platensis is a blue-green microalgae that is classified into Cyanobacteria, single-celled and spiral-shaped [1], which contains protein, vitamins, minerals, and essential amino acids [2]. Arthrospira also has antioxidants, particularly phycocyanin. The body can get its antioxidants from phytocyanin. Children and adults who ingest up to 10 grams of *A. platensis* per day can preserve their physical health [3]. Arthrospira can be supplemented in food products to help the creation of functional foods because of its advantages in preserving health. Arthrospira microalgae have been added as functional food ingredients to a variety of food products, including ice cream [4], jelly [5], cookies [6], and cookies with speck seasoning [7]. Cookies are a form of cake prepared from soft dough that is high in fat and sugar, somewhat crunchy, and when broken into pieces, the cross-section of the pieces has a less thick feel, according to SNI 01-2973-1992. Toddlers, school-age kids, teenagers, and parents frequently eat cookies as a snack [8]. The majority of cookies include a lot of fat and carbs [9]. One sugar replacement that can be made to produce cookie goods that are safe for consumption is switching the dough's sucrose sugar for stevia sugar. Research by [10] stated that the use of *Stevia rebaudiana* water extract as much as 50% in cookie products still obtains good sensory reception, so that higher stevia sugar substitution needs to be carried out in research to obtain low-calorie cookie products.

When A. platensis powder is used, the product becomes green and develops a fishy odor and unpleasant flavor. According to [11] research, adding A. platensis powder to noodles, flakes, or cookies will cause an off flavor to develop. Due to these issues, different alternate dosage forms must be used in place of the arthrospira powder dose form. As an alternative to arthrospira preparations in powder form, [12] research has created granule form, which is anticipated to not impart an off flavor when added to food products. Product color and air resistance, while the granules chemical characteristics are not easily oxidized due to their smaller surface area than that of the powder. Granules are more stable against the effects of air humidity than powders because of their bigger size [13]. This means that even if granules are exposed to air for an extended period of time, they will maintain their original shape. Granulation is a process that can produce *A. platensis* granules. According to [14], granulation is the process of reducing the particle size of a substance or material by forming bigger agglomerates with or without additional additions.

This study is anticipated to provide solutions to issues that have come up in various studies involving cookie items that have been added with arthrospira, including the appearance of a fishy odour and bitter taste, as well as the ability to raise the concentration of adding arthrospira to the product. Due to the granules ability to reduce off-flavor, fortification of arthrospira at larger concentrations can be carried out with improved consumer approval, which is projected to increase the concentration of added arthrospira when used in cookies. The addition of *A. platensis* granules to cookie products is expected to yield cookies that are well welcomed by consumers by retaining the nutritional value of *A. platensis* while not altering the fishy odour and bitter taste of the ingredient.

2. Material and Method

2.1 Material

The ingredients used for making cookies are eggs, mocaf flour (Ladang Lima), instant oats (Haverjoy), baking soda (O-Point), baking powder (O-Point), salt (Refina), vanilla (Cap Bear), butter (Blueband Cake & Cookie), stevia sugar (Stevigrow®), granulated sugar (White sugar), skimmed milk powder (O-Point), cornstarch (Maizenaku) and *A. platensis* granules. The materials used to manufacture *A. platensis* granules include *A. platensis* powder (PT Trans Pangan Spirulindo), tapioca (Rose Brand), cinnamon powder, stevia sugar (Stevigrow) and demineralized water.

2.2 Method

Experimental was designed in Completely Randomized Design, a 3x2 factorial run in this experiment. The treatment were granules (G) as the first factor, and stevia sugar (S) as the second factor. A control group (0%) of no addition and fortified 3%, 6%, 9% (w/w) of arthrospira granules was added to the cookies product. To produce cookies with less calories, stevia is sometimes used in place of sucrose sugar in cookie recipes. These replacements range 30% and 50%.

2.2.1 *Preparation of Arthrospira* Granules

Use 10% (w/w) dry biomass of arthrospira in the dough while making arthrospira granules according to the Candraningsih method (2021). The [6] method uses arthrospira powder to create arthrospira granules, which are then used to make cookies. *Arthrospira platensis* powder, amylum mucilage, cinnamon powder, stevia sugar, tapioca flour, and water are mixed until smooth. The ready wet dough was sieved through an 8-mesh sieve and dried in an oven at 55°C for 24 hours. Furthermore, the dry granules were sifted through a 14-mesh sieve. Amylum mucilage is made by mixing 10% tapioca flour and 90% hot water, then stirring until homogeneous.

2.2.2. *Cookies granule Arthrospira*

Beat the butter, sugar and vanilla for 2 minutes until combined, add the eggs and skim milk and beat again for 2 minutes using a mixer. Mocaf flour, cornstarch, oats, baking soda, baking powder, and salt are added and mixed until smooth. Then added arthrospira granules and mixed using a spatula. The ready dough is stamped and baked in the oven at 160°C for 12 minutes.

2.2.3 *The development of volume cookies determination*

First, measure the volume of the tubular container (V), then weigh the container to be used (W3) and fill the container completely with sesame seeds (W1), cookie weight (W), container + cookie + sesame (W2). The sesame that has been weighed is then removed from the container. Weigh the container filled with cookies and add the pre-weighed sesame while tapping it so that it is compact and at the right volume (W2). The remaining sesame is the volume of the sample cookies (VE). SW is the specific volume of sesame. Cookie development volume value (Sn) is calculated using the following formula:

$$SW = \frac{V}{W1 - W3} \; ; \; VE = SW(W1 - (W2 - W3)) \; ; \; Sn = \frac{VE}{W}$$

2.2.4. *The observation of pore structure of granule cookie Arthrospira*
Observation of the pore structure aims to determine how much effect the treatment of adding A. platensis granules and stevia sugar substitution has on the development of the cookies pores. The procedure for observing the pore structure was carried out by cutting the cookie sample in the middle section without a knife and observing it using a Leica MZ16 stereo microscope. The magnification used is 1.25x.

2.2.5. *Consumer preference using hedonic test*
The hedonic test was carried out to determine the most preferred concentration of arthrospira granules by the panelists. A total of 81 untrained panelists were asked to rate based on their preference for each concentration of granule addition (G) (0, 3, 6, 9% w/w) and stevia sugar substitution (S) (30, 50%). The attributes assessed were appearance, color, aroma, taste, and texture. The rating scale ranges from 1-5, with a score of 1 = really dislike; 2 = dislike; 3 = somewhat like; 4 = likes; and 5 = really like it.

2.2.6. *Quantitative Descriptive Analysis (QDA)*
Prior to the QDA testing, panelist selection was carried out with the aim of obtaining trained panelists who were used for the Quantitative Descriptive Analysis (QDA) and Time Intensity (TI) tests. Panelist selection was carried out using a triangle test, by selecting 1 different sample from 3 samples provided without any reference samples [15]. The samples used were 2 cookies treatments, namely G1S1 treatment (3% granule: 30% stevia sugar) and G1S2 (3% granule: 30% stevia sugar). A total of 24 untrained panelists were asked to differentiate samples based on the intensity of the sweet taste of cookies, carried out with 7 test sets and 3 repetitions. Panelists are declared to pass as trained panelists if in 3 repetitions they are able to answer 5 sets of 7 test sets correctly [16]. The selection of panelists was taken by 7 trained panelists. A total of 7 trained panelists were asked to describe the aroma, taste and aftertaste of the cookies. After that, the trained panelists were asked to conduct a Focus Group Discussion (FGD) to confirm what attributes emerged from the results of the product description. Then a test was carried out using a line scale in the scoresheet, panelists were asked to determine the intensity of aroma, taste and aftertaste by comparing the cookie sample with the FGD results comparison sample. QDA test results data is displayed in the form of a spider web graph.

2.2.7. *Time Intensity*
The TI test was used to determine how strongly the Arthrospira granules and stevia sugar flavor and aftertaste were present in the cookie samples at various times. Seven trained panelists in all were tasked with rating the strength of the flavor and aftertaste that appeared every three seconds for a total of ninety seconds. Data from the TI test results is shown as a curve that comprises the t-ext, T max, I max, and AUC (Area Under The Curve). When the stimulus stops being present, text is the time. T max is the amount of time needed for the stimulus to intensify to its maximum level. The maximal level of sensory stimulation is known as I max. AUC (area under the curve) is the area under the curve which indicates the size of the area detected by sensory sensations.

3. Results and Discussion

3.1. *The Effect of Arthrospira platensis Granules and Stevia Sugar on the Appearance and Development of Cookies*
Figure 1 depicts the cookie product that includes stevia sugar in place of regular sugar and has been enhanced with *Arthrospira platensis* granules. As can be seen from Figure 2, there are enormous voids, tiny cavities, and solids in the pore structure. Indicating poor cookie development and being more

fragile, cookies with a high stevia sugar substitute will have more big cavities. The cookies will become more durable as the sugar concentration rises. According to [6] research, adding a lot of stevia sugar to Arthrospira cookies causes them to create a lot of pores with a huge size.

The treatment with the inclusion of Arthrospira granules and stevia sugar replacement had a significant effect ($p<0.05$) on reducing the swelling volume, as shown by the swelling volume (Table 1). The amount of stevia sugar used to cookies determines how much carbohydrates are present; the more stevia sugar is used, the lower the amount of carbohydrates. These findings are in line with [17] study on the production of red bean and soybean-based cookies with stevia sugar sweetener, which showed that adding stevia sugar decreased the amount of carbohydrates and the crispness of the cookies. The gelatinization process in cookies slows down as the amount of carbohydrates in the dough decreases, making the cookies texture less crispy and their swelling lessened. Another influencing factor is the fat content in the dough, butter as a source of fat will weaken the bonds between starches which causes air to be trapped more easily and the formation of many pores, so that the expansion volume increases [18].

Figure 1. The appearance of cookies granules *Arthrospira platensis*

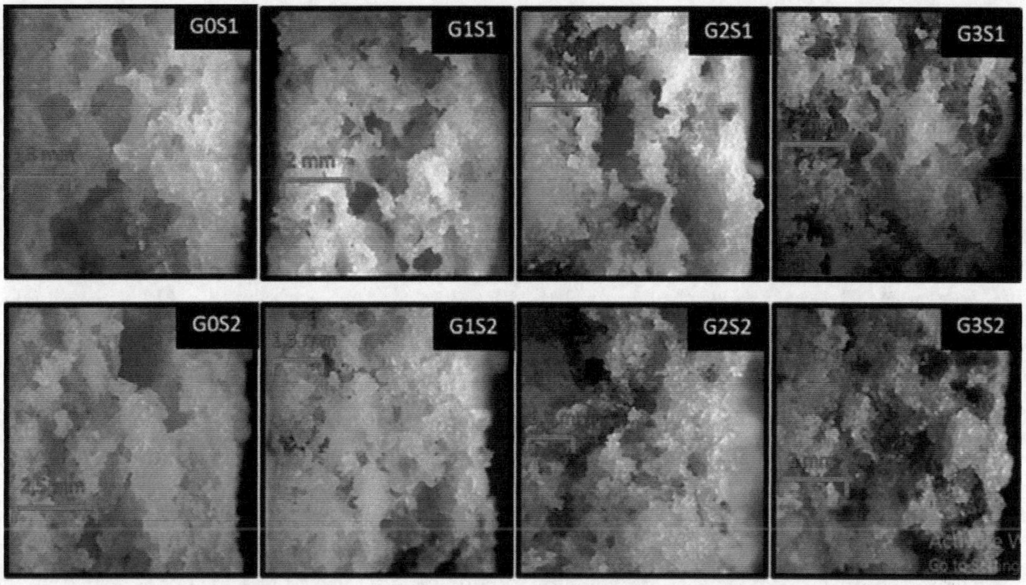

Figure 2. The structure of cookies formation using granule Arthrospira

Table 1. The development of appearance of *cookies Granules Arthrospira*

Treatments	Formulation		Volume of cookies development
	Granules (%)	Stevia (%)	
G0S1	0	30	1.05^{a}
G1S1	3	30	1.04^{a}
G2S1	6	30	1.02^{a}
G3S1	9	30	1.00^{a}
G0S2	0	50	0.80^{b}
G1S2	3	50	0.98^{a}
G2S2	6	50	0.97^{a}
G3S2	9	50	0.93^{ab}

The number that is followed by different letters indicate a significance between treatments ($p < 0,05$)

3.2. *Effect of Arthrospira platensis Granules and Stevia Sugar on Consumer Acceptance*

Table 2 lists the five characteristics of cookies that were assessed using the hedonic test: appearance, color, scent, taste, and texture. Further testing was done on the cookie recipe with the greatest hedonic mean value. Consumer acceptance of cookies made with *A. platensis* granules was higher than that of cookies made with arthrospira powder that had been fortified. Consumer acceptability levels for the cookie formulation with the inclusion of 3% arthrospira granules were 4.15 for aesthetic qualities, 4.12 for color, 4.01 for aroma, 3.86 for taste, and 3.95 for texture, with an average of 4.02. The hedonic value compared to [7],cookies fortified with 1.5% arthrospira powder resulted in a level of consumer acceptance of color attributes (3.48), aroma (3.78), taste (3.84), and texture (3.91) obtained an average of only 3.61. Research by [6] cookies fortified with 3% arthrospira powder resulted in a level of consumer acceptance of appearance (3.39).

Table 2. Consumer preference of Granule cookies Arthrospira

Formulation	Appearance	Colour	Aroma	Taste	Texture	Mean
G0S1	4.10 [a]	4.15 [a]	4.22 [a]	4.04 [a]	3.63 [a]	4.03
G1S1	4.15 [a]	4.12 [a]	4.01 [a]	3.86 [ab]	3.95 [a]	4.02
G2S1	3.68 [c]	3.52 [bc]	3.85 [ab]	3.65 [ab]	3.84 [a]	3.78
G3S1	3.28 [d]	3.22 [c]	3.31 [c]	3.43 [b]	3.77 [a]	3.40
G0S2	4.04 [ab]	4.20 [a]	4.10 [a]	3.83 [ab]	3.89 [a]	4.01
G1S2	4.10 [a]	4.05 [a]	4.09 [a]	3.89 [ab]	3.98 [a]	4.02
G2S2	3.74 [b]	3.74 [b]	3.77 [b]	3.75 [ab]	3.84 [a]	3.77
G3S2	3.22 [d]	3.22 [c]	3.38 [c]	3,47 [b]	3.74 [a]	3.39

The number that is followed by different letters in the same column indicate a significance between treatments (p<0,05)

The analysis of the data revealed that the attributes of appearance, color, scent, and taste were all substantially different (p<0.05), while the texture attribute was not (p > 0.05). The hedonic outcome of the appearance attribute is that consumer preference for the appearance attribute decreases when more grains are added to the dough. The hedonic outcome of the color attribute is that consumer preference for the color attribute decreases when more granules are added to the dough. Because the fishy odour of Arthrospira grains is intensifying, the hedonic outcome of the aroma attribute is that the level of consumer preference for the aroma attribute decreases as more granules are added to the dough. Because the fishy taste of Arthrospira granules seems to be more potent the more granules are added to the dough, the lower the level of consumer preference for the taste attribute is, according to the hedonic outcome of the taste attribute. There is no discernible change between treatments as a hedonic effect of the texture property. The G1S1 formulation (3% Arthrospira granules and 30% stevia) had the highest average value of 4.02 in the sensory average results for Arthrospira granule cookies with 30% stevia sugar substitution. The G0S1 formula average value (30% stevia control) is 4.03. Each attribute value is represented by a number in the G1S1 formula: appearance (4.15), color (4.12), aroma (4.01), taste (3.86), and texture (3.95). The G1S2 formulation (3% Arthrospira granules) had the highest average score of 4.02, indicating the highest degree of desire for Arthrospira granule cookies with 50% stevia sugar substitution. The G0S2, which uses 50% stevia as a control, has an average value of 4.01. The value of each feature, including appearance (4.10), color (4.05), aroma (4.09), taste (3.89), and texture (3.98), is shown by the G1S2 formula. The G1S1 and G1S2 formulations of Arthrospira granule cookies were found to be the best ones when compared to the control cookies, G0S1 and G0S2, based on the findings of the overall average hedonic test.

3.3. *Sensories Characteristics of adding stevia sugar to Arthrospira granule cookies*

3.3.1. *Quantitative Description Analysis*

Three characteristics: aroma, taste, and aftertaste are evaluated by this QDA test. To ascertain the general properties of the test product, the test begins with a description of the four formulations (selected treatments). A Focus Group Discussion (FGD) was used to focus and choose the outcomes of the product descriptions in accordance with the product attributes. Table 3 (aroma profile), Table 4 (taste profile), and Table 5 (aftertaste profile) show the FGD findings. The odour profile, which featured the distinctive odour of sweet, savory milk, savory butter, starchy, and fishy granules of Arthrospira, was validated by the FGD results. The taste profile, which includes the flavors of sweet sucrose sugar, sweet milk, savory butter, and starchy, fishy Arthrospira, was validated by the FGD results. Its characteristics of a bitter aftertaste (not medicinally bitter), sweet sugar, and fishy Arthrospira were validated by the FGD aftertaste profile results.

Table 3. Profile aromas of cookies by selection panelists using Focus Grup Discussion

Formulation	Criteria	Representative
G0S1 (0% Arthrospira granule: 30% stevia)	Sweet, milk tasty aroma	*Skim milk* "O-Point"
	Savory, butter aroma	Margarine cake cookie "Blue Band"
	Starchy	Oatmeal "Haverjoy"
G1S1 (3% Arthrospira granule: 30% stevia)	Sweet, milk tasty aroma	*Skim milk* "O-Point"
	Savory, butter aroma	Margarine cake cookie "Blue Band"
	Starchy	Oatmeal "Haverjoy"
	Fishy-Arthrospira smell	Granules Arthrospira
G0S2 (0% Arthrospira granule: 50% stevia)	Sweet, milk tasty aroma	*Skim milk* "O-Point"
	Savory, butter aroma	Margarine cake cookie "Blue Band"
	Starchy	Oatmeal "Haverjoy"
G1S2 (3% ranula Arthrospira granule: 50% stevia)	Sweet, milk tasty aroma	*Skim milk* "O-Point"
	Savory, butter aroma	Margarine cake cookie "Blue Band"
	Starchy	Oatmeal "Haverjoy"
	Fishy-Arthrospira smell	Granules Arthrospira

Table 4. Profile taste of cookies by selection panelists using Focus Grup Discussion

Formulation	Criteria	Representative
G0S1 (0% Arthrospira granule: 30% stevia)	Sweet, sugar, sucrose	Sugar "Gulaku"
	Sweet, milky taste	*Skim milk* "O-Point"
	Savory, butter taste	Margarine cake cookie "Blue Band"
	Starchy	Oatmeal "Haverjoy"
G1S1 (3% Arthrospira granule: 30% stevia)	Sweet, sugar, sucrose	Gula pasir putih "Gulaku"
	Sweet, milky taste	*Skim milk* "O-Point"
	Savory, butter taste	Margarine cake cookie "Blue Band"
	Starchy	Oatmeal "Haverjoy"
	Fishy-Arthrospira taste	Granula Arthrospira
G0S2 (0% Arthrospira granule: 50% stevia)	Sweet, sugar, sucrose	Gula pasir putih "Gulaku"
	Sweet, milky taste	*Skim milk* "O-Point"
	Savory, butter taste	Margarine cake cookie "Blue Band"
	Starchy	Oatmeal "Haverjoy"
G1S2 (3% Arthrospira granule: 50% stevia)	Sweet, sugar, sucrose	Gula pasir putih "Gulaku"
	Sweet, milky taste	*Skim milk* "O-Point"
	Savory, butter taste	Margarine cake cookie "Blue Band"
	Starchy	Oatmeal "Haverjoy"
	Fishy-Arthrospira taste	Granula Arthrospira

Table 5. Profile aftertaste cookies by selection panelists using Focus Grup Discussion

Formulation	Criteria	Representative
G0S1 (0% granule Arthrospira : 30% stevia)	Bitter (not medicinal bitterness) Sugar, sweet	stevia "Stevigrow" sugar "Gulaku"
G1S1 (3% granule Arthrospira : 30% stevia)	Bitter (not medicinal bitterness) Sugar, sweet Fishy-Arthrospira	stevia "Stevigrow" sugar "Gulaku" Granules Arthrospira
G0S2 (0% granule Arthrospira : 50% stevia)	Bitter (not medicinal bitterness) Sugar, sweet	stevia "Stevigrow" sugar "Gulaku"
G1S2 (3% granule Arthrospira : 50% stevia)	Bitter (not medicinal bitterness) Sugar, sweet Fishy-Arthrospira	stevia "Stevigrow" sugar "Gulaku" Granules Arthrospira

The results of the QDA test for aroma profile (Figure 3) cookies without Arthrospira G0S1 and G0S2 granules had higher intensity of skim milk, butter, and oatmeal aromas than cookies with G1S1 and G1S2 granules. The addition of 3% Arthrospira granules can reduce the intensity of the aroma of other ingredients in cookies, namely reducing the intensity of the aroma of skim milk, butter and oatmeal. The results of the taste profile QDA test (Figure 4) cookies without Arthrospira G0S1 and G0S2 granules had a more dominant flavor intensity of sugar, sucrose and butter. A slight cinnamon spice flavor arises because in the process of making Arthrospira granules cinnamon powder is added which serves to reduce the bitter fishy taste of *A. platensis*. Cinnamon contains cinnamaldehyde [19] which can affect the flavor of food products. Aftertaste profile QDA test results (Figure 5) high stevia sugar substitution cookies produce a more dominant intensity of bitter aftertaste. The treatment of adding granules also strengthens the bitter intensity of stevia sugar, because the sweetener used during the granule making process uses stevia sugar. According to [20], the bitter taste of stevia after consumption is due to the content of tannins, flavonoids, and essential oils. Since the amount of Arthrospira granules added to the cookies was the same, at 3%, the strength of the sweet aftertaste of sugar was lower in cookies with low stevia sugar substitute while it was higher in those with Arthrospira.

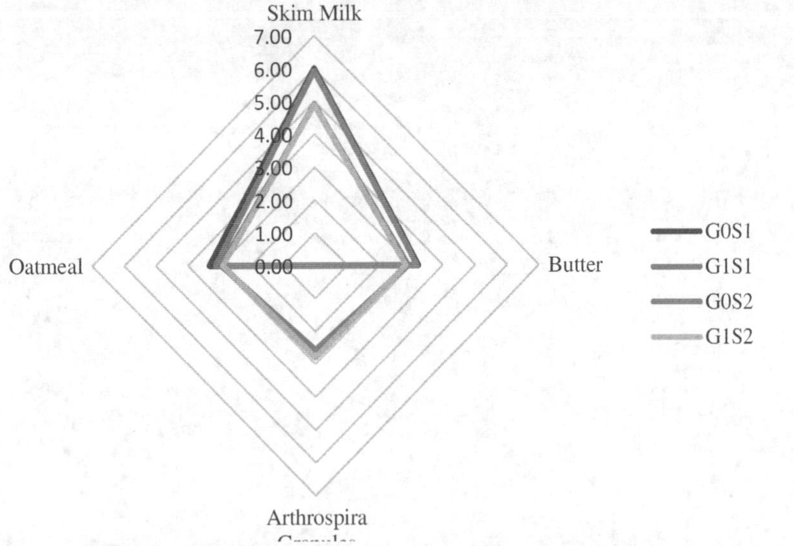

Figure 3. QDA profile aroma of cookie granule arthrospira

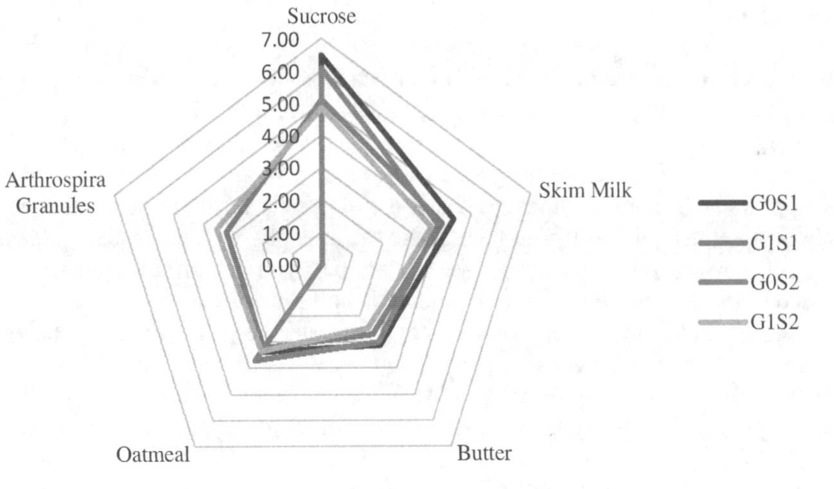

Figure 4. QDA profile taste of cookie granule arthrospira

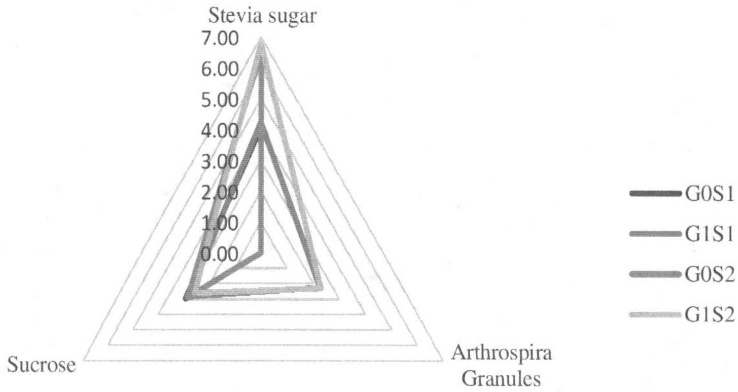

Figure 5. QDA profile of *aftertaste* cookie granule arthrospira

3.3.2. *Time Intensity*

The attributes tested in the TI test were the sweet taste of stevia sugar, the bitter taste of stevia sugar, and the fishy taste of Arthrospira granules. TI measurement aims to determine the pattern of ups and downs of the taste intensity of the attributes being tested from the beginning of the stimulus appears until the stimulus disappears. The results of testing TI cookies can be seen in Table 6.

Based on the sweet taste intensity time curve (Figure 6) and the bitter taste (Figure 7) of stevia sugar that appears in the cookies, it results that the more stevia sugar substitutes added, the greater the area of stevia sweet taste intensity and the greater the area of stevia bitter taste intensity. This shows that stevia sugar will produce a strong sweet taste, but with a high addition it will cause a bitter aftertaste due to the content of tannin compounds [21]. Figure 7 shows that the bitter taste of stevia sugar is experienced after the sweet flavor of stevia has vanished, that it predominates throughout the aftertaste, and that it eventually goes away as the 90-second mark approaches. The fishy taste strength of Arthrospira granules in cookies is affected by the use of stevia sugar replacements, according to the time intensity curve of the fishy taste of Arthrospira granules (Figure 8). The strength of the fishy flavor of Arthrospira granules in cookies was able to be concealed by the addition of more stevia sugar. According to Table 10, the sweet taste of stevia sugar was able to mask the fishy flavor of arthrospira in granula cookies because its AUC value was lower than that of the sweet taste of arthrospira. Because of the same quantity of Arthrospira granules (3%), the AUC area findings between G1S1 and G1S2 were not significantly different, but the fishy flavor of the granules in G1S2 cookies was a little greater because of the smaller volume of cookies development so the texture become denser and the flavor of arthrospira easier to detect.

IOP Conf. Series: Earth and Environmental Science 1289 (2023) 012041 doi:10.1088/1755-1315/1289/1/012041

Table 6. The effect of adding of the granule *Arthrospira platensis* and stevia in Time Intensity

Attribute	G0S1 (0% Granule : 30% Stevia)			G1S1 (3% Granule : 30% Stevia)			G0S2 (0% Granule : 50% Stevia)			G1S2 (3% Granule : 50% Stevia)		
	MS	PS	G	MS	PS	G	MS	PS	G	MS	PS	G
T ext (s)	51	69	0	51	75	48	45	84	0	42	90	54
T max (s)	15	48	0	18	48	18	12	45	0	15	42	18
I max	5.29	3.86	0	4.86	4.29	3.29	6.43	5.29	0	6.00	5.57	3.57
AUC (unit area)	129.4	9.28	0	110.5	99.0	80.7	149.1	165	0	139.1	203.9	94.3

Description: SS = sweet stevia ; BS = bitter stevia ; G = fishy granules of Arthrospira
T ext = time when the stimulus disappeared
T max = time required for the stimulus to reach maximum intensity
I max = the maximum intensity of the stimulus received by the senses
AUC = area under the curve which indicates the area of the area sensory sensation detected

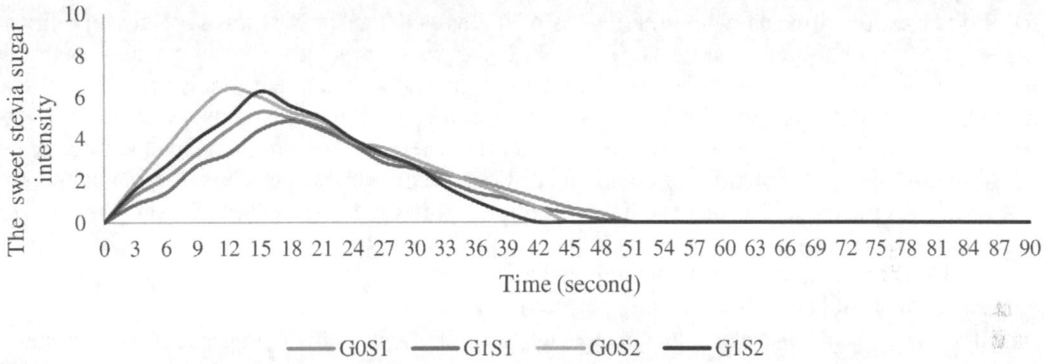

Figure 6. Time intensity curve the taste of the sweet stevia sugar

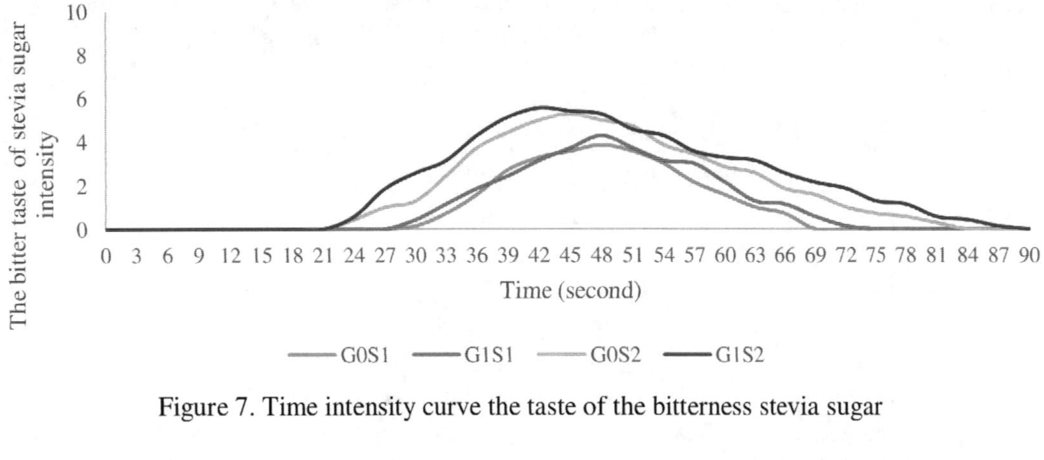

Figure 7. Time intensity curve the taste of the bitterness stevia sugar

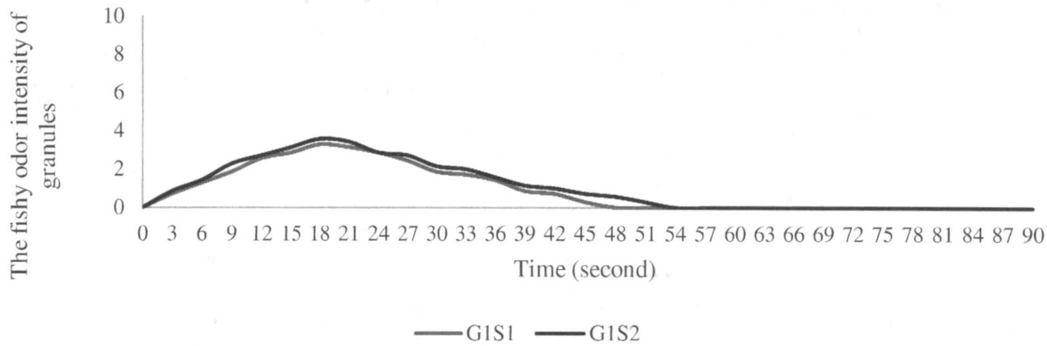

Figure 8. Time intensity curve of the fishy odor Arthrospira granules

3.4. *Characteristics of adding stevia sugar to arthrospira granule cookies on moisture, ash and total sugar content*

The results of the moisture content and total sugar content of cookies were significantly different ($p<0.05$) between treatments, while the results of the ash content of cookies were not significantly different ($p>0.05$) between treatments. The values for water content, ash content and total sugar content can be seen in Table 7. The results of the four cookie treatments moisture contents were statistically different from one another ($p<0.05$) and they complied with the SNI 01-2973-1992 standard for cookie quality, which calls for a moisture content of no more than 5%. Despite meeting the maximum 1.5% [22] criterion for cookie quality, the results of the four cookie ash contents were not significantly different from one another ($p>0.05$). The total sugar content results show that the four cookie formulas have a relatively low sugar content, and the use of Stevigrow® sugar in the dough has no calories, making it suitable for ingestion by diabetics and children [23] states that no more than 50 grams of sugar should be consumed daily per person.

The strength of the fishy odour and bitter taste of *Arthrospira platensis* can be lessened by adding arthrospira granules and stevia sugar to bakery items. Due to the stronger strength of the fishy odour and bitter taste of Arthrospira, the results of the sensory characteristics showed that the aroma and taste attributes increased with the addition of 3% granules and subsequently dropped with the addition of 6% and 9% granules. A larger substitution of 50% will result in a harsh aftertaste of stevia after the cookies have been ingested. A substitution of 30% stevia sugar can provide a strong sweet taste and a somewhat bitter taste of arthrospira granules. The use of stevia as a sugar substitute allowed cookies overall sugar content to be reduced, resulting in the production of alternative low-calorie food products.

Tabel 7. Moisture, ash and total sugar of Arthtrospira granule *cookies*

Treatments	Formulation		Moisture	Ash	Total Sugar
	Granule (%)	Stevia (%)	(%)	(%)	(%)
G0S1	0	30	3.91 ᵃ	0.84 ᵃ	16.57 ᵃ
G1S1	3	30	2.90 ᶜ	0.73 ᵃ	15.22 ᵃ
G0S2	0	50	3.38 ᵇ	0.92 ᵃ	10.60 ᵇ
G1S2	3	50	2.65 ᶜ	0.79 ᵃ	10.04 ᵇ

The number that is followed by different letters in the same column indicate a significance between treatments ($p<0,05$)

The level of consumer approval can be raised by adding granulated *Arthrospira platensis* to cookies in place of regular sugar. In accordance with the findings of the best cookie treatment, G1S1 (3% granule: 30% stevia sugar) has a level of consumer acceptance of the attributes of appearance (4.15), color (4.12), aroma (4.01), taste (3.86), and texture (3.95) with a sensory average of 4.018, better than fortification of 3% arthrospira powder and 50% stevia sugar (appearance attribute (3.39), color (3.42), aroma (3.84) , taste (3.48), and texture (3.92) with a sensory average of 3.61).

4. Conclusion

The best Arthrospira granule cookies recipe combines 3% Arthrospira granules and 30% stevia sugar replacement in cookies. *Arthrospira platensis* fishy odor and unpleasant taste can be lessened by adding the granules to cookie item, while also preserving the product original color. Cookies fortified with *Arthrospira platensis* in the form of granules can increase the level of consumer acceptance. Further investigation is advised, including studies on the shelf life of *Arthrospira platensis* granule cookies items and product mapping on the market to ascertain the product position as a commercial good.

5. Acknowledgements

Authors would like to thank Universitas Gadjah Mada for financial funding through Rekognisi Tugas Akhir (RTA~Final Task Recognition) No: 2488/UN1.P.III/DIT-LIT/PT/2020.

6. References

[1] Budiardi T, Bambang N and Santosa A 2010 Pertumbuhan dan Kandungan Spirulina sp. pada Fotoperiode yang Berbeda *Jurnal Akuakultur Indonesia* **9** 146–56

[2] Shahbazizadeh S, Khosravi-Darani K and Sohrabvandi S 2015 Fortification of Iranian Traditional Cookies With Spirulina platensis *Annu Res Rev Biol* **7** 144–54

[3] Henrikson R 1989 *Earth Food Spirulina* (California: Ronorc Enterprises Inc)

[4] Listyowati T 2017 *Kayu Manis sebagai Agen Masking Aftertaste pada Es Krim Spirulina platensis.* Skripsi (Universitas Gadjah Mada)

[5] Mujiati, S. A. Budhiyanti, N. Ekantari and I. Y. B. Lelana 2010 Kadar β-karoten Minuman Jeli Spirulina pada Berbagai Suhu Pengolahan *Jurnal Perikanan*

[6] Saraswati I G A A I 2019 *Pengaruh Penambahan Stevia Komersial terhadap Oatmeal Cookies yang Difortifikasi sebagai Pangan Fungsional Spirulina platensis* Skripsi (Universitas Gadjah Mada)

[7] Restiningsih M 2020 *Pengaruh Penambahan Bumbu Spekuk terhadap Cookies yang Difortifikasi Spirulina platensis.* Skripsi (Universitas Gadjah Mada)

[8] Moehji S 2000 *Ilmu Gizi dan Penanggulangan Gizi Buruk* (Jakarta: Papas Sinar Sin)

[9] Andriana D 2017 *Uji Daya Terima dan Kandungan Gizi Biskuit dengan Penambahan Tepung Ikan Lele Dumbo (Clarias gariepinus) dan Daun Bangun Bangun (Coleusam boinicus Lour)* (Sumatera Utara: Universitas Sumatera Utara)

[10] Salazar V A G, Encalada S V, Cruz A C and Campos M R S 2018 Stevia rebaudiana : A sweetener and potential bioactive ingredient in the development of functional cookies *J Funct Foods* **44** 183–90

[11] Fitriya W and Alfionita K 2019 The Capability of Cinnamon as an Off-Flavor Masking Agent for Spirulina platensis enriched Food Product *Jurnal Perikanan Universitas Gadjah Mada* **20** 95

[12] Candraningsih M D 2021 *Pengaruh Kombinasi Kayu Manis dan Mucilago Amili terhadap Sifat Sensoris Granula Arthrospira platensis.* Skripsi (Universitas Gadjah Mada)

[13] Nugrahani I, Rahmat H and Djajadisastra J 2006 Karakteristik Granul dan Tablet Propanol Hidroklorida dengan Metode Granulasi Peleburan *Majalah Farmasi Indonesia* **2**

[14] Rajniak P, Mancinelli C, Chern R, Stepanek F, Farber L and Hill B 2007 Experimental study of wet granulation in fluidized bed: Impact of the binder properties on the granule morphology *Int J Pharm* **334** 92–102

[15] Setyaningsih D, Apriyantono A and Sari M P 2010 *Analisis Sensori untuk Industri Pangan dan Agro* (Bogor: IPB Press)

[16] Soekarto S T 1985 *Penilaian Organoleptik untuk Industri Pangan dan Hasil Pertanian* (Jakarta: Bhratara Karya Aksara)

[17] Astuti S D and Agustia F C 2012 *Optimasi Formula dan Karakterisasi Cookies Fungsional Berbasis Kacang Merah dan Kedelai Organik dengan Penambahan Gula Stevia da Kappa Karagenan*

[18] Septieni D 2016 *Mempelajari Pembuatan Cookies Kaya Serat dengan Bahan Dasar Tepung Asia Ubi Jalar* Skripsi (Institut Pertanian Bogor)

[19] Hastuti A M 2014 *Pengaruh Penambahan Kayu Manis terhadap Aktivitas Antioksidan dan Kadar Gula Total Minuman Fungsional Secang dan Daun Stevia Sebagai Alternatif Minuman Bagi Penderita Diabetes Melitus Tipe 2.* Skripsi (Universitas Diponegoro)

[20] Adesh A B, Gopalakrishna B, Kusum S A and Tiwari O P 2012 An Overview on Stevia: A Natural Calorie Free Sweetener *International Journal of Advantages in Pharmacy, Biology and Chemistry* **1** 2277–4688

[21] Kusumaningsih T, Asrilya N J, Wulandari S, Wardani D R T , and Fatikhin K 2015 Pengurangan Kadar Tanin pada Ekstrak Stevia rebaudiana dengan Menggunakan Karbon Aktif *ALCHEMY Jurnal Penelitian Kimia* **11** 81–9

[22] Badan Standarisasi Nasional 1992 *Cookies (SNI 01-2973-1992)* (Jakarta: BSN)

[23] The Indonesian Ministry of Health 2018 How much is the recommended consumption of sugar, salt and fat per day? *https://p2ptm.kemkes.go.id/*

AUTHOR INDEX

Abdullah, C A C ... 228
Adhar, S ... 201
Adi, C A ... 214
Aisyah, A N ... 263
Aisyah, S ... 85
Akbarsyah, Nora ... 99
Alauddin, M H ... 119
Ali, L ... 279
Andriani ... 238
Anwar, M ... 271
Apdillah, Donny ... 21
Apriyana, W ... 271
Arthatiani, F Y ... 109
Arum, S R ... 75
Aynuddin ... 221
Ayuzar, E ... 201
Bardant, TB ... 279
Barokah, G. R. ... 238
Budhiyanti, S A ... 322, 336
Candraningsih, M D ... 375
Castaneda, J I ... 171
Chadijah, A ... 12
Christyandari, DA ... 279
Courtes, F ... 310
Darwanto, Dwidjono Hadi ... 1
Desrita ... 161
Deswati, R ... 109
Ekantari, N ... 336, 375, 389
Elmaghriba, A W ... 375
Fadhilah, A ... 161
Fahmi, A S ... 315
Fauzi, A ... 271
Favian, E ... 360
Febrianto, Try ... 21
Fitriyani, F ... 12
Gavi, M ... 263
Ghazali, M S M ... 228
Gunawan ... 180
Hafidz, AM ... 93
Haji, M T P ... 258
Handaka, A.A. ... 143
Handaka, Asep Agus ... 99
Handoko, Y P ... 119
Hanum, DF ... 201
Harahap, Z A ... 63
Hardianto, E ... 171
Hartami, P ... 201
Hartati, R ... 36
Hartati, Retno ... 41
Hasan, M R ... 228
Hasibuan, J S ... 161
Hatta, M ... 201
Herawati, T. ... 143
Herawati, Titin ... 99
Herjayanto, M ... 85, 93
Hernawan ... 271, 279
Hidayat, M F ... 245
Hidayat, T. ... 134
Hikma, Afida Sabilil ... 192
Hutapea, Jhon H. ... 180
Ibrahim, J ... 12
Ihsan, Y N ... 49
Iskandar ... 143
Islam, MD R ... 171
Jatmiko, TH ... 279
Juligani, B ... 271
Kawinkamraksa, C ... 310
Khasanah, Y ... 271, 279
Kijchavengkul, T ... 310
Kristiany, M G E ... 214
Kumayanjati, B ... 279
Kurniawati, Esty ... 21
Kusumajati, D R S ... 322
Laily, Farkhana ... 192
Lestari, R I ... 155
Luhur, E S ... 109
Ma'mun, Asep ... 21
Mahdaliana ... 201
Mainisa ... 201
Manalu, L.P. ... 134
Mardiana, M ... 119
Marlina, E ... 214
Maulani, A ... 119
Maulita, M ... 119
Mulia, D S ... 161
Muliawan, I ... 109
Mulyo, Jangkung Handoyo ... 1
Mulyono, M ... 119
Mustapha, N A ... 228
Najamuddin ... 63
Nanda, A R ... 12
Nawastuti, Dati ... 1
Netramai, S ... 310
Noranizan, M ... 228
Nugraheni, P S ... 345, 360
Nurfadillah ... 161

Nurhayati, A....143
Nurhayati, Atikah....99
Nurilmala, M. 134
Nurmala, I....119
Pamungkas, Wahyuniar....99
Pangestuti, R....279
Poeloengasih, C D....271
Poeloengasih, CD....279
Prabakusuma, A S....109
Prabowo, G....214
Prasetyo, D J....271
Prasetyo, DJ....279
Pratama, F Y....288
Pratomo, Arif....21
Pujiati, M I....63
Purnamasari, Elly....31
Puspadewi, I J....119
Puspita, A R....49
Putra, M M P....155, 263
Putra, Y....279
Putranti, L N....345
Raharjo, S....389
Rahma, A N F....298
Rahmi, R....12
Rakhmanda, A....221
Rasidi....201
Robin, R....93
Romadhon....315
Rosales, C K....310
Rosanto, A N....155
Rustadi....192
Saleah, F....310
Setiadi, Ananto....180
Setyobudi, E....171
Siahaan, EA....279
Siswanto, A D....75
Slamet, Bejo....180
Suadi....1
Sudarmono, Agnes....99
Sugiharto, Eko....31
Suhana, Mario Putra....21
Suhermanto, A....214
Sukarti, Komsanah....31
Sunarti, T.C.....134
Sunarto....49
Supriyo, Edy....41
Suratno....279
Suryani, R....271, 279
Suwanto, A....271
Suwito, W....238
Syamsia, S....12
Syamsu, K....134
Syamsuddin, M L....49
Syamsudin, F....49
Syamsuri, A S....12
Syarif, A....12
Syarif, AF....85, 93
Syarifah, S M....271
Tahir, I....63
Tamaya, A....271
Triyatmo, B....221
Triyono, H....119
Ustadi....245, 258, 288, 298
Valen, FS....85, 93
Wahyudin, N....279
Wardono, B....109
Wicaksono, A....75
Widianingsih, W....36
Widianingsih, Widianingsih....41
Widiastuti, W....279
Windarsih, A....279
Windrayani, E....336
Wipatanawin, A....310
Yani, F Indah....12
Yuliyanto, P....271
Yuniarti, T....119
Yustiati, A....143
Yusuf, R....109
Zahid, Ahmad....21
Zainuddin, M....49
Zainuri, Muhammad....41
Zulham, A....109